10th Asia Plasma and Fusion Association Conference (APFA 2015)

Journal of Physics: Conference Series Volume 823

Gandhinagar, India
14 – 17 December 2015

Editors:

Ziauddin Khan
Pratibha Semwal

ISBN: 978-1-5108-3962-5
ISSN: 1742-6588

Printed from e-media with permission by:

Curran Associates, Inc.
57 Morehouse Lane
Red Hook, NY 12571

CURRAN ASSOCIATES INC.
proceedings.com

Some format issues inherent in the e-media version may also appear in this print version.

Printed by Curran Associates, Inc. (2017)

Institute of Physics
Dirac House, Temple Back
Bristol BS1 6BE UK

Phone: 44 1 17 929 7481
Fax: 44 1 17 920 0979

techtracking@iop.org

Additional copies of this publication are available from:

Curran Associates, Inc.
57 Morehouse Lane
Red Hook, NY 12571 USA
Phone: 845-758-0400
Fax: 845-758-2633
Email: curran@proceedings.com
Web: www.proceedings.com

Table of contents

Volume 823

10th Asia Plasma and Fusion Association Conference

14–17 December 2015, Institute for Plasma Research (IPR), India

Accepted papers received: 13 March 2017
Published online: 19 April 2017

Preface

011001
OPEN ACCESS
10th Asia Plasma and Fusion Association Conference

011002
OPEN ACCESS
Peer review statement

Papers

Invited lectures

012001
OPEN ACCESS
Indian Test Facility (INTF) and its updates

M Bandyopadhyay, A Chakraborty, C Rotti, J Joshi, H Patel, A Yadav, S Shah, H Tyagi, D Parmar, Dass Sudhir, A Gahlaut, G Bansal, J Soni, K Pandya, R Pandey, R Yadav, M. V. Nagaraju, V. Mahesh, S. Pillai, D. Sharma, D. Singh, M. Bhuyan, H. Mistry, K Parmar, M. Patel, K Patel, B. Prajapati, H. Shishangiya, M. Vishnudev and J. Bhagora.....1

012002
OPEN ACCESS
Upgradation plans of SST-1 Cryogenics system at IPR

V L Tanna, SST-1 Cryo Team and S Pradhan.....7

012003
OPEN ACCESS
Improvements in a Tracer-Encapsulated Solid Pellet and Its Injector for More Advanced Plasma Diagnostics

Naoki Tamura, Shigeru Sudo, Chihiro Suzuki, Hisamichi Funaba, Masaru Takagi, Nakahiro Satoh, Hiromi Hayashi, Hiroya Maeno, Mitsuhiro Yokota and Hideki Ogawa.....13

012004
OPEN ACCESS
Initial results in SST-1 after up-gradation

S Pradhan, Z Khan, V L Tanna, U Prasad, Y Paravastu, D C Raval, H Masand, Aveg Kumar, J R Dhongde, S Jana, B Kakati, K B Patel, M K Bhandarkar, B K Shukla, D Ghosh, H S Patel, T J Parekh, I A Mansuri, K R Dhanani, A Varadharajulu, Y S Khristi, P Biswas, C N Gupta, S George, P Semwal, D K Sharma, H K Gulati, K Mahajan, B R Praghi, M Banaudha, A R Makwana, H H Chudasma, M Kumar, R Manchanda, Y S Joisa, K Asudani, S N Pandya, S K Pathak, S Banerjee, P J Patel, P Santra, F S Pathan, P K Chauhan, M S Khan, P L Thankey, A Prakash, P N Panchal, R N Panchal, R J Patel, G I Mahsuria, D P Sonara, K M Patel, S P Jayaswal, M Sharma, J C Patel, P Varmora, G L N Srikanth, D R Christian, A Garg, N Bairagi, G R Babu, A G Panchal, M M Vora, A K Singh, R Sharma, H D Nimavat, P R Shah, G Purwar, T Y Raval, A L Sharma, A Ojha, S Kumar, N K Ramaiya, V Siju, M V Gopalakrishna, A Kumar, P K Sharma, P K Atrey, SV Kulkarni, K K Ambulkar, P R Parmar, A L Thakur, J V Raval, S Purohit, P K Mishra, A N Adhiya, U C Nagora, J Thomas, V K Chaudhari, K G Patel, S Dalakoti, C G Virani, S Gupta, Ajay Kumar, B Chaudhari, R Kaur, R Srinivasan, D Raju, D H Kanabar, R Jha, A Das and D Bora.....18

Posters

012005
OPEN ACCESS
Wilkinson Type Lumped Element Combiner-Splitter for Indigenous Amplifier Development

Manoj Patel, Akhil Jha, JVS Harikrishna, Rajesh Trivedi and Aparajita Mukherjee.....26

012006
OPEN ACCESS
Experience of 12 kA / 16 V SMPS during the HTS Current Leads Test

P Panchal, D Christian, R Panchal, D Sonara, G Purwar, A Garg, H Nimavat, G Singh, J Patel, V Tanna and S Pradhan.....32

012007
OPEN ACCESS
Performance of Impedance Transformer for High-Power ICRF Heating in LHD

Kenji Saito, Testuo Seki, Hiroshi Kasahara, Ryosuke Seki, Shuji Kamio, Goro Nomura and Takashi Mutoh.....39

012008
OPEN ACCESS
Radio frequency emissions from laser induced material breakdown

Vinoth Kumar, Manikanta Elle and Prem Kiran Paturi.....46

012009
OPEN ACCESS
2D turbulence structure observed by a fast framing camera system in linear magnetized device PANTA

Satoshi Ohdachi, Shigeru Inagaki, Tatsuya Kobayashi and Motoshi Goto.....52

012010
OPEN ACCESS
Improving anti-felting characteristics of Merino wool fiber by 2.5 MHz atmosphere pressure air plasma

Nisha Chandwani, Purvi Dave, Vishal Jain, Sudhir Nema and Subroto Mukherjee.....57

012011
OPEN ACCESS
A Fixed Frequency Reflectometer to Measure Density Fluctuations at Aditya Tokamak

Praveen Kumar Atrey, Dhaval Pujara and Subroto Mukherjee.....63

012012
OPEN ACCESS
Development of heat sink concept for near-term fusion power plant divertor

Sandeep Rimza, Samir Khirwadkar and Karupanna Velusamy.....68

012013
OPEN ACCESS
Radiation effects of the laser ablative shockwaves on aluminium under atmospheric conditions

S Sai Shiva, Ch Leela, P Prem Kiran, C D Sijoy and S Chaturvedi.....74

012014
OPEN ACCESS
Conceptual design of Dump resistor for Superconducting CS of SST-1

Swati Roy, Piyush Raj, Arun Panchal and Subrata Pradhan.....80

012015
OPEN ACCESS
Development, Integration and Testing of Automated Triggering Circuit for Hybrid DC Circuit Breaker

Deven Kanabar, Swati Roy, Chiragkumar Dodiya and Subrata Pradhan.....86

012016
OPEN ACCESS
LOFA analysis in helium and Pb-Li circuits of LLCB TBM by FE simulation

Paritosh Chaudhuri, S Ranjithkumar, Deepak Sharma and Chandan Danani.....92

012017
OPEN ACCESS
Seismic Design of ITER Component Cooling Water System-1 Piping

Aditya P Singh, Mahesh Jadhav, Lalit K Sharma, Dinesh K Gupta, Nirav Patel, Rakesh Ranjan, Guman Gohil, Hiren Patel, Jinendra Dangi, Mohit Kumar and A G A Kumar.....99

012018
OPEN ACCESS
Development of High Voltage and High Current Test Bed for Transmission Line Components

Akhil Jha, Manoj Patel, JVS Harikrishna, P Ajesh, Rohit Anand, Rajesh Trivedi and Aparajita Mukherjee.....105

012019
OPEN ACCESS
Solving the Capacitive Effect in the High-Frequency sweep for Langmuir Probe in SYMPLE

Pramila, J J Patel, R Rajpal, C J Hansalia, V P Anitha and K Sathyanarayana.....111

012020
OPEN ACCESS
The Determination of Plasma Radial Shafranov Shift (ΔR) and Vertical Shift (ΔZ) experimentally using Magnetic probe and Flux loop Method for SST-1 Tokamak

Subrata Jana, Jasraj Dhongde, Harish Masand, Subrata Pradhan and Sameer Kumar.....117

012021
OPEN ACCESS
Observation of MHD phenomenon for SST-1 superconducting tokamak

Manisha Bhandarkar, Jasraj Dhongde and Subrata Pradhan.....122

012022
OPEN ACCESS
Thermal response of actively cooled tungsten monoblocks to inhomogeneous surface heat loading

Y Patil, S S Khirwadkar and D Krishnan.....128

012023
OPEN ACCESS
Preparation of W/CuCrZr mono-block test mock-up using vacuum brazing technique

K Premjit Singh, S Khirwadkar, Kedar Bhope, Nikunj Patel and Prakash Mokaria.....134

012024
OPEN ACCESS
Design and performance of vacuum system for high heat flux test facility

Rajamannar Swamy Kidambi, Prakash Mokaria, Samir Khirwadkar, Sunil Belsare, M S Khan, Tushar Patel and Deepu S Krishnan.....140

012025
OPEN ACCESS
Design of a Prototype Positive Ion Source with Slit Aperture Type Extraction System

Sanjeev K Sharma, Prahlad Vattilli, Bhargav Choksi, Bharathi Punyapu, Rambabu Sidibomma, Sridhar Bonagiri, Deepak Aggrawal and Ujjwal K Baruah.....146

012026
OPEN ACCESS
Sensitivity analysis of upstream plasma condition for SST-1 X-Divertor configuration with SOLPS

M. Himabindu, Anil K. Tyagi, Deepti Sharma, Devendra Sharma and R. Srinivasan.....153

012027
OPEN ACCESS
Ponderomotive potential and second harmonic backward Raman scattering in dense quantum plasmas

Nisha Singh Rathore and Punit Kumar.....158

012028
OPEN ACCESS
Behaviour of TSI in Magnetized Plasmas for different Plasma Oscillation to Debye Wavelength ratio

Jyoti and Rakhee Malik.....166

012029
OPEN ACCESS
Indigenous Manufacturing realization of TWIN Source

R. Pandey, M. Bandyopadhyay, D. Parmar, R. Yadav, H. Tyagi, J. Soni, H. Shishangiya, D. Sudhir Kumar, S. Shah, G. Bansal, K. Pandya, K. Parmar, M. Vuppugalla, A. Gahlaut and A. Chakraborty.....171

012030
OPEN ACCESS
Significance of ITER IWS Material Selection and Qualification

Bhoomi K Mehta, Jigar Raval, Abha Maheshwari, Rahul Laad, Gurlovleen Singh and Haresh Pathak.....177

012031
OPEN ACCESS
Prototyping of radial plates for fusion relevant superconducting magnets

M Ghate, D Bhavasar, A Panchal, S Udgata and S Pradhan.....184

012032
OPEN ACCESS
Assessment of delta ferrite in multipass TIG welds of 40 mm thick SS 316L: A comparative study of ferrite number (FN) prediction and measurements

Ramesh Kumar Buddu, P. M. Raole and B. Sarkar.....189

012033
OPEN ACCESS
ITER ECE Diagnostic: Design Progress of IN-DA and the diagnostic role for Physics

H K B Pandya, Ravinder Kumar, S Danani, P Shrishail, Sajal Thomas, Vinay Kumar, G Taylor, A Khodak, W L Rowan, S Houshmandyar, V S Udintsev, N Casal and M J Walsh.....195

012034
OPEN ACCESS
Preliminary Design of O-mode Radiometer for ITER ECE Diagnostic

S Danani, H K B Pandya, Ravinder Kumar, M E Austin, V S Udintsev and Vinay Kumar.....201

012035
OPEN ACCESS
Development of control system for multi-converter High voltage Power supply using programmable SoC

Rasesh Dave, Jagruti Dharangutti, N P Singh, Aruna Thakar, Hitesh Dhola, Sandip Gajjar, Darshan Parmar, Tanish Zaveri and Ujjwal Baruah.....207

012036
OPEN ACCESS
Integration & Validation of LCU with Different Sub-systems for Diacrode based amplifier

Kumar Rajnish, Sriprakash Verma, Dipal Soni, Hriday Patel, Gajendra Suthar, Hrushikesh Dalicha, Hitesh Dhola, Amit Patel, Dishang Upadhayay, Akhil Jha, Manoj Patel, Rajesh Trivedi, Harsha Machchhar, Raghuraj Singh and Aparajita Mukherjee.....213

012037
OPEN ACCESS
Design of 1 MHz Solid State High Frequency Power Supply

Darshan Parmar, N P Singh, Sandip Gajjar, Aruna Thakar, Amit Patel, Bhavin Raval, Hitesh Dhola, Rasesh Dave, Dishang Upadhay, Vikrant Gupta, Niranjan Goswami, Kush Mehta and Ujjwal Baruah.....218

012038
OPEN ACCESS
Alternate Design of ITER Cryostat Skirt Support System

Manish Kumar Pandey, Saroj Kumar Jha, Girish Kumar Gupta, Avik Bhattacharya, Gaurav Jogi and Anil Kumar Bhardwaj.....226

012039
OPEN ACCESS
Development of electromagnetic welding facility of flat plates for nuclear industry

Rajesh Kumar, Subhanarayan Sahoo, Biswanath Sarkar and Anurag Shyam.....232

012040
OPEN ACCESS
Electron Beam Welding: study of process capability and limitations towards development of nuclear components

Gautam R Vadolia and K Premjit Singh.....238

012041
OPEN ACCESS
Conceptual & Engineering Design of Plug-in Cryostat Cylinder for Super-Conducting Central Solenoid of SST-1

Prabal Biswas, Prosenjit Santra, Kirit Vasava, Snehal Jayswal, Tejas Parekh, Pradeep Chauhan, Hitesh Patel and Subrata Pradhan.....243

012042
OPEN ACCESS
Comparative analysis on flexibility requirements of typical Cryogenic Transfer lines

Mohit Jadon, Uday Kumar, Ketan Choukekar, Nitin Shah and Biswanath Sarkar.....250

012043
OPEN ACCESS
Thermo-mechanical Design Methodology for ITER Cryodistribution cold boxes

Vinit Shukla, Pratik Patel, Jotirmoy Das, Hitensinh Vaghela, Ritendra Bhattacharya, Nitin Shah, Ketan Choukekar, Hyun-Sik Chang and Biswanath Sarkar.....256

012044
OPEN ACCESS
Safety and Environment aspects of Tokamak- type Fusion Power Reactor- An Overview

Bharat Doshi and D. Chenna Reddy.....262

012045
OPEN ACCESS
Indigenously developed bending strain setup for I-V characterization of superconducting tapes and wires

Arun Panchal, Anees Bano, Mahesh Ghate, Piyush Raj and Subrata Pradhan.....270

012046
OPEN ACCESS
Study of Transport and Micro-structural properties of Magnesium Di-Boride Strand under react and bend mode and bend and react mode

Ananya Kundu, Subrat Kumar Das, Anees Bano and Subrata Pradhan.....276

012047
OPEN ACCESS
Design and Development of CRIO based Data Acquisition and Control system for High Voltage Bushing Experiment

Himanshu Tyagi, Sejal Shah, Ratnakar Yadav, Kartik Patel, Hiren Mistri, Deepak Parmar, Dheeraj Sharma, Mainak Bandyopadhyay and Arun K Chakraborty.....282

012048
OPEN ACCESS
Serial Interface through Stream Protocol on EPICS Platform for Distributed Control and Monitoring

Arnab Das Gupta, Amit K Srivastava, S Sunil and Ziauddin Khan.....288

012049
OPEN ACCESS
Development of Data Acquisition Set-up for Steady-state Experiments

Amit K Srivastava, Arnab D Gupta, S Sunil and Ziauddin Khan.....295

012050
OPEN ACCESS
Manufacturing and assembly of IWS support rib and lower bracket for ITER vacuum vessel

R Laad, Y Sarvaiya, H A Pathak, J R Raval and C H Choi.....301

012051
OPEN ACCESS
Assembly & Metrology of First Wall Components of SST-1

Tejas Parekh, Prosenjit Santra, Prabal Biswas, Hiteshkumar Patel, Yuvakiran Paravastu, Snehal Jaiswal, Pradeep Chauhan, Gattu Ramesh Babu, Arun Prakash A, Dhaval Bhavsar, Dilip C Raval, Ziauddin Khan and Subrata Pradhan.....307

012052
OPEN ACCESS
Effect of Geometrical Imperfection on Buckling Failure of ITER VVPSS Tank

Saroj Kumar Jha, Girish Kumar Gupta, Manish Kumar Pandey, Avik Bhattacharya, Gaurav Jogi and Anil Kumar Bhardwaj.....313

012053
OPEN ACCESS
Preparation and analysis of helium purge gas mixture to be used in Tritium Extraction System of LLCB TBM

V Gayathri Devi, Deepak Yadav and Amit Sircar.....318

012054
OPEN ACCESS
Role of Outgassing of ITER Vacuum Vessel In-Wall Shielding Materials in Leak Detection of ITER Vacuum Vessel

A Maheshwari, H A Pathak, B K Mehta, G S Phull, R Laad, M S Shaikh, S George, K Joshi and Z Khan.....323

012055
OPEN ACCESS
Preliminary Design Development of ITER X-ray Survey Spectrometer

Sanjeev Varshney, Siddharth Kumar, Sapna Mishra, Namita Yadav, P V Subhush, T S Chaitanya, Shivakant Jha, Vinay Kumar, Robin Barnsley, Philippe Bernascolle, Natalia Casal, Gunter Bertschinger, Stefan Simrock, Jean-Marc Drevon and Michael Walsh.....328

012056
OPEN ACCESS
Observation of plasma shift in SST-1 using optical imaging diagnostics

Manoj Kumar, Chesta Parmar, Vishnu Chaudhary and Ajai Kumar.....335

012057
OPEN ACCESS
Preliminary Results from Electron Cyclotron Measurements at SST-1

Varsha Siju, Praveena Shukla, Jayesh Raval, Joisa. Y Shankara and S. K. Pathak.....342

012058
OPEN ACCESS
Design and Development of Amplitude and phase measurement of RF signal with Digital I-Q Demodulator

Dipal Soni, Kumar Rajnish, Sriprakash Verma, Hriday Patel, Rajesh Trivedi and Aparajita Mukherjee.....347

012059
OPEN ACCESS
Design of new central solenoid for SST-1

Upendra Prasad, Subrata Pradhan, Mahesh Ghate, Piyush Raj, V L Tanna, Ziauddin Khan, Swati Roy, Prosenjit Santra, Prabal Biswas, A N Sharma, Yohan Khristi, Deven Kanaber and Pankaj Varmora.....353

012060
OPEN ACCESS
Overall behaviour of PFC integrated SST-1 vacuum system

Ziauddin Khan, Dilip C Raval, Yuvakiran Paravasu, Pratibha Semwal, Kalpeshkumar R Dhanani, Siju George, Mohammad Shoaib, Arun Prakash, Gattu R Babu, Prashant Thankey, Firozkhan S Pathan and Subrata Pradhan.....360

012061
OPEN ACCESS
Baking and helium glow discharge cleaning of SST-1 Tokamak with graphite plasma facing components

P Semwal, Z Khan, D C Raval, K R Dhanani, S George, Y Paravastu, A Prakash, P Thankey, G Ramesh, M S Khan, P Saikia and S Pradhan.....366

012062
OPEN ACCESS
Commissioning and experimental validation of SST-1 plasma facing components

Yuvakiran Paravastu, Dilip Raval, Ziauddin Khan, Hitesh Patel, Prabal Biswas, Tejas Parekh, Siju George, Prosenjit Santra, Gattu Ramesh, A ArunPrakash, Prashant Thankey, Pratibha Semwal, Kalpeshkumar R Dhanani, Snehal Jaiswal, Pradeep Chauhan and Subrata Pradhan.....372

012063
OPEN ACCESS
Design and Integration of SMBI System for SST-1 Tokamak

S George, Y Paravastu, M S Khan, K R Dhanani, D C Raval, Z Khan, S Banerjee and S Pradhan.....378

012064
OPEN ACCESS
RF assisted Glow Discharge Condition experiment for SST-1 Tokamak

Dilip Raval, Ziauddin Khan, Siju George, Kalpeshkumar R Dhanani, Yuvakiran Paravastu, Pratibha Semwal, Prashant Thankey, Mohammad Shoaib Khan, Bharat Kakati and Subrata Pradhan.....383

012065
OPEN ACCESS
Gas Fuelling System for SST-1Tokamak

Kalpesh Dhanani, D C Raval, Ziauddin Khan, Pratibha Semwal, Siju George, Yuvakiran Paravastu, Prashant Thankey, M S Khan and Subrata Pradhan.....388

012066
OPEN ACCESS
Nuclear Analyses of Indian LLCB Test Blanket System in ITER

H L Swami, A K Shaw, C Danani and Paritosh Chaudhuri.....394

012067
OPEN ACCESS
Quality control of FWC during assembly and commissioning in SST-1 Tokamak

Hitesh Patel, Prosenjit Santra, Tejas Parekh, Prabal Biswas, Snehal Jayswal, Pradeep Chauhan, Yuvakiran Paravastu, Siju George, Pratibha Semwal, Prashant Thankey, Gattu Ramesh, Arun Prakash, Kalpesh Dhanani, D C Raval, Ziauddin Khan and Subrata Pradhan.....400

012068
OPEN ACCESS
Measurement of electron energy probability function in weakly magnetized plasma

D Kalita, B Kakati, B K Saikia, M Bandyopadhyay and S S Kausik.....405

012069
OPEN ACCESS
Engineering design and integration of in-vessel single turn segmental coil in vacuum vessel of SST-1

Snehal Jayswal, P Chauhan, P Santra, K Vasava, T Perekh, H Patel, P Biswas and S Pradhan.....412

IOP Conf. Series: Journal of Physics: Conf. Series **823** (2017) 011001 doi:10.1088/1742-6596/823/1/011001

APFA 2015

10th Asia Plasma & Fusion Association Conference

14th – 18th December 2015

hosted by

Institute for Plasma Research

Gandhinagar, India

IOP Conf. Series: Journal of Physics: Conf. Series **823** (2017) 011001 doi:10.1088/1742-6596/823/1/011001

Welcome to APFA-2015

It gives me immense pleasure to note that 10th Asia Plasma Fusion Association Conference is being held at Institute for Plasma Research (IPR), Bhat, Gandhinagar 382 428, INDIA during 14-17 December, 2015. This is the second time this conference is being held in India, the first time being 2006. Main aim of the conference is to bring together researchers and students working in the field of fusion plasma science and technology, especially, in the South East Asia region, every alternate year under one umbrella to discuss technology advancement made during this period.

Fusion activities have increased in all the member countries during this period. Smaller countries of this region have also started experimental programmes in plasma physics. Therefore, APFA executive council may look into more participation of these countries and more opportunities for interactions on important fusion science and technology related areas. More Industry participation is also welcome in the meetings. This is being helpful to prepare human resource for the future need in fusion field. Interactions in these meetings are generating interesting regional collaborations.

As India is a full partner in the ITER programme and has now developed and commissioned successfully its own Superconducting Steady state Tokamak (SST-1), research in fusion science and technology needs to increase substantially. APFA is playing an important role in advancing this effort through this meeting.

Few awards for younger colleagues have been instituted which should create a very healthy competition among the participants. We request members to instate more such awards.

IPR colleagues are holding the conference this year to bring the plasma researchers together once again.

I wish APFA, all the best in this endeavour.

Dhiraj Bora
Conference Chair

IOP Conf. Series: Journal of Physics: Conf. Series **823** (2017) 011001 doi:10.1088/1742-6596/823/1/011001

International Advisory Committee

Dhiraj BORA	Chair	IPR
Keeman KIM	Member	NFRI
Yong LIU	Member	SWIP
Jiangang LI	Member	ASIPP
Shaoping ZHU	Member	IAPMC
Toshihiko YAMANAISHI	Member	JAEA
Akio KOMORI	Member	NIFS
Satoru SAKAKIBARA	Member	NIFS
Gyung-su LEE	Member	NFRI
Won NAMKUNG	Member	POSTECH
Predhiman Krishan KAW	Member	IPR
Avinash KHARE	Member	Delhi Univ
Rabindranath PAL	Member	PSSI

International Programme Committee

Dhiraj BORA	Chair	IPR
Damao YAO	Member	ASIPP
Qiang LI	Member	SWIP
Ge ZHUANG	Member	HUST
Shigeru MORITA	Member	NIFS
Mashahiro MORI	Member	JAEA
Atsushi FUKUYAMA	Member	Kyoto Univ
Yong-su NA	Member	SNU
Jin-young KIM	Member	NFRI
Jong-gu KWAK	Member	NFRI
Subrata PRADHAN	Member	IPR
Shishir DESHPANDE	Member	ITER-India
Vinay KUMAR	Member	ITER-India

Local Organizing Committee

Subrata PRADHAN	Chair	IPR
Amita DAS	Member	IPR
Ratneshwar JHA	Member	IPR
Vinay KUMAR	Member	ITER-India
Subroto MUKHERJI	Member	FCIPT-IPR
Srinivasan RADHAKRISHNANA	Member	IPR
A VARADARAJULU	Member	IPR
Ujjwal Kumar BARUAH	Member	ITER-India
Sanjay V KULKARNI	Member	IPR
Tejas PAREKH	Member	IPR
Ziauddin KHAN	Member	LIGO-IPR
Chhaya CHAVDA	Member	IPR
Parveen ATREY	Member	IPR
Hradesh SHARMA	Member	IPR
Ravi A V KUMAR	Member	IPR
Ashlesh SHAH	Member	IPR

IOP Conf. Series: Journal of Physics: Conf. Series **823** (2017) 011001 doi:10.1088/1742-6596/823/1/011001

Exhibitors

I-DESIGN Engineering Solutions Ltd.

INOX India Pvt Ltd

Shell-N-Tube Pvt Ltd.

ATOS Instruments Marketing Services

Larsen & Toubro Limited

Entuple Technologies Pvt. Ltd. (ANSYS)

Blue Sky Spectroscopy Inc.

Pfeiffer Vacuum (India) Private Limited

Strong Below Zero

Aditya High Vacuum Limited

L&T Construction

CAMECA

National Instruments

Air Liquide – Advanced Technologies

Print Vision House

VACUUM TECHNIQUES PVT. LTD.

Continental Electronics Corporation

Linde Kryotechnik AG

Automation System Engineers Pvt. Ltd.

DOTCAD

CryoZone

Editors
Dr. Ziauddin Khan
Mrs. Pratibha Semwal

IOP Conf. Series: Journal of Physics: Conf. Series **823** (2017) 011002 doi:10.1088/1742-6596/823/1/011002

Peer review statement

All papers published in this volume of *Journal of Physics: Conference Series* have been peer reviewed through processes administered by the proceedings Editors. Reviews were conducted by expert referees to the professional and scientific standards expected of a proceedings journal published by IOP Publishing.

Indian Test Facility (INTF) and its updates

M Bandyopadhyay[1], A Chakraborty[1], C Rotti[1], J Joshi[1], H Patel[1], A Yadav[1], S Shah[1], H Tyagi[1], D Parmar[1], Dass Sudhir[1], A Gahlaut[2], G Bansal[2], J Soni[2], K Pandya[2], R Pandey[2], R Yadav[1], M. V. Nagaraju[1], V. Mahesh[2], S. Pillai[1], D. Sharma[2], D. Singh[1] , M. Bhuyan[1], H. Mistry[2], K Parmar[2], M. Patel[1], K Patel[2], B. Prajapati[2], H. Shishangiya[1], M. Vishnudev[1], J. Bhagora[1]

[1]ITER-India, Institute for Plasma Research, Bhat, Gandhinagar 382 428, India.
[2]Institute for Plasma Research, Bhat, Gandhinagar 382 428, India.

E-mail: mainak@ipr.res.in; mainak@iter-india.org

Abstract. To characterize ITER Diagnostic Neutral Beam (DNB) system with full specification and to support IPR's negative ion beam based neutral beam injector (NBI) system development program, a R&D facility, named INTF is under commissioning phase. Implementation of a successful DNB at ITER requires several challenges need to be overcome. These issues are related to the negative ion production, its neutralization and corresponding neutral beam transport over the path lengths of ~ 20.67 m to reach ITER plasma. DNB is a procurement package for INDIA, as an in-kind contribution to ITER. Since ITER is considered as a nuclear facility, minimum diagnostic systems, linked with safe operation of the machine are planned to be incorporated in it and so there is difficulty to characterize DNB after onsite commissioning. Therefore, the delivery of DNB to ITER will be benefited if DNB is operated and characterized prior to onsite commissioning. INTF has been envisaged to be operational with the large size ion source activities in the similar timeline, as with the SPIDER (RFX, Padova) facility. This paper describes some of the development updates of the facility.

1. Introduction

The function of Diagnostic Neutral Beam (DNB) [1] is to diagnose the Helium (He) ash content in the Deuterium – Tritium (D–T) phase of ITER plasma, using the charge exchange recombination spectroscopy (CXRS) [2] technique. DNB, for ITER, is a negative ion source based neutral beam injector (NBI) system. The beam is designed to deliver 18–20 A of 100 keV beam of hydrogen neutrals [3] to the ITER plasma [4], through a narrow beam transmission duct with small blanket aperture opening (H × W = 45 mm × 40 mm). The beam will be operated with a 3 s ON/20 s OFF duty cycle and modulated at 5 Hz to ensure improved signal-to-noise ratio in CXRS diagnostic. There are several challenges involved in DNB development project to achieve its specifications, which are related to transmission of high current (18 – 20 A), low divergence beam (7 mrad or less) over a long distance ~ 20.67 m. In this regard, an accelerated ion beam current of value ~ 60 A, having ~ 35 mA/cm^2 extracted negative ion current density from the ion source of total extraction area (H × W = 1.6 m × 0.6 m) is needed. To transmit the beam through a narrow blanket aperture, placed at distance 20.67 m away from the grounded grid of the ion source, highly focussed beam is required. Such focussing has been achieved by combination of grid segment geometrical shaping, bending and with grid aperture offset setting. The effect on the beam optics due to the bending of the grid segments under heat loads from the incident co-extracted electrons, stripped electrons and back streaming

IOP Conf. Series: Journal of Physics: Conf. Series **823** (2017) 012001 doi:10.1088/1742-6596/823/1/012001

positive ions during the beam operation also needs extensive testing. The DNB beam source, which is under manufacturing, consists of an ion source based on European design [5], and the 100 keV accelerator is designed by India [6]. Functionality of the beam line components (BLCs) like neutralizer, residual ion dump (RID) under different operational scenarios are also matter of concern. INTF is established in IPR, India in order to address these challenges and support ITER in realising a working DNB. The uniqueness of this facility is that, INTF has housed a vacuum vessel of ~23m long to accommodate 20.67 m beam transport experimental setup to characterize the full DNB beamline and establishes itself as the largest NBI beamline testbed. The following sections of this paper describe the facility and some of its R&D activities in detail.

To operate INTF successfully, learning experience to operate negative ion beam source with multi-driver setup operation with a CODAC based data acquisition and control system (DACS) including Cs injection is useful. In this regard, a set of experiments; ROBIN experiment [7], TWIN source experiment [8], High Voltage Bushing experiment [9], Cs oven long delivery tube experiment [10] etc. are being carried out in IPR, to learn different aspects of design, integration and remote operation of large size negative ion source having multiple RF drivers fitted with multiple high power RF generators, HV power supplies, sophisticated diagnostic systems under CODAC based DACS platform. All these experiments are part of IPR's negative ion beam based neutral beam injector (NBI) system R&D program and generate experimental database, useful for ITER NBI activities.

2. Description of INTF and some of its major activities

The INTF facility is in the ITER-India experimental building in IPR [11]. The facility comprises of a large Vacuum vessel (VV), Beamline components (BLCs), Beam source (BS), Cryosorption pumps (CP), HV Bushing, Beam diagnostics, Data acquisition and control system (DACS) and different power supplies with transmission line.

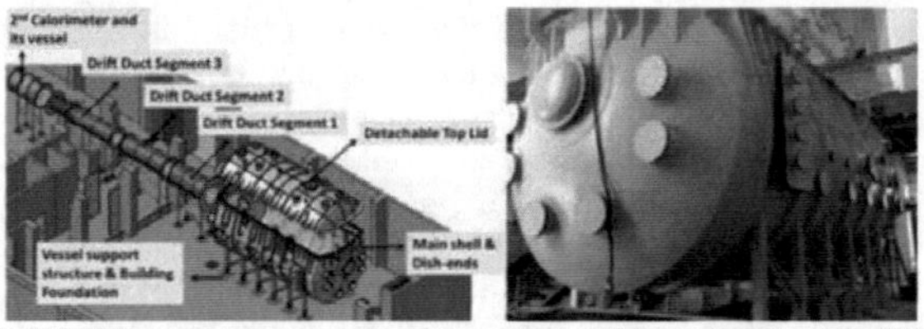

Figure 1. INTV vacuum vessel schematic and its snap inside INTF area.

2.1. Vacuum vessel.

INTF vacuum vessel provides the vacuum confinement to characterize DNB injector system. It serves as the structural support for all in-vessel components like; Beam line components (BLCs) [3], cryo-pumps and the Beam Source (BS). The Vessel has two tori-spherical dish-ends at two sides with large openings for High-Voltage Bushing on one side & 13 m long beam transmission Duct on another side. INTF main Vacuum Vessel which houses all the above components is of cylindrical shape with dimensions of 9 m in length and 4.5 m in diameter and with a detachable top lid configuration for mounting as well as removal of internal components during installation and maintenance phases. Top lid has ~ 30 m long vacuum sealing boundary with double O-ring sealing associated with inter-space pumping provision. The achieved flatness of the top lid after manufacturing is below 2mm over the complete sealing area. At the end of duct, a small vessel is provided which housed the actively cooled beam dump, named "2nd Calorimeter". Total volume of the vessel is ~ 150 m^3 and the vessel is installed and commissioned in INTF lab. The achieved ultimate vacuum is 10^{-6} mbar with maximum deflection on the wall is below 5mm. The global leak tightness is ~ 10^{-6} mbar l/s and corresponding local leak rate is ~ 10^{-9} mbar l/s. The vessel picture inside INTF lab is shown in fig.1. The beamline components (BLCs) and the beam source (BS) are housed inside the vacuum vessel and will be operated under vacuum. Detailed design of the INTF vacuum vessel can be found in ref. [12].

IOP Conf. Series: Journal of Physics: Conf. Series **823** (2017) 012001 doi:10.1088/1742-6596/823/1/012001

2.2. Beamline components

The neutralizer, residual ion dump (RID) and the calorimeter are the beam line components [1]. The layout of the components is identical to DNB with a 1m, 0.75 m and 0.2 m gaps between the grounded grid & neutralizer, neutralizer & RID and RID & calorimeter respectively [3]. In correspondence with the DNB, the 3 m long neutralizer and 1m long RID are 4 channelled structures. Out of five Cu walls in RID for four channels, 2nd and 4th walls are connected to −8 kV potential to generate electric field sufficient to filter out un-neutralized negative ions as well as re-ionized positive ions. Negative hydrogen ions are deflected towards the electrically grounded 1st, 3rd and 5th plates, whereas positive ions are deflected on 2nd and 4th plates respectively. The calorimeter is a movable V – shaped structure made up of two panels with CuCrZr alloy based hyper-vapotron type heat transfer element (HTE) [13] to handle large heat flux when it intercepts the complete beam. The detailed technical specification of each BLC has been prepared by ITER-India and subsequently reviewed and approved by ITER. After successful completion of tendering process, contract has been awarded for procurement of the BLCs. The components are now under initial stage of manufacturing. A parallel prototype manufacturing activity of these BLCs has been also initiated indigenously including material development, high precision machining and electron beam welding for dissimilar metal joining.

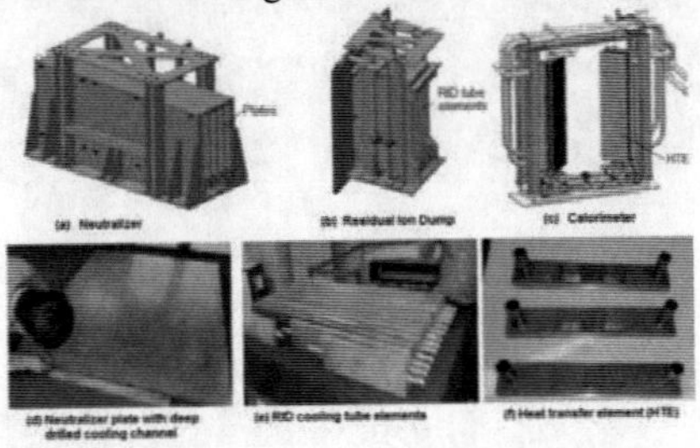

Figure 2. Schematic of beamline components (a, b, c) and its manufactured parts (d, e, f) respectively.

The Beam line components are subjected to several acceptance tests. (i) *Visual inspections* : All the Beam Line Components shall be visually checked according to ASTM Sec. V/ Equivalent EN ISO std. and they shall be free from visual defects like weld spatter , surface cracks, surface porosity etc. (ii) *Dimensional Inspection* : Dimensional control shall be established & verified at different level of manufacturing i.e. after receiving the raw material, during manufacturing of sub-components/sub-assemblies & after assembling of all the sub-components/sub-assemblies to form Main Beam Line Component assembly. The Non-conformance (if any) w.r.t approved drawings observed the same shall be disposed through approved non-conformance report (NCR) procedure. (iii) *Pressure Test*: Cooling circuit of all the BLCs shall be tested for high pressure test at 25 bar. Test pressure is specified according to the recommendation available in the applicable ASME/ EN ISO code. (iv) *Leak Test*: Cooling circuit of all the BLCs shall be subjected to the leak test. Leak test shall be performed at both at ambient temperature and at the maximum and minimum design temperature. During testing, the direction of pressure differential shall be kept in the same direction as during operation. The observed leak shall be less than 1×10^{-10} Pa m^3/s. In addition, some functional tests are also specified for each BLCs. For example, (a) *High voltage Test*: Residual Ion Dump assembly shall be tested for High voltage test at 12 KV for 3 hours. (b) *Flow Test*: One element of each BLCs shall be tested for flow test. (c) *Demonstration of functionality of Movement Mechanism for calorimeter*: The functionality of the Calorimeter Movement Mechanism (CMM) shall be demonstrated in vacuum after installing the Calorimeter assembly along with CMM in the INTF Vessel.

2.3. Beam source

The accelerator system of DNB BS is based on three grid system, consisting of a plasma grid (PG), an extractor grid (EG) with embedded magnets and the grounded grid (GG) [6]. Each grid is made of four segments. A total number of 1280 apertures are arranged on these segments in 16 beam groups in

5×16 matrix, which overall provides a beam extraction area of 0.197 m^2. All the beamlets to be focused at a distance of 20.665 m from the GG to ensure minimum size of the beam near ITER blanket. Stringent focussing requirement is a special design feature of DNB and it is achieved by using a combination of geometrical aiming by placing each segment in a specified angle with respect to the beam axis, aperture offset and geometrical shaping. Apart from that each PG segment is manufactured in a curved profile in horizontal direction to achieve desired horizontal focusing. Prototype of a segment of plasma grid is manufactured and achieved tolerances are ~ 50 micron on flatness over the complete grid segment surface area (1 m × 0.4 m) and ± 0.002 degree in angle tolerance. It is the first time in the world that manufacturing of a six stage angled grid segment is realized, as shown in fig.3a. To hold all the grids and its segments in desired position with respect to the beam axis, maintaining HV isolation requirements, specially designed alumina based ceramic isolators (post insulator) are also manufactured (see fig.3b) as prototypes. These have achieved 140 kV DC voltage isolation for 3hrs with 13 kN cantilever load bearing capability.

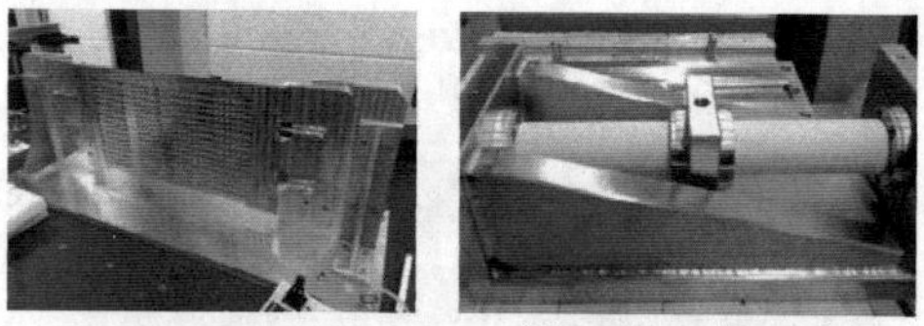

(a) Plasma grid segment (b) Ceramic post insulator

Figure 3. Manufactured angled plasma grid segment and ceramic post insulator on test zig.

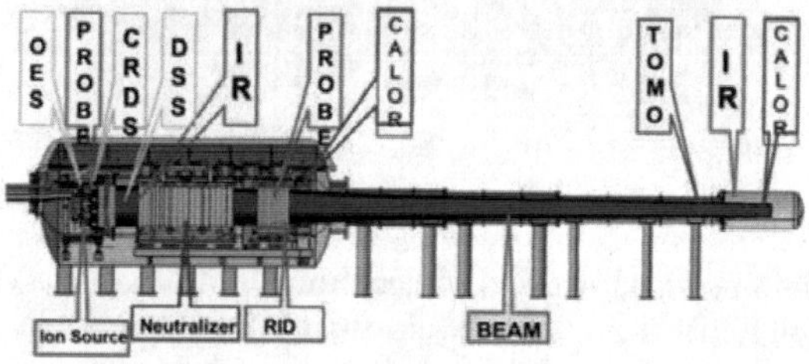

Figure 4. Deployment of characterization diagnostics in INTF

2.4. Diagnostics

Since INTF is envisaged to characterize ITER DNB system and to establish the functionality of its eight inductively coupled RF plasma driver based negative hydrogen ion source with its beam performance including its beam line components, diagnostics plays an important role for its safe and optimized operation. A number of diagnostics are planned in INTF to characterize the ion source and its beam performance. Negative ions and its Cesium contents in the source, will be monitored by optical emission spectroscopy (OES) and cavity ring down spectroscopy (CRDS). Plasma near the extraction region will be studied using standard electrostatic probes. The beam divergence and negative ion stripping losses are planned to be measured using Doppler Shift Spectroscopy (DSS). During initial phase of ion beam characterization, Carbon Fiber Composite (CFC) based infrared (IR) imaging diagnostics will be used. Safe operation of the beam will be ensured by using standard thermocouples and electrical voltage-current measurement sensors. Detailed description of different diagnostic systems corresponding to different beam operation scenarios are depicted and discussed in ref. [14]. The dispersion of all these diagnostics, envisaged in the test facility for characterization, is shown in fig.4. To gain experience, (i) a table top experiment for CRDS is being carried out; (ii) one IR camera is already being procured for CFC based IR diagnostics and will be incorporated in ROBIN [7] beamline for beam characterisation; (iii) Probes, OES, DSS, Calorimetric diagnostics are regularly being utilized in ROBIN test facility to characterize the plasma and the beam respectively.

2.5. Control and Data acquisition system

To ensure successful operation of INTF involving integrated operation of all the constituent plant systems a matured Data Acquisition and Control System (DACS) is required. The INTF DACS is based on Control, Data Access and Communication (CODAC) platform following on Plant Control Design Handbook (PCDH) guidelines. However, the roadmap to learn operation with CODAC platform is through ROBIN experiment [7] and TWIN source experiment [8]. Operation of INTF requires several plant systems like, beam source, gas feed, vacuum, cryogenics, high voltage power supplies, high power RF generators, mechanical systems and diagnostics systems. The experimental phases involve application of HV power supplies (100 KV) and High RF power (~ 800 KW) which will produce energetic beam of maximum power 6MW within the facility for longer durations. Hence the entire facility will be exposed to high heat fluxes and RF radiations. Therefore, to ensure investment protection and to provide occupational safety for working personnel a matured Interlock and Safety system is required for INTF. The Interlock and Safety systems are high-reliability systems devoted completely to the specific functions. These systems will be separate from the conventional DACS of INTF which will handle the conventional control and acquisition functions. Both, the Interlock and Safety systems are based on IEC 61508 and IEC 61226 standards as prescribed by the ITER PCDH guidelines. For complete operation of INTF, approximately 900 numbers of signals are required to be super-intending by the DACS. In INTF, conventional control loop time requirement is within the range of 5–100 ms. It requires sampling rates in the range of 5 sample per second (Sps) to 10 kSps. For high-end diagnostics, the required sampling rates are normally up to 100 MSps. All the corresponding hardware components have been selected from the ITER slow and fast controller catalogues, procured and tested. Combined use of CODAC core software (CCS) and NI-LabVIEW have been finalized due to the fact that full required DAQ support is not available in present version of CCS. Several prototype set-ups have been tested on CODAC-LabView integration. Some details are available in [15]. DACS for ROBIN is already in operation and for TWIN source, it is ready for integration. Different hardware components for INTF DACS are under procurement and testing using developed software.

2.6. High voltage bushing

High voltage bushing (HVB) or feedthrough for INTF is placed on the BS side dish-end of the VV in horizontal configuration and carries all the required feedlines (25 nos.) from transmission line to the beam source, which includes all the cooling lines like, RF transmission lines, HV lines, Instrumentation lines, Gas feed lines, Cesium oven delivery tube along with its heater connections etc. It also forms a primary vacuum boundary. INTF HVB has been designed considering the basic configuration of porcelain based large diameter (~1 m) insulator to provide 100 kV isolation. The design is validated using some analytical calculation and finite element method [16] and prototype experiments. Maximum stress obtained on porcelain ring and epoxy is 2.02 kV/mm and 2.23 respectively. Maximum localize stress obtained on the triple point of EG is 0.7 kV/mm. Maximum stress and the deflection due to vacuum and cantilever load by all the feedlines is determined to be 207 MPa and 3.3 mm respectively. All the stresses and deflections are under acceptable limit. Large diameter (~ 1 m) porcelain ring for HVB is being fabricated at BHEL, India. Several tests on individual components of HVB have been started before going for full-scale manufacturing.

3. Conclusion

The paper describes an overview of INTF facility which will characterize ITER DNB BS and its PSs along with BLCs with full specification before delivery of these components to ITER. Once established and functional, the facility shall provide rich database for the operation of a large size, multi-driver negative hydrogen ion based neutral beam system. The ROBIN and TWIN source experiments are important steps in the development roadmap, to acquire adequate experience on the understanding of operation of negative ion sources and ensuring preparedness for operation of INTF. INTF BS is expected by the end of 2017 and experimentation will initiate following integration of the

BS and experiments on characterisation and operational optimisation shall continue for next 2-3 years before the delivery of the system to ITER site.

References

[1] Chakraborty A, Rotti C, Bandyopadhyay M, Singh M J, Nair R G, Shah S, et al. 2010 *IEEE Trans. Plasma Sci.* **38** 248

[2] M G von Hellermann, Jaspers R, Biel W, Neubauer O, Hawkes N, Kaschuck Y et al. 2006 Proceedings of *21st IAEA Fusion Energy Conference*, Chengdu, China, unpublished.

[3] Bandyopadhyay M, Singh M J, Rotti C, Chakraborty A, Hemsworth R S, Schunke B 2010 *IEEE Trans. Plasma Sci.* **38** 242

[4] Green B J and ITER international team and participant teams 2003 *Plasma Phys. Control Fusion* **45**, 687

[5] Marcuzzi D, Dalla Palma M, Pavei M, Heinemann B, Kraus W, Riedl R 2009 *Fusion Eng. Des.* **84** 1253

[6] Singh M J, H P L De Esch 2010 *Rev. Sci. Instrum.* **81** 013305

[7] Bansal G, Gahlaut A, Soni J et al 2013 *Fusion Eng. Design* **88** 778

[8] Bandyopadhyay M, Pandey R, Shah S, Bansal G, Parmar D, Gahlaut A, Soni J, Yadav R K, Sudhir D, Tyagi H, Pandya K, Parmar K G, Mistri H S, Vuppugalla M and Chakraborty A K, 2014 *IEEE Transactions on Plasma Science* **42** 624

[9] Tyagi H, Shah S, Soni J, Yadav R K, Patel K, Mistri H, Parmar D, Bhagora J, Sharma D, Bandyopadhyay M, Chakraborty A K 2015 *10th Asia Plasma & Fusion Association (APFA-2015) Conference, India*

[10] Bansal G, Mishra S, Pandya K, Bandyopadhyay M, Soni J et al. 2013 *AIP Conf. Proc.* **1515**, 207 (doi: 10.1063/1.4792787)

[11] Singh M J, Bandyopadhyay M, Rotti C, Singh N P, Shah S, Bansal G, Gahlaut A, Soni J, Lakdawala H, Waghela H, Ahmed I, Roopesh G, Baruah U K, Chakraborty A K 2011 *Fusion Engineering & Design* **86** 732

[12] Joshi J, Yadav A, Gangadharan R, Prasad R, Ulahannan S, Rotti C, Bandyopadhyay M, Chakraborty A 2015 *Fusion Engineering and Design* **96** 488

[13] Rotti C, Panda N, Chakraborty A K, Balasubramanian K, Ahmed I, Roopesh G, Bandyopadhyay M, Singh M J, Shah S, Phukan A, Yadav R K 2011 *Proc. 24th Symp. Fusion Engineering (SOFE), Chicago, US*

[14] Bandyopadhyay M, Dass S and Chakraborty A 2016 *Rev. Sci. Instrum.* **87**, 02B906

[15] Soni J, Tyagi H, Yadav R, Rotti C, Bandyopadhyay M, Bansal G, Gahluat A, Dass S, Joshi J, Prasad R, Pandya K, Shah S, Parmar D, Chakraborty A 2015 *Fusion Engineering and Design* **96–97** 961

[16] Shah S, Sharma D, Bandyopadhyay M, Rotti C, Bansal G, Joshi J, Parmar D, Chakraborty A 2014 *Journal of the Korean Physical Society* **65** 1294

IOP Conf. Series: Journal of Physics: Conf. Series **823** (2017) 012002 doi:10.1088/1742-6596/823/1/012002

Upgradation plans of SST-1 Cryogenics system at IPR

V L Tanna[1], SST-1 Cryo Team[1] and S Pradhan[1]

[1]Institute for Plasma Research, Near Indira Bridge, Bhat, Gandhinagar 382 428, India

E-mail: vipul@ipr.res.in

Abstract. Steady State Superconducting Tokamak (SST-1) is India's First Superconducting Tokamak and has Toroidal Field (TF) and Poloidal Field (PF) superconducting coils along with the cold mass support structure weighing about 38 ton of cold mass. A 1.3 kW Helium refrigeration and liquefaction (HRL) at 4.5 K along with its distribution network facilitates the cooling down of the cold mass and cryo-stable operation of SST-1TF magnets. SST-1 experimental campaigns have revealed that the existing HRL is just sufficient for the heat loads acting on it. Further, the SST-1 PF magnets require a higher pressure head and mass flow rate than the nominal values on account of the longer paths of some of the PF magnets. In order to make SST-1 being fully superconducting device, we are introducing superconducting central solenoid (CS). Detailed estimates have been made and it has been found that an additional ~ 900 W at 4.5 K of cryo power is required towards (i) cooling all the PF magnets (ii) the cooling down and the operation of a new Nb_3Sn based CS of SST-1. This paper will elaborate on (i) the experimental heat loads acting on the cryo system (ii) the `thermal runaway amongst the PF magnets observed in the SST-1 campaign' (iii) the robust need of a higher operation pressure up to 2.1 bar (a) (iv) the need of the flow optimizations as per the hydraulic paths (v) the engineering solutions at each of these described (i)-(v) above.

1. Introduction

Several SST-1 cool-down experiments have been carried out for plasma campaigns. The SST-1 has sixteen toroidal field (TF) and nine poloidal field (PF) superconducting coils using NbTi superconductor. The PF coils have non-uniform hydraulic path lengths in the range of 67 m – 130 m. A 1.3 kW at 4.5 K helium refrigerator / liquefier (HRL) system is responsible to cool the TF, PF coils and cold mass support structure at 4.5 K [1]. Experimental data on magnets system have shown that the PF coils (Hydraulic path lengths of 67 m – 130 m) demand more than three times higher pressure head as compared to the TF coils (Hydraulic path length of 48 m) at lower temperature (15 K – 25 K). Till 2014, "Thermal run-away" situation observed amongst the PF coils, as there was only single common control valve for all the PF coils in the return path of distribution having non-uniform lengths. As a part of SST-1 up gradation activities, the PF hydraulic distribution system modified and branched into three groups having uniform path lengths for better control. To cool PF coils and to keep it in superconducting state, additional cold capacity of 450 W at 4.5 K is required [2]. In order to make SST-1 fully superconducting tokamak, Nb_3Sn based superconducting central solenoid (CS) is planned. The heat load for the CS is estimated based on physical sizing, distribution and a pair of current leads. This will demand cooling capacity of about 275 W at 4.5 K. To support the higher pressure head requirement on to the PF coils system, we need to provide the single phase sub-cooled helium at 2.7 bar (a) at 4.5 K. This has been deduced from the inlet and outlet conditions of the helium at rated flow rate of ~ 38 g/s in a single phase considering 100 mbar pressure drop across the heat-exchanger. An additional cold capacity of 175 W is needed to facilitate sub-cooled operation as shown in figure 1.

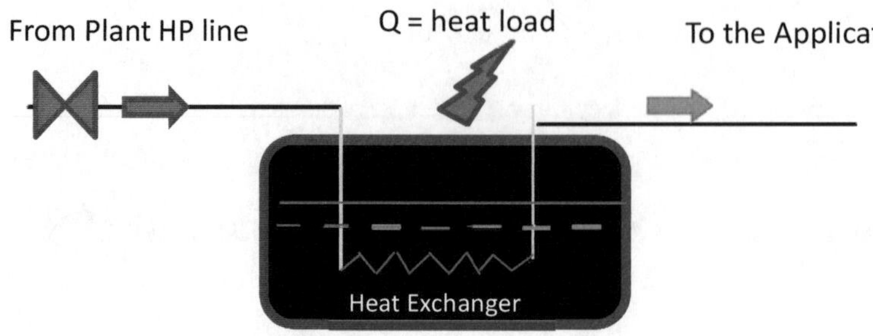

Figure 1. Schematic of a helium sub-cooler

In total, we proposed technical requirement of about 900 W at 4.5 K Helium Refrigeration (HR) system.

2. Cryogenic Heat loads of PF coils system

As mentioned in preceding section 1, 1.3 kW at 4.5 K HRL system is responsible for cooling of the TF, PF coils and cold mass support structure. After several campaigns, it is concluded that the TF coils are cooling down to 4.5 K with existing 1.3 kW HRL for 10-12 days. At present, only 1 pair of vapor cooled current leads are connected. The PF coils are getting cool-down with inlet/outlet conditions of 15 K/24 K with the available pressure head from the HRL system. Estimated values of heat loads on the PF coils system are discussed in Table 1 and corresponding the mass flow rate requirements has been deduced from the actual experimental data on pressure drops and thermo-hydraulics. This has been also verified using established co-relation of H. Katheder [3] relation for CICC pressure drop estimation.

Table 1. Heat loads acting on to the PF magnets system in SST-1 (as installed conditions).

Coils	Nos.	Each heat load from 80 K – 4.5 K (W)	Q_{total} (Watt)
PF-1	01	2.7	2.7
PF-2 (U/L)	02	3.1	6.2
PF-3 (U/L)	02	5.5	11.0
PF-4 (U/L)	02	3.5	7.0
PF-5 (U/L)	02	6.4	12.8
Thermal conduction due to supports		10	25.0
SC bus ducts	18	3 W/bus	54.0
Cryo lines-I	20 m	1.5 W/m	30.0
Cryo-lines-II	20 m		20.0
Valve-Box	01	20	20.0
Total static heat load			189
Transient load			11

IOP Conf. Series: Journal of Physics: Conf. Series **823** (2017) 012002 doi:10.1088/1742-6596/823/1/012002

Total Heat Load @4.5 K	200
Safety margin of uncertainty in heat load estimation (1.5 factor)	300
Cooling capacity sizing (1.5 factor) [6]	450

3. Thermal Runaway in SST- 1 PF coils and its Mitigation

Existing 1.3 kW at 4.5 K HRL along with the IFDCS system used to cool down SCMS of SST-1. During SST-1 campaign, the sixteen TF coils are cooled down from 300K to 4.5K in a uniform and controlled manner. The behaviour of cool-down of individual PF coils is shown in figure 1 for one of the SST-1 campaign. As shown in figure 1, some of the PF coils show thermal runaway behaviour at temperature below ~ 90K and pressure below ~ 6.5 bar (a). Due to thermal runaway the temperature distribution is not uniform in PF coils and the PF coils demands higher pressure heads to get cool-down further. The thermal run away problem is addressed by increasing the Helium pressure head from cold-box of the HRL by optimising the valves of refrigeration stream. The process optimization includes (i) maintaining constant source pressure to refrigeration stream till 125 K and (ii) reducing the source pressure by regulating the refrigeration Joule –Thomson (JT) valve of high pressure (HP) stream and maintaining compressor flow rate within nominal value [4]. This process optimization removed thermal run away as well as reduce the cool down time. The modification within existing helium integrated flow distribution and control (IFDC) system was implemented. In-house development of 3S-2R cryo line has made very important role in realizing the flow distribution upgradation [5]. Successful installation and commissioning of cryo control and on/off valves in IFDCS with process lines layout and provision of proper flexibility have been carried out. During recent SST-1 campaign, IFDC upgraded system was successfully operated. Appropriate precise control and valve adjustment provided in future SST-1 operation will be tried out to make PF coils superconducting as shown in figure 2.

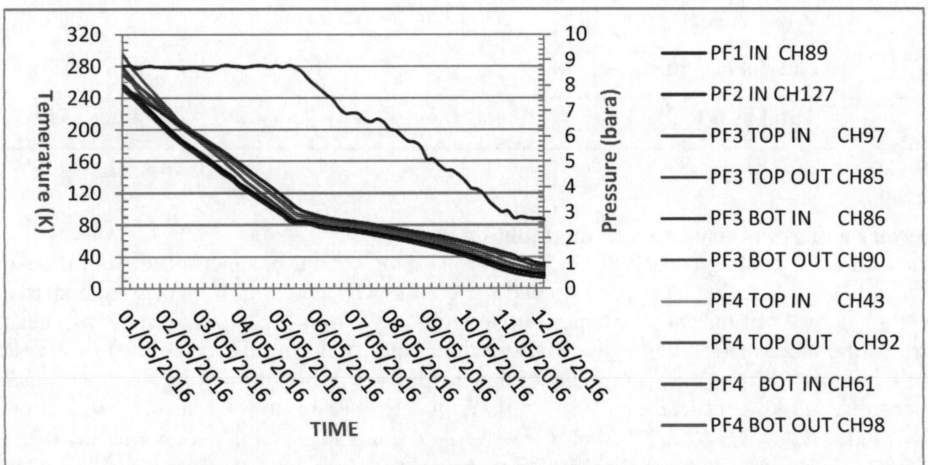

Figure 2. Temperature trend of PF coils for SST-1 Campaign

IOP Conf. Series: Journal of Physics: Conf. Series **823** (2017) 012002 doi:10.1088/1742-6596/823/1/012002

4. Heat loads of the SST-1 Central Solenoid (CS)

In order to make SST-1 as fully superconducting device, as a part of upgradation activities, new central solenoid has been planned using Nb_3Sn superconductors. Therefore, we need an additional cooling power for making the CS to be in a superconducting state. The total heat loads acting on the CS system has been estimated and summarize in Table 2.

Table 2. Heat loads estimation for the CS system.

Sr. No.	Sources of heat load from 80 K – 4.5 K (W)	Quantity	Each heat load estimation (W)	Total heat loads estimation (W)
1	Radiation	01	3.0	3.0
2	Residual gas conduction	01	0.5	0.5
3	Thermal conduction	03	1 W/no	3.0
4	Cryolines	20 m	0.5 W/m	10.0
5	SC feeders + Vacuum Barriers (VB)	5 m	1.6 W/m	8.0
6	Instrumentation losses		9.0	9.0
	Static losses (Total)			**33.5**
	Pulsed / AC losses (During operation)			
7	Joint Joule heating loss	02	0.5 W/joint	1.0
8	AC losses (inclusive of Eddy, Coupling and Hysteresis)			20.8
9	Transient heat load			21.8
	Total heat load			**55.0**

5. Overall cooling capacity sizing as upgradation

In order to size the cooling capacity of helium cryo system, should be determined from the required capacity in both the steady state operation and the cool-down operation. In general, the steady state refrigeration capacity of the cryo system is defined to be 1.5 times of the steady state heat loads of a system. On the other hand, the required refrigeration capacity during the cool-down of a system is used to be estimated from the specified cool-down time and the total mass to be cooled under the conditions of cool-down efficiency of 50% [6]. With reference to above heat load estimation, IPR requires $(300 + 83) \times 1.5W = 575W$ at 4.5 K refrigeration capacity for the cooling of PF and CS systems. There is a pair of MgB_2 based CLs to power the CS at 14 kA. This will have additional cooling capacity demand of 150 W or equivalent to 1.5 g/s at 4.5 K. IPR also needs the high pressure (2.7 bar (a)) cold helium looking at the PF coils pressure head demands with sub-cooled temperature of 4.5 K at the inlet of the PF and CS using heat exchanger bath cooled in a sub-cooled Dewar (SBD). The requirement of cold capacity for sub-cool Dewar is estimated as 175 W equivalent cooling power at 4.5 K for 38 g/s single phase flow at 4.5 K and 2.7 bar (a) (inlet). Therefore, in total, there is a clear need of helium refrigeration plant of 900W at 4.5 K.

6. Cooling Philosophy

As a part of SST-1 up gradation activities, the cryo-distribution of PF system will be modified such that uniform path lengths of PF coils are grouped together with combination of more or less equal path lengths. These modifications have been already made in existing IFDCS. These three groups of PF coils will be cooled with new helium refrigeration (HR) system. The PF coils require much higher pressure head of 2.7 bar(a). The new HR system must supply cold power at 4.5 K, 2.7 bar(a) in sub-cooled single phase form. The return from PF system is at 1.1 bar (a) or equal to suction pressure of new HR system. In distribution system, 3 – supply, 2 – return (3S-2R lines) cryo lines are fabricated to provide cold power from IFDCS system to SST-1. Intended new HR plant will be completely separate from the existing 1.3 kW at 4.5 K HRL system. However, they will use common buffer as well as recovery systems. They will be interfaced at IFDC and SST-1 systems as shown in figure 3.

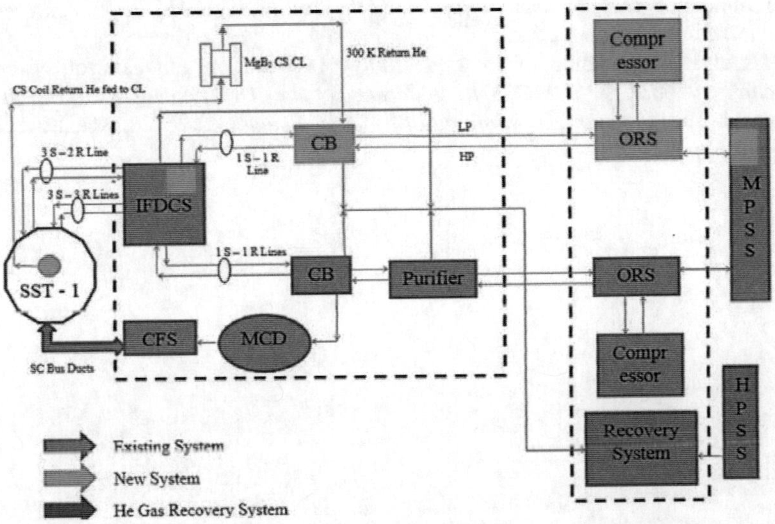

Figure 3. Cooling Philosophy of existing and upgraded helium cryo system at IPR

The CS will be made from Nb_3Sn and have a pair of MgB_2 based current leads (CL) rated at 14 kA. It will be cooled by new HR system with sub-cooled single phase helium at 4.5 K, 2.7 bar (a) as shown in figure 3. Similar to the PF system, the CS will have CICC path lengths of 200 m approximately, which require higher pressure head of 1.2 bar in comparison to PF & TF system. To take the advantage of MgB_2 based CL, the return cold helium from the CS, the helium flow will be fed to the bottom of individual CL equally. The helium gas is warmed up to 300 K along the MgB_2 CL het exchangers and fed back to the suction of compressor of HR plant at room temperature. Considering this aspect, as this return enthalpy is not used in cold-box, equivalent cold flow taken as liquefaction load in calculation of plant capacity. For this purpose 1.5 g/s at 4.5 K, 2.7 bar(a) single phase flow is considered to remove static and dynamic (transient) heat load of CS. The equivalent additional refrigeration power for CL is 150 W at 4.5 K needs to be considered while sizing of HR capacity. A sub-cool Dewar (SBD) with heat exchanger in bath is required to mitigate the longer path lengths with inlet of 2.7 bar(a), 4.5 K and mass flow rate of ~ 38 g/s. The plant outlet must offer stable 4.5 K after SBD.

IOP Conf. Series: Journal of Physics: Conf. Series **823** (2017) 012002 doi:10.1088/1742-6596/823/1/012002

7. Conclusions

As a part of SST-1 cryo system upgradation, we have made the estimation of cryo heat loads and which was compared with the actual SST-1 experimental data from the recent campaigns. It is clear that we need an additional cold capacity of about 900 W at 4.5 K for facilitating the PF coils and central solenoid to be in a superconducting state at higher pressure head requirements as part of SST-1 upgradation plans. In-house efforts such as increasing the pressure head from existing HRL by optimising the process of refrigeration stream of cold-box of HRL and IFDC upgradation has facilitated the uniform cool-down of PF coils system without any thermal run away situation.

References

[1] Tanna V L et al Cryogenics related Development and Test activities for the SST-1 refurbishment, *ICEC-ICMC-2012 at Fukuoka, Japan May 2012.*

[2] Pradhan S and SST-1 Team 2015 *Nuclear Fusion* **55** 104009

[3] Zanino R et al *Cryogenics* 2006 **46** 541

[4] Panchal P et al Process optimization of Helium Cryo Plant Operation for SST-1 superconducting magnet system *presented in 26thICEC-ICMC-2016 conference at New Delhi during 7-11 March 2016*

[5] Panchal R et al Upgradation of Integrated Flow Distribution and Control system of SST-1 *presented in 26thICEC-ICMC-2016 conference at New Delhi during 7-11 March 2016*

[6] Akiyama M 1991 *Design Technology of Fusion Reactors* (World Scientific Publication) Chapter 5 pp 289-290.

IOP Conf. Series: Journal of Physics: Conf. Series **823** (2017) 012003 doi:10.1088/1742-6596/823/1/012003

Improvements in a Tracer-Encapsulated Solid Pellet and Its Injector for More Advanced Plasma Diagnostics

Naoki Tamura[1], Shigeru Sudo[2], Chihiro Suzuki[1], Hisamichi Funaba[1], Masaru Takagi[3], Nakahiro Satoh[3], Hiromi Hayashi[1], Hiroya Maeno[1], Mitsuhiro Yokota[1] and Hideki Ogawa[1]

[1]National Institute for Fusion Science, 322-6 Oroshi-cho, Toki-shi, Gifu 509-5292, Japan
[2]Chubu University, 1200 Matsumoto-cho, Kasugai-shi, Aichi 487-8501, Japan
[3]Hamamatsu Photonics Inc., 1820 Kurematsu-cho, Hamamatsu-shi, Shizuoka 431-1202, Japan

E-mail: ntamura@lhd.nifs.ac.jp

Abstract. A Tracer-Encapsulated Solid Pellet (TESPEL) has been developed for promoting a precise study of the impurity transport in a magnetically-confined high-temperature plasma. This paper gives a brief report of the recent improvements in the TESPEL and its injector for more advanced plasma diagnostics. The TESPEL can be considered as a double-layered impurity pellet. This structure enables us to produce a both poloidally and toroidally localized "tracer" impurity source in the plasma, and to specify the total amount of the tracer impurity deposited in the plasma precisely. Recent experiments on the Large Helical Device by using the TESPEL suggest that the importance of the impurity source location in the impurity transport study. Thus we have developed new type TESPELs, which are greatly improved in regard to the above-mentioned unique features. In addition, we also developed a new TESPEL injector, which enables us to inject the TESPEL obliquely into the plasma. This injector can also contribute to a further shallower penetration of the TESPEL into the plasma.

1. Introduction

Although we are approaching the age of the International Thermonuclear Experiment Reactor, many physics issues related to the confinement of magnetically-confined toroidal plasma still remain to be clarified. For example, under some conditions, impurities inside the magnetically-confined toroidal plasma tend to accumulate in the core region of the plasma [1, 2]. This will cause a dilution of fusion fuel. Moreover, radiation loss from the core plasma will be enhanced due to the impurity accumulation, and then the temperature in the core region will be decreased dramatically. Consequently, fusion plasma performance will be degraded below the acceptable level. In order to develop strategy for obviating and suppressing impurity accumulation, it is important to gain a full understanding of impurity transport in the magnetically-confined toroidal plasma.

In the study of impurity transport in the magnetically-confined toroidal plasmas, a spatio-temporal behavior of the impurity injected into the plasma whether intentionally or accidentally has been utilized [3, 4]. The most common method for the external injection of impurities is a bare impurity pellet injection method [5]. The impurities injected by this conventional method will be deposited inevitably even in the edge ergodic region (outside the last closed flux surface). This would become a

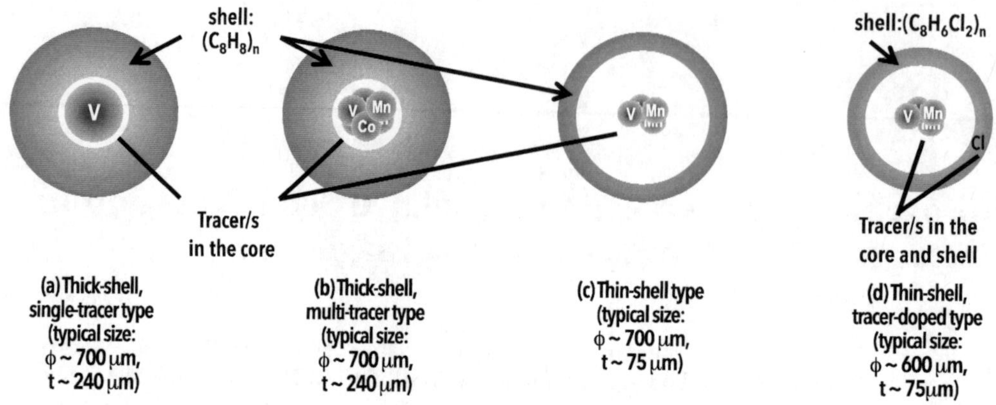

Figure 1. Summary of the tracer-encapsulated solid pellets (TESPELs) developed to date. (a) Thick-shell single-tracer type (typical size: OD = 700 μm, shell thickness ~ 240 μm), (b) Thick-shell multi-tracer type (typical size: OD = 700 μm, shell thickness ~ 240 μm), (c) Thin-shell type (typical size: OD = 700 μm, shell thickness ~ 75 μm), (d) Thin-shell tracer-doped type (typical size: OD = 600 μm, shell thickness ~ 75 μm). The development of the TESPEL has proceeded in the same order of this itemized list. The material for the shell is a polystyrene $(C_8H_8)_n$ polymer, except for the thin-shell tracer-doped type, whose shell material is made of poly-2,6-dichlorostyrene $(C_8H_6Cl_2)_n$ polymer. The tracer impurities in the TESPEL core used recently are shown in figure: vanadium (V), manganese (Mn) and cobalt (Co).

problem in estimating the total amount of the externally-injected impurity deposited in the confined plasma. In order to overcome this disadvantage, a Tracer-Encapsulated Solid Pellet (TESPEL) [6, 7, 8] has been developed at the National Institute for Fusion Science (NIFS) in Japan. This paper reports recent improvements in the TESPEL and its injector for more advanced plasma diagnostics.

2. Improvements in the Tracer-Encapsulated Solid Pellet (TESPEL)

The TESPEL can be considered as a double-layered impurity pellet. The TESPEL basically consists of a hydrocarbon polymer as an outer shell and a tracer impurity as an inner core. This configuration enables the following unique features: a) the TESPEL injected into the magnetically-confined toroidal plasma can produce both a toroidally and a poloidally localized tracer impurity source in the core region of the plasma; b) the total amount of the tracer impurity deposited in the core plasma can be identified clearly; since the size of the inner core of TESPEL can be measured during the TESPEL production process; c) various elements can be selected as the tracer impurity; and d) variable penetration depth of the tracer impurity can be obtained owing to the flexible TESPEL size. When the TESPEL enters into the magnetically-confined toroidal plasma, the outer shell of the TESPEL is first ablated, and then is followed by the ablation of the tracer impurity in the core of the TESPEL. Therefore, the control of the penetration depth of the TESPEL corresponds to that of the deposition location of the tracer impurity by the TESPEL. The detailed production procedure of the TESPEL is summarized in ref. [8].

The first version of the TESPEL is a thick-shell, single-tracer type, as shown in Fig. 1(a). The typical outer diameter and shell thickness of this thick-shell, single-tracer type TESPEL is ~ 700 μm and ~ 240 μm, respectively. The outer diameter can range widely from 400 to 900 μm. Consequently, the feature of this type of TESPEL is a wide range of the total amount of the tracer impuirty. The second version of the TESPEL is a thick-shell, multi-tracer type, as shown in Fig. 1(b). This type of TESPEL has been developed for the purpose of estimating precisely the difference of the behavior between the intrinsic (chromium (Cr), iron (Fe) and nickel (Ni)) and the tracer impurities (vanadium

(V), manganese (Mn) and cobalt (Co)) [9]. In the case of the above-mentioned types of the TESPEL, the penetration depth of the TESPEL is mostly affected by the outer diameter of the TESPEL. That is, the region where the tracer was deposited can be controlled almost by the size of the TESPEL. Recent TESPEL experiments on the Large Helical Device (LHD) suggest the importance of the impurity source location in the impurity transport study [10]. Thus when the region of interest is the outer region of the plasma (e.g., that near the last closed flux surface), the required TESPEL size becomes inevitably smaller, and correspondingly the amount of the tracer impurity also becomes smaller. This becomes a problem that confronts the impurity diagnostics such as a spectrometer. In order to solve this problem, we developed a thin-shell type TESPEL, which is based on the fabrication technology of a polystyrene polymer shell [11]. The typical outer diameter and shell thickness of this thin-shell type TESPEL is ~ 700 μm and ~ 75 μm, respectively. Thus the shell thickness of this thin-shell type TESPEL is 70 % thinner than that of the thick-shell type TESPEL. Another advantage of the thin-shell type TESPEL is that the amount of the tracer impurity can be increased more than that of the thick-shell, single-tracer type TESPEL. For obtaining a further shallower penetration than even with the thin-shell type TESPEL, we developed a thin-shell, tracer-doped type TESPEL [12]. For this type of TESPEL, the shell is made of poly-2,6-dichlorostyrene $(C_8H_6Cl_2)_n$. Thus the chlorine (Cl) in the shell behaves as a tracer impurity. The typical outer diameter and shell thickness of this thin-shell, tracer-doped type TESPEL is ~ 600 μm and ~ 75 μm, respectively. As can be seen from Fig. 1(d), the thin-shell, tracer-doped type TESPEL can also contain the tracer impurities in the core.

The deflection effect of the TESPEL flight path due to the fast ions originated from the neutral beam injection for the plasma heating could affect the deposition location of the tracer impurity by the TESPEL. However, the deflection effect of the TESPEL flight path has no appreciable effect on the impurity transport study, because the impurity transport study generally does not need the deep deposition location of tracer impurity in the confined plasma, in contrast to the solid hydrogen isotope ice pellet injection for the plasma fueling. And in the case for the shallow penetration with the thin-shell type TESPEL, the deflection effect due to the fast ions could occur further inner side of the target region with the thin-shell type TESPEL [13], and thus the deflection effect of the TESPEL flight path has little impact on the study of interest.

3. New TESPEL Injector

As described above, the recent LHD experiments using the TESPEL suggest that the importance of the impurity source location in the study of impurity transport. Therefore, it is very important to put more control into the placement of the tracer impurity deposition for further promoting the impurity transport study. One solution is the development of the new-type TESPELs described above. Another solution is the development of an injector, which can control more precisely the location of the tracer impurity deposition. In the development of such an injector, we decided to build a new TESPEL injector capable of making an oblique injection of the TESPEL into the LHD plasma, because the existing TESPEL injector has no room for the addition of such a function.

The basic design of the new TESPEL injector is almost the same as the current injector, which consists of the TESPEL injection system and the differential pumping system. In the TESPEL injection system, the TESPEL will be accelerated through the gun barrel (with the inner diameter of 1 mm and its length of ~ 410 mm) by a helium gas with a typical pressure of several MPa. The differential pumping scheme works for preventing the high-pressure helium gas for the TESPEL acceleration from penetrating into the LHD vacuum vessel. Here, a three-stage differential pumping system is installed. The volume of the expansion chambers is 0.05 m^3 (first, furthest from LHD), 0.06 m^3 (second) and 0.06 m^3 (third, nearest to LHD), respectively. The first expansion chamber is evacuated by a helical groove pump (0.27 m^3/s for He) and the second and third expansion chambers are evacuated by turbo-molecular pumps (0.16 m^3/s for He and 0.32 m^3/s for He, respectively). The first expansion chamber is connected with the second expansion chamber through an ultrafast shutter valve (closing time, less than 10 milliseconds), and then the second expansion chamber is connected with the third expansion chamber through a gate valve (closing time, ~ 1 second). After the ejection of

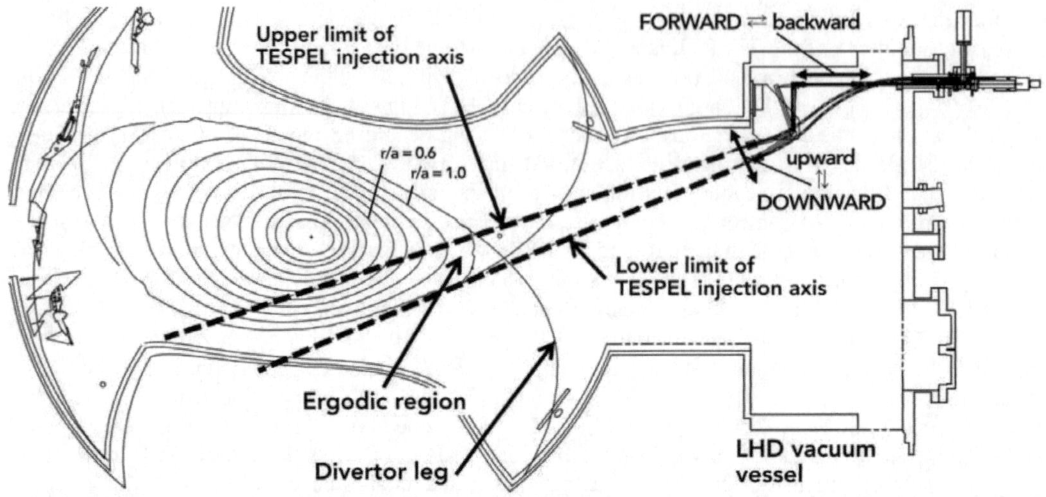

Figure 2. Cross-sectional view of the LHD with the plane of the TESPEL injection by the new injector. The remote-controlled mechanism can move the final guide tube of the new injector upward and downward. Consequently, the expected location of the tracer deposition with the new injector is ranged from the location at $r/a \sim 0.6$ to that inside the ergodic region (outside the last closed flux surface).

the TESPEL, these valves will be closed immediately. In order to realize effectively the oblique injection of the TESPEL, the final guide tube (i.e., nearest to the LHD plasma) for the TESPEL is installed in the vacuum vessel of LHD. Figure 2 shows a cross-sectional view of the LHD with the plane of the TESPEL injection by the new injector. In figure 2, the magnetic surfaces with the magnetic axis R_{ax} = 3.75 m are also depicted for reference. The remote-controlled mechanism can move the slightly-bended final guide tube of the new injector upward and downward. Consequently, the expected location of the tracer deposition with the new injector ranges from the location at $r/a \sim 0.6$ to that inside the ergodic region. The TESPEL injection axis only in the edge ergodic region will be extremely useful for the study of an edge-radiation-enhanced operation scenario. The TESPEL entering LHD plasma will be observed by the existing spectroscopic diagnostics [14] and a stereoscopic ultra-fast camera system [15], which enable us to confirm the TESPEL penetration depth and study the TESPEL ablation cloud physics. The detailed comparison of the TESPEL injection on between the straight path with the conventional injector and the oblique path with the new injector is a future work.

4. Conclusion

Important improvements in the TESPEL and its injector have been recently performed for more advanced and precise study of the impurity transport. Concerning the TESPEL, two new types of the TESPEL, the thin-shell type and the tracer-doped-thin-shell type, have been recently developed. The thin-shell structure enables us to achieve the shallower penetration of the TESPEL. And the tracer-doped-thin-shell structure allows us to deposit the tracer impurity around the last closed flux surface. As a consequence, the possible location for the tracer impurity deposition by using the TESPEL technique becomes more flexible. Concerning the TESPEL injector, the new TESPEL injector has been recently installed on the LHD for realizing the oblique injection of the TESPEL. The new TESPEL injector empowers us to achieve the even shallower penetration of the TESPEL, in particular to deposit the tracer impurity only in the ergodic region outside the last closed flux surface.

Acknowledgement
The authors acknowledge all of the technical staff of NIFS for their excellent support. They also would like to thank Prof. Y. Takeiri (Director General of NIFS) for his continuous encouragement. This work is supported by a Grant-in-Aid for Young Scientists from the Toray Scientific Foundation, JSPS KAKENHI Grant Numbers 23360415, 15H04234 and a budgetary Grant-in-Aid (No. ULHH012) of the National Institute for Fusion Science.

References
[1] Matthews G F et al. 2013 *J. Nucl. Mater.* **438** S2
[2] Nakamura Y et al. 2014 *Plasma Phys. Control. Fusion* **56** 075014
[3] Zurro B et al. 2011 *Nucl. Fusion* **51** 063015
[4] Sertoli M et al. 2015 *Nucl. Fusion* **55** 113029
[5] Kaneko H et al. 1987 *Nucl. Fusion* **27** 1075
[6] Sudo S et al. 1993 *J. Plasma Fusion Res.* **69** 1349
[7] Khlopenkov K V et al. 1998 *Rev. Sci. Instrum.* **69** 3194
[8] Sudo S and Tamura N 2012 *Rev. Sci. Instrum.* **83** 023503
[9] Sudo S et al. 2012 *Nucl. Fusion* **52** 063012
[10] Sudo S et al. 2013 *Plasma Phys. Control. Fusion* **55** 095014
[11] Takagi M et al. 2002 *Fusion Sci. Technology* **41** 278
[12] Tamura N et al. 2015 *Plasma Fusion Res.* **10** 1402056
[13] Mishra J S et al. 2011 *Nucl. Fusion* **51** 083039
[14] Tamura N et al. 2008 *Rev. Sci. Instrum.* **79** 10F541
[15] Tamura N ea al., 2016 *Rev. Sci. Instrum.* **87**, 11D615

IOP Conf. Series: Journal of Physics: Conf. Series **823** (2017) 012004 doi:10.1088/1742-6596/823/1/012004

Initial results in SST-1 after up-gradation

S Pradhan[1], Z Khan[1], V L Tanna[1], U Prasad[1], Y Paravastu[1], D C Raval[1], H Masand[1], Aveg Kumar[1], J R Dhongde[1], S Jana[1], B Kakati[1], K B Patel[1], M K Bhandarkar[1], B K Shukla[1], D Ghosh[1], H S Patel[1], T J Parekh[1], I A Mansuri[1], K R Dhanani[1], A Varadharajulu[1], Y S Khristi[1], P Biswas[1], C N Gupta[1], S George[1], P Semwal[1], D K Sharma[1], H K Gulati[1], K Mahajan[1], B R Praghi[1], M Banaudha[1], A R Makwana[1], H H Chudasma[1], M Kumar[1], R Manchanda[1], Y S Joisa[1], K Asudani[1], S N Pandya[1], S K Pathak[1], S Banerjee[1], P J Patel[1], P Santra[1], F S Pathan[1], P K Chauhan[1], M S Khan[1], P L Thankey[1], A Prakash[1] , P N Panchal[1], R N Panchal[1], R J Patel[1], G I Mahsuria[1], D P Sonara[1], K M Patel[1], S P Jayaswal[1], M Sharma[1], J C Patel[1], P Varmora[1], G L N Srikanth[1], D R Christian[1], A Garg[1], N Bairagi[1], G R Babu[1], A G Panchal[1], M M Vora[1], A K Singh[1], R Sharma[1], H D Nimavat[1], P R Shah[1], G Purwar[1], T Y Raval[1], A L Sharma[1], A Ojha[1], S Kumar[1], N K Ramaiya[1], V Siju[1], M V Gopalakrishna[1], A Kumar[1], P K Sharma[1], P K Atrey[1], SV Kulkarni[1], K K Ambulkar[1], P R Parmar[1], A L Thakur[1], J V Raval[1], S Purohit[1], P K Mishra[1], A N Adhiya[1], U C Nagora[1], J Thomas[1], V K Chaudhari[1], K G Patel[1], S Dalakoti[1], C G Virani[1], S Gupta[1], Ajay Kumar[1], B Chaudhari[1], R Kaur[1], R Srinivasan[1], D Raju[1], D H Kanabar[1], R Jha[1], A Das[1], D Bora[1] and SST-1 Team[1]

[1]Institute for Plasma Research, Bhat Village, Gandhinagar-382 428, India

E-mail: pradhan@ipr.res.in

Abstract. SST-1 Tokamak has recently completed the 1st phase of up-gradation with successful installation and integration of all its First Wall components. The First Wall of SST-1 comprises of ~ 3800 high heat flux compatible graphite tiles being assembled and installed on 132 CuCrZr heat sink back plates engraved with ~ 4 km of leak tight baking and cooling channels in five major sub groups equipped with ~ 400 sensors and weighing ~ 6000 kg in total in thirteen isolated galvanic and six isolated hydraulic circuits. The phase-1 up-gradation spectrum also includes addition of Supersonic Molecular Beam Injection (SMBI) both on the in-board and out-board side, installation of fast reciprocating probes, adding some edge plasma probe diagnostics in the SOL region, installation and integration of segmented and up-down symmetric radial coils aiding/controlling plasma rotations, introduction of plasma position feedback and density controls etc. Post phase-I up-gradation spanning from Nov 2014 till June 2016, initial plasma experiments in up-graded SST-1 have begun since Aug 2016 after a brief engineering validation period in SST-1. The first experiments in SST-1 have revealed interesting aspects on the `eddy currents in the First Wall support structures' influencing the `magnetic Null evolution dynamics' and the subsequent plasma start-up characteristics after the ECH pre-ionization, the influence of the first walls on the `field errors' and the resulting locked modes observed, the magnetic index influencing the evolution of the equilibrium of the plasma column, low density supra-thermal electron induced discharges and normal ohmic discharges etc. Presently; repeatable ohmic discharges regimes in SST-1 having plasma currents in excess of 65 KA ($q_a \sim 3.8$, $B_T = 1.5$ T) with a current ramp rates ~ 1.2 MA/s over a duration of ~ 300 ms with line averaged densities ~ 0.8×10^{19} and temperatures ~ 200 eV with copious MHD signatures have been experimentally established. Further elongation of the plasma duration up to one second or more with position and density feedback as well as coupling of Lower Hybrid waves are currently being persuaded in SST-1 apart from increasing the core plasma parameters with further optimizations and with wall conditioning.

1. Introduction

The Steady State Supercondcuting Tokamak (SST-1) is a 'working' experimental superconducting device since late 2013. In the long term, SST-1 envisages steady state operation in both the single null and double null configurations in large aspect ratio configurations [1]. The primary magnetic configurations and plasma shaping magnetic requirements are provided by Superconducting Magnet Systems (SCMS) comprising of sixteen superconducting D-shaped Toroidal Field (TF) magnets and nine superconducting Poloidal Field (PF) magnets together with a pair of resistive PF coils inside the vacuum vessel. An air-core Ohmic transformer together with three pairs of resistive compensating coils along with a 500 kW, 500 ms Gyrotron are used for the pre-ionization, initial break-down and initial current start-up. A pair of resistive vertical field magnets outside the cryostat placed symmetrically around the mid plane provides the initial equilibrium. Figure 1 shows the various components of SST-1 machine shell.

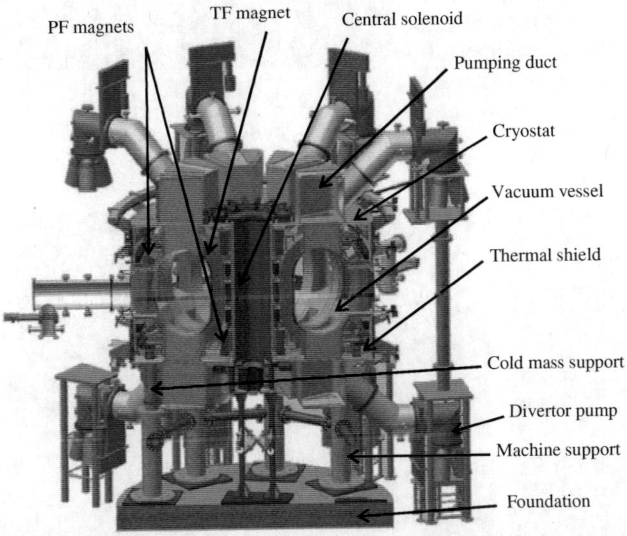

Figure 1. 3D cut view of SST-1 Machine

Table 1. Major SST-1 machine parameters (PFC related)

Parameters	**Values**
Major radius	1.1 m
Minor radius	0.2 m
SS surface area of VV	75 m^2
Exposed surface area of PFC	40 m^2
Plasma species	Hydrogen
Volume enclosed by PFC	16 m^3
Steady State Heat Flux	
Main Baffle	0.25 MW/m^2
In / Outboard Passive Stabilizer	0.25 MW/m^2
In / Outboard Divertor Plate	0.6 MW/m^2
In / Outboard Poloidal Limiter	1.0 MW/m^2

After obtaining a plasma current of ~ 75 kA at a central field of 1.5 T for ~ 500 ms in a limiter equipped SST-1 in 2014, as part of phase-1 up-gradation of SST-1 Plasma Facing Components (PFC) were installed successfully during Oct 2014 – May 2015 (Table-1). Subsequently, PFC equipped SST-1 under gone a rigorous 'engineering validation' to validate leak tightness, thermal and electrical isolations, meteorology consistencies etc. as described in section-2 of this paper. Initial experiments also quantified eddy currents influencing the NULL, the plasma break-down and start-ups, the equilibrium field index and the initial equilibrium etc. in the PFC equipped SST-1 as described in section-3 of the paper. The MHD aspects, Disruption, Edge Turbulence and Radial transport observed have been elaborated in section-4. The future plans inclusive of engineering up-gradation and physics on SST-1 have been briefly outlined in section-5 of this paper along with a summary.

2. SST-1 Up-gradation (Phase-1)

During 2014-15, the major up-gradations in SST-1 has been carried out involving PFC integration, cryogenic flow distribution and control (IFDC) system and SMBI system for gas fuelling.

IOP Conf. Series: Journal of Physics: Conf. Series **823** (2017) 012004 doi:10.1088/1742-6596/823/1/012004

2.1. PFC Up-gradation

SST-1 Tokamak was successfully commissioned in 2012 and the first plasma was obtained in June 2013 with poloidal limiters having SS 304L as vessel wall material. In the phase-1 up-gradation [2], Graphite based PFCs designed to withstand an input heat load of 1.0 MW/m^2 have been mounted on a back plate made of Cu-Zr & Cu-Cr-Zr alloys embedded with SS 304L piping to carry the heat away. Approximately 3800 tiles have been mounted on 132 numbers of copper alloys back-plates on a surface area of 40 m^2 on a vacuum vessel volume of ~ 16 m^3. PFC components are baked up to 250 °C for twenty four hours under UHV conditions using hot nitrogen baking system while keeping the vacuum vessel at 150 °C.

SST-1 Prototype is used for first installation of plasma facing components along with supports using combination of ECDS (Electronic co-ordinate determining system) and Photogrammetric measurements. Mechanical templates and Go and No-go gauge were made to ensure the alignment and space between each module during assembly. Then after, full assembly of copper modules was carried out in SST-1 main vessel. Appropriate mechanical spacers and shims were used to correct alignment of the components with respect to other modules and machine axis.

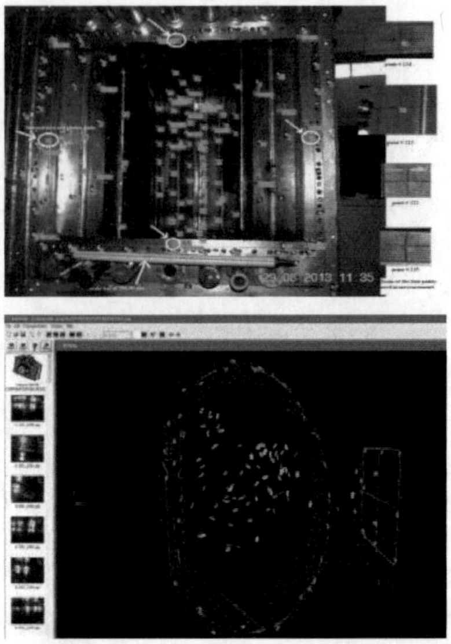

Figure 2. Coded and un-coded targets mounted on prototype vacuum vessel and co-ordinate generation in AUSTRALIS 7.0

Figure 3. Full assembly of copper module in SST-1 vessel

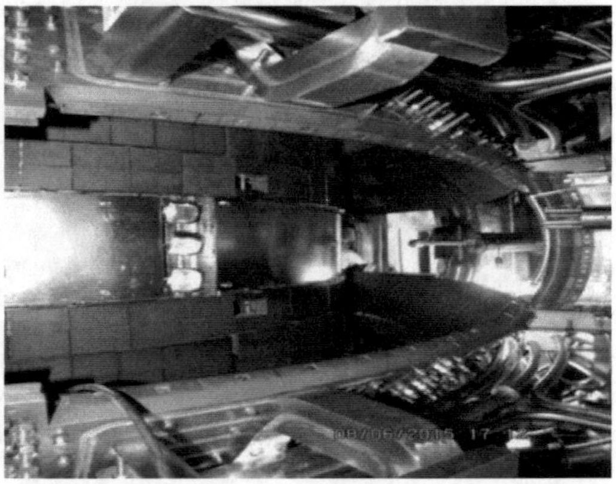

Figure 4. Final assembly of PFC modules (with graphite tiles) in SST-1 vessel

Graphite tiles are mechanically bolted on the back plate with a graphoil sheet in between to ensure good thermal contact whereas the temperature of the back plate is monitored with thermocouples. In continuation to PFC assembly, the interconnecting pipes are also laid to connect different modules with supply and return headers for baking and cooling. As a part of PFC conditioning, baking of components is carried out at 250 °C for extended duration (figure 5). Afterwards, prior to plasma experiments hydrogen discharge cleaning followed by extended helium discharge cleaning are usually carried out.

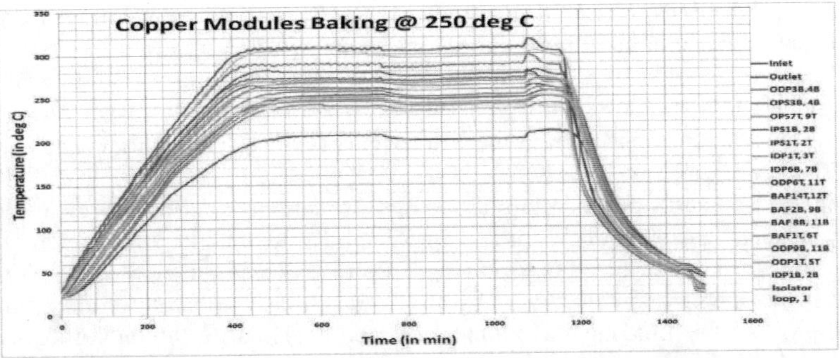

Figure 5. Performance of Baking of Copper Modules at 250 °C.

2.2. *IFDC Up-gradation*

Hydraulic imbalances and thermal run way had been observed amongst the different hydraulic path lengths of PF coils systems. The original PF distribution scheme had a single control valve for the entire PF magnets. The flow distribution scheme has been subsequently modified (figure 6 and figure 7) balancing the hydraulic impedances to various unequal PF flow paths. With these modifications and up-gradations, the cooling down in PF magnets has now been devoid of thermal run away.

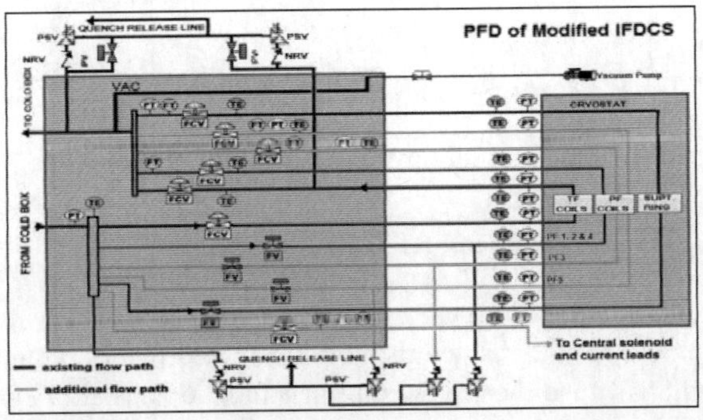

Figure 6. Process flow and instrumentation diagram of upgraded IFDCS

Figure 7. Snapshots of Installation of upgraded IFDCS and 3S-2R transfer line.

2.3. *SMBI system for SST-1 Tokamak*

The currently adopted fueling techniques in Tokamak devices include conventional gas puffing (GP), supersonic molecular beam injection (SMBI) and pellet injection (PI) etc. Among these methods, SMBI is found to be better as compared to GP and is technically simpler in comparison to PI. The fueling efficiency of SMBI is around 30 – 60 %, about three to four times higher than the conventional gas puffing (GP). SMBI can also enhance the penetration depth and the fueling efficiency. A higher fueling efficiency, reduced gas-wall interaction and therefore a lower wall saturation limit have been observed in L-mode plasma experiments in limiter tokamaks. In divertor tokamak experiments with the supersonic gas jet, a fueling efficiency of 10 – 30 % has been reported. In SST-1 machine, two Laval nozzles made up of SS 304L with throat diameters of 0.5 mm and 1.0 mm are installed at high field side (HFS) at the port locations 7 and 10 respectively. The taper angle is 12°–15° for these two nozzles having the exit diameters of 5.0 mm & 10.0 mm with lengths of 12.6 mm and 30 mm respectively.

The initial characterization of the nozzle for flow rate and injection rate was carried out in the laboratory using a chamber of 50 mm diameter and 400 mm length (figure 8). The cylinder is connected to a turbo-molecular pump having an effective pumping speed of 150 l/s. The pressure inside the chamber was maintained at ~ 1.0×10^{-2} Pa. A laser source (405 nm, 5 mW) was mounted horizontally with a fast imaging camera mounted perpendicular to the line of sight of laser as shown in figure 9.

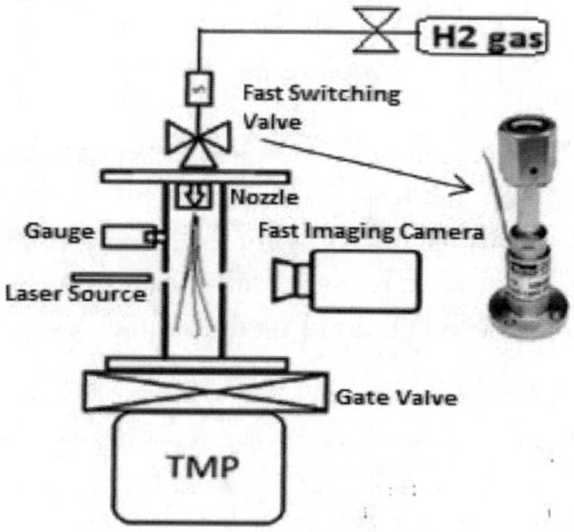

Figure 8. Schematic of experimental set-up

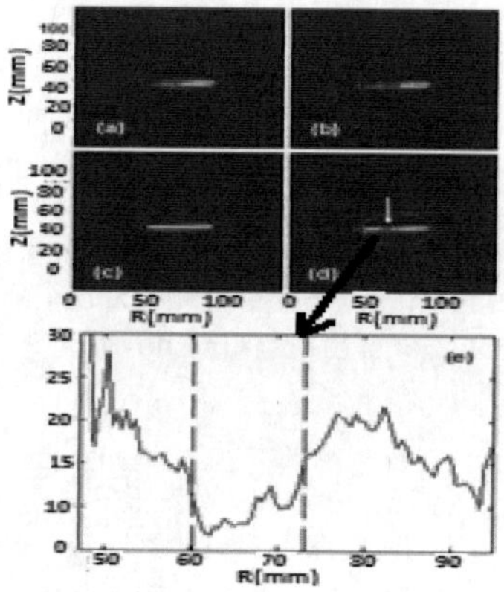

Figure 9. Fast imaging data.

3. Initial validations of PFC equipped SST-1 circular plasma

3.1. Eddy current Aspects and Equilibrium Re-construction

Eddy currents temporal and spatial characteristics have altered in PFC equipped SST-1 compared to earlier limiter equipped SST-1 and thereby was influencing the transient field errors and subsequently the magnetic NULL and break-down characteristics. A detailed quantitative distribution of the eddy current in the electrically continuous SST-1 vacuum vessel have been computed employing the array of internal voltage loops (flux loop) and with the help of simple circuit model [3-4]. The computed results from this model have been benchmarked against experimental measured integrated eddy current. The computed values and the experimentally measured values are in good agreement. The calculated eddy current based on flux loop signal and circuit equation model have been extended towards the reconstruction of the vessel magnetic field line (B-field) for SST-1. Comparison between the field line of plasma and vacuum shot explains the plasma current contribution from all other sources also.

IOP Conf. Series: Journal of Physics: Conf. Series **823** (2017) 012004 doi:10.1088/1742-6596/823/1/012004

The eddy current (figure 10) and magnetic field (figure 11) for eddy has been computed through a MATLAB code using simulated data calculated from each vessel sectional current and location of the segment and cryostat position. The inductance and resistive effect of vessel segments have also been included here. Results show under representative conditions, the maximum Eddy current has been computed to be ~ 15 kA. The rate of change of current in the central solenoid ($(dI_{OT})/dt$) and the sudden disappearance of plasma current during the disruptions are the primary sources of eddy currents apart from sudden plasma movements. The computed eddy current distributions including their temporal behaviour shall be extended to plasma equilibrium and electromagnetic modelling [5] as well as plasma discharge process in future.

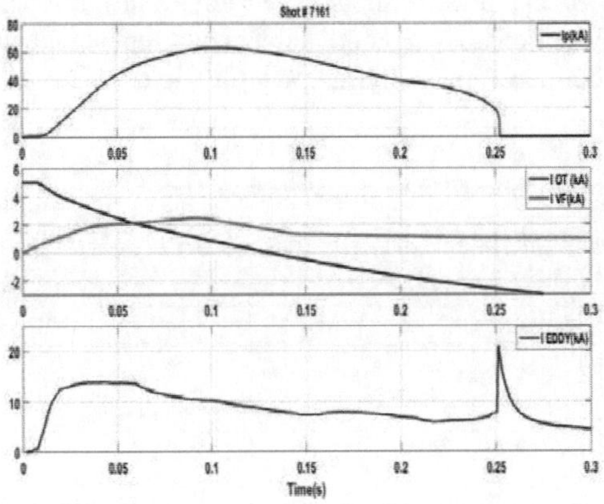

Figure 10. Total vessel eddy current profile and Ip , IOT, IVF (red) for plasma shot no 7161

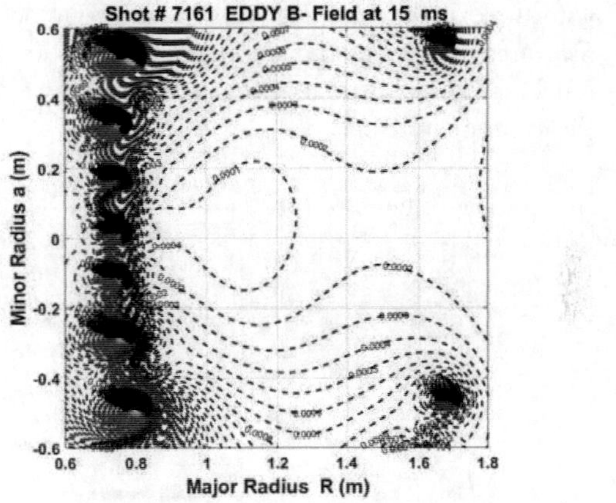

Figure 11. The eddy B-field contours of plasma shot 7161

3.2. Magnetic flux surfaces and Radial Shafranov shifts in SST-1 Tokamak Plasma

Magnetic flux surface contours of SST-1 tokamak plasma have been computed from the magnetic probes, flux loops experimental data and from the analytical solution [6] of Grad-Shafranov equation (GSE). SST-1 plasma, at the present phases of operations is circular in shape and leans against the limiters. The radial Shafranov shift (ΔR) [7] from the above formulations has been calculated from experimental measurements.

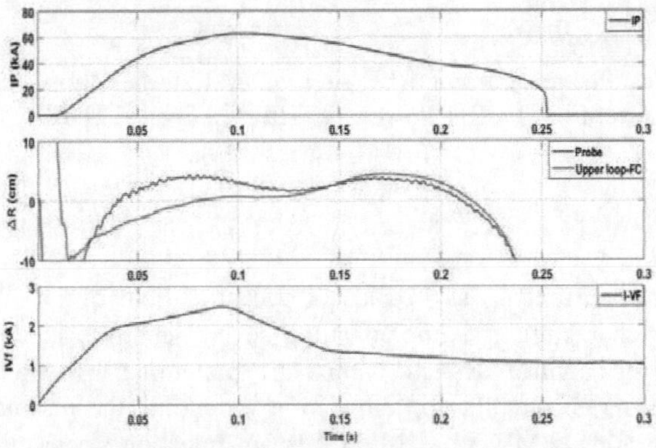

Figure 12. Plasma current (I_P), Plasma Radial shift (ΔR) & Vertical Field current (I_{VF}) for SST-1 Shot No: 7161.

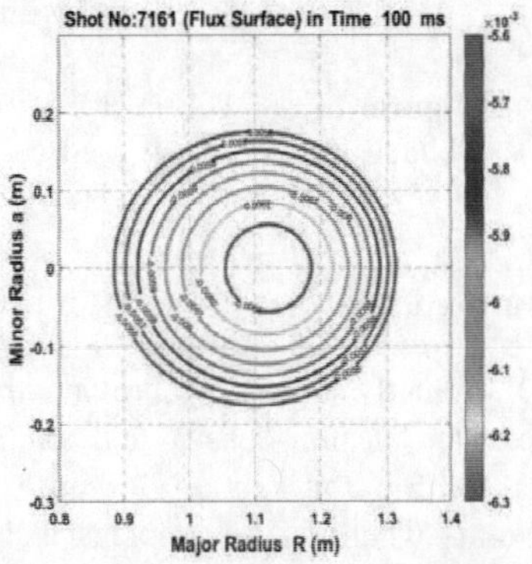

Figure 13. Magnetic flux surface for SST-1 shot no 7161 at time instant 100ms at maximum plasma current (I_P).

Since the control of plasma position plays an important role in plasma confinement and optimized tokamak operations, this ΔR would be used as a plasma position feedback control parameter in long duration SST-1 plasma experiments later. The plasma flux surface contours computed (figure 13) as above have also been approximately compared with the time synchronized fast imaging signals of the SST-1 plasma. These comparisons shows a good agreement between the real time trends of plasma position shift at the equilibrium regions with those seen with fast imaging.
The plasma current and applied vertical field current (I_{VF}) have been plotted at top and bottom of figure 12.

The Radial Shafranov shift (ΔR) has been measured [8-9] using the described flux loops and magnetic probes.

The flux surface contours computed following these prescriptions have been in good agreement with the experimental data from a large number of SST-1 plasma shots. The radial Shafranov shift measured using the flux loops and magnetic probes are also in good agreement, repeatable and reliable. Plasma currents up to 65kA have been obtained repeatedly at a central field of 1.5 T as shown in figure 14.

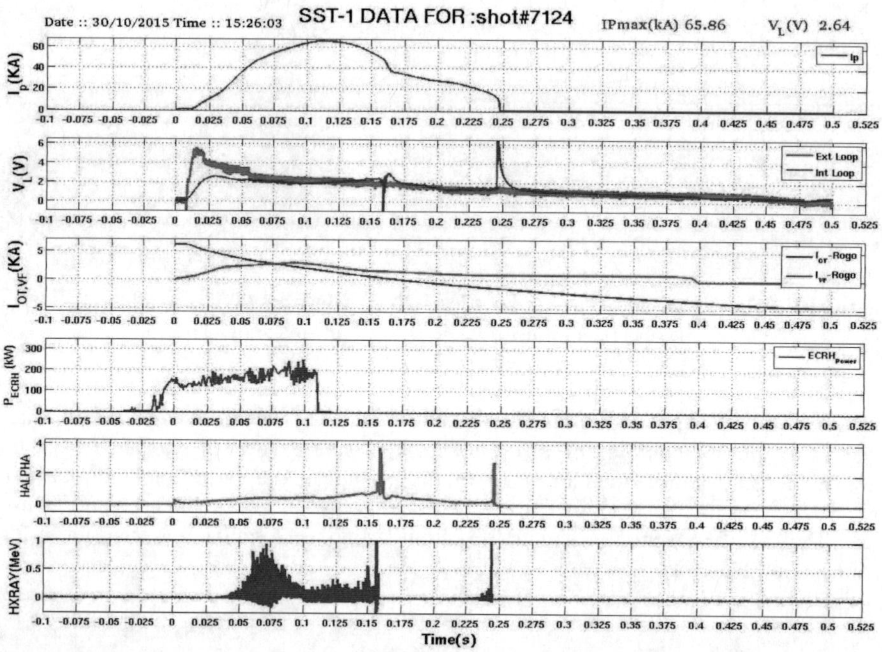

Figure 14. SST-1 shot no. 7124, (a) Plasma current, (b) Internal & External loop voltage, (c) Ohmic & Vertical field current (d) ECRH Power, (e) Hα (f) Typical Hard X-ray signal

4. Future Plans

SST-1 presently aims at improving the plasma current up to 110 kA at a central field of 1.5 T, improve the plasma density and temperature characteristics, implementing several controls in the current profile through an in-house developed control system within the prevailing hardware constraints. The near future plans include plasma current flat-top elongation, extending the plasma duration assisted with LHCD as well as integrating a MW level Neutral Beam Injection system to SST-1.

5. Summary

SST-1 has been upgraded with graphite first wall and is now a compatible Tokamak towards long pulse plasma in both circular and elongated cross sections. Initial experiments have begun in SST-1 and plasma currents in excess of 65 kA with typical core density ~ 0.8×10^{19} m^{-3} and core electron temperatures ~ 200 eV having duration in excess of 250 ms corresponding to q_{edge} ~ 3.8 have been obtained. The pre-ionization induced break down, eddy current induced NULL dynamics, subsequent plasma start-up, copious MHD activities, mode locking and other signatures of standard Tokamak plasma have been obtained. The circular equilibrium has been reconstructed from experimental signals and has been validated with imaging measurements.

SST-1 experiments observe the MHD instabilities such as saw teeth, m/n = 1/1 and m/n = 2/1 tearing modes, similar to the observations reported by other Tokamaks. Study of such MHD instabilities and disruption characteristics will be useful for the subsequent disruption mitigation and avoidance strategies in long pulses of SST-1.

References

[1] Pradhan S, Khan Z, Tanna V L et. al. 2015, *Nuclear Fusion* **55** 104009
[2] Jacob S et al. 1996, *Journal of Nuclear Materials* vol. **233-237** pp. 655-659
[3] Jana S , Pradhan S et. al (accepted 2016), *Fusion engineering and design*
[4] Gates D A, Menard J E et. al 2004 , *Review of scientific instruments* 75(12) 5090.
[5] Leuer J A et al., 2010 *Fusion science and technology* **38(3)** 333
[6] Solovev L S et al, 1968 *Soviet Physics JETP* **26(2)** 400.
[7] Mukhovatav V S and Shafranov V D, 1971 *Nuclear Fusion* **11** 605.
[8] SalarElahi A and Ghoranneviss M, 2012 *Journal of Nuclear and Particle Physics* **2(6)**142.
[9] Rahimirad A, Emami M et al 2010, *Journal of Fusion Energy* **29** 73.

IOP Conf. Series: Journal of Physics: Conf. Series **823** (2017) 012005 doi:10.1088/1742-6596/823/1/012005

Wilkinson Type Lumped Element Combiner-Splitter for Indigenous Amplifier Development

Manoj Patel[1], Akhil Jha[1], JVS Harikrishna[1], Rajesh Trivedi[1] and Aparajita Mukherjee[1]

[1]ITER-India, Institute for Plasma Research, Bhat, Gandhinagar - 382428, Gujarat, India

E-mail: manoj.patel@iter-india.org

Abstract. Development of wideband solid state power amplifier in ITER [1] ICH&CD frequency range is ongoing. A 12 kW/CW Solid State Power Amplifire (SSPA) is being developed in house. 1 × 16 wideband splitter/combiner will be used at input/output side. Study has been carried out on two options i.e. coaxial type & lumped element based Wilkinson splitter/combiner. Tentative power level of both input N-Type ports of combiner is ~ 1 kW. Design and simulation for coaxial type Wilkinson combiner is done. Quarter wave length for centre frequency is ~ 1500 mm. To reduce mechanical dimension of combiner, PTFE dielectric is used with sophisticated arrangement. Coaxial combiner required unique fabrication process. Alternate option is proposed as a lumped element based Wilkinson combiner with reduced size, cost & development time. Design and simulation was carried out. Required PCB design & fabrication was done accordingly. Same design will be implemented for splitter as well. Design scheme for the splitter/combiner will be finalized depending on the achieved performance of both the designs. In this paper, detailed design, simulation and test results are presented for both types of combiners. A detailed comparison of combiners is provided.

1. Introduction

The Ion Cyclotron Heating and Current Drive (ICH &CD) system has to couple 20 MW Radio Frequency power to ITER [1] plasma for heating and driving plasma current, in the frequency range of 35-65 MHz [2]. There will be 8 RF sources to generate total 20 MW of RF power at VSWR 2:1[3]. Each RF source is made up of two RF amplifier chains combined at end stage through high power combiner. Each amplifier chain has three cascaded amplifiers, where final stage (~ 1.7 MW) is driven by driver stage (~ 150 kW) which is further driven by a pre-driver stage (~ 10 kW). The driver and final stages are tube based (Diacrode®/Tetrode) amplifiers and pre-driver is a solid state power amplifier (SSPA). SSPA is a wideband solid state amplifier using MOSFET as an active device. There are total 16 pallet modules each having maximum output power of ~1 kW. Each pallet module contains one MOSFET and input/output matching circuit. The goal is to achieve power level of ~ 12 kW/CW. 16 pallet amplifier modules will be combined using 16 × 1 wideband combiners. 16 RF signals, with equal phase, will be required to drive each pallet module. 1 × 16 wideband splitter will be used at input side. Figure 1 shows the overall scheme of SSPA.

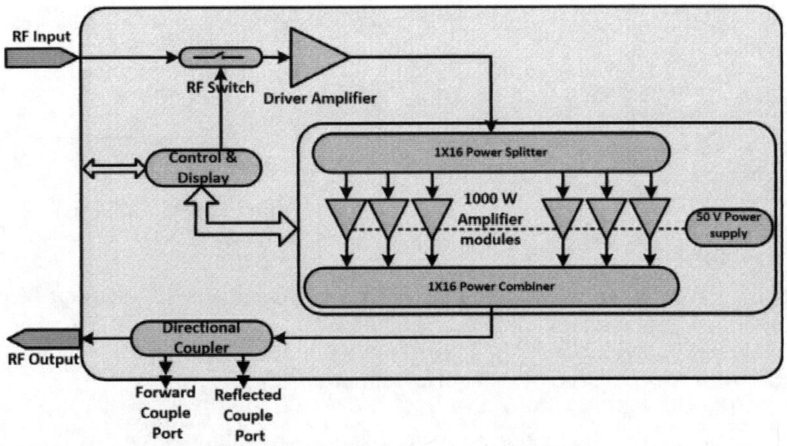

Figure 1. Solid State Power Amplifire (SSPA) block diagram

Wilkinson type combiner/splitter [4] is found as a good choice for development of the input splitter and output combiner for SSPA.

2. Wilkinson combiner/splitter:

Major technical specification for the 2 × 1 Wilkinson combiner /splitter is as given in table 1.

Table 1: Major technical specifications

Parameters	Values
Operating frequency	-65 MHz
Flatness (at edge frequency w.r.t centre)	Within 0.5 dB
Return loss (at input port)	Better then - 20 dB
Power handling	~ 2 kW
Connector type	N female (50 ohm)

Wilkinson combiner/splitter is a 3 port (in 2 × 1 scheme) lossy, matched and reciprocal network. Figure 2a shows the scheme of a 2 × 1 Wilkinson combiner/splitter. Port 3 of characteristic impedance Z_o is split in two arms with each arm having a quarter wave (λ/4) impedance transformer of characteristic impedance $\sqrt{2}Z_o$. Port 1 and 2 are connected at the other end of transformer. Isolation between the port 1 and 2 is achieved by connecting a resistor of impedance $2Z_o$ between the two ports.

For characteristic impedance $Z_o = 50\ \Omega$ and design frequency $f_o = 50$ MHz
Quarter wave transformer characteristic impedance is [4]

$$Z_{trans} = \sqrt{2} \times 50 = 70.7\ \Omega \qquad (1)$$

The length of transformer is [4]

$$L_{trans} = \frac{\lambda_o}{4 \times \sqrt{\varepsilon_r}} = \frac{6}{4 \times \sqrt{2.1}} = 1.035\, m \qquad (2)$$

where wavelength (λ_o) = 6 m, PTFE dielectric constant (ε_r) = 2.1

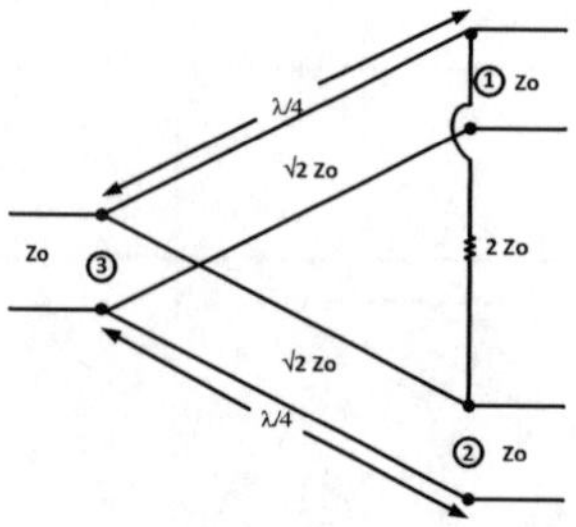

(a)

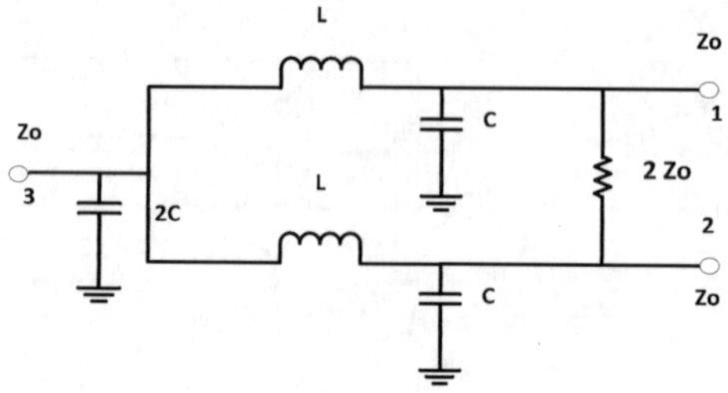

(b)

Figure 2. 2 × 1 Wilkinson combiner/splitter (a) layout (b) corresponding lumped equivalent

The lumped circuit equivalent of 2 × 1Wilkinson combiner/splitter is shown in figure 2b. The quarter wave transformer is replaced by an equivalent pi (C-L-C) circuit.

Inductive reactance [4] $\omega_o L = Z_{trans}$

$$L_{trans} = \frac{Z_{trans}}{2\pi f_o} \tag{3}$$

From equation (1) and (3) at design frequency 60MHz; calculated inductance L = 187 nH

The resonance frequency is defined as [4] $f_o = \frac{1}{2\pi\sqrt{LC}}$ (4)

From equation (4) at design frequency 60 MHz; calculated capacitance C = 37.6 pF. Value of resistance connected between port 1 and port 2 is 100 Ω.

3. Modelling and Simulation:

3.1. Coaxial 2x1combiner/splitter:

Modeling and simulation of coaxial type combiner/splitter was performed using CST MWS software. Inner conductor diameter was constrained on the available copper wire diameter (a = 2.85mm) while corresponding outer conductor diameter (for transformer b = 15.74 mm and for ports b = 6.56 mm) is found from equation (5) [4].

$$Z_{trans} = \frac{138}{\sqrt{\varepsilon_r}} \log_{10} \frac{b}{a} \tag{5}$$

Where $\varepsilon_r = 2.1$ (PTFE)

The outer conductor is simulated by cylindrical grooving on two aluminum plates placed back to back as shown in figure 3a. Copper inner wire is supported coaxially inside the grooves using PTFE rods. The effective length of quarter wave transformer is reduced to 1.035m. Overall size of the coaxial combiner is reduced by providing unique shape to the transformer section. Coaxial combiner/splitter shows transmission S31or S32 value -3.04 dB at band center (50 MHz) and - 3.15 dB at band edged (35 and 65 MHz). Return loss S11 and S22 are better then - 30 dB for the whole band and isolation between port 1 and 2, S12 or S21 is better than -15 dB for whole band.

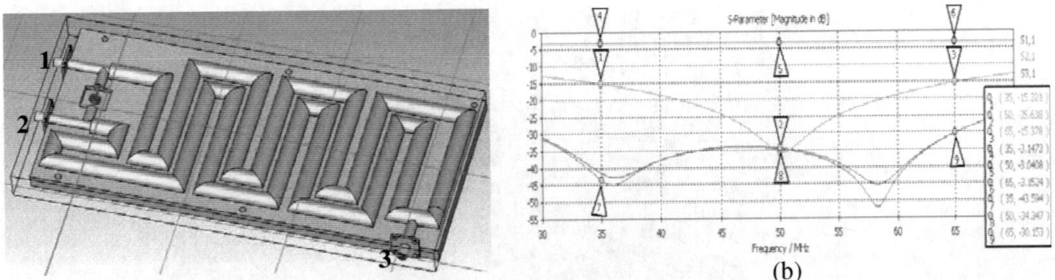

(a) (b)

Figure 3. (a) CST model of 2 × 1coaxila Wilkinson combiner (b) S-parameter results of the model

3.2. Lumped 2 × 1 combiner/splitter

Lumped type combiner designed for 50MHz shows a very quick fall in the transmission response ~ 1.0dB when we go above 60MHz (up to 65 MHz); to make the response more flat at higher frequency the design frequency was shifted to 60 MHz. The value of inductor and capacitor were calculated as given in equation (3) and (4). The closest value found practically for the inductor and capacitor were 192nH and 37.5pF and they were taken for simulation. Figure 4a shows the simulated circuit of lumped combiner/splitter. The transmission S31 or S32 are - 3.01 dB at 60 MHz. The transmission reduces to - 3.37 dB and - 3.09 dB at band edges 35 and 65 MHz respectively. Return loss S11 or S22 is better than - 23 dB for the whole band. Isolation is better than -10 dB for the whole band.

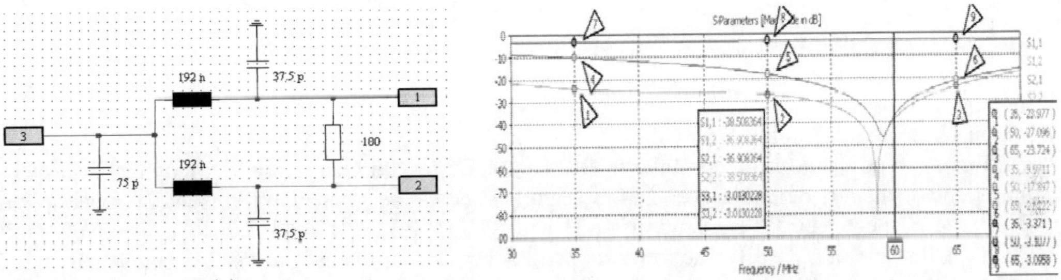

(a) (b)

Figure 4. (a) Simulated circuit for 2 × 1 Wilkinson combiner/splitter (b) S parameter responses

4. Fabrication and Testing

4.1. Coaxial 2 × 1combiner/splitter

Coaxial type combiner/splitter is machined over an aluminium plate, with PTFE rods as dielectric to support the inner wire. Two aluminium plates were machined to achieve the half circular groove and then bolted back to back to with PTFE rods within the groove. N type connectors were mounted at the ports and inner pin was soldered with the inner conductor wire. Figure 5 shows the fabricated and assembled coaxial type combiner/splitter with VNA setup for testing. The Insertion loss S31 and S32 is - 3.20 dB and - 3.01 dB respectively at band centre (50 MHz). Transmission at band edge (35 and 65 MHz) has better than 0.11 dB flatness. Return loss S22 or S11 is better than - 20 dB for whole band. Figure 6(a) and 6(b) shows the S-parameter response of coaxial combiner/splitter.

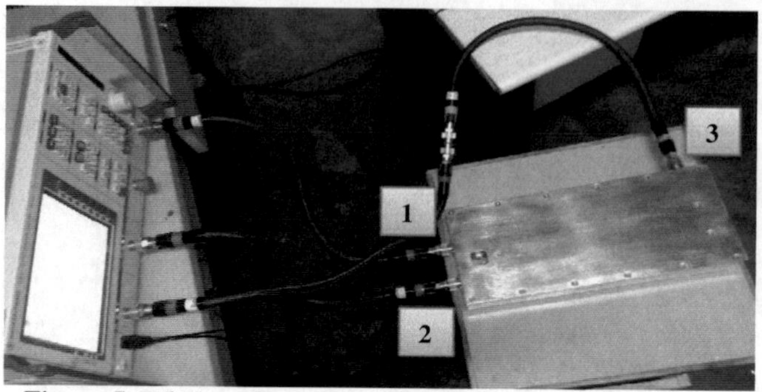

Figure 5. Fabricated coaxial splitter/combiner VNA setup

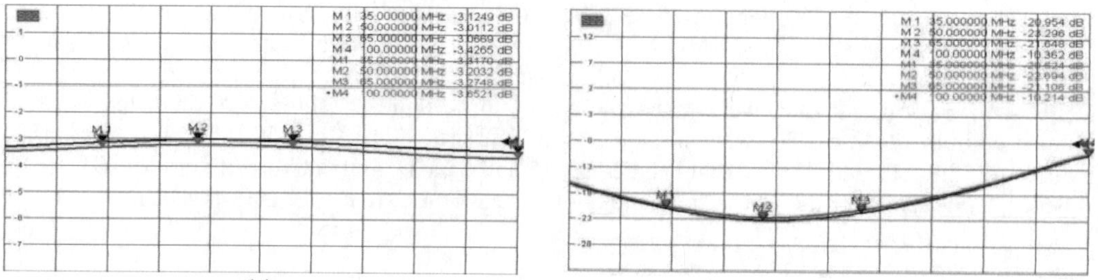

(a) (b)

Figure 6. S parameter response (a) Insertion Loss (S31 and S32) (b) Return loss (S11 and S22)

4.2. Lumped 2 × 1 combiner/splitter

The Lumped type Wilkinson combiner/splitter is built by lumped circuit components (resistor, inductor and capacitor) mounted over a PCB. Figure 7 shows the lumped Wilkinson combiner/splitter with N type connector ports. Transmission S31 and S32 are - 3.0 dB and - 3.2 dB respectively at 50 MHz and a flatness of 0.31 dB is achieved at band edge. Return loss S11 or S22 is better than -17dB for whole frequency range. Isolation S12 or S21 is - 10 dB at 35 MHz and improves up to - 16 dB at 65 MHz. Figure 8(a) and 8(b) shows the S-parameter response of lumped combiner/splitter

Figure 7. Lumped 2 × 1 Wilkinson splitter/combiner

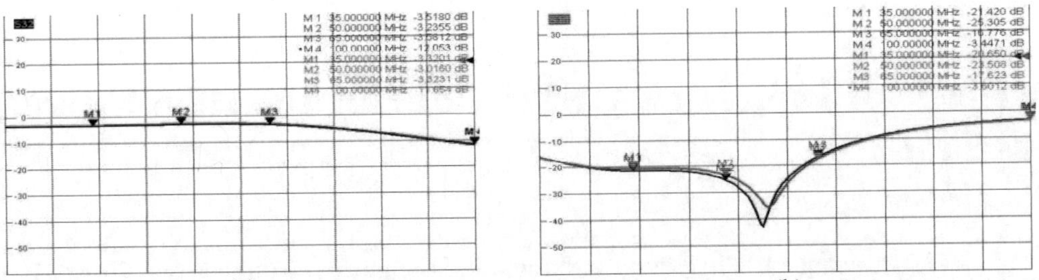

(a) (b)

Figure 8. S parameter response (a) Insertion Loss (S31 and S32) (b) Return loss (S11 and S22)

5. Conclusion

Detailed study of Wilkinson type 2 × 1splitter/combiner was done. Two options for development of combiner i.e. coaxial and lumped element were chosen. Calculation for coaxial type combiner transformer impedance and corresponding lumped equivalent is provided. Both the options were simulated using CST MWS and fabrication was done and test results were obtained for comparison. A comparison for the two options is provided in table 2.

Table 2. Comparison of lumped and coaxial 2 × 1 Wilkinson type combiner/splitter

Properties	Lumped type	Coaxial type
Transmission flatness over whole band	~ 0.3 dB	~ 0.11 dB
Return loss over whole band	Better than - 17 dB	Better than - 20 dB
Isolation between input ports	- 10 dB at 35 MHz 16 dB at 65 MHz	Better then - 20 dB
Power handling	Depends upon lumped element dissipation limit (~ 1.5 kW)	Restricted by the N type connecter power handling (~ 3 kW)
Size	Smaller (PCB mount)	385 × 170 × 40 mm
Fabrication method	PCB development and soldering of lumped elements	Machining of aluminium plate and bolted
Cost	Comparatively less	Machining cost is high

Lumped type combiner/splitter is easier to fabricate and has lower fabrication cost but it has lower isolation, lower return loss and lower transmission flatness between input ports, smaller bandwidth and lower power handling capability (depending on components used) with respect to the coaxial type combiner/splitter. Coaxial type combiner developed here has larger dimension and relatively higher fabrication cost.

References

[1] ITER-Physics Expert group 1999 *Nuclear Fusion* **39** 2137
[2] Lamalle P U et al. 2013 *Fusion Eng. Des.* **88** 517
[3] Mukherjee A et al 2015 *Fusion Engineering and Design* **96–97** 542
[4] Pozar D M 2012 *Microwave Engineering* fourth edition (John Wiley & Sons, Inc), chapter 7 p 328-329

IOP Conf. Series: Journal of Physics: Conf. Series **823** (2017) 012006 doi:10.1088/1742-6596/823/1/012006

Experience of 12 kA / 16 V SMPS during the HTS Current Leads Test

P Panchal[1], D Christian[1], R Panchal[1], D Sonara[1], G Purwar[1], A Garg[1], H Nimavat[1], G Singh[1], J Patel[1], V Tanna[1] and S Pradhan[1]

[1]Institute for Plasma Research, Near Indira Bridge, Bhat, Gandhinagar 382 428, India

E-mail: pradip@ipr.res.in

Abstract. As a part of up gradation plans in SST-1 Tokamak, one pair of 3.3 kA rated prototype hybrid current leads were developed using Di-BSCCO as High Temperature Superconductors (HTS) and the copper heat exchanger. In order to validate the manufacturing procedure prior to go for series production of such current leads, it was recommended to test these current leads using dedicated and very reliable DC switch mode power supply (SMPS). As part of test facility, 12 kA, 16 VDC programmable SMPS was successfully installed, commissioned and tested. This power supply has special features such as modularity, N+1 redundancy, very low ripple voltage, precise current measurements with Direct Current Current Transformer, CC/CV modes with auto-crossover and auto-sequence programming. As a part of acceptance of this converter, A 5.8 mΩ water-cooled resistive dummy load and PLC based SCADA system is designed, developed for commissioning of power supply. The same power supply was used for the testing of the prototype HTS current leads. The paper describes the salient features and experience of state-of-art of power supply and results obtained from this converter during the HTS current leads test.

1. Introduction

Steady-state Superconducting Tokamak (SST-1) superconducting magnet system (SCMS) consists of sixteen Toroidal Field coils and nine Poloidal Field coils [1]. The SCMS is cooled by1.3 kW at 4.5 K Helium Refrigerator cum Liquefier [2]. The power supply at ambient temperature is connected to the SCMS at 4.5 K through a pair of current lead to minimize the heat leaks [3]. Presently SST-1 has 10 kA rated vapour cooled conventional current leads, which has been indigenously designed, developed and fabricated at IPR. A test facility was planned to characterize superconducting current leads up to 12 kA at 4.5 K [4]. To limit AC losses in these current leads, a dedicated power supply with very low ripple voltage at output terminal is required. 12 kA / 16 V DC programmable Switch Mode Power Supply (SMPS) has been procured from Kemppi Oy, Finland as per SST-1 cryogenic requirement to test the current leads. It has very low ripple voltage (< 8 mV_{rms}), modular architecture with N+1 redundancy, wide operating current and voltage range, precise current measurements with Direct Current Current Transformer (DCCT) and constant current (CC) / constant voltage (CV) mode with auto-crossover.

As a part of up gradation plans in SST-1, one pair of 3.3 kA rated prototype hybrid Current Leads (CL) were developed using Di-BSCCO as High Temperature Superconductors (HTS) and copper heat exchanger. Consequently the HTS CL saves a substantial cryogenic operation cost. This12 kA SMPS was used to test the prototype HTS CL. The first part of the paper describes salient features of power supply with installation and commissioning experience. The second part describes the test set-up of prototype HTS CL and results.

2. Modular Architecture of Power Supply

The power supply is divided into seven sub-converters for fully modular concept and better availability. Each sub-converter is a basic constant current source with rating of 1714 A, 16 VDC. Each sub-converter is controlled by phase shift peak current controlled zero voltage switching (ZVS) type soft switching method. However, the complete power supply is controlled by central control unit with Proportional Integrating Derivative (PID) control algorithms in CC/CV. The power supply including bus-bar at top is actively water-cooled to dissipate heat effectively. Basic sub-converter [1714 A, 16 V] is split in two types of modules i.e. one number of input module and three number of output module for compactness and easy maintenance. The schematic of sub-converter is shown in figure 1.

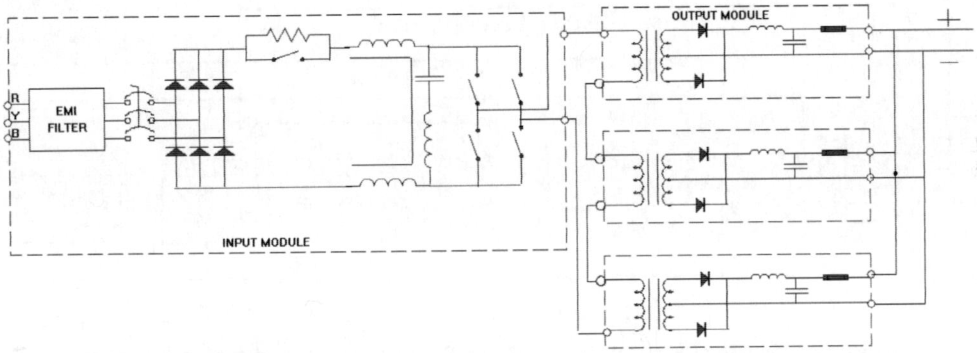

Figure 1. Basic Building Block of Sub-converter.

2.1. Input Module

Input module consists of EMC filter, power contactor on the AC mains (3 × 415V, 50 Hz), in-rush limiting or pre-charge circuits, a 3-phase full wave rectifier with a damped L-C passive filter, full bridge phase shift zero voltage switching inverter operating at 19.4 KHz. The control of the IGBT in phase shift full-bridge inverter is achieved by introducing phase shift between two legs of the bridge. This topology was combined with the peak current control, introducing a feed forward parameter from the intermediate DC voltage, to assist the low frequency performance.

2.2. Output Module

Output module consists of high frequency transformers (with ratio of 2:1 for each transformer), rectifier stage, output filter, blocking diode and fuse. For ease of operation and maintenance, output module is physically split into three modules of 570 A rated current each. Several capacitance filter banks are fitted at output terminals to reduce the high frequency noise. Secondly free-wheel diodes are provided at the output bus-bars for the protection of power supply for inductive load.

2.3. Sub-converter Control Circuit

This circuits control the output of sub-converter according the current reference received from the common control electronics. The control loop is conventional Proportional Integrating (PI) controller with addition of peak current control, i.e. feedback of the output voltage AC components is added to current feedback. The control circuit also handles various types of interlocks and diagnostics for proper operation of sub-converter and communicates to central control electronics.

3. Control Philosophy of Power Supply

Central control unit performs all functions concerning complete converter operation and have interface for remote automation systems. It performs accurate output control by sending the current reference to each sub-converter in the form of 0-10 V signal. For current measurement, highly precise 16 kA DCCT type transducer is used. This type of transducer gives precise measurement in the range of 10 ppm and preferable over shunt or Hall Effect transducers. Each sub-converter acts as independent current source. A conventional PID algorithm is used to implement the analog or digital control loops as shown in figure 2. A digital control loops are programmed in micro-controller and references / feedback signals are routed through high precision ADC and DAC. The analog control loops are developed using operational amplifier. For analogy control both I_{ref} & V_{ref} signals are required for proper operation of CC/CV. For digital control I_{ref} & V_{ref} signals are fed by either local display or remote master terminal through digital communication.

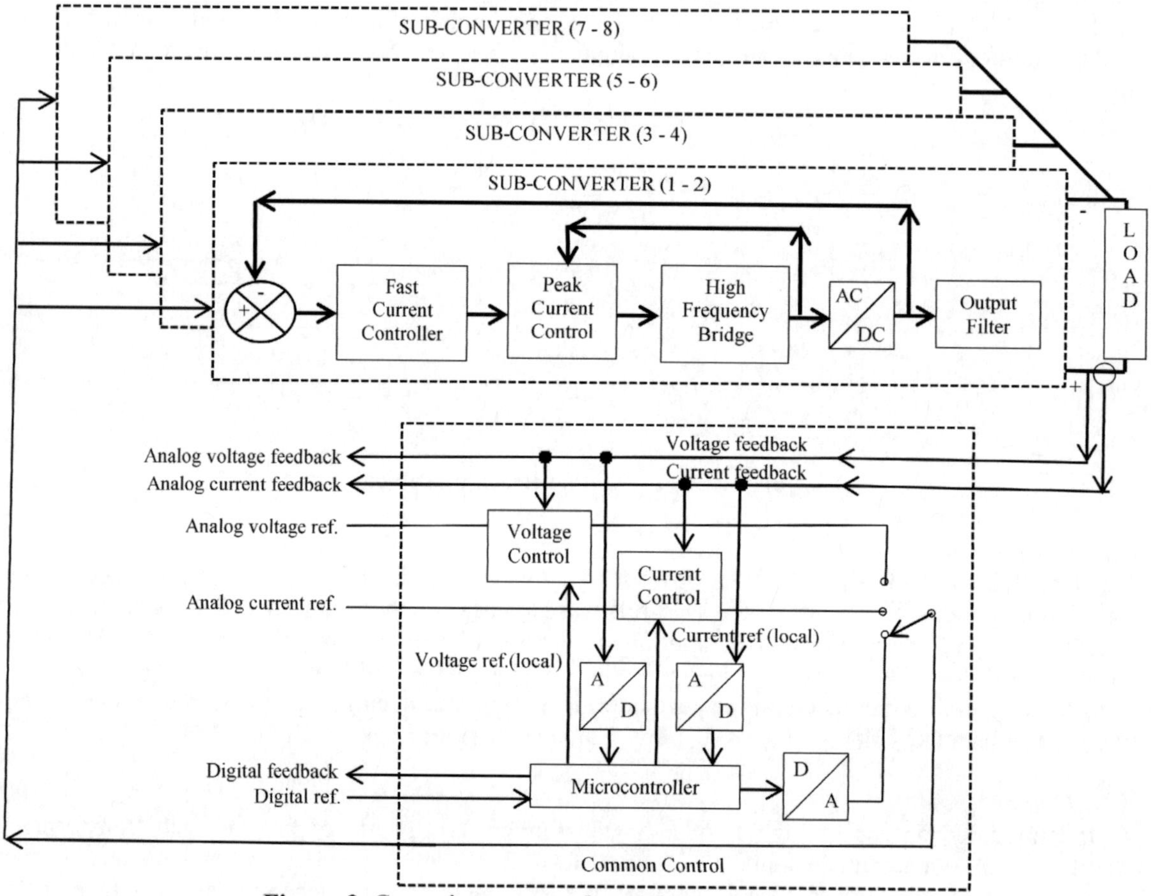

Figure 2. Control structure of complete power supply.

Normally seven sub-converters are used for nominal output power, but for reasons of redundancy, extra sub-converter can be operated with others in parallel. Under normal condition, all sub-converters are operating in parallel in relation to the necessary power. If one or more sub-converter fails, on-state sub-converter reference will increase, so that output current and voltage do not change.

4. Installation and Commissioning Experience

Performance and validation test were carried out at different operating points in CC / CV modes at site. To meet this requirement, an in-house 5.8 mΩ water-cooled dummy load is developed. For full functionality test, Supervisory Control and Data Acquisition (SCADA) application with Programmable Logic Controller (PLC) automation is developed.

IOP Conf. Series: Journal of Physics: Conf. Series **823** (2017) 012006 doi:10.1088/1742-6596/823/1/012006

4.1. Development of 5.8 mΩ, 40 kW Water-cooled Resistive Dummy Load

Generally low resistance with high wattage dummy load is required to test such high current low voltage power supply. Forced flow water-cooled resistance made of stainless steel 302 is preferred due to small size, weight and its higher resistivity compared to copper. Due to limitation of only 50 litres per minute chilled water available from source at installation site, a 2.5 m length and 300 sq. mm of cross-section made of SS-302 pipe is taken as resistive load. Four copper terminals at regular interval are brazed to have different resistance (1.67, 2.46, 4.13 and 5.82 mΩ respectively). A dummy load is shown in figure 3. During commissioning of power supply, 2.5 kA current have been circulated through it with maximum output voltage of 16 VDC at output end. The temperature of the load is maintained to 70 °C. The dummy load performed satisfactorily in CV mode. The CC / CV crossover functionality of power supply is checked satisfactorily with this dummy load.

4.2. Development of SCADA Application with PLC Automation

The SMPS is interfaced with PLC of test facility. A detailed PLC ladder logic is developed to acquire analog signals and digital command / status signals from SMPS. A new mimic screen and real-time / historical trend are developed in Wonderware's InTouch SCADA to control SMPS remotely. Figure 4 shows the mimic screen developed for remote operation of power supply. During commissioning the interface between power supply and PLC based SCADA system is checked and found satisfactory operation.

Figure 3. Water-cooled dummy load with connection

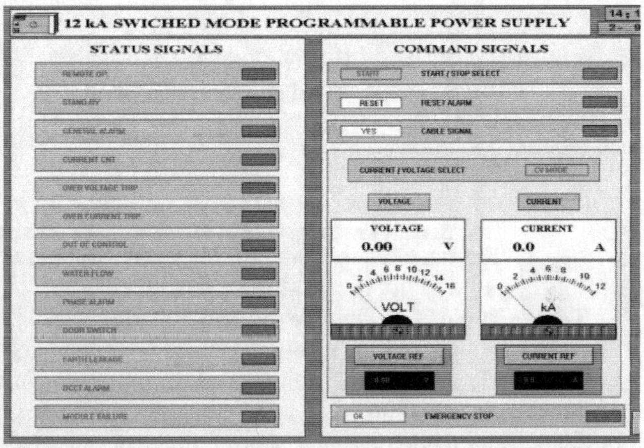

Figure 4. Mimic screen for power supply operation in remote mode

4.3. Performance Results

Initially, all pre-commissioning activities were carried out to check electrical and water connections as per standard practice. Control loops in central control unit were tuned in CC and CV mode with water-cooled cables and dummy load respectively. The main performance data of the SMPS are as below:

- Less than 8 mV_{rms} ripple voltage for the voltage range of 1 to 16 VDC at output terminals
- Auto-sequencing with user defined number of steps and duration in CC/CV mode
- Less than 50 ms transient response
- Less than 5 % current sharing between sub-converters
- Remote analog /digital communication with IPR automation system and master application

Voltage ripple is measured with precise true RMS voltmeter, while current ripple is measured with Rogowski coil for complete range. Other electrical parameters were found within the specifications. Figure 5 shows the operation of SMPS at full rated specification in CC and CV mode. Figure 6 shows the historical trend of power supply voltage and current during auto-sequence testing of SMPS with dummy load. This feature is very useful for testing superconducting resistive load.

IOP Conf. Series: Journal of Physics: Conf. Series **823** (2017) 012006 doi:10.1088/1742-6596/823/1/012006

Figure 5(a). Rated Current in CC mode

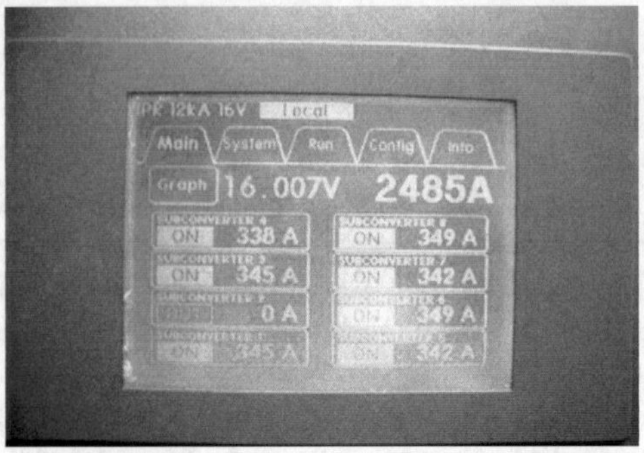

Figure 5(b). Rated Voltage in CV mode

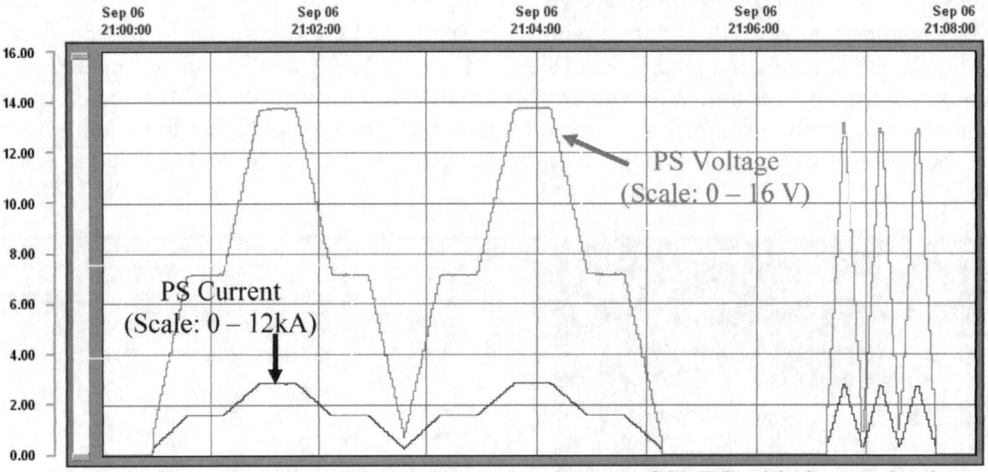

Figure 6. Demonstration of Auto-sequencing feature of SMPS with dummy load

5. Test set-up of Current Leads and Results

The HTS CL has mainly two parts namely copper heat exchanger (300 K – 77 K) and HTS module (77 K – 4.5 K) of Di-BSCCO material. In real application, the lower end of HTS module is connected to the superconducting bus-bar of magnets. For test purpose, lower ends of both leads are connected through 1 mΩ feeder system (electrical resistance between two leads). A special test set-up with 12 kA SMPS as shown in figure 7 was prepared to characterize the HTS CL.

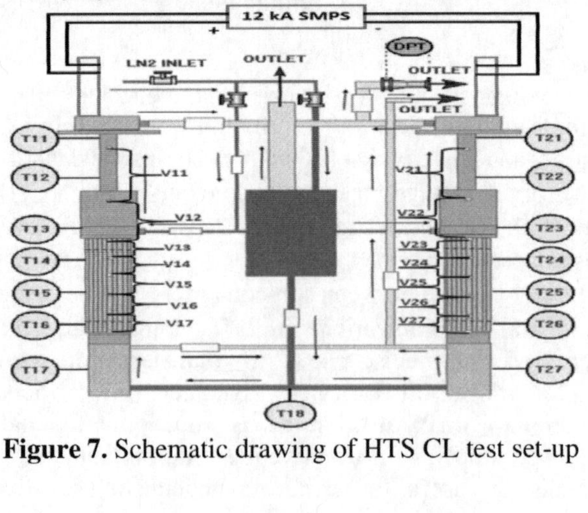

Figure 7. Schematic drawing of HTS CL test set-up

The test set-up consists of cryostat with vacuum pump and gauges. Several sensors were mounted on CL to measure its electrical and thermal performance. These sensors are for voltage, temperature, flow, pressure and vacuum. A 12 kA SMPS is connected to the top end of leads through flexible copper cables. Before cooling down the leads, several electrical tests have been performed on CL for its integrity and isolation to ground. After cooling down the CL and stabilizing the temperature along HTS module, current leads are energized at different currents from 12 kA power supply in steps. All measurements from various sensors were acquired in data acquisition system for analysis.

During the tests, the current in leads is ramped with different rates and kept constant for several hundred of seconds. Only four sub-converters were switched ON for HTS CL test and performance of power supply with water cooled cable is found satisfactory during test. Figure 8 and 9 show the ramp profiles of current having highest flat top current and longest flat top duration respectively during test.

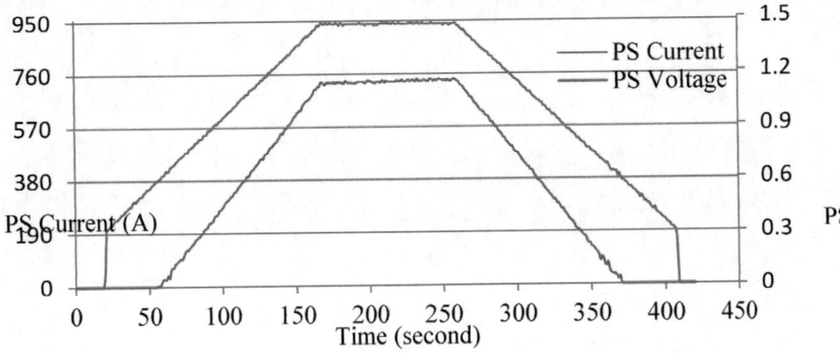

Figure 8. The ramp profile having highest flat top current.

Flat Top
Time:- 94 sec
Current:-940 A

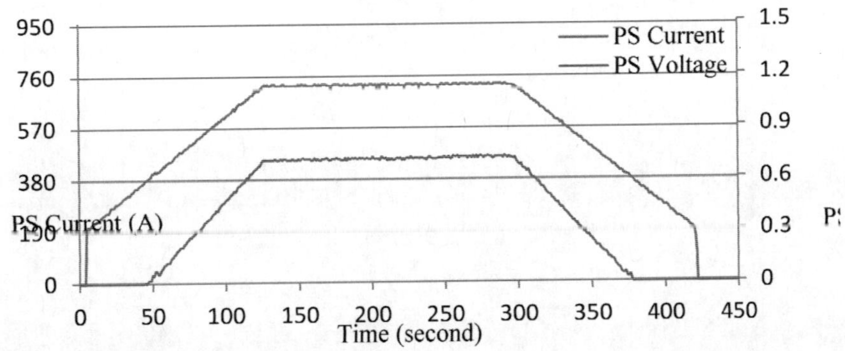

Figure 9. The ramp profile having longest flat top duration.

Flat Top
Time:- 171 sec
Current:- 730 A

6. Conclusion

The 12 kA / 16 V DC programmable switch mode power supply was successfully installed, commissioned and validated at IPR as part of low temperature test facility to characterize the superconducting current leads. It has special features of low ripple voltage at output terminals, CC / CV crossover and auto sequence programmability, which is essential for testing superconducting current leads. The performance of 3.3 kA prototype HTS current lead is validated up to 940 A using this power supply with different ramp rates and other scenario.

Acknowledgement

The authors are very thankful to A. Varadharajulu for their technical inputs during procurement of SMPS for test facility and to Y. Khristi and SST-1 magnet division for providing data acquisition system for voltage measurements across HTS current leads.

References

[1] Pradhan S and SST-1 Team 2015 *Nuclear Fusion* **55** 104009
[2] Pradhan et al. 2015 *Physics Procedia* **67**756

IOP Conf. Series: Journal of Physics: Conf. Series **823** (2017) 012006 doi:10.1088/1742-6596/823/1/012006

[3] Patel R, Panchal P et al. 2015 *Physics Procedia* **67** 170
[4] Tanna V L et al. 2012 *Cryogenics and Superconductivity Society of Japan* **932** 619

IOP Conf. Series: Journal of Physics: Conf. Series **823** (2017) 012007 doi:10.1088/1742-6596/823/1/012007

Performance of Impedance Transformer for High-Power ICRF Heating in LHD

Kenji Saito[1], Testuo Seki[1], Hiroshi Kasahara[1], Ryosuke Seki[1], Shuji Kamio[1], Goro Nomura[1] and Takashi Mutoh[1]

[1]National Institute for Fusion Science, National Institutes of Natural Sciences, Toki, Gifu 509-5292, Japan

E-mail: saito@nifs.ac.jp

Abstract. There are two types of ion cyclotron range of frequencies antennas in the Large Helical Device. The handshake form (HAS) antenna has high heating efficiency. However, its loading resistance is small, and injection power is limited by the voltage of the transmission line. On the other hand, the field-aligned-impedance-transforming (FAIT) antenna has higher loading resistance than the HAS antenna despite having a smaller antenna head. However, the high voltage on the transmission line is again the bottleneck for high-power injection, as with the HAS antenna. We developed an ex-vessel impedance transformer for the HAS and FAIT antennas to decrease the voltage on the transmission lines by increasing loading resistance. The estimated enhancement factors of loading resistance were 1.65 and 2.50 for the HAS and the FAIT antennas, respectively, and the experimental result for the HAS antenna was consistent with the estimation. Therefore, higher power injection will be possible.

1. Introduction

In the Large Helical Device (LHD) [1], there are presently two pairs of ion cyclotron range of frequencies (ICRF) antennas. They are referred to as the handshake form (HAS) antennas [2] and the field-aligned-impedance-transforming (FAIT) antennas [3]. The heads of the HAS antennas are aligned in the direction of the magnetic field lines to excite waves with a large wavenumber component parallel to the magnetic field lines. Therefore, the HAS antenna has high performance in heating efficiency of minority ion heating at the 0-π current phase [4], where heating efficiency is defined as the power absorbed by plasma divided by the injected power from the antenna. However, the loading resistance R, defined by $R = 2P\,(z_c/V_{max})^2 = z_c/VSWR$, was small, and the maximum injection power was limited by the voltage on the transmission line, where P is the forward power minus the reflected power, and V_{max} is the maximum voltage on the line with the characteristic impedance z_c, which is 50 Ω in the LHD. In this paper, loading resistance is defined on the transmission line not on the antenna head, and it is equivalent to the impedance at the voltage node on the transmission line. We limit the V_{max} to 30 kV in normal operation to prevent breakdowns on the transmission line, and the interlock level of V_{max} is normally set to 35 kV. The power feedback system [5] decreases the injection power if the voltage is increased beyond 30 kV, and the interlock system turns off the power immediately if the voltage reaches 35 kV. The typical plasma loading resistance R_p of the HAS antenna was only 2 Ω, where the loading resistance of vacuum injection was subtracted to eliminate loading resistance caused by the circuit loss. The maximum injection power calculated with the plasma loading resistance of 2 Ω and the voltage limit of 30 kV is only 360 kW.

The heads of the FAIT antennas are aligned in the direction perpendicular to the magnetic field lines; however, heating efficiency is comparable to that of the HAS antennas [6]. Meanwhile, although the area of the Faraday-shield of the FAIT antenna is approximately half of that of the HAS antenna, it has a higher plasma loading resistance of typically 5 Ω due to the optimized in-vessel impedance transformer between the antenna head and the ceramic feed-through [3,7]. It was deduced from the equation $P = \frac{1}{2} R\ (V_{max}/z_c)^2$ that an injection power of 1.8 MW or power density (injection power normalized by the area of Faraday-shield) of 15 MW/m^2 would be possible for several seconds. This was because a voltage of 38.5 kV was once applied as a trial on the transmission line without arcing on the antenna head, and the plasma loading resistance $R_p > 6\ \Omega$ was realized with a line-averaged electron density of more than 2×10^{19} m^{-3} at the standard antenna position [3]. However, application of voltage of 38.5 kV on the transmission line may cause breakdowns on the transmission line. Once a breakdown occurs in the transmission line, high voltage cannot be applied to the transmission line again. Therefore, to reduce the voltage without decreasing the injection power, we developed an ex-vessel impedance transformer (EVIT).

In section 2, we present the detailed design of the EVIT. An electromagnetic simulation of the designed EVIT is conducted in section 3, and section 4 demonstrates its high performance experimentally. Section 5 presents the summary.

2. Design of ex-vessel impedance transformer

In order to decrease the maximum voltage on the transmission line by increasing the loading resistance, conversion of impedance near the antenna port is necessary. A pre-stub tuner was one of the candidates, however, space is limited around the antenna port. Moreover, the in-vessel impedance transformer for the FAIT antenna worked well [3]. Therefore, we decided to develop a compact ex-vessel impedance transformer (EVIT) for the HAS and FAIT antennas inserted in the transmission line outside the vacuum vessel at the port of the ceramic feed-through. The drawback of the impedance transformer is that the frequency is fixed. The frequency for the EVIT was set to 38.5 MHz, which is the design frequency for the FAIT antennas, since this frequency is favourable for minority ion heating at the standard magnetic field of 2.75 T on the magnetic axis at the major radius of 3.6 m [8,9]. This frequency is also favourable for the second-harmonic heating of deuterium plasma.

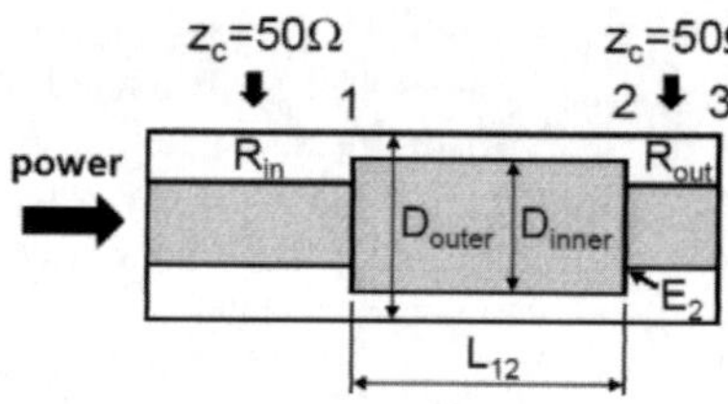

Figure 1. Simple model of the EVIT. D_{outer} is the diameter of the outer conductor; D_{inner} and L_{12} are the diameter and length of the inner conductor between points 1 and 2, respectively. R_{in} is the converted loading resistance on the transmission line on the input side with the characteristic impedance of 50 Ω, and R_{out} is the loading resistance between points 2 and 3, where the characteristic impedance is also 50 Ω.

In order to determine the dimensions of the EVIT, simulations were conducted with the simple model shown in figure 1, where the length and diameter of the inner conductor between points 1 and 2 were variable. The diameter of the outer conductor was ϕ 241 mm, which is the same as that of the transmission line. The direction of the electric field was assumed to be radial. Point 3 is the port of the ceramic feed-through, and a bellows is inserted between points 2 and 3 in order to absorb the deformation of the ceramic feed-through. Therefore, a 100 mm length is necessary between points 2 and 3. The following equation was used for the rough estimation of the electric field E:

$$E = \frac{V}{r \ln(D_{outer}/D_{inner})} \quad (1)$$

where V is the voltage applied on the inner conductor, r is the distance from the axis, and D_{outer} and D_{inner} are the diameters of the outer and inner conductors, respectively. The value of impedance at point 2 must be obtained in order to calculate the converted loading resistance and the maximum strength of the electric field in the impedance transformer.

The length L is defined as the distance between a voltage node in the transmission line and the impedance matching device, including the uncertainty of a half wavelength multiplied by an integer. The loading resistance R corresponds to the impedance at the node. For the upper and lower port FAIT antennas, the maps of plasma loading resistance R_p (= $R-R_{vacuum}$) and the shift of voltage node ΔL (= $L-L_{vacuum}$) were obtained before the installation of the EVIT, as shown in figures 2(a) and (b), where the subscript 'vacuum' means injection to the vacuum without plasma. The typical values of R = 5.6 Ω (R_p = 5 Ω, R_{vacuum} = 0.6 Ω) and L = L_{vacuum}-0.05 m shown by '+' were used for the design of the EVIT. For the HAS antenna, R = 2.3 Ω (R_p = 2 Ω, R_{vacuum} = 0.3 Ω) was used as the typical loading resistance. Data of ΔL for the HAS antennas is insufficient. Therefore, we assumed L = L_{vacuum} as the typical value. The typical impedances at point 2 for the FAIT and HAS antennas were calculated from the values of R and L as 6.6+21.3j Ω and 4.7+50.7j Ω, respectively. Whereas the differences in impedance between upper and lower port antennas were small, therefore, they were averaged.

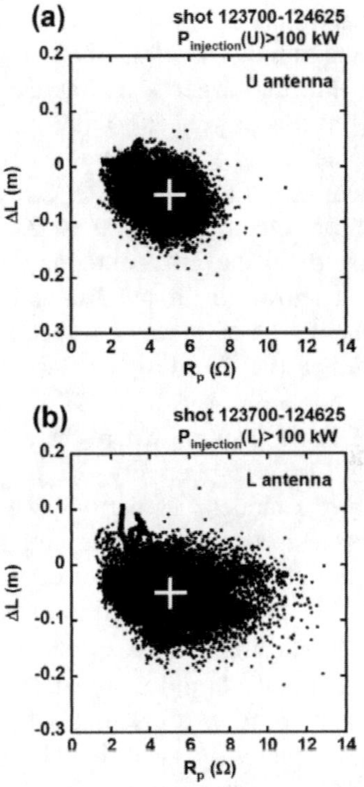

Figure 2. R_p-ΔL maps for (a) upper and (b) lower port FAIT antennas.

Figure 3(a) shows the enhancement factor of loading resistance R_{in}/R_{out} for the FAIT antenna calculated with the obtained typical impedance. The high-enhancement factor

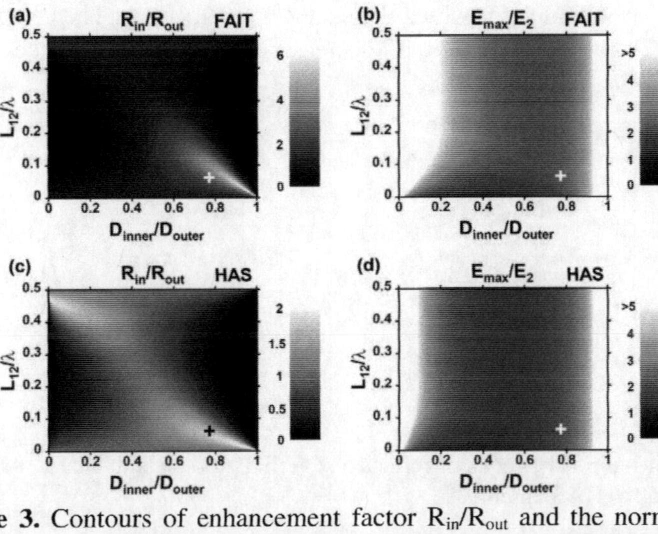

Figure 3. Contours of enhancement factor R_{in}/R_{out} and the normalized maximum strength of the electric field E_{max}/E_2. The ordinate is the length of the inner conductor normalized by the wavelength λ, and the abscissa is the ratio of the diameters of the inner and outer conductors. (a) R_{in}/R_{out} for the FAIT antenna; (b) E_{max}/E_2 for the FAIT antenna; (c) R_{in}/R_{out} for the HAS antenna; (d) E_{max}/E_2 for the HAS antenna.

region was located in the short L_{12} and large D_{inner}. The short L_{12} was favourable for the design of a compact EVIT. Figure 3(b) shows the maximum electric field normalized by E_2 shown in figure 1. We determined L_{12} = 495 mm and $D_{inner} = \phi$ 186 mm as shown by the '+', since the value of L_{12} was the maximum for the limited space, and the larger D_{inner} increased the electric field, although the larger L_{12} and D_{inner} increased loading resistance. At the design values of L_{12} and D_{inner}, the enhancement factor was 2.82, and the normalized maximum electric field was 2.13. They were also calculated for the HAS antenna with scanning L_{12} and D_{inner}, as shown in figures 3(c) and (d). In this case, two regions of high-enhancement factor appeared, and it was determined that the design values of L_{12} = 495 mm and $D_{inner} = \phi$186 mm for the FAIT antenna were also useful for the HAS antenna, where the enhancement factor and the normalized maximum electric field were 1.64 and 1.88, respectively.

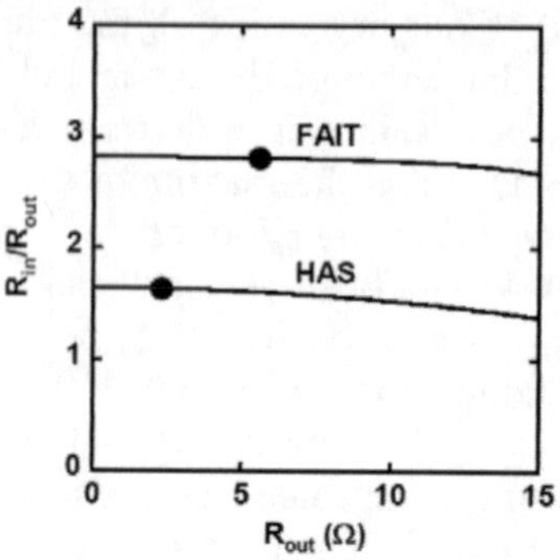

Figure 4. Dependency of enhancement factor on the loading resistance in the output side of EVIT. Closed circles denote the typical discharge.

From the model calculations, we also found that the dependency of the enhancement factor on the loading resistance R_{out} at the output port was weak, as shown in figure 4, where the position of the voltage node in the antenna side was fixed. Therefore, the difference in enhancement factor between the FAIT and HAS antennas was mainly caused by the position of the voltage node. With a characteristic impedance z_{ct} of 15.7 Ω calculated from the diameters of inner and outer conductors between points 1 and 2, the reflection coefficient Γ on the transmission line with a characteristic impedance of z_c was calculated by changing the position of point 1, and plotted on Smith charts, as shown in figures 5(a) and (b). The equation of the locus is given as follows:

$$\frac{|\Gamma + (z_c - z_{ct})/(z_c + z_{ct})|}{|\Gamma + (z_c + z_{ct})/(z_c - z_{ct})|} = \frac{|\Gamma + 0.52|}{|\Gamma + 1.92|} = \text{const.} \quad (2)$$

This equation describes the circle of Apollonius for which the center is on Im(Γ)=0. The center of the locus is located in the negative Re(Γ) region, therefore, |Γ| decreases and as a consequence, loading resistance increases. With the obtained values of L_{12} and D_{inner}, we designed the compact EVIT. Figure 6 shows the upper port FAIT antenna with the EVIT. The outer conductor of the EVIT was made with aluminum and the inner conductor was coated with silver on stainless steel to reduce power loss. The flange-to-flange length was only 628 mm, and the edge of the

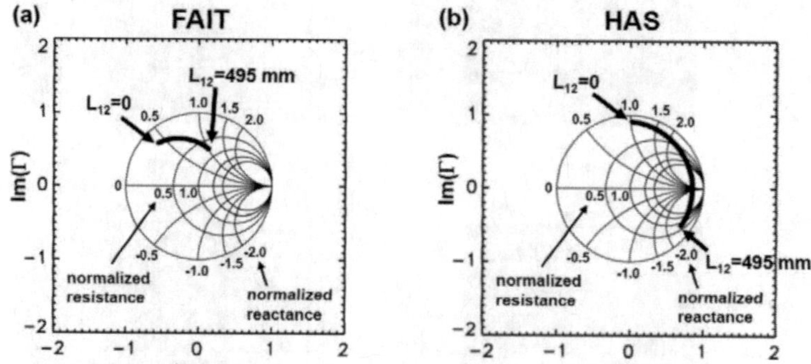

Figure 5. The locus of reflection coefficient on Smith charts for (a) FAIT and (b) HAS antennas.

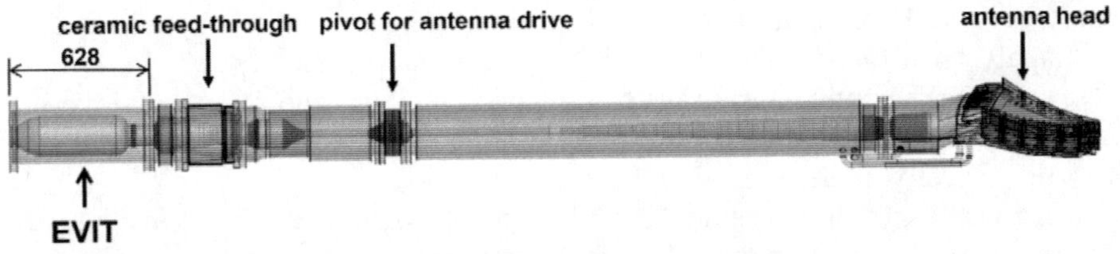

Figure 6. The upper port FAIT antenna with the EVIT.

thick inner conductor was rounded in order to avoid the convergence of electric field lines.

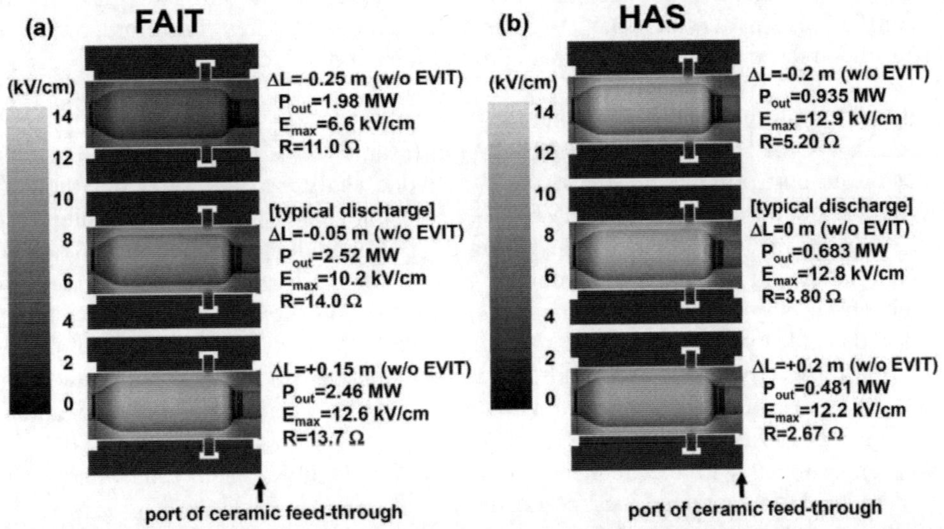

Figure 7. The strength of the electric field in the EVIT for (a) FAIT and (b) HAS antennas depending on the voltage node position ΔL without the EVIT. The maximum voltage in the transmission line was set to 30 kV. The output power P_{out}, the maximum electric field strength E_{max}, and the loading resistances R with the EVIT are also shown. The loading resistances of the FAIT and HAS antennas without the EVIT are 5.6 Ω and 2.3 Ω, respectively.

3. Electromagnetic simulation of ex-vessel impedance transformer

Electromagnetic simulation was performed with HFSS, as shown in figures 7(a) and (b), to obtain detailed estimation of the enhanced loading resistance and the strength of the electric field. The frequency was 38.5 MHz and the maximum voltage in the transmission line connected to the EVIT was set to the operational limit of 30 kV by adjusting output power. The estimated enhancement factor of loading resistance was 2.50 and 1.65 for FAIT and HAS antennas, respectively, at the typical discharge. These were consistent with the calculations by the simple model shown in figure 1. We also changed the voltage node position on the antenna side by ±20 cm, for the enhancement factor strongly depends on this position, as mentioned in section 2. The range of ±20 cm was enough for the FAIT antennas according to figures 2(a) and (b), although data for the HAS antennas was insufficient. In the case of the FAIT

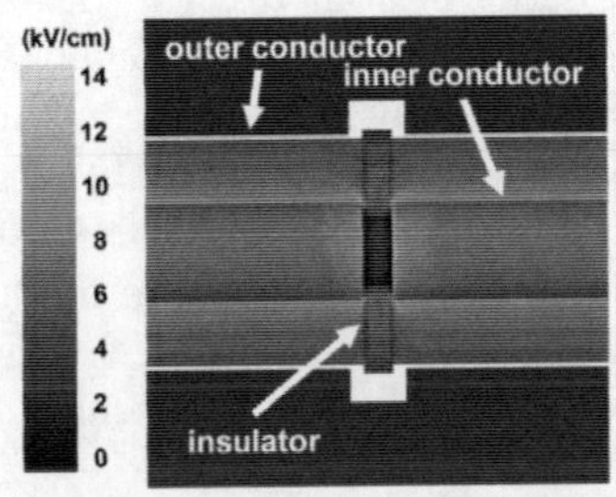

Figure 8. The strength of the electric field near the insulator. The applied voltage was 35 kV.

antenna, the loading resistance was kept high. On the other hand, in the case of the HAS antenna, the enhancement factor was small when the voltage node shifted to the antenna side by 20 cm, but still larger than unity ($R_{in}/R_{out} = 1.16$). The power loss in the transmission line was approximately inversely proportional to the loading resistance. Therefore, the power loss decreased to 40% and 60% for the FAIT and HAS antennas, respectively, in the typical discharge.

The electric field in the transmission line around the PTFE (polytetrafluoroethylene) insulator for the support of the inner conductor was also simulated in order to investigate the breakdown condition as shown in figure 8. The frequency was 38.5MHz and the applied voltage was 35 kV, which is the interlock level, and breakdowns sometimes occurred at the insulator beyond this voltage, although nitrogen was filled at a pressure of 0.3 MPaG in order to prevent breakdowns. The simulated maximum electric field was 13.5 kV/cm at the edge of the inner conductor. As shown in figures 7(a) and (b), the maximum strength of the electric field in the EVITs, where nitrogen is also filled at the pressure of 0.3 MPaG, is smaller than 13.5 kV/cm in all cases in the figures. Therefore, breakdowns in EVITs will be avoided if the voltage in the transmission line is limited to 30 kV.

4. Enhancement of loading resistance in the HAS antenna

EVITs were attached to the HAS antennas in 2014. We measured the reflection coefficient of the HAS antenna at the lower port without plasma using a network analyser, and compared the reflection coefficient to that before the installation of EVIT. At a frequency of 38.5 MHz, the loading resistance without plasma R_{vacuum} was 0.296 Ω before installing the EVIT, and it increased to 0.537 Ω after the installation. The enhancement factor in vacuum injection was 1.81, which almost agreed with the simulated result of 1.71 by HFSS. The voltage node position on the transmission line L_{vacuum} was shifted to the antenna side by 0.764 m, which also agreed well with the simulated result of 0.752 m. At the frequency of 77 MHz, which corresponds to the frequency of the second-harmonic heating of hydrogen, the loading resistance decreased from 0.441 to 0.213 Ω. Therefore, in the HAS antenna, the EVIT was not suitable for this frequency.

In the plasma experiments with the frequency of 38.5 MHz, loading resistance with and without the EVIT for the lower HAS antenna was compared changing the distance Δ between the Faraday-shield and the last closed flux surface, as shown in figure 9. The upper port antenna was turned off in order to avoid the mutual coupling effect. The range in line-averaged electron density was from 0.8 to 1.5 × 10^{19} m^{-3}, and the major radius of the magnetic axis and strength of the magnetic field on the axis were 3.6 m and 2.75 T, respectively, in both experiments. As a reference, data from the lower port FAIT antenna were plotted. These data showed that the conditions were nearly the same in both experiments, since the EVIT was not attached in the FAIT antenna in both experiments. The loading resistance of the HAS antenna was approximately doubled by the EVIT. The shift of voltage node ΔL for the HAS antenna was negligible in both experiments. In this case, the simulated enhancement factor was 1.65, which was consistent with the experiments. Higher power injection was enabled by the enhancement of loading resistance in the HAS antennas.

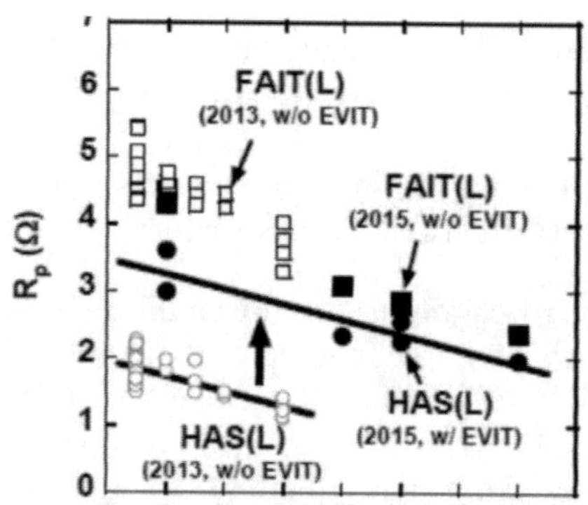

Figure 9. Comparison of R_p with and without EVIT in various antenna positions.

5. Summary

Compact EVITs were developed and installed in the ICRF antennas in the LHD to increase loading resistance in order to increase the ICRF injection power. This was because high voltage on the transmission line had been the bottleneck in high-power operation. High loading resistance is also

IOP Conf. Series: Journal of Physics: Conf. Series **823** (2017) 012007 doi:10.1088/1742-6596/823/1/012007

favourable for the reduction of power loss in the transmission line. Plasma experiments were conducted with and without the EVIT using a HAS antenna. In the experiments, the loading resistance increased, and the experimental enhancement factor of approximately 2 was consistent with that obtained by the simulation (1.65). According to the simulation, the EVIT will also increase the loading resistance of the FAIT antennas by 2.5 times in the typical discharge and as a consequence, a high-power injection of 1.8 MW from one FAIT antenna will be possible.

Acknowledgement
The authors would like to thank the staff of the LHD Experiment Group at the National Institute for Fusion Science for their helpful support during this work. This work was supported by NIFS budget ULRR703.

References

[1] Kaneko O, Yamada H, Inagaki S, Jakubowski M, Kajita S, Kitajima S, et al., 2013 *Nucl. Fusion* **53** 104015.

[2] Kasahara H, Seki T, Saito K, Kumazawa R, Mutoh T, Shimpo F, et al., 2010 *Plasma and Fusion Research* **5** S2090.

[3] Saito K, Seki T, Kasahara H, Seki R, Kamio S, Nomura G, et al., 2015 *Fusion Eng. Des.* **96-97** 583-588.

[4] Kasahara H, Saito K, Seki T, Kumazawa R, Nomura G, Shimpo F, et al., 2011 *Proc. 38th EPS Conf. Strasbourg* P2.099 and http://ocs.ciemat.es/EPS2011PAP/pdf/P2.099.pdf.

[5] Kamio S, Kasahara H, Seki T, Saito K, Seki R, Nomura G, et al., 2015 *Fusion Eng. Des.* **101** 226-230.

[6] Kamio S, Seki T, Saito K, Kasahara H, Seki R, Nomura G, et al., 2015 *Proc. 42nd EPS Conf. Lisbon* P5.154 and http://ocs.ciemat.es/EPS2015PAP/pdf/P5.154.pdf.

[7] Saito K, Seki T, Kasahara H, Seki R, Kumazawa R, Nomura G, et al., 2013 *Fusion Eng. Des.* **88** 1025-1029.

[8] Saito K, Kumazawa R, Mutoh T, Seki T, Watari T, Torii Y, et al., 2001 *Nucl. Fusion* **41** 1021-1035.

[9] Saito K, Kumazawa R, Seki T, Kasahara H, Osakabe M, Isobe M, et al., 2010 *Fusion Science and Technology* **58** 515-523.

IOP Conf. Series: Journal of Physics: Conf. Series **823** (2017) 012008 doi:10.1088/1742-6596/823/1/012008

Radio frequency emissions from laser induced material breakdown

Vinoth Kumar[1], Manikanta Elle[1] and Prem Kiran Paturi[1]

[1]Advanced centre of research in high energy materials (ACRHEM), University of Hyderabad, Gachibowli, Hyderabad 500 046, India

E-mail: premkiranuoh@gmail.com

Abstract. Observation of radio frequency (RF) emissions from laser induced breakdown (LIB) of stainless steel (SS304) and the studies on the origin of these emissions are presented. The emitted frequencies were observed to be in the range of the estimated ion-plasma frequencies, indicating the contribution of the ion-neutral collisional interactions in the radiation of RF emissions from plasma. These studies can be extended to develop an RF based tool to study the plasmas of varying electron temperatures and densities.

1. Introduction

The low frequency emissions from plasma and their applications have been widely investigated for several decades [1]. The RF and microwave radiations generated from plasmas due to laser induced breakdown (LIB) [2,3] have been studied to make them feasible for applications like laser ground penetrating radars (LGPR) [4] that has several advantages when compared with the conventional GPR. From our studies, the spectral selective RF emissions from nanosecond (ns) and picosecond (ps) LIB of different targets, that can be developed into a sample identification tool, was understood [5]. These emissions were mainly observed due to the ion plasma frequencies due to the atomic and molecular ionized species [6]. By varying the input laser parameters, plasmas with varying electron temperatures and densities hence varying plasma ion frequencies can be created, making the study of RF emissions a generic tool for plasma diagnostics. For example, in tokomak plasmas [7], the α particles, emitted during the fusion process, due to their cyclotronic gyration in the confining magnetic field are predicted to give rise to radio frequency (RF) emissions in 10-500 MHz band [8]. Experimental observation of RF emissions due to fast ions situated in the core of the discharge were reported to be useful in extracting the information concerning the fusion products [9]. Though the plasma is actively held away from the walls of the tokamak container by using powerful magnetic fields, sometimes the plasma can interact with the inner walls of the reactor which might result in plasma contamination. Therefore, understanding the RF emissions from tokamak plasmas pave way to study the thermal contaminants present in the plasma system over long duration. In this paper, we report the observation and the origin of RF emissions from the LIB of stainless steel (SS304), which is one of the important materials considered for the tokamak walls.

2. Experiment

Second harmonic (532 nm) of Nd:YAG lasers of different pulse durations (7 ns and 30 ps) were focused, in the focal geometry (f/D) ~ 10 and 6.5 in the ns and ps regimes respectively, on the target (SS304) to generate LIB. The laser intensities in case of ns LIB and ps LIB are of the orders of ~10^{11}

W/cm^2 and ~10^{13} W/cm^2 respectively. The RF emission detection was carried out with a broadband mini biconical antenna (Lindgren 3182) of frequency range 30 MHz – 1 GHz while the waveforms were recorded with a spectrum analyzer (Agilent PSA E4448A) which can scan over 3 Hz – 50 GHz, giving rise to the frequency evolution over the entire band of the detecting antenna at a single shot. The polarization state of ns and ps laser pulses was vertical and horizontal, respectively. As depicted in figure 1(a), the antenna polarization is termed as vertical (V-pol), whenever it is oriented parallel to the x-axis. When it is rotated parallel to the y-axis, it is termed as horizontal antenna polarization (H-pol). The antenna was rotated to measure the H-pol and V-pol components of emissions. Single shot-LIB was ensured with the target mounted on an electronically controlled XY translation stages (M-443, LTA-HA, and ESP-300, M/s. Newport). In order to cut down the contribution of ambient RF noises, the spectra without plasma formation (noise) was subtracted from the LIB data that contains noise and the signals of interest. Besides, the average RF spectrum from the LIB of atleast 100 laser shots was considered for data analysis. The spectrum analyzer was externally triggered by the laser pulse to ensure better signal to noise ratio of the recorded spectra [10].

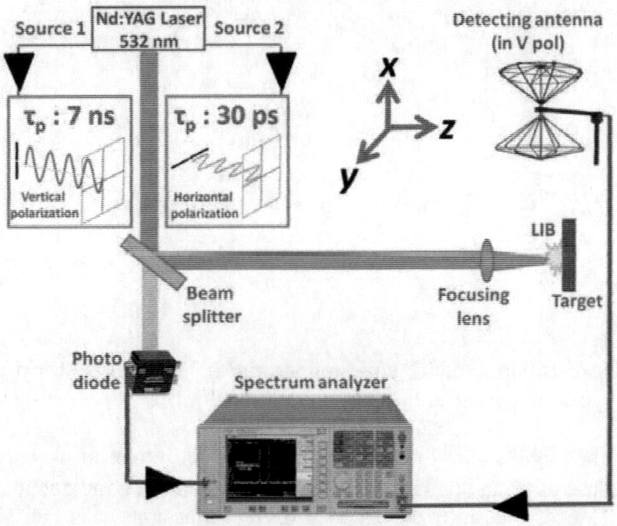

Figure 1. (a) Experimental setup with laser propagating along z direction.

Before proceeding with the experiments, it is important to position the detector at the point where maximum radiation is received. Therefore, the angular distribution of RF emissions from the LIB of a solid target (copper) was studied by detecting the RF emissions from LIB at different angles between the source and the detector, by keeping the distance fixed. The amplitude of RF emissions from ns LIB and ps LIB were observed to be increasing with the angle between the source and the detector with the maximum when the source and the detector are perpendicular to each other (90º and 270º). In order to arrive at the integrated power of RF emissions from LIB of the targets, the frequency domain RF spectra were converted into the corresponding time domain spectra by inverse fast Fourier transformation (IFFT). The area under the curve of the time domain spectrum gives the total energy output of the emitted RF radiation. The total output energy is normalized with respect the solid angle subtended by the detector. This is termed as the normalized RF output. The fit of normalized RF with respect to of angle of detection is given in figure 2.

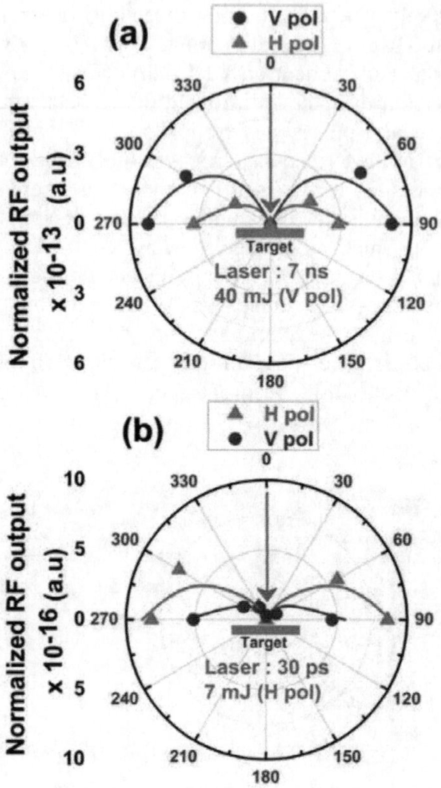

Figure 2. (a) Angular distribution of RF emissions from (a) 7 ns LIB at 40 mJ and (b) 30 ps LIB at 7 mJ of solid target (copper) under different antenna polarization.

The best fits for the angular distribution of normalized RF emissions were turned out to vary as $\sin^{1.6}\theta$ and $\sin^{1.5}\theta$ in case of ns and ps LIB, respectively, for the laser and antenna polarizations used in the present study. Hence, the $\sin^2\theta$ dependence of RF emissions was confirmed in case of solid targets. The dipolar emission pattern ($\sin^2\theta/r^2$) [11] was confirmed from the RF emissions from LIB of atmospheric air [6]. Thus, in our studies, antenna was positioned, closer to the plasma source (~ 1 metre), perpendicular to the laser propagation direction. It is also observed that the emissions with same laser and antenna polarization were observed to be stronger than those with cross polarization. Therefore, the analyses were emphasized with same laser and antenna polarization conditions.

3. Results and discussion

The RF emissions from ns and ps LIB of SS304, at different input laser energies, are given in Figure 3. The emissions were observed to be strongly dependent on the laser parameters (pulse duration, input laser energy etc.). The dominant RF emissions from ns and ps LIB were observed to be spread within 300 MHz barring few relatively less intense peaks at higher frequencies. From the positioning of the frequency peaks, it is confirmed that no harmonic signals are observed from the LIB. This shows that the frequency lines represent the RF emissions from the LIB of the target medium. Both in ns and ps LIB, the RF emissions at lower input laser energies were observed to be higher than those at higher input laser energies. When compared with the emissions from ns LIB, the emissions from ps LIB were observed to be 2-3 orders lesser.

IOP Conf. Series: Journal of Physics: Conf. Series **823** (2017) 012008 doi:10.1088/1742-6596/823/1/012008

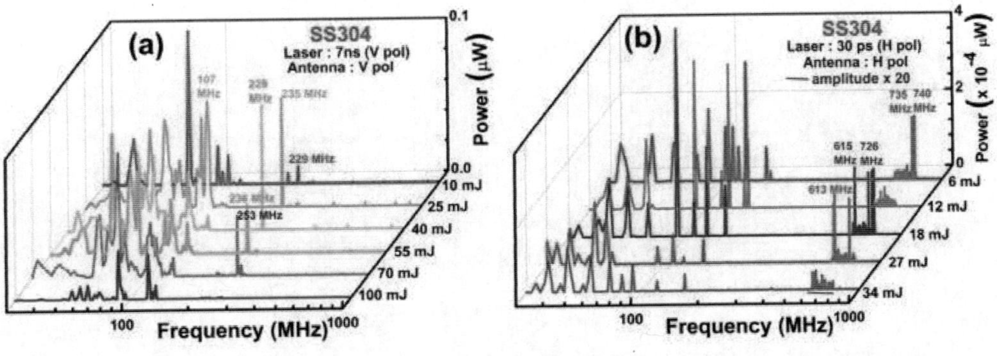

Figure 3. RF emissions from (a) 7 ns and (b) 30 ps LIB of SS304 under same laser and antenna polarization. *The line in 3 (b) represents the amplified portion of the spectrum for the better view of the frequencies*

The interaction of the plasma constituents, as depicted in figure 4, will result either in the growth (further ionization) or decay (recombination) of plasma and the corresponding energy release [12]. The RF emissions from LIB depend on the degree of ionization that determines the plasma number density (n_e, n_i) which in turn depends on the laser parameters (energy, pulse duration etc.) and the plasma recombination. In underdense plasmas as in our case, the interaction of charged particles with atomic clusters will be dominant [13]. These charge-neutral interactions give rise to emissions in the RF and microwave range of frequencies [1, 14] as shown in figure 4. Therefore, the RF emissions take place as long as there is interaction between the charged particles and the neutrals in the plasma system. Evidently, the RF emissions were observed to be relatively higher at lower input laser energies where the electron-neutral interaction remains higher (figure 3). As the ionization is higher with the ps pulses, due to relatively higher peak power when compared to the ns pulses, resulting in lower probability of charge-neutral interactions. Moreover, the rate of recombination is relatively higher in plasma due to ps LIB. Thus the emitted RF powers from ps LIB are lower than those from ns LIB. Besides the emissions due to charge-neutral interactions, RF emissions also take place owing to the oscillations of the dipoles that are formed due to the difference in the drift velocities of the electrons (~1500 m/s) and the ions (~5 m/s) [4]. However, after a certain value of input laser intensity, the damping of plasma brings down the emission of radiation. The damping of the plasma oscillations is predominantly due to collision/recombination and electron-attachment [13]. According to the recombination model, the loss of plasma by recombination is proportional to the product of electron and ion number densities ($n_e \times n_i$). At moderate input laser intensities, the number of charged species and their velocities in the underdense plasma is relatively lower and hence the charge particle-neutral collisions are dominant. In this case, the relaxation time for the kinetic energy of the electrons, which is proportional to the electron temperature, is longer than that for the average velocity (of the ions and neutrals) by a factor of (m_n/m_e) [12,13]. Therefore, the RF emissions were observed to be higher at moderate input intensities. At higher laser intensities, the higher value of n_e (n_i) and also the temperature of the electrons leads to the interaction between the plasma constituents. In this case, recombination will take place at a faster rate, when compared to that in plasmas due to lower input intensities, reducing the plasma density more drastically. Thus the RF emissions come down at higher input intensities. This is reflected in the data presented in figure 3.

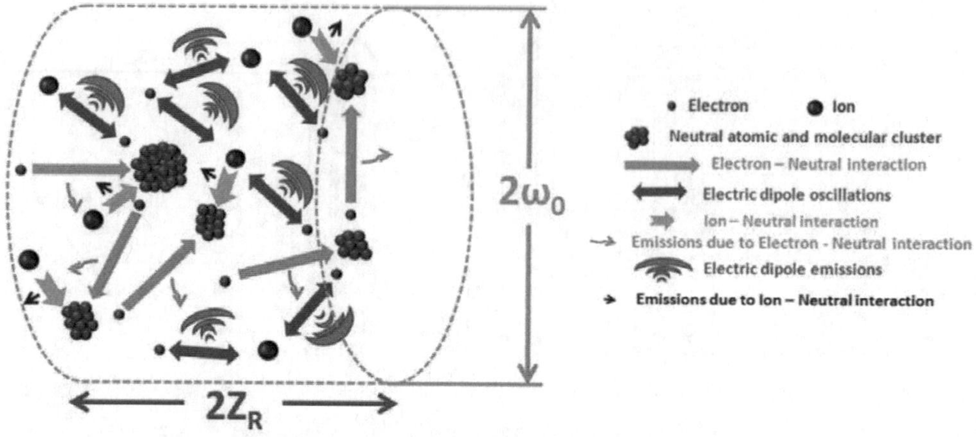

Figure 4. RF emission mechanisms from LIB of target material

Further, to arrive at the origin of these emissions, the ion plasma frequencies (ω_{pi}) were estimated using the following equation.

$$\omega_{pi}^2 = \frac{n_i Z^2 e^2}{m_i \varepsilon_0} \quad (1)$$

where n_i is the ion number density, Z and m_i are the charge state and mass of the ions respectively. In our case, the atoms are considered to be singly ionized. Therefore, Z is taken to be one and hence $n_e = n_i$. The target material (SS304) was studied to be composed of Fe (~ 72%), Cr (~ 18%), and Ni (~ 8%) and Mn (~1%). The value of $n_i = n_e$, under the studied range of laser intensities were obtained from literature [15]. The ω_{pi} values corresponding to the different possible atomic and molecular species in the LIB of SS304 were estimated and given in table 1.

Table 1. Estimated ion-plasma frequency (ω_{pi}) values

Laser	Intensity range (W/cm^2)	Electron density range (m^{-3})	Estimated ion-plasma frequency values (MHz)	
			Atomic species	Molecular species
7 ns	3 - 7×10^{10}	1 - 3×10^{17}	Fe^+: 235 Mn^+: 236 Cr^+: 243 Ni^+: 229	$(Fe_2O_3)^+$: 110 $(Mn_3O_4)^+$: 110
30 ps	1 - 4×10^{12}	7 -10×10^{18}	Fe^+: 1247 Ni^+: 1216 Mn^+: 1257 Cr^+: 1292	$(Fe_3O_4)^+$: 612 $(Fe_2O_3)^+$: 737 $(Mn_3O_4)^+$: 616 $(Mn_2O_3)^+$: 742 $(Ni_2O_3)^+$: 725

It is evident that the emitted radio frequencies, shown in figures 3(a) and (b), correspond to the estimated ion-plasma frequencies shown in table 1. This shows the importance of the ionization and ion-neutral interactions in the RF emissions from the underdense plasma due to LIB. The estimated ω_{pi} values are observed with ± 3 MHz frequency error, which is the minimum frequency that the spectrum analyzer can resolve. The ω_{pi} values, corresponding to the atomic species, in ps LIB were observed to be in GHz range which is beyond the detection range of the antenna. However, some of the peaks in the MHz region can be accounted for the contributions from the oxides (molecular species) of the elements present in the system. Owing to relatively smaller velocity of the charged particles (electrons and ions) that collide with the neutrals, the RF power of high frequency emissions (250-760 MHz) is relatively lower than that of the low frequency emissions (30-250 MHz). Moreover, in case of ps LIB, the charge-neutral interaction will be relatively low. This is reflected from the low amplitude of emissions in case of ps LIB. Therefore, the RF emissions from LIB can be useful in extracting the elemental information of the target.

4. Conclusion

The studies on RF emissions from LIB of SS304 were carried out. The emitted frequencies from LIB were observed to be in the range of estimated ion-plasma frequency ω_{pi}. Therefore, the emissions can be attributed to the contribution of ion-neutral interactions to the radiation emissions from the plasma source. These studies can be extended to develop an RF based tool to study the plasmas of varying electron temperatures and densities. By varying the input laser parameters (energy, polarization, pulse duration), plasmas with varying electron temperatures and densities hence varying plasma ion frequencies can be created, making the study of RF emissions a generic tool for plasma diagnostics.

Acknowledgement

The authors thank DRDO, Government of India, for the grants-in-aid support.

References

[1] Bekefi G 1966 *Radiation processes in plasmas* (Wiley)
[2] Pearlman J S and Dahlbacka G H 1978, *J. Appl. Phys.* **49**, 457
[3] Gerdin G and Tanist M J 1986, *Plasma Phys. Control.* Fusion **28**, 527
[4] Nakajima H, Shimada Y, Somekawa T, Fujita M and Tanaka K A 2009, *IEEE Geosci. Remote Sens. Lett.* **6**, 718
[5] Vinoth Kumar L, Manikanta E, Leela Ch and Prem Kiran P 2014, *Appl. Phys. Lett.* **105**, 064102
[6] Vinoth Kumar L, Manikanta E, Leela Ch and Prem Kiran P 2016, *J. Appl. Phys.* **119**, 214904
[7] McCracken G and Stott P 2005, *Fusion the energy of the universe* (Elsevier academic press)
[8] Kadomtsev B B 1992 *Tokamak plasma: A complex physical system* (Taylor & Francis group)
[9] Fraboulet D 1996 *Ion cyclotron emission in tokamak plasmas* (PhD dissertation, Grenoble Univ)
[10] Vinoth Kumar L, Leela Ch, Manikanta E, Tewari S P and Prem Kiran P 2012, *Proc SPIE.* **8434**, 84340V1
[11] Jackson J D 1998 *Classical Electrodynamics* (Wiley)
[12] Chen F F 1984 *Introduction to plasma physics and controlled fusion vol 1: plasma physics* (Plenum press)
[13] Bittencourt J A 2004 *Fundamentals of plasma physics*(Springer)
[14] Akcasu A Z and Wald L H 1967, *Phys. Fluids* **10**, 1327
[15] Singh J P and Thakur S N 2007 *Laser-Induced Breakdown Spectroscopy* (Elsevier Science)

2D turbulence structure observed by a fast framing camera system in linear magnetized device PANTA

Satoshi Ohdachi[2,3], Shigeru Inagaki[1] , Tatsuya Kobayashi[2,3], and Motoshi Goto[2,3]

[1]Research Institute for Applied Mechanics, Kyushu University, Kasuga-shi, Fukuoka-ken, Japan

[2]National Institute for Fusion Science, Toki-shi, Gifu-ken, Japan

[3]SOKENDAI (The Graduate University for Advanced Studies), Toki-shi, Gifu-ken, Japan

E-mail: ohdachi@nifs.ac.jp

Abstract. Two dimensional turbulence observed in the linear magnetized device PANTA is studied using a visible high speed camera system. When the fluctuation component are decomposed with Fourier-Bessel expansion, complex density fluctuations are recognized as the superposition of the several mode having low mode number, e.g., m = 1, 2, and 3. The phase relations between waves indicate that they are non-linearly coupled.

1. Introduction

Mesoscale structures, such as the zonal flow and the streamer play important role in the drift-wave turbulence. The interaction of the mesoscale structure with the turbulence is a key to understand the structural formation of turbulence and the turbulence-driven transport in the magnetically confined plasmas. In a cylindrical magnetized device, the interaction of the streamer and the drift wave has been found [1]. In that study the spatial shape of the streamer-like structure were obtained by using the bi-spectrum analysis with a fixed probe array and a movable probe system. Statistical nature of the mesoscale structure can be studied by this method. However, if the dynamic wave-wave interaction can be observed directly, our understanding of the turbulence will be improved further. For this purpose, direct visualization of the turbulence with a high-speed camera system was performed. The analytical procedure and initial results are presented in this article.

2. Experimental Setup

A fast framing camera system observing the cylindrical plasma through an end port of the PANTA device is schematically shown in figure 1(A). If the perturbation is aligned to the magnetic field lines, two dimensional turbulent structures can be observed from this arrangement. It is noted that when we observed the plasma column perpendicularly, the mode structure is well aligned to the magnetic field lines. The parameters of the target plasma is the following; T_e ~ 3 eV, n_e ~ 1 × 10^{19} m^{-3}, T_i ~ 0.3 eV, B = 900 G, the neutral pressure P_n = 0.8 m Torr, the minor radius a ~ 6 cm and L = 4 m. Helicon source (7 MHz, 3 kW) is used to generate plasma. No filter was used for the camera. Since the electron temperature is several eV, line emission of Ar I is the dominant component of the visible image. Its intensity is sensitive to the local electron density fluctuations and local electron temperature fluctuations. A sample image is shown in figure 1(B). The plasma is surrounded by azimuthally aligned 64-pin Langmuir probe array. The contour plot with white lines shows the coherence between

IOP Conf. Series: Journal of Physics: Conf. Series **823** (2017) 012009 doi:10.1088/1742-6596/823/1/012009

imaging data and the ion saturation current measure by a probe tip of 64 pin system. High-coherence region is well localized so that the local density / temperature perturbation can be estimated from the visible emission intensity. It is confirmed that the spatial / frequency spectra measured by ion saturation using the probe system is quite similar to that by 2D visible measurement described in detail in Section 3. It is thus reasonable we assume that the fluctuation of the light emission density fluctuations having low mode number. .

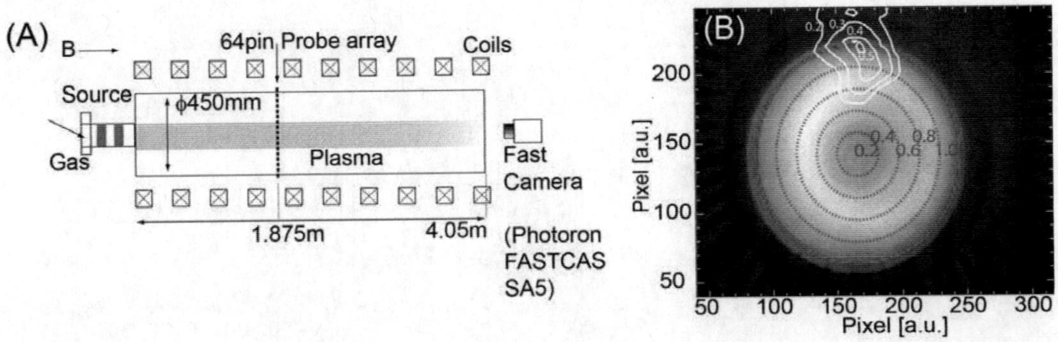

Figure 1. (A) Schematic drawing of the high-speed-camera measurement. (B) A typical emission profile (Blue-White gradation) and the contour of the correlation coefficient between the ion saturation current of a probe and the emission data are shown with the white lines. The region where the Fourier-Bessel decomposition is performed is shown by the red circles.

3. Analysis and Initial Results

Fluctuating component of the visible emission *g(r, θ)* is decomposed to orthogonal components by the Fourier-Bessel expansion method [2] as,

$$g(r,\theta) = \sum_{m=0}^{m_{max}} \sum_{l=0}^{l_{max}} \left(a_m^l \cos(m\,\theta) + b_m^l \sin(m\theta)\right) J_m\left(\lambda_m^{l+1} r\right) \quad (1)$$

where $\lambda_m^l$is the *l*-th zero point of the *m* th order Bessel function $J_m(z)$.

Decomposition is performed where emission light is strong. The boundary is shown by the red *r*=1.0 circle in figure 1(B). Since the Fourier-Bessel series are orthogonal, it is quite easy to obtain the coefficients a_m^l and b_m^l. Error in the decomposition is estimated to less than 5% when $m_{max} = 4$ and $l_{max} = 10$ are assumed. If we combine coefficients a_m^l and b_m^l to complex coefficient $z_m^l = a_m^l + i\,b_m^l$, z_m^l represents the magnitude of mode with phase information. Complex amplitude at a minor radius *r* with a mode number *m* can be written from the integral as,

$$z(t,r,m) = \sum_{l=0}^{l_{max}} z_m^l J_m\left(\lambda_m^{l+1} r\right) \quad (2)$$

Using the Fourier transform $Z(\omega, r, m) = \int_{-\infty}^{\infty} z(t,r,m) e^{-i\omega t} dt$, rotary spectrum *S(ω, m)* [3] can be defined as $S(\omega, m) = \langle Z^* Z \rangle$.

Rotary spectrum with *m* =1, 2 and 3 are shown in figure 2. There are many sharp peaks, labelled by f_1, f_2, ⋯, in broad spectra. It is noted that the *m* = 2 and *m* = 3 components are not strong in the core region. Those spectra in the edge region (e.g., $\rho = 0.8$) are quite similar to the spectra estimated with the multi-pin probe array data [1].

Figure 3 shows time evolutions of the emission pattern. Though the changes of the raw image are quite complicated, the dominant movement is the rotation of the m = 1 mode in the electron diamagnetic drift direction with f = f_2 and f_3, shown by the red curve in figure 3(A). Interaction with the other component, e.g., (m = *1*, f =f_1) which rotates in ion diamagnetic drift direction, shown by the blue dotted curve in figure 3(A), disturbs the constant rotation of the modes.

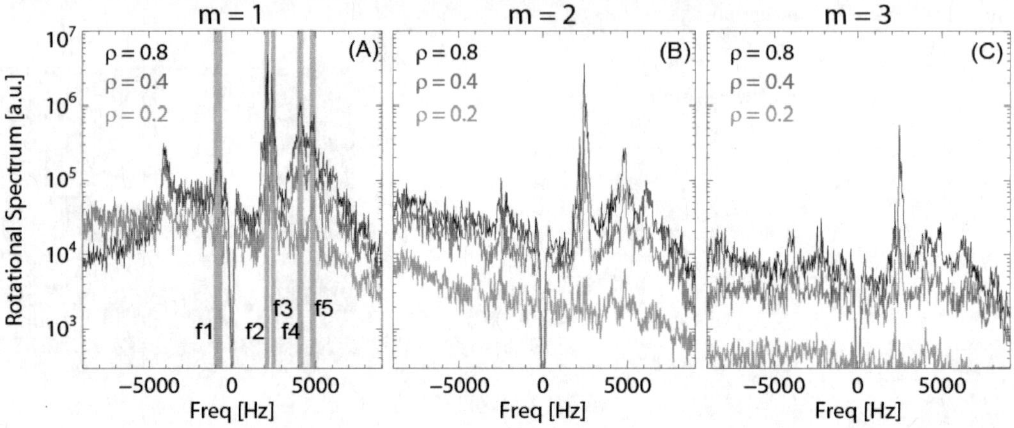

Figure 2. (A) Rotary spectra of fluctuations with the m=1(A), 2(B) and 3(C) are shown.

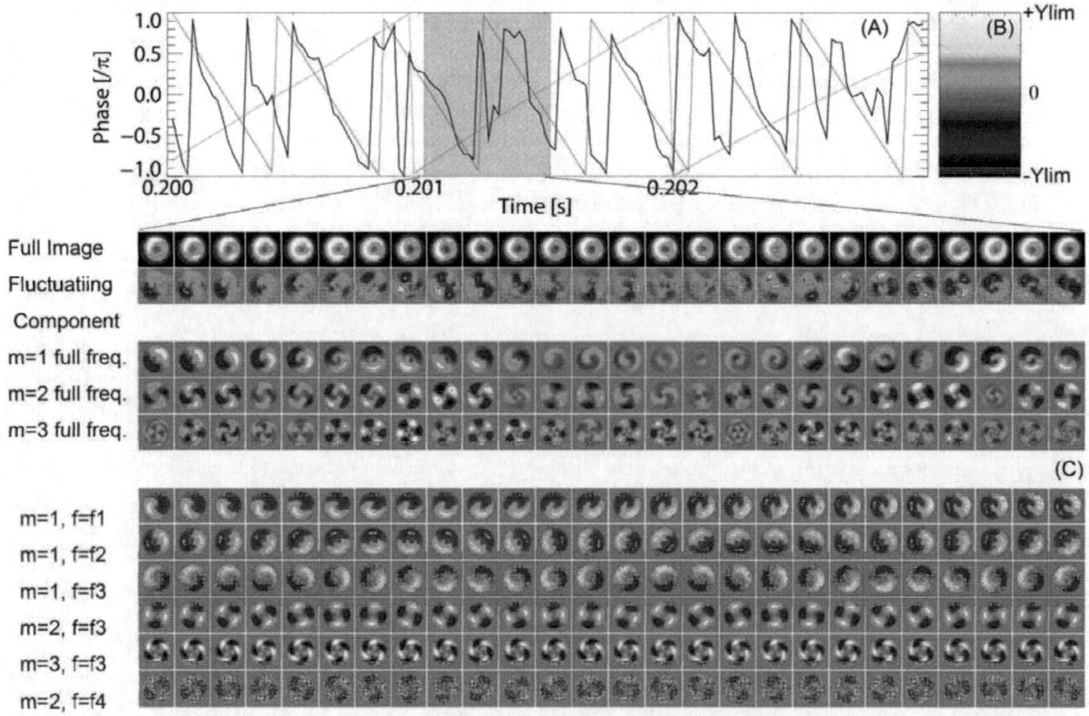

Figure 3. Time evolution of the rotation angle of the fluctuation is shown (A). Phase of total m=1 component, component with m = 1 and f= f_2, and component with m = 1 and f = f_1 is drawn by red b line, black line and blue dotted line, respectively. The color bar for the 2D images is shown in (B).

IOP Conf. Series: Journal of Physics: Conf. Series **823** (2017) 012009 doi:10.1088/1742-6596/823/1/012009

Sequential images of the emissions are shown in (C). Here, [+Ylim, -Ylim] = [22500, 0] (Full image), [0.18, -0.18] (Fluctuating components), [0.09, -0.09](m=1, full frequency), [0.09, -0.09](m=2, full freq.), [0.05, -0.05] (m = 3, full freq.). The image with frequency filter (6 bottom rows) is drawn with auto-scale in order to emphasize the spatial structure.

Wave-wave interaction of the each component is then investigated to study the non-linear nature of the turbulence. Among the possible wave-wave coupling pair, interaction of mode A ($m = 1$ and $f = f_2$) and mode B ($m = 2$ and $f = f_4$, ($\sim 2 \times f_2$,)) is quite obvious. The phase of the Mode B, ϕ_B and the two times of the phase of A, 2 ϕ_A are shown in figure 4(A).

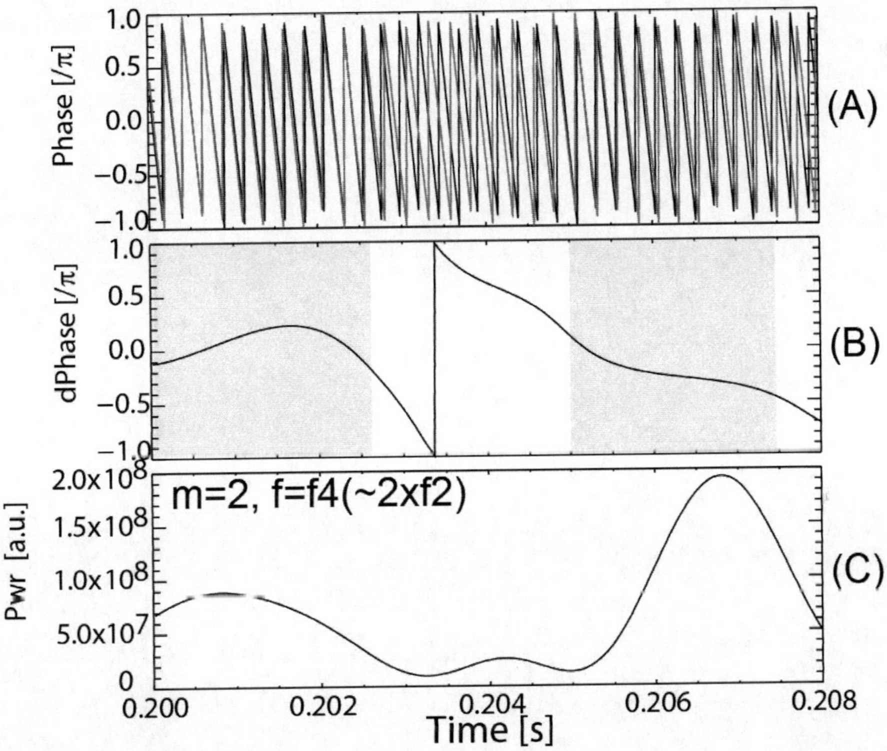

Figure 4. Time evolution of two times of the phase of the Mode A ($m = 1$ and $f = f_2$) ϕ_A (black line)) and the phase of the Mode B ($m = 1$ and $f = f_2$) (red line)) are shown in (A). The difference of the two phases is shown in (B). The amplitude of the Mode B is shown in (C).

There is a time period where the two phases are locked between each other (Shaded region of figure 4(B)). That is the reason finite bi-coherence between mode A and mode B is observed. The amplitude of the mode B is larger when the two phases are locked (figure 4(C)). It is possible that the mode B is produced from the mode coupling of the mode A. Many peaks in the rotary spectra might be caused by this kind of wave-wave coupling of the fundamental mode. It is noted that from the time evolution of the phase difference (figure 4(B)), there is a period where the two waves are unlocked, as well. Therefore, the wave-wave coupling process is found to be not simple steady process but a kind of intermittent process.

In summary, Fourier-Bessel expansion method is shown to be an effective tool to analyse two dimensional turbulences observed in the linear device. Non-linear coupling of the waves are clearly observed. It is found that coupling process is not constant in time and intermittently activated.

Acknowledgement

This work was supported by the collaboration programs of the RIAM of Kyushu University. and is also partially supported by the Ministry of Education, Science, Sports and Culture, Grant-in-Aid for Scientific Research 26249144, by the JSPS-NRF-NSFC A3 Foresight Program (NSFC: No.11261140328), and by NIFS/NINS under the project of Formation of International Scientific Base and Network.

References

[1] Yamada T et. al. 2010 *Phys. Rev. Lett.* **105** 225002
[2] Nagayama Y 1987 *J. Appl. Phys.* **62** 2702
[3] Gonella J 1972 *Deep Sea Research and Oceanographic Abstracts* **19** 833

IOP Conf. Series: Journal of Physics: Conf. Series **823** (2017) 012010 doi:10.1088/1742-6596/823/1/012010

Improving anti-felting characteristics of Merino wool fiber by 2.5 MHz atmosphere pressure air plasma

Nisha Chandwani[1], Purvi Dave[1], Vishal Jain[1], Sudhir Nema[1] and Subroto Mukherjee[1]

[1]Institute for Plasma Research, Near Indira Bridge, Bhat, Gandhinagar 382 428, India

E-mail: nisha@ipr.res.in

Abstract. The present work investigates the effect of high frequency (2.5 MHz) Dielectric Barrier Discharge (DBD) in air on surface characteristics of Merino wool as a function of plasma exposure time (5s to 15s). The FE-SEM (Field Emission Scanning Electron Microscopy), EDS (Energy Dispersive X-ray spectrum) and Derivative ATR-FTIR (Attenuated Total Reflection- Fourier Transform Infrared) Spectroscopy are used to study physio-chemical changes induced by plasma. These physio-chemical properties of fibers can be co-related with the felting behaviour of the wool fiber, which leads to shrinkage and pilling of garments while laundering. Felting occurs mainly because of presence of outermost hydrophobic cuticle layer having sharp scales. The FE-SEM analysis of wool fiber surface reveals that cuticle scales on wool fiber become blunt after plasma processing. The ATR-FTIR analysis along with second order derivative spectroscopy demonstrates the cleavage of di-sulphide bonds of cuticle and formation of sulphur-oxygen groups such as Cystine Sulphonate (-S-SO3-), cysteic acid (-SO3-), cystine monoxide(-SO-S-), cysteine di-oxide (-SO2-S-). A possible explanation about how the combined effect of morphological and chemical changes induced by plasma results in minimizing the felting of wool fibers is discussed.

1. Introduction

Wool is one of the natural fiber, having a number of advantageous characteristics. However, one of the drawback of this fiber is having felting property. Felting is the process of the entanglement of fibers, when exposed to aqueous medium. The felting results in shrinkage and pilling of woolen products, while laundering. Wool fiber consists of multiple layers with unique properties. The outer-most layer of wool fiber called cuticle is hydrophobic and scaly in nature. When fiber is immersed in water, each layer absorbs water more or less and swells. The outer layer being hydrophobic swells less as compared to inner layer cortex which is hydrophilic. This difference in expansion leads to uprising of scales and thus entanglement of fibers [1-4]. Conventionally, chlorination process is used to impart anti-felting or shrink proof property to wool. However, worldwide research has been increasing for use of low-temperature plasmas as an eco-friendly tool for surface modification of wool. A number of previous researches have been reported for low pressure non-thermal plasmas for shrink proofing of wool [5-8]. However, industrial requirement for wool processing can be fulfilled only by atmospheric pressure plasmas, as the processing cost is much lower as compared to low pressure plasma and also inline treatment in continuous mode is possible. Atmospheric pressure non-thermal plasmas can be generated by Dielectric Barrier Discharge (DBD) technique. Few of the reported work for the use of atmospheric pressure air plasmas for shrink proofing of wool are discussed here. Jayant et.al 2011 [9] has used 50 Hz air DBD system. In their experimental results, the descaling of the wool is not

IOP Conf. Series: Journal of Physics: Conf. Series **823** (2017) 012010 doi:10.1088/1742-6596/823/1/012010

achieved even when plasma exposure time was quite more (40 to 60 minutes). The present work comprises of investigating effect of high frequency (2.5 MHz) Dielectric Barrier Discharge (DBD) air plasma on surface characteristics of Merino wool for exposure time of few seconds (0 to 15s) only. This high frequency power source has been chosen, as it generates uniform glow dielectric barrier discharge (DBD) plasma in air at atmospheric pressure with higher density (~ 2.5W/cm^2) over the large surface of 70 cm^2 [18],which makes plasma processing significantly faster as compared to other DBD configurations.

2. Experimental Details

The DBD set-up consists of a pair of cylindrical electrodes made up of SS wire mesh (#325) and a polyester film (thickness 125 micron) separated by about 2 mm gap and high current driving power supply. A uniform glow like discharge in air at atmospheric pressure is generated in the inter-electrode gap as shown in figure 1. This power supply unit provides 2.5 MHz damped oscillations with the voltage magnitude of 5.8 KV (first cycle) and current magnitude of 50-100A (first cycle) for a duration of 0.25 μs [18]. The wool fibers are inserted manually between pair of cylindrical electrodes and discharge is generated by switching on the power supply for different exposure times from 5s to 15s, at a constant power density of 2.5 watt/cm^2.

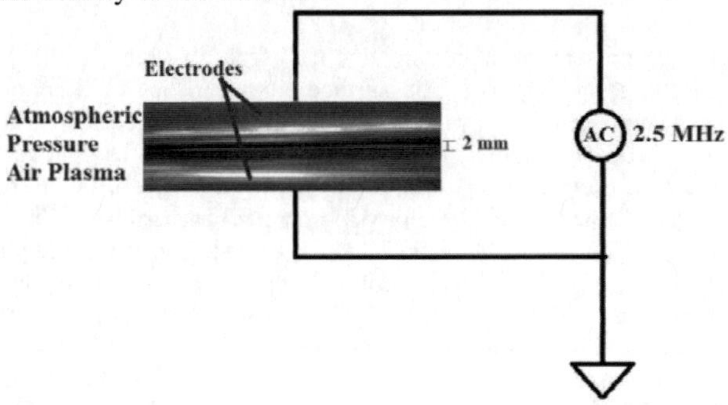

Figure 1. Schematic diagram of experimental set up for dielectric barrier discharge

3. Results & Discussion

3.1. Surface Morphology

In the present work FESEM (Field Emission Scanning Electron Microscope) Model: Merlin VP by Carl Zeiss has been used for surface analysis. The SEM images of Merino wool fiber at magnification of 10.0 KX are shown in Figure 2 (a) to (d) with plasma exposure time of 0, 5s, 10s and 15s, respectively. It can be clearly observed that the scales of untreated fiber are sharp and plasma exposure

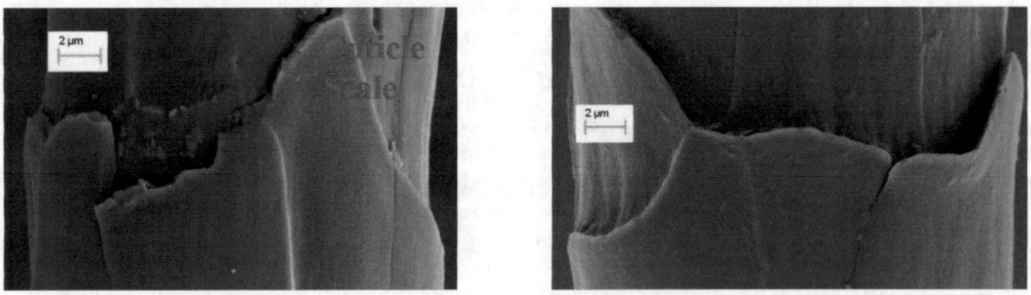

(a) 0s (b) 5s

IOP Conf. Series: Journal of Physics: Conf. Series **823** (2017) 012010 doi:10.1088/1742-6596/823/1/012010

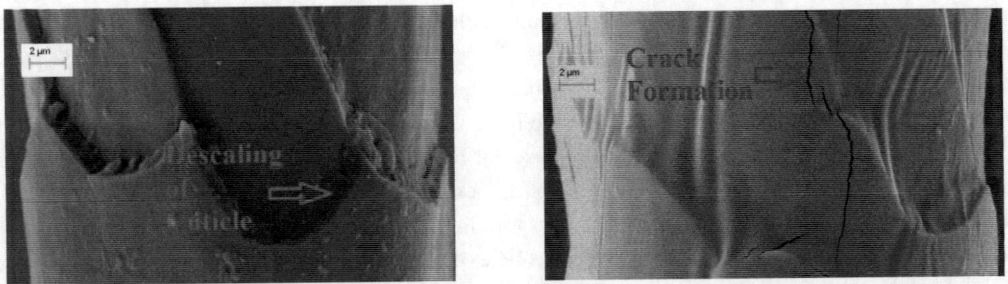

(c) 10s (d) 15s

Figure 2. SEM image of Merino Wool Fiber at 10.00 KX for plasma exposure time 0 to 15s

results in making the cuticle scales blunt. As we increase the exposure times from 5s to 15s, the scales become more and more blunt. However, crimps and cracks are observed on the fiber surface at exposure time of 15s, which means fiber gets damaged when exposed for higher time. The FE-SEM analysis of wool fiber reveals that after plasma exposure the overlapping cuticle scales on the fiber surface become blunter. The blunting of scales results in reducing the directional friction among the fibers. This indeed lowers the chances of inter-locking of the fibers, due to uprising of scales in aqueous medium and hence the felting can be prevented [1 and 4].

3.2. FTIR Study

The IR spectra of wool fiber have been taken with the Nicolet 6700 FTIR instrument having a resolution 4 cm^{-1} and multi bounce HATR accessory. The FTIR spectrum of untreated and plasma

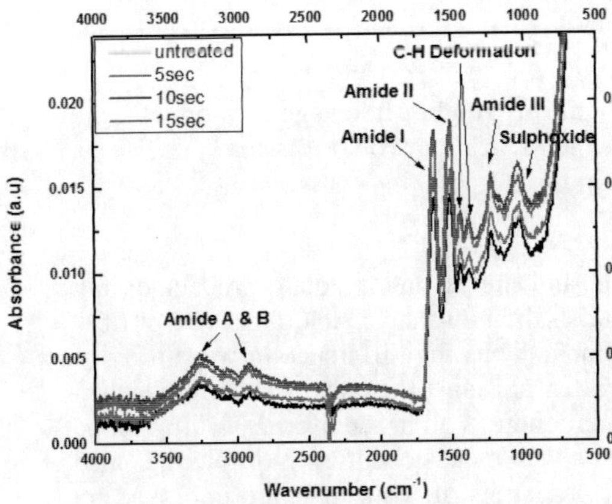

Figure 3. FTIR-ATR spectrum of untreated and plasma treated Merino wool

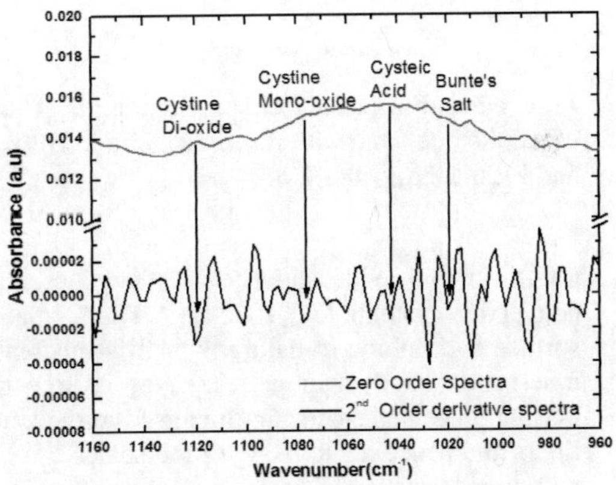

Figure 4. Sulphoxide band of Merino wool and its second order derivative spectrum

treated Merino wool fiber is shown in Figure 3.There are no significant changes between untreated and plasma treated fiber. Similar results have been previously reported by Masukuni Mori et.al [8]. However, it is known that exposure of wool fiber to air plasma leads to oxidation of fibers. In this process the protein cystine is converted to cysteic acid (-SO3-) [7, 13] .This phenomenon leads to formation of intermediate products such as Cystine Mono-Oxide(-SO-S-), Cystine Di-Oxide (-SO2-S-) and Cystine-Sulphonate (-S-SO3-) or Bunte's Salt. We cannot identify these individual sulphur – oxygen groups in unprocessed spectrum. Hence, we have performed second derivative analysis on IR spectra for resolving the bands and also for obtaining quantitative results by estimating concentration of these cystine oxides. A portion of experimentally obtained FTIR-ATR spectrum of Merino wool fiber and its second order derivative spectrum is shown in figure 4. We can easily identify peak positions of individual sulphur-oxygen group's bunte's salt (1020 cm^{-1}), cysteic acid (1043 cm^{-1}), cystine monoxide (1076 cm^{-1}) and cystine dioxide (1121 cm^{-1}). The concentration or absorption ratio is obtained by dividing absorbance value of these individual sulphur-oxygen groups in derivative spectrum with the absorbance value of Amide III (1232 cm^{-1}), which is the reference band. The results are shown in Figure 5, which are average of three reading. We have observed that concentration of bunte's salt and cysteic acid is increasing with plasma exposure time, as higher exposure time leads to more oxidation of the fiber. The concentration of both Cystine Mono-oxide and Cystine – dioxide seems to slightly increase at 5s exposure time and as we increase exposure time further, we find different results and hence error obtained is relatively higher. This may have happened as

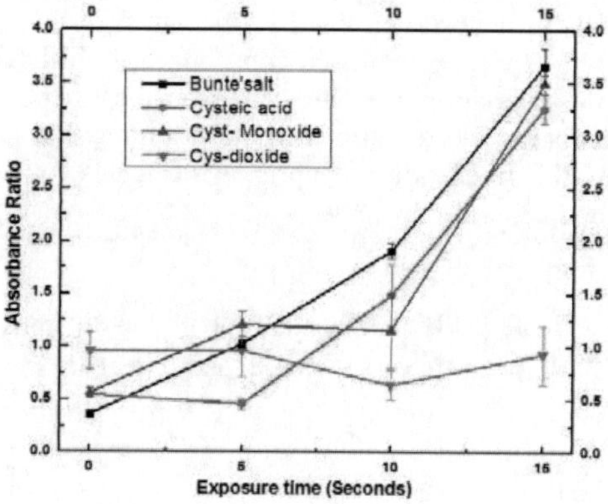

Figure 5. Absorbance ratio of functional groups of sulphoxide band with respect to Amide-III band and Plasma Exposure Time

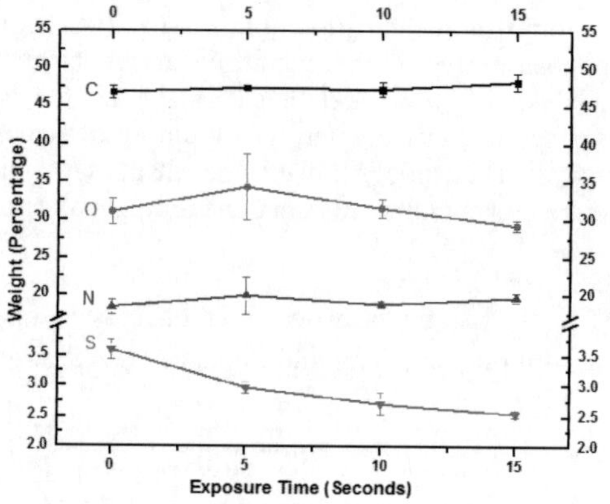

Figure6. Weight percentage of various elements as a function of Plasma Exposure time

Cystine mono-oxide and Cystine di-oxides are not stable and are an intermediate product during conversion cysteine to cysteic acid. The formation of Bunte's salt and Cysteic Acid makes wool fiber surface hydrophilic in nature, which is inherently hydrophobic. Thus the difference in wettability of inner cortex and outer cuticle layer is now minimized after plasma processing. Therefore, when plasma processed fibers are immersed in the aqueous medium, there is no non-uniform swelling of the fibers and hence the uprising of the scales can be prevented which means occurrence of felting is also minimized. Further, Gomez et.al [14] has reported shrink properties of wool fiber are improved by formation of Bunte's Salt. Wakida et.al [17] described cysteic acid along with the bunte's salt on polypeptide chain improves the wettability of the fabric considerably. Maclaren & Kirkpatrick [10] have explained formation of Cystine mono-oxide and cystine di-oxide generates more reactive substrate which provides suitable sites for dyes and softeners.

3.3. EDX Analysis

The EDX (Energy Dispersive X-ray spectrum) spectrum of Merino wool fiber shows mainly Carbon, Oxygen, Nitrogen and Sulphur peaks. The weight percentage of these elements is shown in figure 6 as a function of plasma exposure time. The results show the percentage of sulphur has decreased significantly after 5s plasma exposure time. As we increase the exposure time further, the decrease in sulphur percentage is relatively slow. The sulphur in the wool is present in amino acid cysteine as well as thioester linkages of lipids on the epicuticle surface [13]. The decrease in sulphur percentage can be attributed to removal of outer most layer epicuticle [15]. It is possible that in 5s exposure time, the major portion of epicuticle is removed and hence higher reduction in sulphur percentage is observed. The nitrogen percentage has slightly increased in 5s plasma exposure time, while for 10s and 15s exposure time; there is no significant change in nitrogen percentage values. This can be explained as the outer lipid layer containing sulphur is removed; the underlying nitrogen content from proteneous part of wool in the cortex is exposed and hence shows relative increase in weight percentage. We have observed a small increase in oxygen percentage value, which may be due to oxidation of the fiber. There, is no significant change in weight percentage of carbon, as it is present in entire fiber.

3.4. Shrink Proof Test

The Aachen felt ball method has been used to examine the changes in the shrink resistance of wool fibers. This test has been carried out at Wool Research Association (WRA) with standard IWTO-20-69. The felt ball diameter for untreated fiber was found to 2.509 cm while for 10s plasma treated sample, diameter slightly increased to 2.597 cm. Hence felting density of untreated fiber is found to be approximately 0.12 gm/cc while plasma treated fiber has a value of about 0.10 gm/cc. These results as obtained in our study by using DBD are quite similar to as obtained for low pressure glow discharge system by German Wool Research Institute [16]. However to achieve a felting density value of 0.04 gm/cc as obtained by conventional chlorination treatment , additionally a resin treatment may be required.

Atmospheric Pressure Plasma consists of energetic electrons, ions, neutrals, radicals, etc. When these species interact with the wool fiber surface, the fibers are etched due to bombardment of electrons and ions and also fiber surface is oxidized. The cuticle which is the outer layer of the wool fibers mainly consists of cystine linkages having di-sulphide bonds. When high energy electrons interact with the fiber there is cleavage of di-sulphide bonds (Dissociation energy 2.6 eV) which leads to formation of highly reactive radical anion SS. These anions react in oxygen and form cystine Sulphonate (-S-SO3-) or cysteic acid (-SO3-). Cystine mono-oxide (-SO-S-) and cystine di-oxide (-SO2-S-) are intermediate product when cystine is converted to cysteic acid. The evidence of these interactions is given by EDS (Energy Dispersive X-ray spectrum) and Derivative ATR-FTIR (Attenuated Total Reflection- Fourier Transform Infrared) Spectroscopy analysis techniques in the present work. The physio-chemical changes induced on the fiber surface can be co-related with the felting behavior of the wool fiber [2]. The SEM study indicates the plasma exposure time of 15s leads to damage of fiber and for 10s significant de-scaling is observed. The FTIR and EDS analysis also shows removal of epicuticle layer and significant oxidation of the fiber for 10s exposure time. It seems for the present experimental set-up, a plasma exposure time of 10s will do optimum surface modification of the fiber required to improve its anti-felting property. The result of shrink resistance test by Aachen Felt Ball method confirms the improvement in anti-felting property of Merino wool fiber by atmospheric pressure air plasma with the felting density value of about 0.10 gm/cc.

4. Conclusion

In the present work, we have used high frequency (2.5 MHz) atmospheric pressure dielectric barrier discharge in air to modify surface properties of Merino wool fiber and physio-chemical properties of wool fiber are studied as a function of plasma exposure time. The surface morphology is well examined by FE-SEM and results revealed how morphological changes with exposure time. The FTIR analysis with second-derivative spectroscopy explained the wool fiber oxidation by plasma and cleavage of di-sulphide bonds. The elemental study by EDS substantiates the results obtained by FTIR. The changes induced in physio-chemical properties of fibers by plasma processing are co-related with its anti-felting properties. Further, the Aachen Felt Ball method has validated the improvement in felting characteristics. The present study has revealed that exposure time of 10s resulted in optimum surface modification to achieve anti-felting properties. A higher exposure time has shown damage of the fiber and hence it may have impact on tensile strength of the fiber.

The experimental parameters reported in present study such as exposure time of 10s, use of ambient air at atmospheric pressure and simple electrode set-up makes it quite suitable for industrial use. The present work clearly demonstrates the potential of using 2.5 MHz atmospheric pressure air DBD for improving anti-felting property of wool fiber in an environmental friendly manner.

Acknowledgement

Authors are thankful to Wool Research Association (WRA), Mumbai for Felt-Ball Analysis of Merino wool fiber.

References

[1] Liu et.al 2007 *Textile research journal* **77** 12 957
[2] Mercer et.al 1947 *Journal of Textile Institute* **38** 227
[3] Makinson and Sydney K R 1979 *Marcel Dekker Inc. New York*
[4] Udakhe J et.al 2011 *Journal of the Textile Association* p171-179
[5] Hesse A et.al 1995 Text.Res.Journal **65** 6 p355-361
[6] P.erra et.al 1999 *Text. Res. Journal* **69** 11 p811-815
[7] Kan et.al 2003 *Autex Research Journal* **3** 4
[8] Mori M et.al 2006 *RJTA* **10** 1
[9] Jayant et.al 2011 *Manmade Textiles in India* **XXXIX No. 4** SASMIRA p137
[10] Maclaren J A et.al 1968 *Journal of Society of Dyers and Colourists* **84** p564
[11] Odlyha M et.al 2007 *Autex Research Journal* **7** 1
[12] Douthwaile et.al 1994 *Journ. of the society of dyers & colors* **110** 9 p304-307
[13] Hacke 2006 *John Rylands University Library of Manchester*
[14] Gomez et.al 1994 *Journal of Textile Institute* **85** 2
[15] Restivo A et.al 2014 *Microscopy and Microanalysis* **07** 20 p5
[16] Shishoo R 2007 (Woodhead Publishing Limited) **Chapter 9** p228-244
[17] Wakida et.al 1993 *Textile Research Journal* **63** 8 p438-442
[18] Jain V et.al 2016 *IPR/RR-779/2016*

IOP Conf. Series: Journal of Physics: Conf. Series **823** (2017) 012011 doi:10.1088/1742-6596/823/1/012011

A Fixed Frequency Reflectometer to Measure Density Fluctuations at Aditya Tokamak

Praveen Kumar Atrey[1,2], Dhaval Pujara[2] and Subroto Mukherjee[1]

[1]Institute for Plasma Research, Near Indira Bridge, Bhat, Gandhinagar 382 428, India
[2]Institute of Technology, Nirma University, Ahmedabad, Gujarat, India

E-mail: atrey@ipr.res.in

Abstract. Amongst modern diagnostics of fusion plasmas, microwave methods, both passive and active, play an important role. Microwave Reflectometer is used to measure the plasma density and its fluctuations in fusion research device like tokamak. A fixed frequency (O – mode) microwave reflectometer at 22 GHz (cut – off density $n_c = 6 \times 10^{12}$ cm^{-3}) has been designed, developed and used to measure the critical density layer and its fluctuations in Aditya tokamak. It can measure the plasma density fluctuations from r = 11 to 22 cm for central electron density 7.5×10^{12} cm^{-3} and more, respectively. The output signal of reflectometer is analyzed and compared with the density measurement from the microwave interferometer. When the density measured by interferometer is constant, then the fluctuations of local density are seen from the reflectometer signal. Analysis of initial results show that density fluctuation at r = 21 cm in the main plasma has correlation time of about 8 µsec and frequency spectrum is broad. Use of 22 GHz incident wave allows the observation of density fluctuation with wave number in the range of 0 – 9.2 cm^{-1} from the reflecting region at the receiving horn. Radial variation of the fluctuation level is observed from 5% to 22% for minor radius 11 to 22 cm, respectively.

1. Introduction

The tokamak is a device designed to magnetically confine hot plasmas in a toroidal configuration, with the eventual purpose of producing controlled thermonuclear fusion. The fixed frequency reflectometry diagnostic is widely used to determine the electron density fluctuations in tokamak [1-3]. Its principle is based upon the radar technique and determines the phase shift of microwave propagating into the plasma and reflected at cut-off layer. Physically, it possesses several advantages: a good spatial and temporal resolution (δr < 1 cm and δt < 5 µs) [4]. Technically, the reflectometry needs a very small access to the vacuum vessel. These principal characteristics make reflectometry an important diagnostic for the present and future tokamaks like SST-1 [5] and ITER [6]. In this paper, design, development and use of a fixed frequency reflectometer as a diagnostic to study the electron density fluctuations is reported.

2. Reflectometry

2.1. Basic Principle

Reflectometry relies on the total reflection of electromagnetic waves by plasma when the local refractive index μ equals zero. For the ordinary mode waves (**E** || $\mathbf{B_T}$), this occurs when the wave frequency f = f_p given by

$$f_p = \left(\frac{n_e e^2}{4\pi^2 \epsilon_0 m_e}\right)^2 \quad (1)$$

where, e = charge of the electron, m_e = mass of the electron and ε_0 = permittivity of free space. This corresponds to a critical reflecting density n_c, where

$$n_c = f^2\left(\frac{4\pi^2 \epsilon_0 m_e}{e^2}\right) = \frac{f^2}{80.6}\ m^{-3} \quad (2)$$

The phase change that a wave will undergo in propagating through plasma to cut-off layer, being reflected at the cut-off layer, and propagating back out of the plasma is given by

$$\emptyset = \left(\frac{2\omega}{c}\right)\int_r^{r_c} \mu(r)dr - \frac{\pi}{2} \quad (3)$$

This equation shows that the phase change can result from three possible sources:

1. Movement of the cut-off layer r_c
2. Changes in the refractive index μ(r)
3. Changes in the probing frequency ω

Thus if the probing frequency of reflectometer is fixed and plasma refractive index is constant, then the change in the phase will be due to the movement of cut-off density layer which is known as density fluctuations.

A fixed frequency (O – mode) microwave reflectometer at 22 GHz (cut-off density $n_c = 6 \times 10^{12}$ cm^{-3}) has been designed and developed to measure the critical density layer and its fluctuations at tokamak ADITYA. The plasma density fluctuations has been measured from r = 11 to 22 cm for central electron density 7.5×10^{12} cm^{-3} and more, respectively.

2.2. Design

The block diagram of a 22 GHz reflectometer is shown in figure 1. A Gunn oscillator at 22 GHz is used as a microwave source, whose output is about 70 mW. The source along with power supply is kept 4 meter away from the center of the tokamak Aditya. At this distance, magnetic fields effect on the isolator is negligible. The output of the source is connected to an isolator whose isolation is more than 35 dB. The output of isolator is divided by using 10 dB directional coupler (D.C) in such a way that 90% power should go to plasma path and 10% should go to reference path. The K – band waveguide of 4 m length is used to carry the power output to the horn antenna (Gain = 20 dB) and reflected power to the power combiner (Magic Tee). The same antenna is used for transmitting and receiving purposes. A circulator is used to get reflected signal from the antenna. A magic tee is used to see the interference between reference path and plasma path signal and output of this will go to the detector. An attenuator (A) and a phase shifter (P.S) are used in reference path to make power levels equal and set initial phase shift to zero between two paths, respectively. Homodyne detection technique is used to measure the cut-off density layer and its fluctuations. Electrical breaks are provided at regular intervals in the waveguide and the supporting structure to avoid close loop formation. A Teflon lens (diameter = 100 cm) is used in front of horn to make a parallel beam of microwave before it enters to the plasma through a 100 CF vacuum window of Kodial glass. The sensitivity of the detector is 1000 mV/mW. The output of the detector is amplified with a gain of 10 and digitized by a digitizer at the sampling rate of 1 MHz.

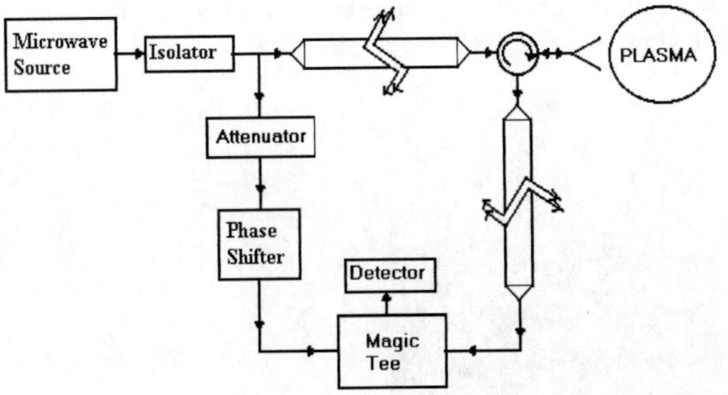

Figure 1. The block diagram of 22 GHz reflectometer

3. Results and Discussion

For this experiment, the Aditya operating parameters are the toroidal magnetic field $B_T \sim 0.75$ T, the neutral base pressure P ~ 3.5– 5.0 × 10^{-5} Torr, the plasma current Ip ~ 60 – 80 kA, the discharge duration **t** ~ 60 – 120 ms and the chord averaged density $n_e \sim (0.5 - 1.5) \times 10^{13}$ cm^{-3}.

The time evolution of the plasma current, the chord averaged plasma density and the reflectometer signal is shown in figure 2. Initially (1.5 – 3.0 ms), a few small fringes are seen in reflectometer output, which is due to the phase shift introduced by the low density plasma which allows the wave to go up to the inside wall of the vessel and get reflected. When the plasma density reaches to the cut off density, the wave is reflected from the cut off layer and we start getting the higher amplitude signal.

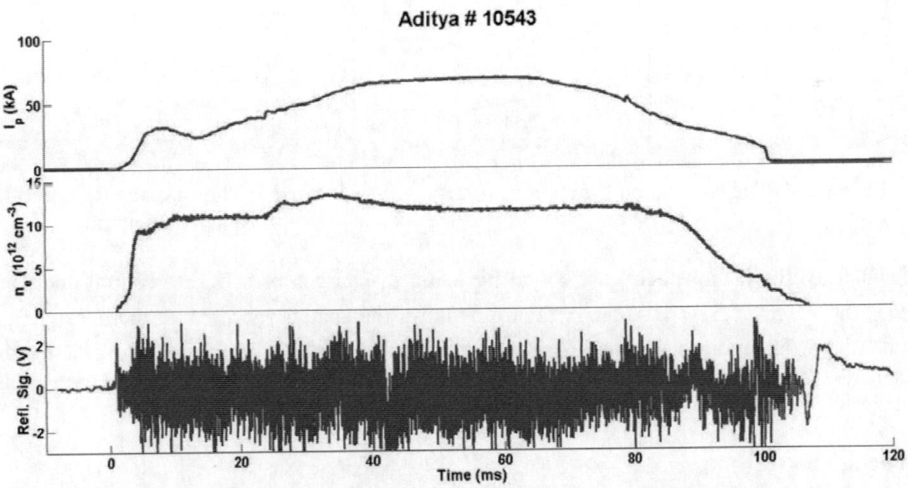

Figure 2. The plasma current, the plasma density and the reflectometer signal vs time

As the density evolves, cut off layer starts moving outside, and fringes of increasing amplitude due to the layer movement are seen. Chord averaged density at the center channel remains constant between 45 – 80 ms at about 1.2 × 10^{13} cm^{-3}. The cut-off density layer for 22 GHz ($n_c = 6 \times 10^{12}$ cm^{-3}) appears at r = 21 cm. The reflectometer signal between 45 and 80 ms carry the information of the fluctuation of the cut-off density layer.

The power spectrum analysis (figure 3) shows that the frequency spectrum is broadband. Figure 4 shows the auto correlation function vs time delay. The auto-correlation time is ~ 8 μs. The use of 22

GHz incident wave allows the observation of density fluctuations with wave number in the range of 0 – 9.2 cm^{-1} from the reflecting region at the receiving horn.

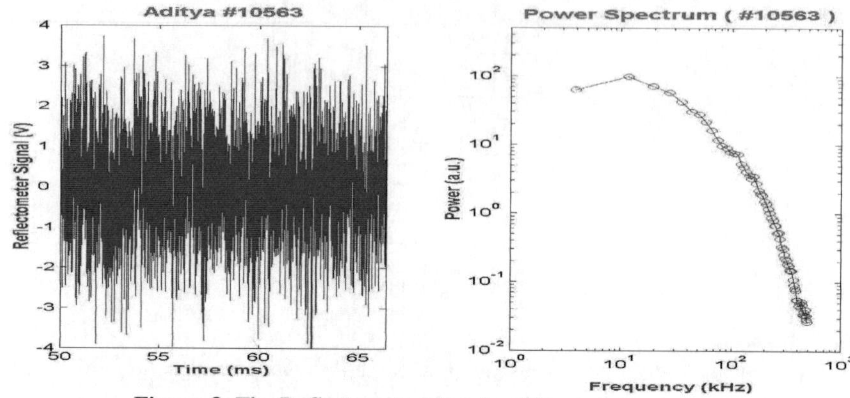

Figure 3. The Reflectometer signal and its power spectrum

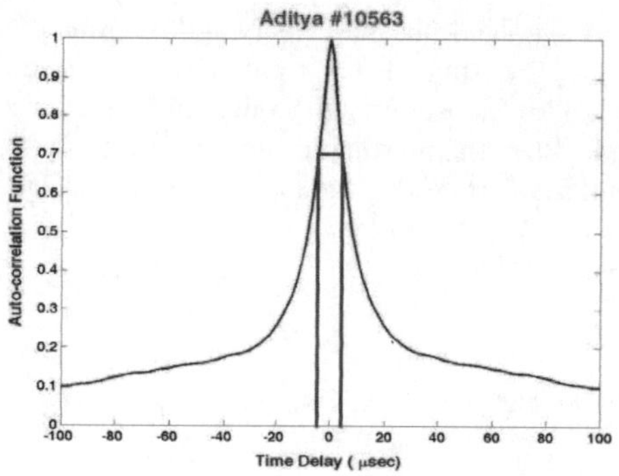

Figure 4. The correlation function vs time

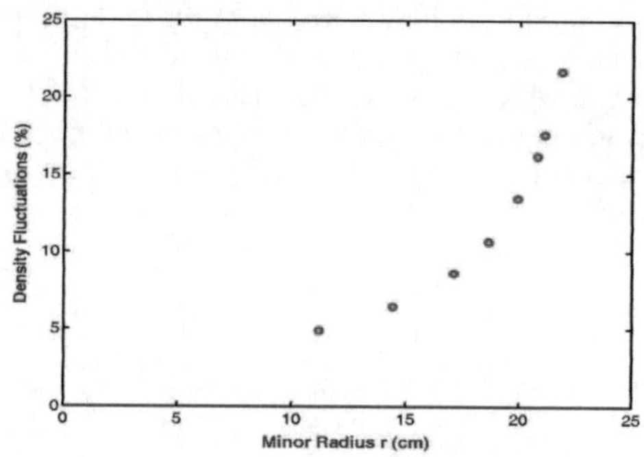

Figure 5. The spatial variation of the density fluctuation

The density fluctuation levels are found by using discharges of different central chord averaged densities (n_e ~ 0.5 – 1.5) × 10^{13} cm^{-3}). Figure 5 shows the spatial variation of the density fluctuation level at Aditya. The radial variation of the fluctuation level is observed from 5% to 22% for the minor radius 11 to 22 cm, respectively. The amplitude of the fluctuation level increases exponentially from the center of the plasma to the edge of the plasma.

4. Conclusion

A fixed frequency (O – mode) microwave reflectometer at 22 GHz ($n_c = 6 \times 10^{12}$ cm^{-3}) is designed, developed and used to measure the plasma density fluctuations. The plasma density fluctuations from r = 11 to 22 cm for different central electron densities is measured. The evolution of reflectometer output signal is studied and compared with microwave interferometer signal. The measured auto-correlation time of the reflectometer signal is ~ 8 ms. The power spectrum analysis shows that the frequency spectrum is broadband. The use of 22 GHz incident wave allows the observation of density fluctuations with wave number in the range of 0 – 9.2 cm^{-1} from the reflecting region at the receiving horn. The radial variation of the fluctuation level is observed from 5% to 22% for the minor radius 11 to 22 cm, respectively.

IOP Conf. Series: Journal of Physics: Conf. Series **823** (2017) 012011 doi:10.1088/1742-6596/823/1/012011

References

[1] M. A. Heald and C. B. Wharton 1978 *Plasma Diagnostics with Microwaves* (New York) **193**.
[2] TFR Group 1985 *Plasma Physics and Cont. Fusion* ***27(11) 1299***
[3] W. A. Peebles etal 1990 *Rev. Sci. Instrum.* ***61(11) 3509***
[4] C. Laverne, A. J. H. Donne etal 1996 *Plasma Physics and Cont. Fusion* ***38 905***
[5] Y. C. Saxena and SST – 1 Team 2000 *Nuclear Fusion* ***40 1069***
[6] G. Vayakis, N. Bretz, etal 1997 *Rev. Sci. Instrum.* ***68 435***

IOP Conf. Series: Journal of Physics: Conf. Series **823** (2017) 012012 doi:10.1088/1742-6596/823/1/012012

Development of heat sink concept for near-term fusion power plant divertor

Sandeep Rimza[1], Samir Khirwadkar[1], Karupanna Velusamy[2]

[1]Divertor & First wall Technology Development Division, Institute for Plasma Research (IPR), Gandhinagar, India
[2]Mechanics and Hydraulics Division, Indira Gandhi Centre for Atomic Research (IGCAR), Kalpakkam, India

E-mail: sandeepr@ipr.res.in, sandeeprimza13@gmail.com

Abstract. Development of an efficient divertor concept is an important task to meet in the scenario of the future fusion power plant. The divertor, which is a vital part of the reactor has to discharge the considerable fraction of the total fusion thermal power (~15%). Therefore, it has to survive very high thermal fluxes (~10 MW/m^2). In the present paper, an efficient divertor heat exchanger cooled by helium is proposed for the fusion tokamak. The Plasma facing surface of divertor made-up of several modules to overcome the stresses caused by high heat flux. The thermal hydraulic performance of one such module is numerically investigated in the present work. The result shows that the proposed design is capable of handling target heat flux values of 10 MW/m^2. The computational model has been validated against high-heat flux experiments and a satisfactory agreement is noticed between the present simulation and the reported results.

1. Introduction

Helium cooled divertor design is one of the potential options for future fusion reactors [1]. Since, helium is chemically and neutronic inert which is favorable from a safety aspect. On the other hand, due to deprived thermo physical property of helium a highly efficient design and cooling technique is desirable for the reactor divertor to sustain the heat flux ~10 MW/m^2. The present paper describes the detailed design study of a new divertor heat exchanger concept for the tokamak application. The effects of critical thermal hydraulic parameters on the heat transfer characteristics of the module are presented as a function of Reynolds number (*Re*).

Different type of helium cooled concept have been reported for the plasma-facing surface for the power plant application. Pizzuto et al. [2] studied the High Efficiency Thermal Shield (HETS) concept for tokamak design. The result showed that the HETS solution is feasible in order to achieve a good heat transfer at the low cost of pressure drop. Diegele et al. [3] investigated the modular divertor for future fusion power plant with an improved heat transfer technique. In their investigation, they found that modular concept has the potential of removing high heat flux (~10 MW/m^2) at reasonable pumping power. Experimental and numerical investigation on divertor finger mock up was investigated by Rimza et al. [4-6]. They discovered that, the potential of divertor to handle heat flux increase largely, by the use of the extended surfaces. High Efficiency Thermal Shield (HETS) is based on a finger mock-up concept for tokamak design. In the present study, a new efficient divertor heat exchanger design concept has been proposed to provide a larger scale divertor concept that would minimize the number of units. Hence, the cost of manufacturing will also decrease. The effect of critical hydraulic parameter on the thermal hydraulic performance of divertor heat exchanger has been investigated using Fluent 14.0 [7].

2.0. Solution Methodology

2.1. Proposed Design

Plasma-facing surface of divertor is fabricated of tungsten with sacrificial layers (4 mm) as shown in figure 1. Tile is joined on the thimble by brazing and fabricated by high ductility tungsten alloy. A high temperature filler material Pd-Ni 40 (1511 K) has been used as a brazing filler between the mating surfaces [8]. To cooled the hot thimble, a jet cartridge carrying the multiple jets is placed concentrically inside the thimble. All the geometrical parameters of a divertor heat exchanger are shown in the Table 1.

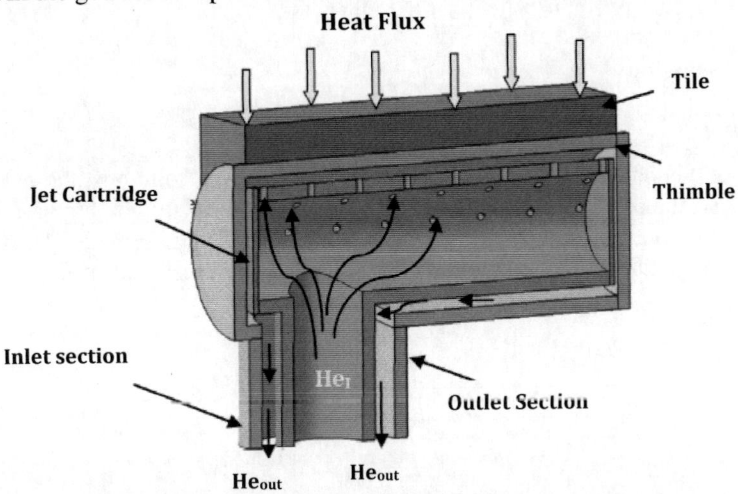

Figure 1. Diagram of proposed divertor heat exchanger.

Helium (He) enters through the inlet section of the cartridge, and accelerates through the multiple jets (Ø 0.6) and flows radially outwards through outlet section as shown in figure 1. The most important design criterion is to keep thimble temperature above ductile brittle transition temperature (~ 600°C) and below the brazing filler temperature of the mating surface (1511 K). The pumping power ratio should be lower than ~10 % of incident power [4].

Table 1. Detailed geometrical parameters of a divertor heat exchanger.

Divertor	Dimension (mm)		Material
	Diameter	Thickness	
Tile	16.8	4	Tungsten (W)
Thimble	16	1.25	Tungsten Alloy (WL-10)
Jet Cartridge	11.5	1.25	Indian RAFMS

2.2. Initial Boundary Conditions

The computational domain of interest considered for the present numerical simulation is depicted in figure 1. Due to symmetry, only 180° sector model has been considered for the present simulation. The domain consists of three solid volumes viz., tile, thimble, cartridge and one fluid volume. The material properties of the tile and thimble are modeled as temperature-dependent [9]. Helium is specified as an ideal gas. The boundary conditions for 3D simulations are as follows:

- Adiabatic boundary conditions are applied on outer sides of the domain.
- The top surface of the domain is applied with constant heat flux ~10 MW/m^2.
- Specified mass flow rate, temperature and pressure (600 °C and 10 MPa) are assigned at the inlet manifold [4].

Pressure-velocity coupling between the incompressible Navier-Stokes and continuity equations are resolved using the SIMPLE algorithm [10]. The effect of accompanied by adapting the k-ε realizable model. The convective and diffusive fluxes are combined using the second order upwind scheme. To declare convergence, the errors and discretized momentum and continuity equation, are set to a value below 10^{-4}, whereas for energy equation it is set to below 10^{-6}.

3.0. Results and Discussion

3.1. Model Validation

To test the validity and enhance the level of confidence in numerical solutions, the present simulation has been validated against the results of Konc˘ar et al. [11]. Systematic studies for various mass flow rates with different heat flux, inlet pressure and temperatures have been performed. The predicted results are compared with experimental as well as numerical results and are presented in figure 2, for a fixed flow rate of 13.7 g/s.

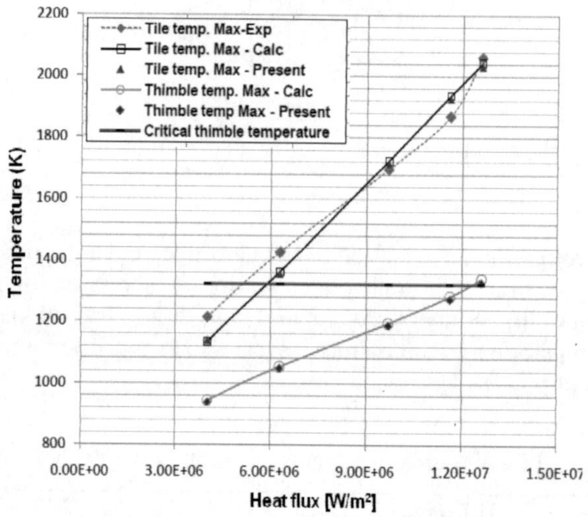

Figure 2. Comparison of present maximum tile and thimble temperature and reported results as a function of heat flux.

The temperature increase more or less linearly with heat flux, as expected. The temperature distributions along the tile, thimble, helium temperature difference of the present simulation at different heat fluxes are in satisfactory agreement with the reported results.

3.2. Assessment of Thermal-hydraulic Performance

The prime motive of this study is to find the required mass flow rate at a reasonable pressure drop for cooling of heat exchanger. Towards this, numerical studies have been performed at various mass flow rates, viz., 10, 13, 15, 20 and 25 g/s, corresponding to the *Re* numbers is 1.2×10^4, 1.5×10^4, 1.8×10^4, 2.4×10^4 and 3.0×10^4. Temperature distributions on the surface of tile and thimble for various Re Numbers for design heat flux (10 MW/m^2) is depicted figure 3(a). It can be noticed that, the tile and thimble temperatures decrease with *Re* numbers. This implies that increasing the *Re* Numbers makes eddies stronger, and increase the fluid mixing causing the highest heat transfer rate in the modular heat exchanger. The results show that tile and thimble mating surface filler temperature are within design limits for *Re* numbers of $\sim 1.58 \times 10^4$. The temperature distributions in the modular heat exchanger, as predicted by the numerical model are depicted in figure 3(b). The temperature across the left corner of tile is slightly varied, presumably due to impact of jet velocity.

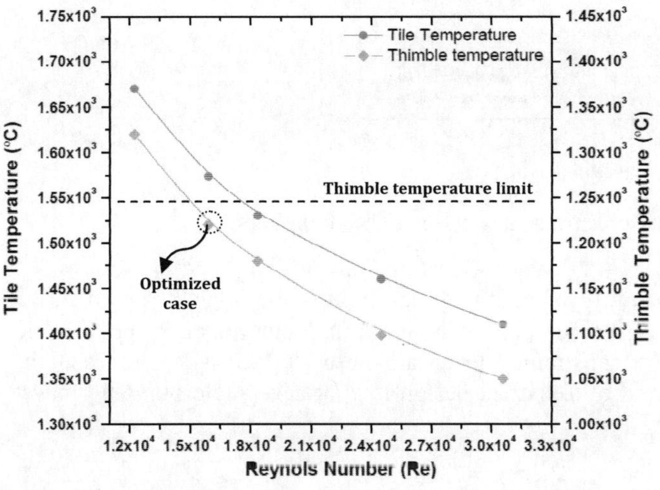

(a)

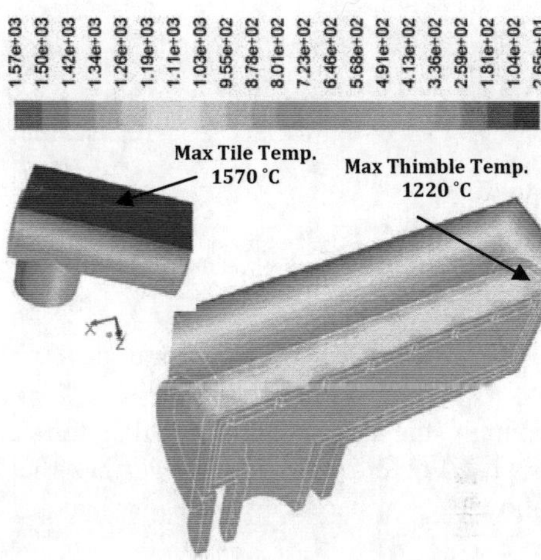

(b)

Figure 3. Comparison of (a) maximum tile & thimble temperature as a function of Re Numbers and (b) predicted temperature distribution over the heat exchanger at $Re \sim 1.58 \times 10^4$.

Pumping power ratio *(J)* is one of the most significant factors that reflect the hydraulic performance of the heat exchanger. It is the ratio of required pumping power to circulate helium in the heat exchanger to cool the hot target surface to the total incident power [4]. The Pumping power ratio for the modular heat exchanger is given by,

$$J = W/Q_T \tag{1}$$

where W is the pumping power for modular heat exchanger of divertor ($W = m \times \Delta P / \rho$), m is the mass flow rate of coolant, ΔP is the pressure drop in in divertor heat exchanger, ρ is the density of coolant, and Q_T is the total incident power.

Figure 4 presents the variation in the pumping power ratio as a function of *Re* numbers. It is clear that pumping power ratio increase continuously with *Re* numbers. The possible reason is that, the formation of vortex and high velocity, turbulence level increases, leading to the associated increase in heat transfer. However, eddies and flow acceleration caused by high velocity, increase the pressure loss in the heat exchanger.

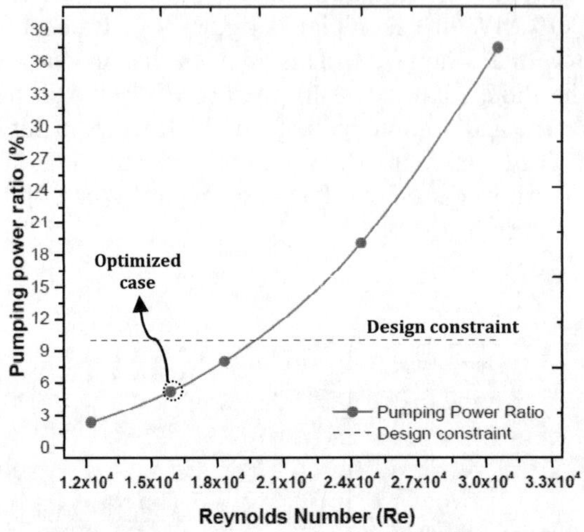

Figure 4. Variation in pumping power ratio as a function of Re Numbers

The results show that pumping power ratio is well within the design limit (< 10 %) at *Re* number 1.58 $\times 10^4$. It is observed that only ~ 5 % pumping power ratio is required at prototypical *Re* numbers to maintain the desired thimble filler temperature constraints. From all these studies, it is clear that the proposed divertor heat exchanger has the capability to extract design load at acceptable pumping power ratio.

4.0. Conclusions

The three dimensional heat transfer characteristics of a new divertor heat exchanger proposed for fusion reactor are investigated through jet impinging technique. The main purpose of the present investigation is to evaluate thermal hydraulics performance of proposed heat exchanger. Numerical investigations showed that the proposed design is capable of handling target heat flux values of 10 MW/m^2. The present analysis shows that proposed heat exchanger design should be accepted for Re numbers less than ~ 1.58×10^4 (13 g/s) to make sure desired pumping power ratio and thimble temperature limits.

Acknowledgement

This research work was carried out at the Institute for Plasma Research (IPR) and supported by the Department of Atomic Energy (DAE), India.

References

[1] Maisonnier D et al. 2006 *Fusion Eng. Des.* **81** 1123
[2] Pizzuto A et al. 2005 *Fusion Eng. Des.* **75-79** 481
[3] Diegele E et al. 2003 *Fusion Eng. Des.* **66-68** 383
[4] Rimza S et al. 2014 *Fusion Eng. Des.* **89** 2647
[5] Rimza S et al. 2015 *Appl. Therm. Eng.* **82** 390
[6] Rimza S et al. 2015 *Fusion Eng. Des.* **100** 581

IOP Conf. Series: Journal of Physics: Conf. Series **823** (2017) 012012 doi:10.1088/1742-6596/823/1/012012

[7] ANSYS FLUENT 2011 Release 14.0, User Guide, Ansys, Inc., Lebanon, US
[8] Roth A 1996 *Vacuum Technology* (Amsterdam: Elsevier) p 359
[9] ITER, 2001 *Material Properties Handbook*, **ITER Document No. G. 74 MA 900-11-10W 0.1**
[10] Patankar S V 1980 *Numerical Heat Transfer and Fluid Flow* (New York: Hemisphere) p12
[11] Končar B et al. 2010 *Appl. Therm. Eng.* **30** 697

IOP Conf. Series: Journal of Physics: Conf. Series **823** (2017) 012013 doi:10.1088/1742-6596/823/1/012013

Radiation effects of the laser ablative shockwaves on aluminium under atmospheric conditions

S Sai Shiva[1], Ch Leela[1], P Prem Kiran[1], C D Sijoy[2] and S Chaturvedi[2]

[1]Advanced Centre of Research in High Energy Materials (ACRHEM), University of Hyderabad, Hyderabad 500046, India
[2]Computational Analysis Division, Bhabha Atomic Research Centre (BARC), Visakhapatnam, India

E-mail: premkiranuoh@gmail.com

Abstract. Numerical investigation of laser induced shock wave (SW) propagation into bulk aluminium target with and without the effects of electron thermal radiation (ETR and No-ETR) is demonstrated using MULTI-fs 1D-code over intensity range 10^{10} – 10^{11} W/cm^2. The radiation emitting from the plasma is observed to show negligible effects on the SW propagating into aluminium target for low energy (25 mJ) and significant effects at high energy (175 mJ) which was found to be dominant up to 50 ns of time. The observations show that two SW have been launched on to the target surface: one during the pulse duration termed as primary SW (PSW) and the other immediately after the laser pulse termination termed as secondary SW (SSW). The effects of ETR were found dominant on SSW compared to that on PSW for 175 mJ. The PSW and SSW found to coalesce at around 30-40 ns and move as a single SW after these time scales. The resultant pressure after coalesce is higher than the individual ones before coalesce for 175 mJ. The PSW pressures at 25 mJ and 175 mJ were found to be ~1.5 GPa and ~7 GPa, respectively that were launched at 10 ns and 7 ns.

1. Introduction

When a laser beam of sufficiently high intensity is focused on to the target surface, the breakdown of the medium occurs and subsequently formation of plasma takes place due to coupling of laser energy to the target [1]. The plasma then ablates from the target and expands into background gas by absorbing the remaining incoming laser energy until the laser pulse terminates. Due to the ablation, the shock wave (SW) or compression wave (CW) is launched into the target in order to conserve the momentum created by the ablated material [1, 2]. Due to transient laser intensity pulse, the SW launched onto the target also has a transient nature making it a very efficient tool to study the dynamic response of materials. Moreover, with the laser driven shock waves (LDSW) very high pressures over very short durations of time (pulse durations) can be generated unlike the flyer plate impact technique. Due to very high pressures generated at very short times, the impulses will be very high making it efficient technique to generate the equations-of-state [3] of the material. The response of the material to LDSW loadings can give rise to obtain the parameters such as pressure (P), particle (u_p) and shock velocity (U_{sh}), density (ρ) and specific internal energy (E_{sp}) inside the target. The SW pressures launched onto the target depends on various parameters such as target properties, incident laser intensity, wavelength, pulse duration, background gas and also on the radiation emitted from the ablated plasma. The effects of ETR on the ablated plasma and SW expanding into the background medium for different laser intensities is investigated [4,5] and observed that the radiation plays an important role on the plasma and SW dynamics with the effects becoming significant with increasing laser energy. Many efforts have been made on the LDSW propagating into solid targets [2, 3, 6] and found some of the important aspects [2, 3, 7-9].

In this paper, the numerical simulation of LDSW propagation through bulk aluminum target whose thickness taken to be ~ 2 mm is presented. The intensities ranging between 10^{10} – 10^{11} W/cm^2 were used to launch the SW on to the target surface and the SW dynamics was investigated over the time scales of up to 1000 ns considering with and without electron thermal radiation effects (ETR and No-ETR). The numerical simulations are carried out using modified 1D-radiation hydrodynamic code MULTI-fs [5, 10] for planar, cylindrical and spherical geometries.

2. Simulation methodology

The LDSW in the experiments were launched using the second harmonic of Nd:YAG laser with the excitation wavelength of 532 nm and pulse duration of 7 ns (FWHM). The detailed setup of the experiments presented elsewhere [11, 12]. The simulations have been performed using modified MULTI-fs [5] 1D-RHD code. The laser ablative shock waves (LASW) from Al target propagating into atmospheric air obtained from the simulations are validated [5,11] with the experimental results. Hence the numerical results presented here are the SW propagation into Al target obtained with same input parameters as that in experiments considering the effect of ablation of Al in ambient air. In figure 1, the origin represents the target surface (air-target interface) and the left portion of surface represents the Al target. Similarly the right side represents the background air. The laser is allowed to incident from right to left on to the target surface (origin).

3. Results and Discussion

3.1. *Effects of ETR on SW propagating into Al*

Figure 1 compares the pressure of shock wave propagating into Al target considering ETR and No-ETR effects for 25 mJ and 175 mJ at 14 and 30 ns, respectively. The origin in figure represents Al-air interface where the laser is incident onto the target surface from right to left.

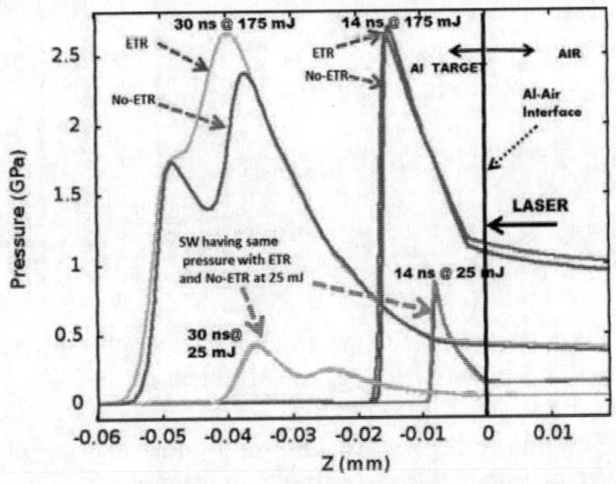

Figure 1. Comparison of shock wave propagating into Al target with ETR and No-ETR effects for 25 mJ and 175 mJ input laser energies at 14 and 30 ns, respectively.

The peak pressures at 25 mJ with ETR and No-ETR found to have same values at both time scales. Whereas with 175 mJ, the pressures are observed to differ slightly at 14 ns, but differ largely at 30 ns. The percentage of laser absorption by the ablated plasma increases with increasing laser energy. Due to increase in the absorption, the electron number density and electron temperatures increases largely within the ablated plasma, also simultaneously the radiation emitting from the plasma increases due to collisions with the ions. The emitted radiation propagates into ambient air and also into the target. The influence of the ETR effect on the SW propagating in to ambient air validated with experimental data is reported in [5] and observed that the ETR effect become dominant with increasing input laser energy. It is reported [13] that the radiation emitted by the plasma close to the target surface is mostly due to free-free transitions. Since the temperature in the laser affected region within the target increases abruptly, the radiation penetrating into the target increases to over few orders of laser wavelength. This is due to the opacities (Rosseland and Planck) [10] of the material being dependent on temperature and mass density. Since the emissions coming out of the plasma increases in case of 175 mJ, the radiation influence on the SW propagating into targets also becomes prominent at these energies. Hence the pressures were found to have higher values with ETR (~ 2.75 GPa) compared No-ETR (~ 2.3 GPa) at 30 ns. However, the influence of the radiation diminishes quickly at around 50 ns and the SW pressures become similar with ETR and No-ETR.

3.2. Formation and coalescence of the primary and secondary shock wave

In figure 2(a & b) it is clearly visible that two SW emanate inside the Al target. The first one represents the primary shock wave (PSW) that is originated during the initial breakdown of the target within the laser pulse duration. The second one represents the secondary shock wave (SSW) originated immediately after the termination of the laser pulse. During the initial breakdown of the material the plasma expands into the ambient air creating a momentum onto the surface that propagates in the form of PSW into the target. The PSW in case of 25 mJ (figure 2(a)) is observed launching at 10 ns during trailing edge of the laser pulse. Whereas at 175 mJ (figure 2(b)) it occurs at 7 ns during leading edge of the pulse. After the breakdown occurs, the plasma absorbs the remaining laser energy and expands continuously into ambient air until laser pulse terminates. Since the laser interaction time with the ablated plasma is more in case of 175 mJ, the plasma attains very high pressures and temperatures leading to the generation of SSW pressures higher than the PSW. The PSW and SSW pressure at 20 ns with 25 mJ found to be ~ 0.2 GPa and ~ 0.6 GPa, respectively.

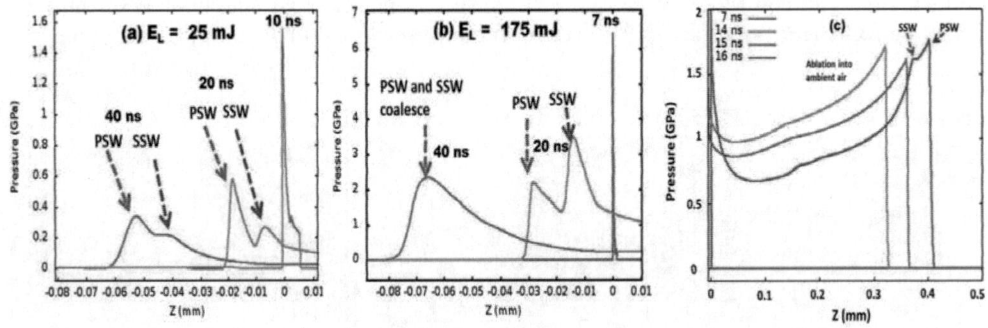

Figure 2. Propagation of the primary and secondary shock wave pressure into Al target at (a) 25 mJ, (b) 175 mJ over the time scales 7 to 40 ns and (c) into ambient air for 175 mJ.

Similarly, with 175 mJ it is found to be ~2 GPa and ~3.5 GPa, respectively. At both the energies it is observed that the two SW (PSW and SSW) coalesce at some point of time during their propagation into target. In the case of 25 mJ, since the strength of SSW is weak compared to PSW, the SW coalesces weakly at around 40 ns. While at 175 mJ, since the SSW strength is more than the PSW the propagation speed will be higher than the PSW. Hence the SSW quickly catches the PSW and coalesces at little early times of around 30 ns. During the coalescence time with 25 mJ, the resultant peak pressure is found have same value as that of PSW before coalescence, whereas with 175 mJ it is found to have higher value than the individual SW before coalescence. Since the two SWs have been generated in to the target correspondingly the SWs will also be generated in the ambient air (shown in figure 2(c)) in order to conserve the momentum. The SWs propagating in air also coalesce at some point of time where during this time the SW loses its planarity and transfers to the cylindrical nature [5]. The laser ablative shock wave from Al propagating into ambient air is also observed to show an increased velocity at latter time scales [11].

In figure 3 (a & b), the negative particle velocities (u_p) signifies the propagation direction of the SW and particles propagation into Al target. It is obvious from figure 3 (a & b) that, as the pressure increases the particles behind the shock front gain high velocities which is due to high compression of the material particles by the SW. The particle motion follow similar trend as that of the pressure (PSW and SSW) observed in figure 2 (a & b). The particle velocities (u_p) behind the PSW at 25 mJ (figure 3 (a)) were observed to decrease from ~ 0.28 - 0.07 km/s and behind the SSW from ~ 0.06 - 0.05 km/s over 10 – 40 ns and 20-40ns, respectively. Similarly, at 175 mJ (figure 3 (b)) u_p behind PSW it is decreased from ~ 0.85 - 0.4 km/s over 7 – 40 ns and behind SW from ~ 0.6 - 0.4 km/s over 20 – 40 ns.

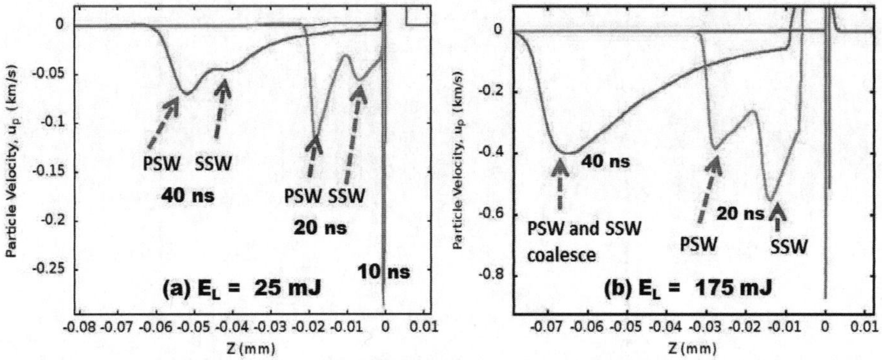

Figure 3. Particle velocities behind the primary and secondary shock wave at (a) 25 mJ and (b) 175 mJ over the time scales 7 to 40 ns.

3.3. Temporal evolution of pressure and particle velocity

In figure 4 (a & b) the temporal evolutions of the pressure and particle velocities is presented for 175 mJ over the time scales of 7 – 1000 ns considering the effects of ETR.

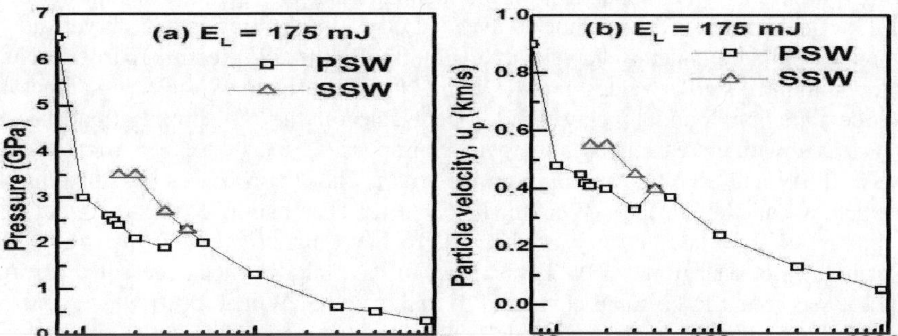

Figure 4. (a) pressure of the SW and (b) particle velocity behind the SW propagating into Al target for the input laser energy of 175 mJ over the time scales of 7 to 40 ns.

The PSW pressure (figure 4 (a)) decay from ~6.5 -1.8 GPa over the time scales 7 – 30 ns and rises to over ~2.4 GPa at 40 ns due to coalescence of PSW and SSW. Similarly, the SSW decays from ~ 3.5 - 2.3 GPa over 16 – 30 ns. The pressure then decreases from ~ 2.3 - 0.25 GPa over 40 – 1000 ns. The particle velocities (figure 4 (b)) behind the PSW decay from ~ 0.9 - 0.3 km/s over the time scales 7 – 30 ns and rises to over ~ 0.4 km/s at 40 ns due to coalescence of PSW and SSW. Similarly, the particle velocities behind SSW decay from ~0.58 - 0.4 km/s over 16 – 30 ns. Finally, the particle velocities decrease from ~ 0.4 - 0.1 km/s over 40 – 1000 ns.

3.4. P - u_p Hugoniot

Figure 5 corresponds to the P - u_p Hugoniot curve of the Al target obtained for two laser energies 25 mJ and 175 mJ, respectively. As observed from figure that as the shock pressure increases the particle velocities also increases linearly. The maximum particle velocity at 1 GPa observed to be ~ 0.18 km/s (at 25 mJ) similarly, at 6.5 GPA observed to be ~ 0.9 km/s. The highest SW velocity (U_{sh}) achieved with 25 mJ and 175 mJ found to be ~2.0 km/s and 2.6 km/s. So the maximum particle velocities attained at pressures 1 and 6.5 GPa are very small compared to the maximum shock speeds achieved.

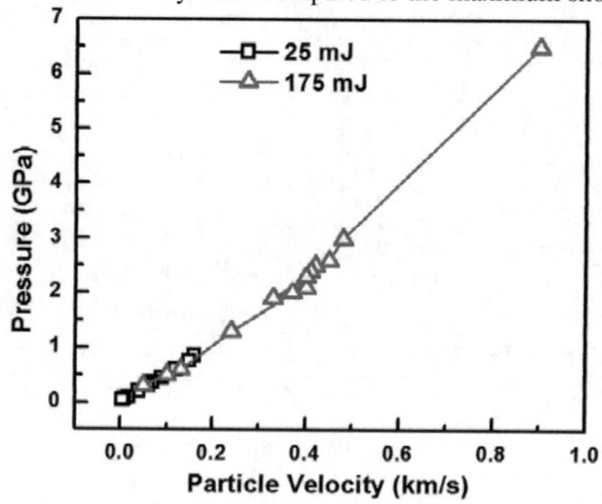

Figure 5. P-u Hugoniot curve for the input laser energies of (a) 25 mJ and (b) 175 mJ.

4. Conclusion

Numerical simulations performed with 1D-RHD code show that the laser driven shock waves (LDSW) propagating into Al target are dependent on the incident laser energy. At low input laser energy (25 mJ) the ETR effects on laser driven shock waves (LDSW) propagating into Al target show negligible influence on the pressure and particle velocity evolution. So the SW pressure, particle velocity behind SW was found to have similar values with ETR and No-ETR effects. While at high input laser energy (175 mJ) the radiation found to play a significant role on the SW propagation. The shockwave pressure increased with increasing laser energy. One important aspect observed from the simulations is that two SW (PSW and SSW) emanating into the target. The first SW was launched during the laser pulse interaction and the second SW immediately after termination of the laser pulse. The PSW pressure in case of low laser energy is observed to have higher pressure than the SSW pressure whereas at high laser energy the SSW is observed to have higher shock pressure. The role of ETR effects on PSW is found to be small compared to that of the SSW at 175 mJ which confirms that the radiation emitting from ablated plasma is small during the laser pulse interaction process and becomes significant after the laser pulse termination due to expansion and cooling of the ablated plasma. The ETR effects are found to play a significant role up to 50 ns time scales and later on show negligible effects on the SW propagation. The two shock waves were observed to coalesce at around 30 – 40 ns and finally propagate as single shock wave. The two SWs observed to be generated in ambient air and found to coalesce at around 50 ns where during this time SW was observed to transit from planar to cylindrical nature. This was observed to lead to laser ablative shock propagating with varying velocities at latter time scales [11]. During the coalescence time the resultant SW is found to have higher pressure than the individual one before coalesce. The P - u_p Hugoniot of Al target is presented from 1- 6.5 GPa SW pressure.

Acknowledgement

Authors from ACRHEM (S Sai Shiva, Ch Leela and P Prem Kiran) thank DRDO for Grants-in-Aid funding.

References

[1] Radziemski L J, Cremers D A 1989, *Laser-induced plasmas and applications* (New York:Marcel Dekker Inc)
[2] Benuzzi-Mounaix A, Koenig M, Ravasio A et al. 2006 *Plasma Phys. Control. Fusion* **48** B347
[3] Remington B A, Drake R P, Takabe H, Arnett D 2000 ,*Phys. Plasmas* **7** 1641
[4] Laville S, Vidal F, Johnston T W, Chaker M, Le Drogoff B, Barthélemy O, Margot J, Sabsabi M 2004 ,*Phys. Plasmas* **11** 2182
[5] Sai Shiva S, Leela C, Prem Kiran P, Sijoy C D, Chaturvedi S 2016 ,*Phys. Plasmas* **23** 053107
[6] Batani D, Stabile H, Ravasio A, Lucchini G, Strati F, Desai T, Ullschmied J, Krousky E, Skala J, Juha L, Kralikova B, Pfeifer M, Kadlec C, Mocek T, Präg A, Nishimura H, Ochi Y 2003 ,*Phys. Rev. E* **68** 067403
[7] Luo S N, Swift D C, Tierney IV, Paislet D L, Kyrala G A, Johnson R P, Hauer A A, Tschauner O and Asimow P D, 2004 *High Pressure Res.* **24** 409
[8] Thibaut de R, Didier L, Andre D and Emilien L 2014 ,*Metals* **4** 490
[9] Atzeni S, Ribeyre X, Schurtz G, Schmitt A J, Canaud B, Betti R and Perkins L J 2014 , *Nucl. Fusion* **54** 054008
[10] Ramis R, Eidmann K, Meyer-ter-Vehn J, Hüller S 2012, *Comput. Phys. Commun.* **183** 637
[11] Leela C, spatio temporal imaging of laser induced shock waves and plasma plume from ambient air, metals, periodic structured surfaces and sub-micron sized compacted powders 2014 Ph. D. thesis University of Hyderabad India
[12] Leela C, Venkateshwarlu P, Singh R V, Verma P, Kiran P P 2014 ,*Opt. Express* **22** A268
[13] Rezaei F, Tavassoli S H 2013 ,*Phys. Plasmas* **20** 013301

IOP Conf. Series: Journal of Physics: Conf. Series **823** (2017) 012014 doi:10.1088/1742-6596/823/1/012014

Conceptual design of Dump resistor for Superconducting CS of SST-1

Swati Roy[1], Piyush Raj[1], Arun Panchal[1] and Subrata Pradhan[1]

[1]Institute for Plasma Research, Near Indira Bridge, Bhat, Gandhinagar 382 428, India

E-mail: swati@ipr.res.in

Abstract. Under upgradation activities for SST-1, the existing resistive central solenoid (CS) coil will be replaced with Nb_3Sn based superconducting coil. Design of Central solenoid had been completed and some of the initiative has already taken for its manufacturing. The superconducting CS will store upto 3 MJ of magnetic energy per operation cycle with operating current upto 14 kA. During quench, energy stored in the coils has to be extracted rapidly with a time constant of 1.5 s by inserting a 20 mΩ dump resistor in series with the superconducting CS which is normally shorted by circuit breakers. As a critical part of the superconducting CS quench protection system, a conceptual design of the 20 mΩ dump resistor has been proposed. The required design aspects and a dimensional layout of the dump resistor for the new superconducting CS has been presented and discussed. The basic structure of the proposed dump resistor comprises of stainless steel grids connected in series in the in the form of meander to minimize the stray inductance and increase the surface area for cooling. Such an array of grids connected in series and parallel will cater to the electrical as well as thermal parameters. It will be cooled by natural convection. During operation, the estimated maximum temperature of the proposed dump resistor will raise upto 600 K.

1. Introduction

One of the critical applications of superconducting magnets is in the field of fusion grade reactors and tokamak devices. They are cooled to their appropriate cryogenic temperatures during their operation. These superconducting magnets have the ability to support a very high current density with a vanishingly small resistance. Owing to large operating current and magnet's inductance, magnitude of the stored energy of these magnets is also large, rising upto the order of Mega Joules to Giga Joules. Thus, handling and operation of these superconducting magnets becomes critical and off-balancing in operating parameters may lead to magnet quench. The portion of the magnet which quenches becomes resistive and hence becomes a heat source due to Joule heating at that location. In order to prevent further permanent damage to the superconducting coil, the quench protection system of the magnet should come into the picture immediately. The quench protection system safely extracts the inductively stored energy of the magnet. This extracted energy is dumped into external dump resistor in the form of heat energy [1] and [2].

In SST-1, the resistive central solenoid is to be replaced with a new superconducting central solenoid (SC-CS). Design of the new CS based on Nb_3Sn Cable in Conduit conductor (CICC) has been completed. The new SC CS coil will have 576 turns with operating current upto 14 kA [3]. The major electrical parameters of the coils have been tabulated in Table 1.

Table 1. Parameters of SC-CS quench protection system.

SC-CS Parameters	Values
CS Inductance, Lmagnet	30 mH
Operating current, I	14 kA
Energy stored in the coil, E	~3 MJ
Hot spot temperature, T	150 K
U(θ)	6.140×10^{16} A^2 s m^{-4}
Time constant, τ	1.5 s
Dump Resistor, Rd	20 mΩ
Fast Decay time	7.5 s

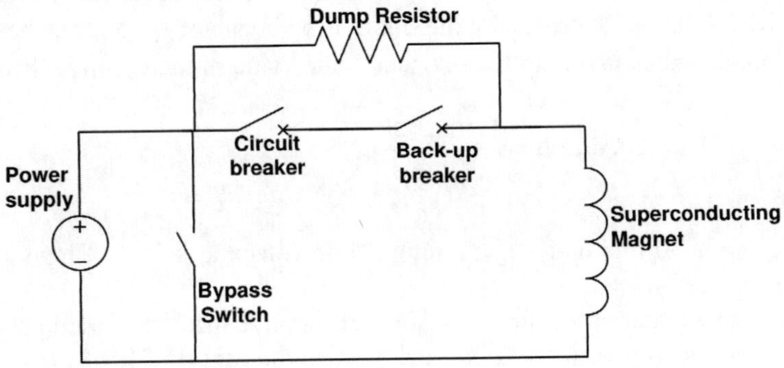

Figure 1. Protection system for superconducting magnets.

A typical quench protection system for superconducting magnet has been shown in figure 1. Under normal operating conditions, circuit breaker (*CB*) remains closed and it carries the operational current. During quench, CB is opened the current commutated to the dump resistor (R_d) connected across it, where the coil current decays exponentially with a defined time constant (L_{magnet}/R_d). Thus, magnet stored energy is extracted and dissipated in the external resistor kept outside cryostat at room temperature, leading to less damage or no damage to the magnet [4]-[7]. This type of quench protection is termed as *active protection.* Thus dump resistor acts as energy extraction unit for superconducting CS.

The design of dump resistor must be carried out considering with following important points [1] and [2]:

i. The hot spot calculations comprising of SC-CS parameters such as coil inductance (L_{magnet}), operating current, and magnet insulation strength, specific heat, density, etc.
ii. The peak voltage rise across the dump resistor, which is limited by the insulation level of superconducting coil;
iii. The physical parameters of the dump resistor has to be decided such that it should be able to absorb the stored magnetic energy ($1/2*L_{magnet}*I^2$), without getting damaged;
iv. It must withstand the thermal as well as electro-magnetic stress generated during energy dumping process

Considering the above criteria and constraints, a dump resistor for superconducting has been designed. The design philosophy and calculations have been presented in this paper. The selection of material for the dump resistor and suitable thermal design based on operating parameters has been outlined in subsequent sections. Finally, the dimensional layout of the dump resistor is presented.

2. Parameters for Dump resistor design

The dump time constant has been calculated using adiabatic hot spot formula as shown in Eq. 1 [3]

$$\int_0^t J^2(t)\,dt = \int_{Tb}^{T} \frac{S(T)}{\rho(T)}\,dT \qquad (1)$$

where $J(t)$ is the operating current density, 2.21×10^8 $A^2\ m^{-4}$, $S(T)$ and $\rho(T)$ are temperature dependent physical properties of the superconducting magnet as volumetric heat capacity and effective longitudinal resistivity respectively. T_b is operating temperature, 4.5 K and T is the maximum hot spot temperature, 150 K for SC-CS.

From the SST-1 SC-CS magnet design calculations, RHS of Equation (1) has been obtained in order of 6.140×10^{16} $A^2\ s\ m^{-4}$ and thereby the dump time constant τ_{dump} has been obtained as 1.5 s [3]. From magnet inductance and dump time constant, the value of the dump resistor is obtained as:

$$\text{R dump} = \frac{\text{L magnet}}{\tau \text{ dump}} \qquad (2)$$

For the SC-CS L_{magnet} is 30 mH and τ_{dump}, dump time constant is 1.5 s. Hence, the value of dump resistor has been obtained as 20 mΩ.

The criteria to select material for the fabrication of dump resistor are based on thermal response as well as change in electrical characteristics with respect to thermal profile of the conductor. Following the requirement, it should have high resistivity, high density, high strength and low temperature coefficient and non-magnetic. In the view of same, the material selected for dump resistor is SS304L due to high resistivity, high strength and heat resistant characteristics and hence design of the dump resistor has been done considering SS304L.

3. Design of Dump Resistor

3.1. Calculation for overall mass of dump resistor

Maximum allowable temperature of the dump resistor during energy dump has been limited to T_{max} and mass has been calculated.

Electromagnetic forces are directly proportional to the square of the operating current, hence during magnet discharge, huge forces will be offered on the structural material of the dump resistor, may leading to deformation. Thus, to reduce the magnitude of electromagnetic forces, it is proposed to divide 20 mΩ resistors into 5 parallel units as shown in the figure 2. The 100 mΩ has been split in series combination of two units of each 50 mΩ. A detailed calculation for electromagnetic force during operational scenario has been mentioned in the subsequent section.

IOP Conf. Series: Journal of Physics: Conf. Series **823** (2017) 012014 doi:10.1088/1742-6596/823/1/012014

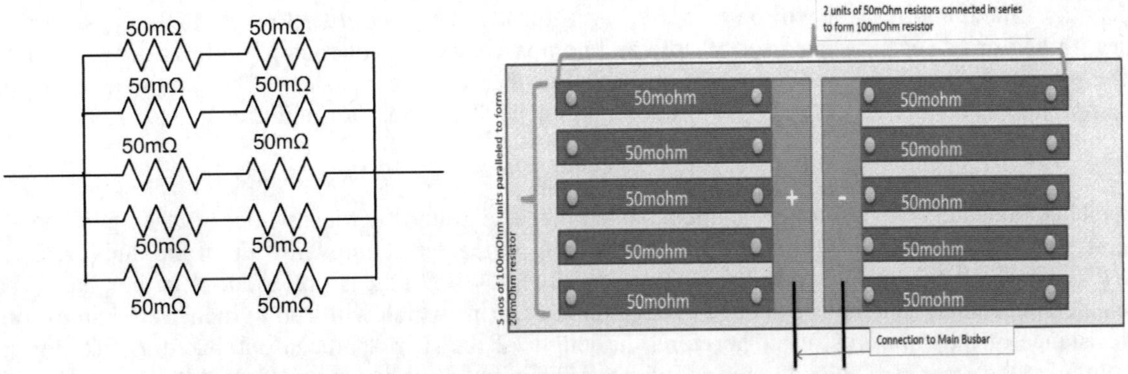

Figure 2. Schematic and layout for 20mΩ Dump resistor for SC-CS

During quench, dumping ~ 3 MJ of magnet energy will lead to rise in the temperature of the dump resistor. With the proposed configuration of resistor units, the total energy to be absorbed by the dump resistor (3 MJ) will also get divided in 5 parallel branches as 0.58 MJ each and current in each parallel branch will be 14 kA/3 = 2.8 kA.

The mass of single 100 mΩ resistor is calculated considering a safe upper limit for the maximum allowable temperature for SS material as 600 K

$$\int_{300}^{T\,max} \mathrm{m\,C(T)\,dt} = 0.58\,\mathrm{MJ} \tag{3}$$

where, m = mass of 100 mΩ resistor, $C\,(T)$ = temp dependent specific heat for SS 304L, specific heat profile for SS304L in the range of 250 K– 650 K, taken from CRYOSOFT ® [9] as,

$$\mathrm{C(T)} = -2.7747\,\mathrm{E}-9T^4 + 6.1484\,\mathrm{E}-6T^3 - 0.0050T^2 + 1.9741T + 199.4746 \tag{4}$$

Hence, the required mass for 100 mΩ resistor is ~ 4 kg (limiting T_{max} to 600 K). Since there are 5 parallel branches, the total mass of the dump resistor will be ~20 kg.

3.2. Structural Layout of the Dump Resistor

The required dump resistor shall be suitably compact as well as the layout should allow natural cooling of the resistor.

As discussed earlier, the 100 mΩ resistor is made of series combination of 2 arrays of 50 mΩ in series. Further, each 50 mΩ is made of several strips of SS304L connected in series in the form of meander. Thus, the dimensional calculation for 50 mΩ resistor unit has been carried out. The required length for 50 mΩ is calculated as follows:

$$\mathrm{R} = \rho\frac{\mathrm{l}}{A}.\ \text{Hence,}\quad \frac{\mathrm{l}}{A} = \frac{\rho}{R} \tag{5}$$

$$\text{Density} = \frac{\mathrm{m}}{A \times l} \tag{6}$$

where, R= resistance of one unit (50 mΩ), ρ = resistivity of conductor ($0.72 \times 10^{-6}\Omega$ m), l = length of the conductor, A= cross sectional area of the conductor, *density* = Density of conductor (7900 kg/m^3), m ~ 2 kg.

Solving Equation (5) and Equation (6), the required length for 50 mΩ resistor has been obtained as l ~ 4085 mm, with cross sectional area, A ~ 58.8 mm^2. The required length is large, and since the 50 mΩ resistor unit will be connected in series with another 50 mΩ unit to make 100 mΩ, the overall

IOP Conf. Series: Journal of Physics: Conf. Series **823** (2017) 012014 doi:10.1088/1742-6596/823/1/012014

length of the dump resistor will be increased even further. Thus, in order to meet the requirement of length and cross sectional area for 50 mΩ and compact size, the conductor length is converted into sections of length 550 mm, connected in series in the form of meander. Thus, number of strips required to be connected in series (N) to form one unit of 50 mΩ resistor unit is calculated as

$$N = \textit{total length of conductor for resistor/ length of single strip} = 7.42$$

This value of N is not whole number so we have to round it. Meanwhile, another point to be considered is that in order to have power connections at the centre-outward side of the dump resistor as seen in Figure 3, an even number of strips in single 50 mΩ unit is preferred. Hence, N has to be evened-out to meet the requirement of even number of strips which will lead to higher error in overall resistance of the dump resistor. Therefore, the length of the strip has been optimized as 510 mm to minimize the error as well as to comply with requirement of even number of strips for proper power connections. The number of strips, N calculated with length of single strip as 510 mm is 8.01, hence the overall error in resistance value reduces to well below 0.14 %.

While calculation, the thickness of the strips has to be deliberated because the cross sectional area of the strip $A \sim 58.8$ mm^2 is constant, so we have to define either thickness or width of the strip. Thickness of the strip has a trade-off with electro-magnetic forces between two adjacent strips. Since the strips are connected in the form of meander, the direction of the current is opposite to each other, thereby causing a repulsive force between two strips. The electro-magnetic forces will appear at all the strips of the array. These electro-magnetic forces get neutralized in the middle portions of the array, but strips at the ends will experience maximum force.

The electro-magnetic force is expressed as a function of current I, dimensions of the strip (thickness) a and spacing between two strips d as seen in Equation below [10].

$$\mathrm{F} = \frac{\mu 0\, I^2}{\pi a}\left\{\arctan\left(\frac{a}{d}\right) - \frac{d}{2a}\ln\left(1+\frac{a^2}{d^2}\right)\right\} \tag{7}$$

Since, the electro-magnetic force between the adjacent strips, varies as square of operating current; the current has been reduced to one-fifth value (as discussed in Section 3.1) to reduce these forces by 1/25. Further, the thickness of the strip and distance between the strips has also been varied and optimized, resulting in electro-magnetic forces value of ~ 35 N with 3 mm thickness of single strip of 50 mΩ resistor unit.

The overall optimized designed values have been summarized in the Table 2. The dimensional layout is represented in figure 3.

Table 2. 20 mΩ Dump resistor design and layout details.

Dump resistor	20 mΩ
Material	SS304
Resistivity	0.72 uΩ -m
Density	7900 kg/m^3
Energy to be dumped	3 MJ
Maximum temp	600 K
Overall mass	20 kg
Overall Size	1.09 × 0.51 × 0.5 m

Structural design and layout	
Nos. of units connected in parallel	5
Resistance of each parallel unit	100 mΩ
Nos. of arrays in series for 100mΩ	2
Resistance of each array	50 mΩ
Nos. of strips in an array	8
Dimensions of single strip	510 × 20 × 3 mm
Resistance of single strip	6.25 mΩ
EM force (maximum)	35 N

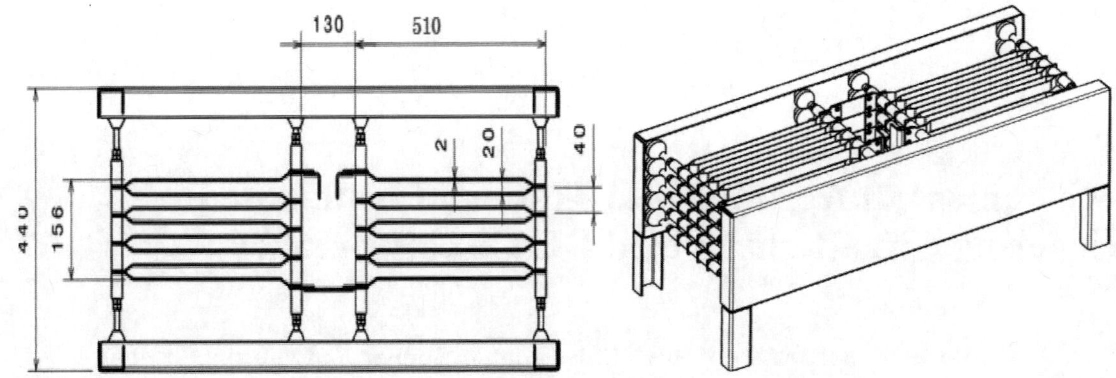

Figure 3(a). Top view of 20 mΩ dump resistor (dimension in mm); **3(b)**: Isometric view image of 20 mΩ dump resistor

4. Conclusion

Dump resistor for the quench protection system of SC-CS for SST-1 has been obtained from dump time constant derived from thermal and electrical parameters of the SC- CS. 20 mΩ dump resistor has been designed using SS304L for operating at 14 kA operating current capable of absorbing 3 MJ of coil energy during magnet quench. The dimensional layout including component layout has also been proposed based on safe thermal as well as electro-magnetic parameters. The mass of the proposed dump resistor will be ~ 20 kg and the maximum temperature during energy dissipation of 3 MJ is expected to be 600 K. The electromagnetic forces are expected to be ~ 35 N.

References

[1] Wilson M N 1983 *Superconducting Magnets* (Clarendon: Oxford U.K.).
[2] Iwasa Y 1994 *Case Studies in Superconducting Magnets: Design and Operational Issues* (Plenum: New York).
[3] Prasad U et.al. 2015, *Design of New Superconducting Central Solenoid of SST-1 Tokamak Proc. Int. Conf. on Asia Pacific Fusion Assosciation (India)*
[4] Le D K et.al 2012, *Quench Protection System for the KSTAR Toroidal Field Superconducting Coil J Elec Eng & Tech* **7**.
[5] Fu P et.al 2006 *Quench protection of the poloidal field superconducting coil system for the EAST tokamak Nucl. Fusion,* **46** S85.
[6] Chikaraishia H et al. 1998, *Quench protection system using a novel d.c. circuit breaker for LHD Fusion Eng. and Des.* **41** 259.
[7] Takahata K. et.al 2007 *Proc. Int Toki Conf/Int Stellarator Helicon Workshop (Japan) Internal energy dump for superconducting magnet protection of the LHD-type fusion reactor FFHR* .
[8] Lawrence D 1995 *Stability of Superconductor* (Kluwer Academic Publishers: New York) p 123.
[9] Kedia S, Roy S, Pradhan S, 2010 *IEEE Trans. on applied Superconductivity,* **20** 6 p 2354.
[10] Moon F C 1984 *Magneto-Solid Mechanics* (Wiley: New York).

Development, Integration and Testing of Automated Triggering Circuit for Hybrid DC Circuit Breaker

Deven Kanabar[1], Swati Roy[1], Chiragkumar Dodiya[1] and Subrata Pradhan[1]

[1]Institute for Plasma Research, Near Indira Bridge, Bhat, Gandhinagar 382 428, India

E-mail: deven@ipr.res.in

Abstract. A novel concept of Hybrid DC circuit breaker having combination of mechanical switch and static switch provides arc-less current commutation into the dump resistor during quench in superconducting magnet operation. The triggering of mechanical and static switches in Hybrid DC breaker can be automatized which can effectively reduce the overall current commutation time of hybrid DC circuit breaker and make the operation independent of opening time of mechanical switch. With this view, a dedicated control circuit (auto-triggering circuit) has been developed which can decide the timing and pulse duration for mechanical switch as well as static switch from the operating parameters. This circuit has been tested with dummy parameters and thereafter integrated with the actual test set up of hybrid DC circuit breaker. This paper deals with the conceptual design of the auto-triggering circuit, its control logic and operation. The test results of Hybrid DC circuit breaker using this circuit have also been discussed.

1. Introduction

Superconducting magnets are one of the critical components of modern tokamaks. Magnet quench has been defined as a major threat to these superconducting coils where a portion of superconductor starts becoming resistive. Under such scenario, it is essential to rapidly extract the stored energy of the coil and dump it externally [1]. For large superconducting coils where stored energy is of the order of Giga Joules, the process of energy extraction is accomplished by inserting a dump resistor in series with the superconducting coil. DC circuit breaker (DCCB) and pyrobreaker are important components of quench protection system for superconducting magnets. DCCB acts as main opening switch and an explosive triggered opening switch (pyro-breaker) is provided as a back-up switch for DCCB [2]. The pyrotechnic based current interrupter is very useful for protecting electrical systems with high continuous currents (> 5 kA) when rapid interrupting is required. However, there are certain limitations of pyrobreaker such as the recovery time is long and the cost of the replacement is high [3]. Opening of these switches causes load current to commutate into dump resistor, thus bringing dump resistor in series with the superconducting coil leading to required energy extraction during quench.

Conventional method of current commutation involves mechanical circuit breakers due to their low on-state resistance, high dielectric strength and capability of carrying high currents. But these breakers do have considerable disadvantages too. During current interruption, their contacts experience arcing which leads to erosion of the contact material, thereby degrading life expectancy and performance of the breaker and in turn affecting the reliability of entire quench protection system [4]. In many of the industrial applications, static breakers have been employed for fast and arc-less operation due to their high reliability and reduced maintenance. But such solutions also have certain critical issues. High on-

IOP Conf. Series: Journal of Physics: Conf. Series **823** (2017) 012015 doi:10.1088/1742-6596/823/1/012015

state loss requires effective cooling for static breaker. This is the reason why static switches have not been so popular against conventional breakers [5].

Hybrid DC circuit breaker (DCCB) is a combination of mechanical switch and static switch which provides advantages of both types of switches. In this switching technique, the mechanical switch is generally a conventional mechanical breaker which carries all the continuous current to the load and the static switch will operate only at the final stage of current commutation process [6-8].

A prototype scale IGBT based hybrid DCCB has been conceptualized and tested at 1.5 kA current [9]. The triggering of mechanical and static switches in Hybrid breaker can be automatized which can effectively reduce the current commutation time. With this view, a dedicated control circuit (auto-triggering circuit) has been developed which can decide the timing and pulse duration for mechanical switch as well as static switch from the operating parameters. The methodology of auto-triggering circuit (ATC) has been elucidated using flowchart in the first section of this paper. This circuit has been fabricated and tested with simulated signals from function generator. Thereafter, it has been integrated and tested with 1.5 kA Hybrid DCCB set up. The testing scheme and obtained results are discussed in the subsequent sections.

2. Control scheme of auto-triggering circuit

Initially, the mechanical switch signal is kept ON i.e. mechanical switch is closed and carrying the continuous current to the load, whereas the static switch is in OFF condition. Once quench signal is received, the mechanical switch is turned OFF and simultaneously the static switch signal is turned ON. When the contacts of the mechanical switch begin to separate, the current starts commuting from mechanical switch to the static switch branch. The current in static switch branch is continuously monitored and as it crosses a predefined set current, static switch is turned OFF after a small intentional delay. This delay assures full commutation of current from mechanical switch to static switch. Hereafter, the load current commutes finally into the dump resistor. The flowchart of ATC has been outlined in figure 1.

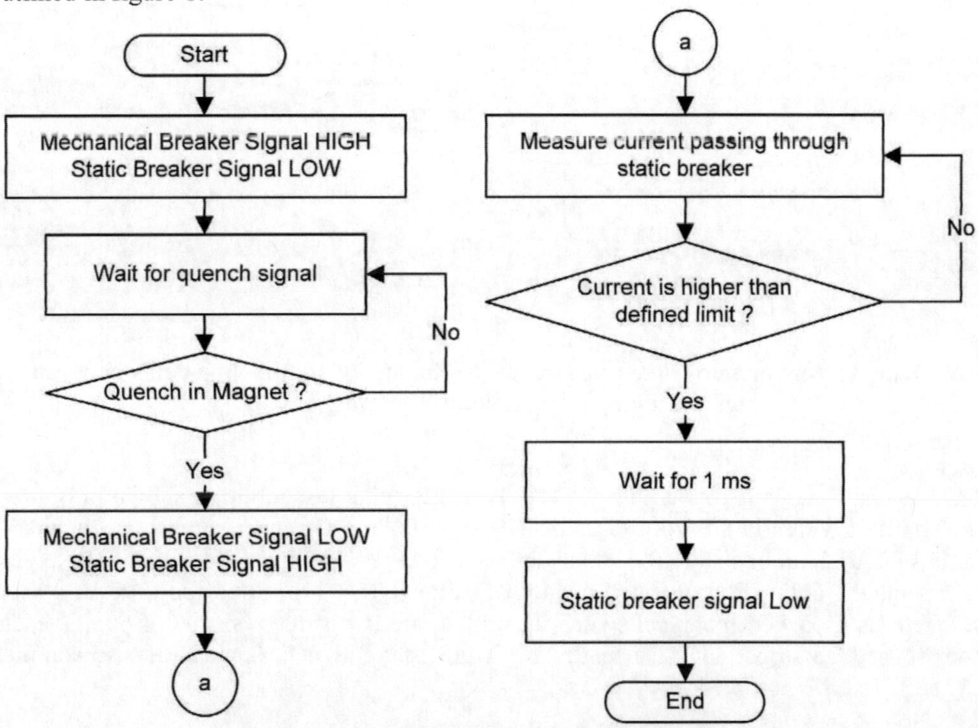

Figure 1. Methodology of auto-triggering circuit.

3. Testing of auto-triggering circuit

An auto-triggering circuit has been designed based on the mentioned flowchart in figure 1. The ATC has been fabricated using logic gates, flip-flops and comparators as shown in figure 2.

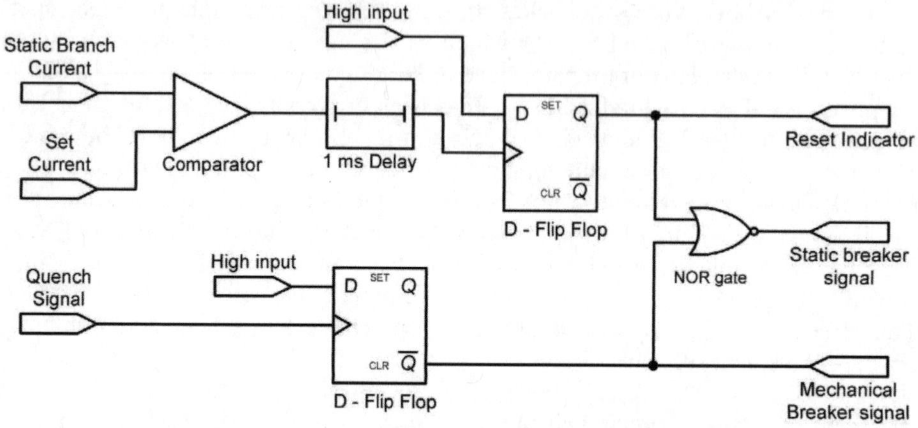

Figure 2. Schematic of auto-triggering Circuit

The fabricated ATC has been tested with function generator. The input signals such as quench and static switch current signals were generated by function generator and the output of the ATC (i.e. signal for the static switch) was examined. Figure 3(a) & (b) shows the test results in this simulated scenario.

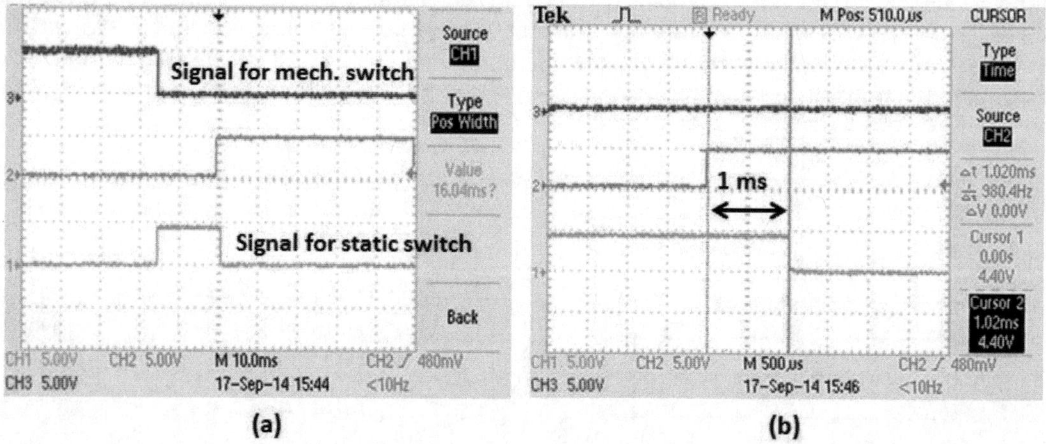

Figure 3. (a) Testing of auto-triggering circuit. (b) Zoomed view showing delay between current signal and the output of auto-triggering circuit.

4. Test set up

The tested ATC has been integrated with 1.5 kA Hybrid Breaker test set up as shown in figure 4. The test set up for 1.5 kA hybrid CB comprises of 220 V, 1500 A DC power supply, mechanical switch (MS), static switch, dump resistor (R_d), freewheeling diode (FWD) and load resistor (R_L). The static switch is a parallel combination of four SKM400GB12E4 IGBTs [10] rated for 1200 V, 400 A each. Power contactor is used as mechanical switch. R_d and R_L are low value resistors of 60mΩ, each made of meander shaped SS strips [11]. Current through MS, static switch, R_d and R_L has been measured with LEM DCCTs LTC 505-S/SP-5 [12].

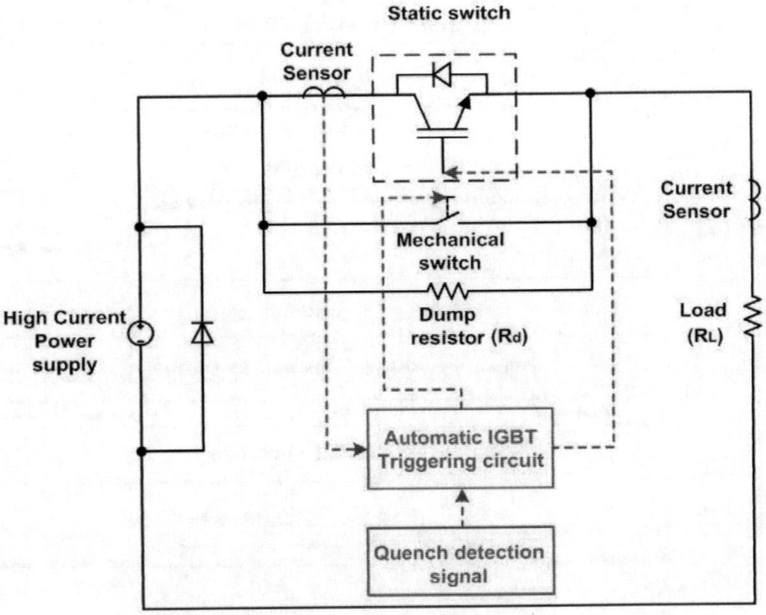

Figure 4. Schematic of auto-triggering Circuit.

For the hybrid CB operation, the static and mechanical switches are triggered in a defined sequence [13]. During quench, the mechanical switch is triggered to open and simultaneously static switch is closed. The load current therefore transfers from mechanical switch to static switch. As soon as the mechanical switch is opened completely and all the current is transferred to the static switch, static switch is also opened. ATC will decide the time at which static switch has to be opened, thus it will reduce the overall current commutation time of hybrid DCCB operation.

5. Results & Discussion

The hybrid breaker testing has been carried out at various current levels up to 1.5 kA and the obtained waveforms are shown in figure 5 and figure 6. At t = 0 ms, quench signal is received, hence the ATC turns OFF the mechanical switch and turns ON the static switch simultaneously, so the current starts flowing in static switch branch after t = 0 ms. At t = 13.19 ms, the contacts of the mechanical switch start separating, hence current from mechanical switch branch starts commuting in static switch branch. ATC is continuously monitoring the current in static switch branch and comparing it with a set reference current of 1 kA. As the current in static switch crosses this reference value, the circuit issues turn OFF command to the static switch after a set delay (1 ms). As shown in figure 5 at t = 14.36 ms, the static switch is turned OFF and current is commuted from static switch branch to dump resistor branch. At this point, the current in static breaker branch becomes zero and the load current also decreases as dump resistor is inserted in series with the load.

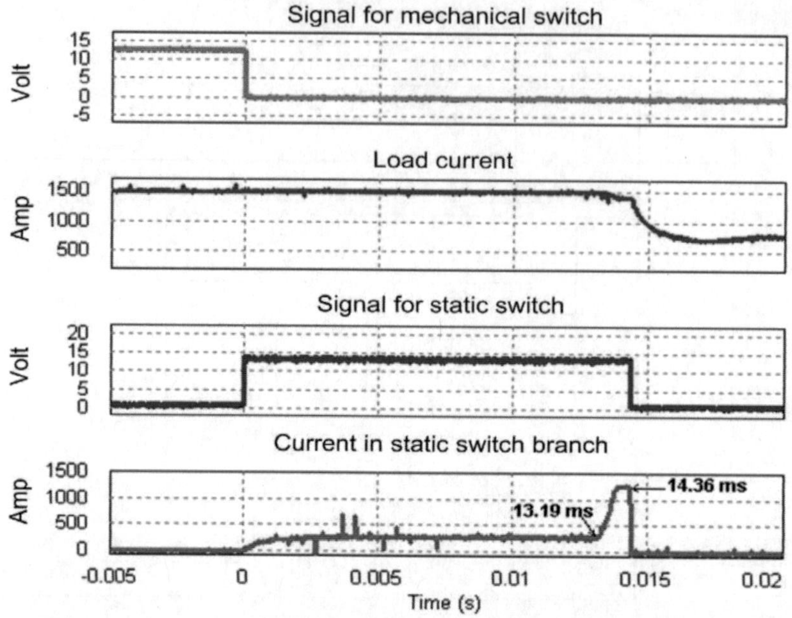

Figure 5. Hybrid DCCB testing at 1500 A with auto-triggering circuit

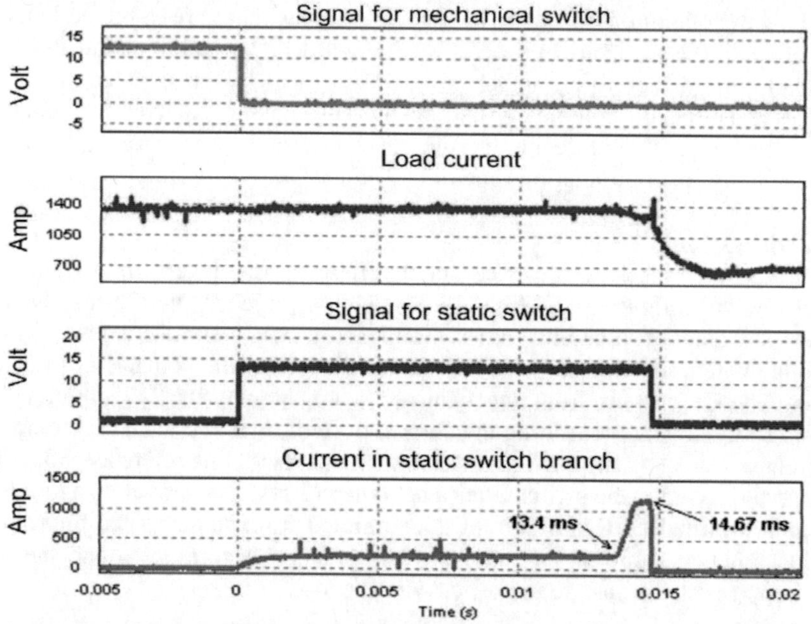

Figure 6. Hybrid DCCB testing at 1360 A with auto-triggering circuit

It is generally observed that the opening time of mechanical switch is not always constant. Thus, the static switch has to be kept ON for longer period of time to ensure that the current is fully commuted from the mechanical switch to the static switch. By implementing this circuit, the need for manual tuning can be eliminated because it senses the opening of mechanical switch. Hence, ATC makes the operation of hybrid DCCB independent of opening time of mechanical switch.

IOP Conf. Series: Journal of Physics: Conf. Series **823** (2017) 012015 doi:10.1088/1742-6596/823/1/012015

6. Summary

In this paper, an auto-triggering circuit to control the operation of Hybrid DCCB has been conceptualized where the triggering pulses of mechanical as well as static switches are automatically decided from the circuit parameters itself. The circuit has been successfully tested at lab scale and thereafter it has been implemented into actual 1.5kA Hybrid DCCB test set up. The control logic, circuit layout and operation of auto-triggering circuit have been presented in this paper. The test results of Hybrid DCCB using this circuit have also been discussed. Further, this auto-triggering circuit makes the operation of hybrid DCCB independent of opening time of mechanical switch, thus it reducing the overall current commutation time.

References

[1] Iwasa Y, 1994 *Case Studies in Superconducting Magnets: Design and Operational Issues* (New York Plenum).
[2] Pradhan S et. al, 2004 Proceedings of 20th IAEA fusion energy conference **36** 35.
[3] Atmadji A M S, "Direct current hybrid breakers: A design and its realization", Ph.D dissertation, Dep. of Electrical Eng., Eindhoven University of Technology, Eindhoven, The Netherlands, May 2000.
[4] Yang F, et. Al 2013 *Journal for Physics D: Appl. Phys.* **46 (27)** 19
[5] Atmadji A M S and Sloot J G J 1998 *IEEE-Proceedings of EMPD'98* **vol. 2** pp. 683-688
[6] Gaio E et. al 2008 *Proceeding of the 25th Symposium on Fusion Technology (SOFT-25)* pp 804–809.
[7] Novello L et. Al 2009 *IEEE Trans. on Applied Superconductivity*, **19(2)** 76
[8] Novello L, Analyses and experimental tests for the development of the Quench Protection System for the superconducting magnets of the Satellite Tokamak JT-60SA, Ingegneria Industriale Energetica.

IOP Conf. Series: Journal of Physics: Conf. Series **823** (2017) 012016 doi:10.1088/1742-6596/823/1/012016

LOFA analysis in helium and Pb-Li circuits of LLCB TBM by FE simulation

Paritosh Chaudhuri[1], Ranjithkumar S[1], Deepak Sharma[1], Chandan Danani[1]

[1]Institute for Plasma Research, Bhat, Gandhinagar – 382428, Gujarat, India

E-mail:paritosh@ipr.res.in

Abstract. One of the main ITER objectives is to demonstrate the feasibility of the breeding blanket concepts that would lead to tritium self-sufficiency and the extraction of a high-grade heat for electricity production. India has developed the LLCB TBM to be tested in ITER for the validation of design concepts for tritium breeding blankets relevant DEMO and future power reactor. LLCB concept has the unique features of combination of both solid (lithium titanate as packed pebble bed) and liquid breeders (molten lead lithium). India specific IN-RAFMS is the structural material for TBM. The First Wall is actively cooled by high-pressure helium (He) gas [1]. It is important to validate the design of TBM to withstand various loads acting on it including accident analysis like LOCA, LOFA etc. Detailed thermal-hydraulic simulation studies including LOFA in helium and Pb-Li circuits of LLCB TBM have been performed using Finite Element using ANSYS. These analyses will provide important information about the temperature distribution in different materials used in TBM during steady state and transient condition. Thermal-hydraulic safety requirement has also been envisaged for the initiation the FPPS (Fusion Power Shutdown System) during LOFA. All these analysis will be presented in detail in this paper.

1. Introduction

India is developing Lead-Lithium cooled Ceramic Breeder (LLCB) TBM for testing in one half of the ITER port no.2. It contains lithium meta-titanate (Li_2TiO_3) ceramic breeder (CB) as the tritium breeder material in the form of packed pebble beds. Molten Pb-16Li eutectic acts as tritium breeder, neutron multiplier and coolant for the CB zones. India specific Reduced Activation Ferritic Martensitic Steel (IN-RAFMS) is the structural material for TBM. The First Wall (FW) of LLCB TBM is actively cooled by high-pressure helium (He) gas. One of the functions of FW of is to remove surface heat flux and energy from plasma during normal and off-normal ITER operational conditions [1]. During the various ITER operating conditions, the LLCB TBM set will be subjected to a complex load scenario. It is necessary to perform a detailed analysis corresponding to different operating conditions by a very systematic approach based on the load specifications [2]. It is important to validate the design of TBM to withstand various loads acting on it including accident analysis like LOCA, LOFA etc. Detailed thermal-hydraulic simulation studies of loss of flow analysis (LOFA) in both helium and Pb-Li circuits of LLCB TBM have been performed using Finite Element (FE) using ANSYS. These analyses will provide important information about the temperature distribution in different materials during the accident condition. Thermal-hydraulic safety requirement has also been envisaged for the initiation the

FPPS (Fusion Power Shutdown System) during LOFA. All these analysis will be presented in detail in this paper.

2. LLCB TBM Description

LLCB TBM set consists of LLCB TBM and Shield Block (SB) as shown in figure 1 along with its dimensions. The TBM set is welded to the flange, which is bolted to the port plug frame. Figure 2-2 shows the 3D and 2D view of LLCB TBM and its internals. The LLCB TBM has both the features of solid breeder and liquid breeder blankets. It contains lithium meta-titanate (Li_2TiO_3) as ceramic breeder (CB) material in the form of packed pebble beds and Pb-Li eutectic as tritium breeder material. Pb-Li also acts as neutron multiplier and coolant. India specific Reduced Activation Ferritic Martensitic steel (IN-RAFMS) is used as the structural material for the LLCB TBM. The shield block is located behind the LLCB TBM, for shielding from high energy neutrons to magnets and other ITER components behind the TBM set and port plug. The shield block is connected to the TBM by attachments as shown in figure 2(a). The Pb–Li flows around the ceramic breeder pebble bed compartments as shown in figure 2(b) in low velocity to extract the volumetric heat in Pb-Li and also ceramic breeder zones due to fast neutrons.

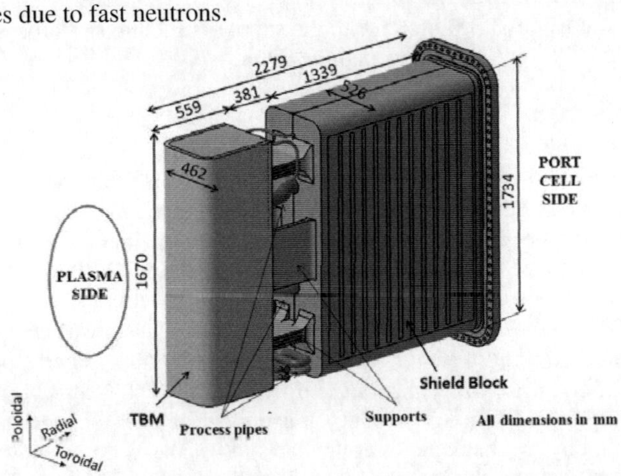

Figure 1. LLCB TBM with shield block and attachment plates

During plasma operation LLCB TBM is cooled by two process fluid: (i) 8 MPa helium is used to cool the FW structures and (ii) 1.2 MPa Pb-Li cools all CB zones and also the heat generated in Pb-Li. Apart from the two coolant circuits, low pressure helium purge gas at 0.12 MPa absolute pressure is flown through the ceramic breeder box to extract tritium generated within the ceramic breeder canister box. Tritium generated in Pb–Li circuit is extracted separately by low-pressure purge helium gas mixed with hydrogen and external detriation system.

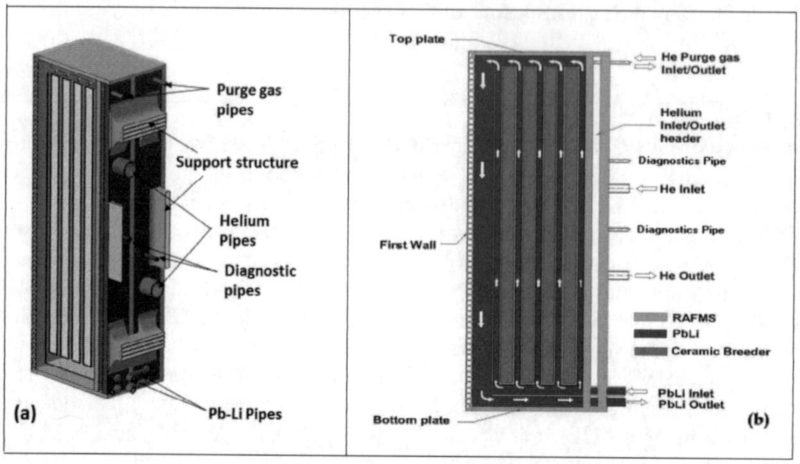

Figure 2. (a) 3D view of LLCB TBM along with the support structure and process pipes, (b) Pb-Li flow path in 2D view

2.1. First Wall of LLCB TBM

The First wall (FW) assembly of LLCB TBM is designed to withstand the heat flux from the plasma and neutronics heat generation on the FW structure to maintain its temperature below the allowable limits. The typical dimension of the FW is 1670 mm (poloidal) × 462 mm (toroidal) × 559 mm (radial) which is connected helium inlet and outlet manifolds as shown in figure 3(a). FW of LLCB TBM is made of 28 mm thick U-shaped structure made of low-activation IN-RAFMS, having 65 square cooling channels. The 65 coolant channels are divided into 13 circuits. Each circuit is having 5 channels of 20 mm × 20 mm cross section as shown in figure 4. The flow direction in five channels is also shown in this figure. Velocity of helium in each channel of ~ 50 m/s corresponds to a mass flow rate of 0.125 kg/s which gives rise to a temperature difference of ~ 40 °C between inlet and outlet for the incident heat flux of 0.3 MW/m^2. The inlets and outlets of each channel are connected to the respective headers formed by two back plates at the backside of the TBM as shown in figure 3(b). For all 13 circuits, helium enters from the inlet header side and after passing through the five passes it is collected in the outlet header.

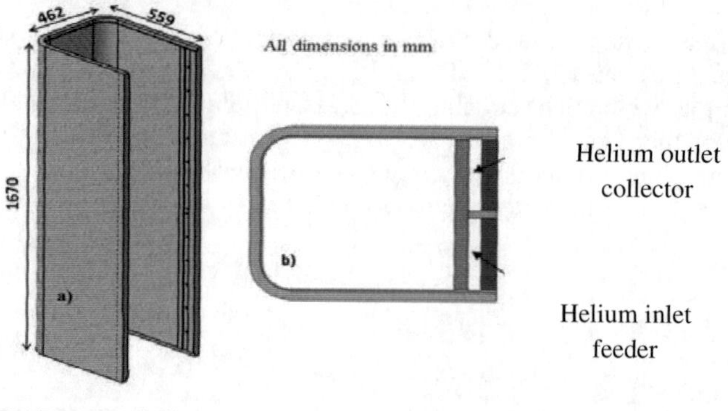

Figure 3. (a) FW of LLCB TBM and (b) inlet and outlet manifolds of helium (remove box).

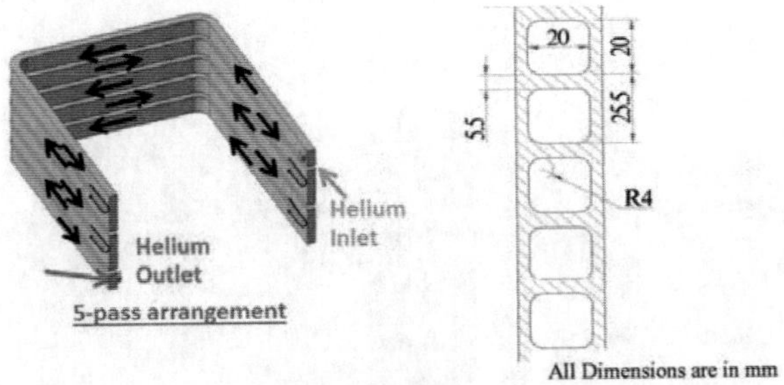

Figure 4. Five pass channel arrangement and dimension of cooling channels in FW

2.2. Internal components in LLCB TBM
The internals of LLCB TBM consisting of four ceramic breeders (CB) zones, Pb-Li flow paths and the Pb-Li separator plate as shown in the top view of LLCB TBM in figure 5. It shows the nomenclature of different ceramic breeder (CB) canisters and lead-lithium (LL) channels. All CB canisters are cooled by Pb-Li and its flow arrangement is shown in figure 2(b). Pb-Li enters at 300 °C and the flow is forced upwards through a set of five parallel channels. Pb-Li removes the heat generated in CB zones and also within it and finally reaches the top header. The accumulated Pb-Li flow through first Pb-Li channel (LL1) and exits the TBM. All CB canisters have the same thickness of 50 mm. All Pb-Li parallel flow paths have the same widths of 23 mm.

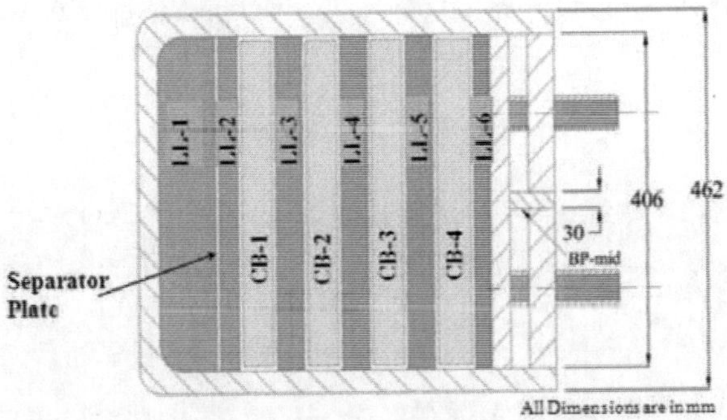

Figure 5. Top view of Pb-Li flow channels and CB zones

3. Steady state analyses
Steady state thermal analysis has been carried out using ANSYS Finite element (FE) code to estimate the temperature of different components of LLCB TBM. Thermal loads and boundary conditions(BCs) are applied on the FE model include the surface heat flux 0.3 MW/m^2, neutron heat generations (volumetric heating) in different components of LLCB TBM based on the ITER neutron wall loads of 0.78 MW/m^2. Convective BCs (heat transfer coefficient (HTC) and bulk temperature of the fluid) have been applied on the coolant channel surfaces [3]. Optimized value of HTC of helium has been used to keep RAFMS material within its allowable temperature window. Figure 6 shows the temperature profile of LLCB TBM for ITER normal operation. The maximum temperature of 920.25 °C occurs in first CB zone (CB1). FW structure temperature ranges from 302 to 474 °C which is below the maximum allowable limit of 550 °C as shown in figure 6.

IOP Conf. Series: Journal of Physics: Conf. Series **823** (2017) 012016 doi:10.1088/1742-6596/823/1/012016

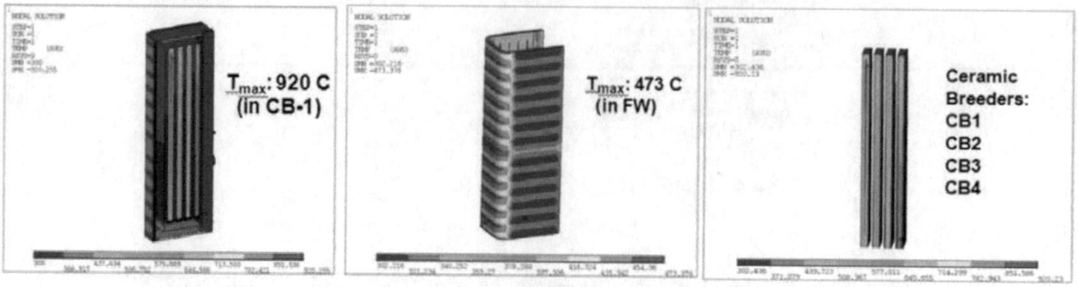

Figure 6. Steady state Temperature profile in LLCB TBM.

4. Transient analyses

Transient thermal analyses were performed, in 2D model for different ITER heat load conditions i.e., cyclic loads defined as inductive and non-inductive operations by the typical ITER pulses. Both these operations have two scenarios of normal and fusion power excursions. Figure 7 shows the Temperature-Time histories in different components in LLCB TBM during inductive normal operation. It has been observed that after each pulse cycle, the temperature inside CB bed appreciably increases. FW temperature doesn't rise with each pulse unlike ceramic breeder temperature. During power excursion the FW temperature reaches to 499 °C which also below its limiting value. The details of the transient analyses for inductive and non-inductive operation can be found in ref. 3.

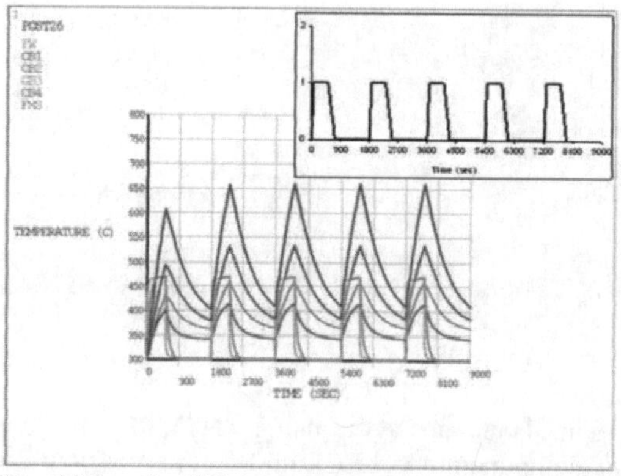

Figure 7. Temperature-Time history in different components in LLCB TBM during inductive normal operation.

5. Loss of Flow Accident (LOFA)

5.1 LOFA in First Wall cooling circuit

LOFA in FW Helium Cooling System (FWHCS) occurs due to helium circulator trip causing loss of flow of helium. During this event, it is assumed that the Pb-Li flow in LLCS loop is maintained at its nominal operating parameters. In the safety design of LLCB TBS, it is assumed that on the low flow signal from FWHCS, Fusion Plasma Shutdown System (FPSS) is triggered [4]. Failure of FPSS will lead to failure of TBM FW and subsequent in-vessel helium coolant leak. The temperature evolution TBM FW during LOFA is shown in figure 8. LOFA in He circuit initiated after 60 s full power flat top and the FW reaches 550 °C in 17 s. Since the time required to initiate FPPS is 3-4 s. therefore, sufficient margin for fast plasma shutdown. The same has been analyzed using RELAP/MOD4.0 and found to be in very good agreement [5].

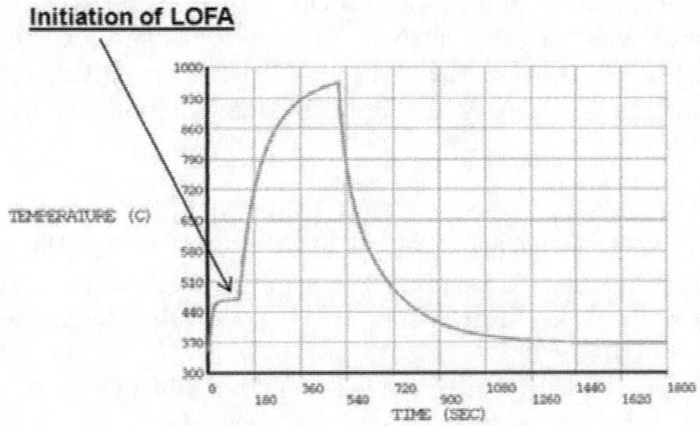

Figure 8. Temperature-Time history in FW during FWHCS LOFA.

5.2. LOFA in Pb-Li cooling circuit

LOFA in Lead Lithium Cooling System (LLCS) occurs due to pump trip and will lead to stagnant Pb-Li in TBM. This case is studied with helium flow in FWHCS maintained at its nominal operating parameters. The increase in Pb-Li temperature will cause the temperature of CB canister to rise because of no heat transfer from Pb-Li to them. Temperature evolution of internal components of LLCB TBM during LLCS LOFA is shown in figure 9. The time duration is long enough to go for slow plasma shut down (instead of FPSS) in order to bring down the Pb-Li and CN canister structure zone temperatures within the allowable limits i.e. 480 °C (above which the corrosion in structural material increases rapidly). There is no significant temperature variation in TBM FW during LOFA in LLCS.

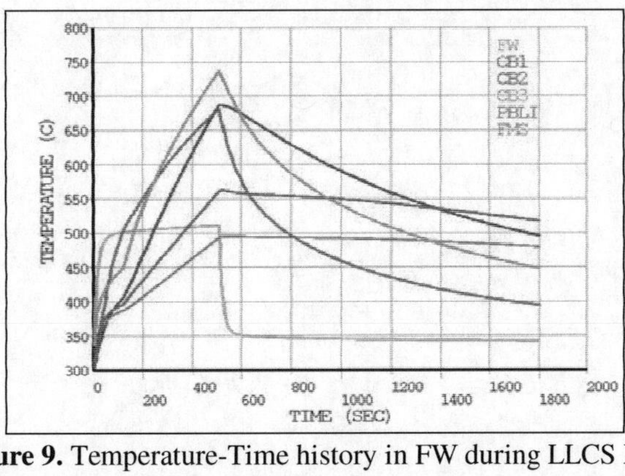

Figure 9. Temperature-Time history in FW during LLCS LOFA

IOP Conf. Series: Journal of Physics: Conf. Series **823** (2017) 012016 doi:10.1088/1742-6596/823/1/012016

6. Summary

Temperatures in LLCB TBM are well within the allowable limits in ITER inductive and non-inductive (normal and power excursion) operations. During LOFA in FW cooling circuit, the FW reaches its limiting temperature of 550 °C in 17 s. Actuation of FPSS is recommended to keep the TBM FW temperature below design limit during LOFA in FWHCS. Slow plasma shutdown following the pump trip (LOFA) in LLCS will be enough for the temperatures to remain under design limit. Time duration long enough to go for slow plasma shut down (instead of FPSS) in order to keep the TBM temperatures within the allowable limits (< 480 °C for LLCS operation) The results obtained from FE analyses are compared with accident analysis using RELAP/MOD4.0 and found to be in good agreement. LOFA and LOHS (Loss of Heat Sink) in FWHCS and LLCS are analyzed, the objective of these analyses is to provide the safety design inputs for I&C safety functions.

References

[1] Chaudhuri P, Ranjithkumar S, Sharma D, Danani C et al. 2014 *Fusion Eng. Des.* **89** 1362

[2] Ranjithkumar S, Sharma D, Chaudhuri P et al. 2016 *Fusion Eng. Des.***109-111** 1581

[3] Chaudhuri P, Ranjithkumar S, Sharma D, Danani C et al. 2016 *Fusion Eng. Des.* **109-111** 906

[4] "I&C safety function specification for Fusion Power Shutdown System (FPSS)", IDM UID 9TF3Y8, Aug. 2012.

[5] Sandeep T, Vilas Chaudhari V 2016 *Fusion Eng. Des.* **109-111** 1534.

IOP Conf. Series: Journal of Physics: Conf. Series **823** (2017) 012017 doi:10.1088/1742-6596/823/1/012017

Seismic Design of ITER Component Cooling Water System-1 Piping

Aditya P Singh[1], Mahesh Jadhav[1], Lalit K Sharma[1], Dinesh K Gupta[1], Nirav Patel[1], Rakesh Ranjan[1], Guman Gohil[1] , Hiren Patel[1], Jinendra Dangi[1], Mohit Kumar[1] and A G A Kumar[1]

[1]ITER-India, Institute for Plasma Research, Block A, Sangath Skyz, Bhat-Motera Road, Koteshwar, Ahmedabad, India 380005

E-mail: aditya.singh@iter-india.org

Abstract. The successful performance of ITER machine very much depends upon the effective removal of heat from the in-vessel components and other auxiliary systems during Tokamak operation. This objective will be accomplished by the design of an effective Cooling Water System (CWS). The optimized piping layout design is an important element in CWS design and is one of the major design challenges owing to the factors of large thermal expansion and seismic accelerations; considering safety, accessibility and maintainability aspects. An important sub-system of ITER CWS, Component Cooling Water System-1 (CCWS-1) has very large diameter of pipes up to DN1600 with many intersections to fulfill the process flow requirements of clients for heat removal. Pipe intersection is the weakest link in the layout due to high stress intensification factor. CCWS-1 piping up to secondary confinement isolation valves as well as in-between these isolation valves need to survive a Seismic Level-2 (SL-2) earthquake during the Tokamak operation period to ensure structural stability of the system in the Safe Shutdown Earthquake (SSE) event. This paper presents the design, qualification and optimization of layout of ITER CCWS-1 loop to withstand SSE event combined with sustained and thermal loads as per the load combinations defined by ITER and allowable limits as per ASME B31.3, This paper also highlights the Modal and Response Spectrum Analyses done to find out the natural frequency and system behavior during the seismic event.

1. Introduction

Piping subjected to high seismic acceleration needs to be designed stiff enough to maintain its structural integrity. At the same time it is also flexible enough to allow the expansion of pipe due to thermal gradient. It is very challenging task to optimize the design which has a balance between rigidity and flexibility to qualify the pipe as per the code requirement. All the piping design activities has been performed with the compliance and guidelines of international and industrial codes & Standards as well as the law and regulations of respective local authority.

2. System Classifications

The CCWS-1 piping is classified as below briefly according to the related documents of ITER [1]. This table represents principal relationship between safety, seismic and quality class. The classification of each system and component has something to do with most of all activities such as design, selection of materials, fabrication, installation, examination, testing, inspection and equipment qualification, etc.

IOP Conf. Series: Journal of Physics: Conf. Series **823** (2017) 012017 doi:10.1088/1742-6596/823/1/012017

Table 1. System classification

System Name	Safety Class	Seismic Class	Quality Class
CCWS-1Inside Tritium Building	SIC-2	SC-1	QC-1
CCWS-1 Outside Tritium Building	SR	SC-2	QC-2

3. Design Input

Pipe size and thickness has been selected as per the process requirement and hydraulic analysis. Preliminary pipe routing has been done with in available space. Flexibility of piping system is mainly dependent on the layout. While finalizing the location of equipment's the connecting piping flexibility is also to be considered along with the process flow, accessibility to valves, instruments, equipment maintenance, cleaning, operational safety, headroom clearance and aesthetics. The piping layout designer has to undergo number of iterations to reach to a final layout. Pipe Routing is always decided based on the equipment layout. The best possible pipe routing is achieved by knowing the process flow and the above criterion for layout.

Valve and support location has been provided as per the standard engineering practice. Pressure, temperature, insulation, materials requirement has been given in the table below [3].

Table 2. Pressure and temperature of process branch connection and main header

Clients	Design Pressure (MPa)	Design Temperature °C	Material	Pipe Nominal Size,DN
IBED PHTS-B1 HX	5	270	A358 GR 304L Cl1	500
IBED PHTS-B2 HX	5	270	A358 GR 304L Cl2	500
IBED PHTS-B3 HX	5	270	A358 GR 304L Cl3	500
IBED PHTS-B4 HX	5	270	A358 GR 304L Cl4	500
IBED PHTS Baking heat exchanger	5	270	A358 GR 304L Cl5	350
Auxilliary Cooling Header	1	90	A312 TP 304L	150
NBI- PHTS	3.5	120	A358 GR 304L Cl7	600
VV-PHTS& Volume Control Heat Exchanger for VV-PHTS	3.8	180	A312 TP 304L	200
CVCS HX for IBED PHTS	5	275	A358 GR 304L Cl9	250
Sampler Coolers	5	275	A358 GR 304L Cl10	250
TBM Primary Cooling System Components outside Port Cells	10	100	A312 TP 304L	200
CCWS-1 Main Header	16	78.1	A358 GR 304L Cl12	1600

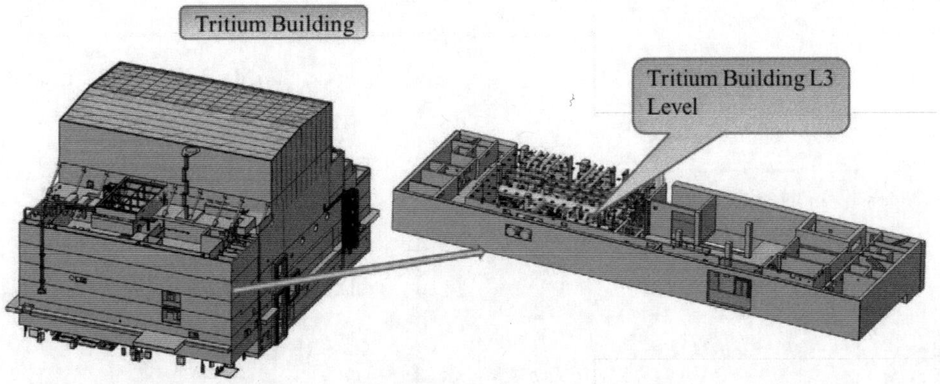

Figure 1. Location of piping in Tokamak Building

3.1. Seismic Input

Tritium building is multi story building; CCWS-1 pipes are entering into the Tritium building at L3 level. Seismic Response Spectra has been given at each floor levels and at definite node locations [5]. If the pipe is running in multi nodal area, the envelope of all the nodal spectrum values has been taken for seismic qualification of pipe otherwise specific node will be considered. Below graphs (see figure 2, 3 and 4) showing the acceleration vs frequency values in X, Y and Z direction respectively. From the graph it is clear that peak acceleration for 4% coming in X, Y & Z directions are 8.0529, 8.133 & 54.475 m/sec^2 respectively.

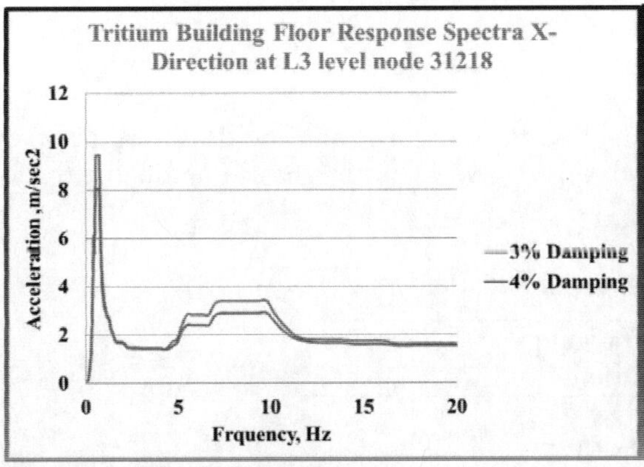

Figure 2. FRS X-Direction

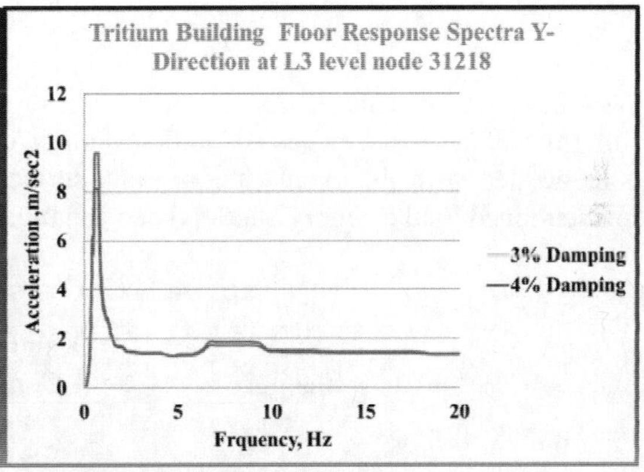

Figure 3. FRS Y-Direction

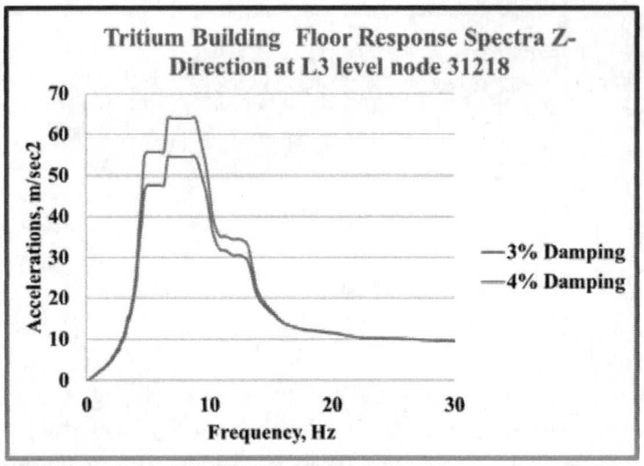

Figure 4. FRS Z-Direction

Table 3. Seismic Control Parameters

Seismic control parameters	Values/Method
Frequency Cutoff (Hz)	33
Closely Spaced Mode Criteria	0.1
Load Duration (sec)	20
Damping (DSRSS) (ratio of critical)	0.04
ZPA (Reg. Guide 1.60/UBC -- g's)	0.5
Spatial or Modal Combination First	MODAL
Spatial Combination Method	SRSS
Modal Combination Method	DSRSS
Include Missing Mass Components	Z PA X Missing
Missing Mass Combination Method	SRSS
Directional Combination Method	SRSS
Mass Model	LUMPED
Frequency Cutoff (Hz)	33

3.2. Load Cases and Combination

To full fill the service requirement and structural integrity, the system should be design as per the loads associated with the event category and service levels. CCWS-1 piping has to be qualified for below mentioned load category and load combination [1].

Table 4. Load Cases and Combination

Event Category	Design Loading Conditions	Service Level
I	P+Dw	A,B
	P+Dw+T	
II	P+Dw+SL-1	
	P+Dw+SL-1+T	
III	P+Dw+SMHV+T	C
IV	P+Dw+SL-2	D
	P+Dw+SL-2+T	

P= Design pressure, Dw = Dead weight of pipe, T=operating temperature, SL-1 = Seismic level-1, SL-2 = Seismic level-2, SMHV = Maximum Historically Probable Earthquake

4. Modeling and Analysis

Preliminary CAD model has been modeled in CAESAR II. Initial analysis run has been done to decide the support location as per the sustained loading condition which correspond to the event category I and service level A. After preliminary supports, stress check for the secondary loading condition corresponds to event category I service level B has been performed. During thermal stress check it has been found that there was very less flexibility in between the main header branch to the client connections which causes the failure of branches at client anchor points, tee connections and bend locations as shown in figure 5. All these locations except anchor points are failing due to high bending

IOP Conf. Series: Journal of Physics: Conf. Series **823** (2017) 012017 doi:10.1088/1742-6596/823/1/012017

stress caused by the thermal expansion of pipes. To avoid the failure due to the thermal expansion, branch connection from main header to the interface points has been modified as shown in figure 5. After modification all the branches has been qualified in thermal loading conditions as described in table 2. Piping has to be qualified for event category IV & service level D also, which includes SL-2 type events. For dynamic qualifications of pipes response spectrum method has been followed.

Spectrum values has been applied [5] and it is found that due to more flexibility in the branch connections pipes are made to free and the natural frequencies of system are coming into the range of corresponding peak acceleration in the spectrum. So due to peak acceleration the model exceeding the code specified allowable limit at many locations. To make system more rigid, support locations and configurations has been modified. Pipe route also modified as per the requirement to qualify.

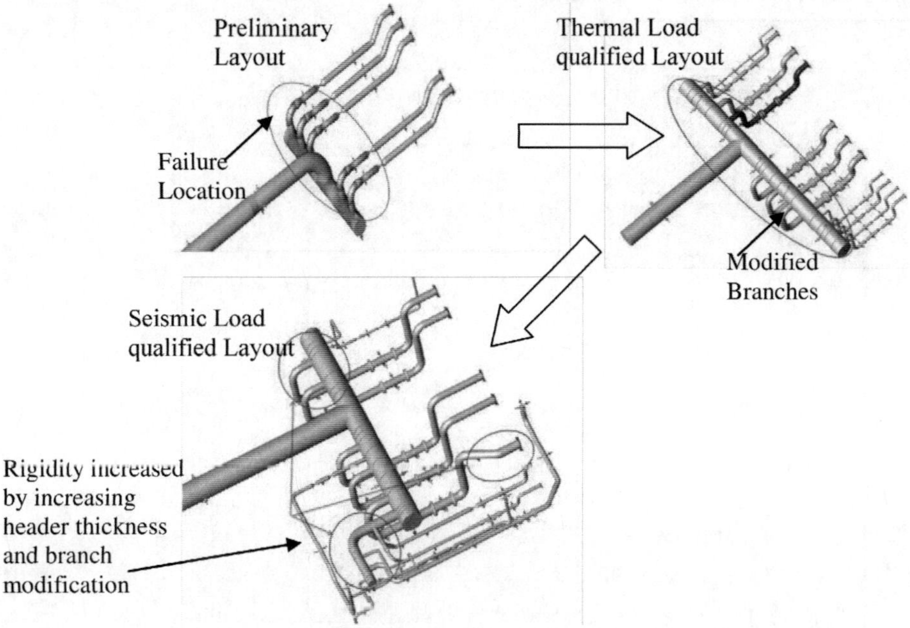

Figure 5. Design phases of cooling layout

5. Results and Conclusion

To optimize the routing of pipe between flexibility which is required in stress qualification in thermal loading condition and rigidity required in stress qualification in seismic loading condition many iterations has been performed with different configuration of supports and pipes. Code Compliance report shows that the all the code stresses are well below the code specified limits. Comparison of natural frequencies of thermal and dynamic qualified model and code compliance results are shown in figure 6 and 7 respectively.

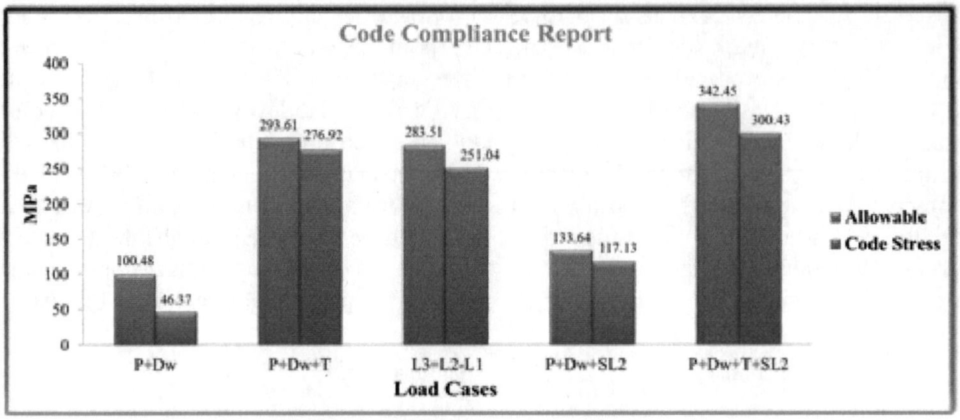

Figure 6. Code Compliance Report

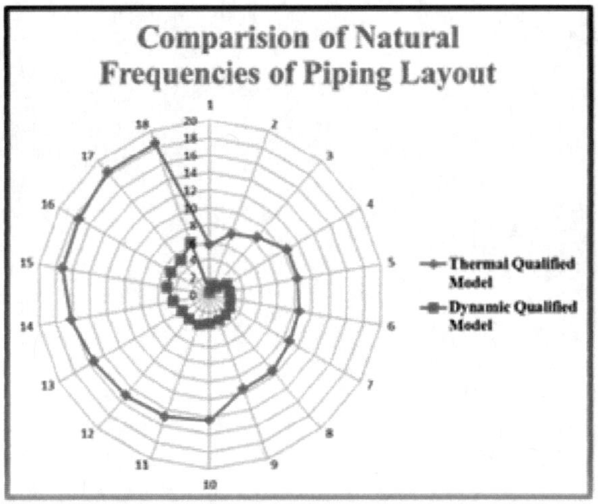

Figure 7. Comparison of Natural Frequencies of Piping Layout

References

[1] Load Specification for Cooling Water System_3YGYH7_v5.2
[2] Load Specifications _222QGL_v6.0
[3] Piping Material Specification for CCWS CHWS and HRS-II-M9RF9YQ_v2.5
[4] ASME B31.3 2012
[5] Tender Design- Tokamak Complex PBS 62.11 PBS 62.14 and PBS 62.74- Generation of the Floor Response Spectra_3Z5N6P_v1.3

IOP Conf. Series: Journal of Physics: Conf. Series **823** (2017) 012018 doi:10.1088/1742-6596/823/1/012018

Development of High Voltage and High Current Test Bed for Transmission Line Components

Akhil Jha[1], Manoj Patel[1], JVS Harikrishna[1], Ajesh P[1], Rohit Anand[1], Rajesh Trivedi[1] and Aparajita Mukherjee[1]

[1]ITER-India, Institute for Plasma Research, Near Indira Bridge, Bhat, Gandhinagar 382 428, India

E-mail: akhil.jha@iter-india.org

Abstract. A test bed for testing of MW level transmission line components, based on the concept of standing wave resonator is being developed at ITER-India, Ion Cyclotron Heating and Current Drive (ICH&CD) lab. This test bed can be configured and operated for various lengths of the resonator for optimum requirement. Estimated 32 kV peak voltage and 650 A peak current can be achieved inside the resonator during operation with an input power level of ~ 20 kW. In this paper, detailed design and simulation results of test bed using high frequency simulator Microwave Studio (MWS) is presented. A brief description on implementation plan is described as well.

1. Introduction

The Ion Cyclotron Heating and Current Drive (ICH & CD) system has to couple 20 MW of Radio Frequency (RF) power to ITER [1] plasma for heating and driving plasma current, in the MHz frequency range [2]. The same system will also be used for wall conditioning at lower power level in between plasma shots. India is responsible to deliver total 9 number of RF sources to ITER project, 8 will be used for plasma operation and 1 will be spare. Each RF source will provide 2.5 MW of RF power at Voltage Standing Wave Ration (VSWR) 2:1 in the frequency range of 35 to 65 MHz [3]. A large number of high power transmission line components (3", 6" and 12") are required for connecting various stages of RF source. Analytical calculation shows that peak RF voltage and current in transmission line can reach up to 22.5 kV and 450 A during operation at 2.5 MW power level with VSWR 2:1 [4].

To test the passive transmission line components at high voltage and large current, prior to integration with main RF source system, a test bed is required. RF test facilities comprising of stub and phase shifter combination has been developed elsewhere [5, 6] for testing vacuum transmission line (especially vacuum feedthrough) and coaxial transmission line components at high voltages (~ 30 kV or higher). Resonant ring concept has been used at ORNL which utilizes directional coupler to build up high circulating power with very little reflection [7]. Such test stands are usually bulky and require more space. A compact test bed based on standing wave resonator is designed which can be configured and operated for quarter wave ($\lambda/4$), half wave ($\lambda/2$) and three quarter wave ($3\lambda/4$) resonator length. It is basically a "Transmission line resonator" which uses transmission line sections with various lengths and terminations (usually open- or short circuited) to form resonators [4]. Input impedance of the resonator is matched with external RF source (50 Ω) using a tunable matching L-C circuit, which provides impedance matching for different operating conditions at resonance frequency.

IOP Conf. Series: Journal of Physics: Conf. Series **823** (2017) 012018 doi:10.1088/1742-6596/823/1/012018

The Device Under Test (DUT) (i.e. transmission line components for testing) needs to be connected in-line during operation.

2. Standing Wave Setup Configuration:

Proposed resonator test bed are in three configurations, namely quarter wave (λ/4) open circuit stub, half wave (λ/2) short circuit stub and three quarter wave (3λ/4) open circuit stub. All three resonator configurations can be represented with a series resonator equivalent circuit [4]. A lossy quarter wave or three quarter wave open circuit (half wave short circuit) stub and their series equivalent circuit are shown in figure 1.

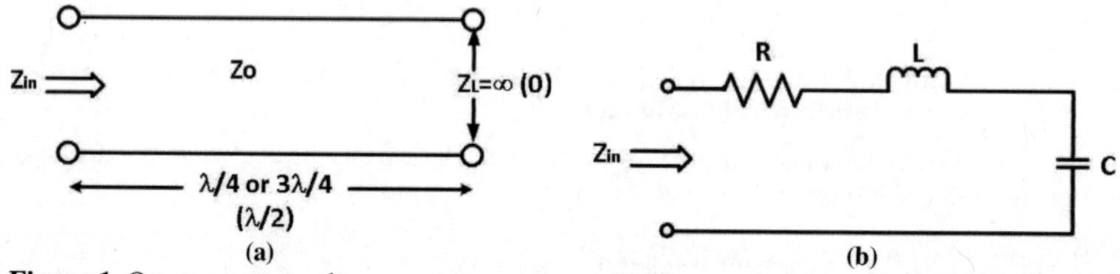

Figure 1. Quarter wave or three quarter wave open (Half wave short circuit) stub (a) layout (b) equivalent circuit.

The stub has characteristic impedance Z_o and load impedance $Z_L = \infty$ (open stub). Input impedance of low loss ($\alpha l \ll 1$ $i.e.$ $tanh\alpha l \sim \alpha l$) coaxial quarter wave (λ/4) stub near resonance is given as [4]:

$$Z_{in} = Z_o\left[\alpha l + i\frac{\pi\Delta\omega}{2\omega_o}\right] \tag{1}$$

where ω = ω_o+ Δω where $\Delta\omega$ is small, α attenuation constant, β propagation constant and l is distance from load (open circuit). Comparing equation (1) with input impedance of series LCR circuit provides resistance $\boldsymbol{R} = \boldsymbol{Z_o \alpha l}$ and inductance $\boldsymbol{L} = \frac{\pi Z_o}{4\,\omega_o}$. Similarly for three quarter wave (3λ/4) open circuit stub and half wave (λ/2) short circuit stub inductance are $\frac{3\pi Z_o}{4\,\omega_o}$ and $\frac{\pi Z_o}{2\,\omega_o}$ respectively with resistance $\boldsymbol{Z_o \alpha l}$. Unloaded Q for stub is [4]

$$Q_U = \frac{\beta}{2\alpha} \tag{2}$$

where for quarter wave (λ/4) stub$\beta = \frac{\pi}{2\,l}$, for three quarter wave (3λ/4) stub$\beta = \frac{3\pi}{2\,l}$ and for half wave (λ/4) stub$\beta = \frac{\pi}{l}$.

3. Modelling and Simulation

3.1. Quarter wave (λ/4) and Three Quarter wave (3λ/4) open circuit stub:

Modelling and simulation of 12inch Standing Wave Resonator was performed with CST Microwave Studio (MWS) software. The detail cut view of CST model of 12inch coaxial open circuit resonator is shown in figure 2. Open circuit stub is connected with port 1 (indicated in figure 2) of 12inch coaxial TEE (denoted in figure as TEE1). Length of stub (***lo***) is decided depending on the quarter wave (λ/4) and three quarter wave (3λ/4) mode of operation. Port 2 of coaxial TEE 1 is extended and terminated with a short circuit plunger. Port 3 of TEE 1 connected with a coaxial capacitor. Capacitor is formed by introducing a discontinuity in the inner conductor. The inner conductor in capacitor is made in two parts; one with larger diameter provides a 29.6 Ω section which is fixed and the other 107.7 Ω section with smaller diameter slides in and out of the larger diameter section. The overlapping area between two inner conductor sections decides the resultant capacitance of the coaxial capacitor. Inner conductor of 107.7 Ω section is extended out of the resonator using Teflon rod (denoted in the bottom right of figure) which will be connected to an external drive for tuning the capacitor. To fulfil the above purpose a coaxial TEE (denoted as TEE2) is used. Port 4 of the TEE 2 is used for extending out the Teflon rod and port 5 is reduced in 3-1/8inch port using a 12inch to 3-1/8 inch reducer. RF power is fed to the resonator through 3-1/8 inch input port.

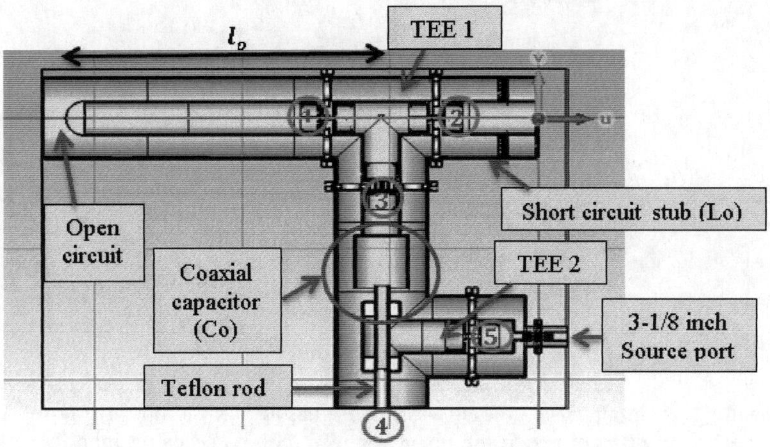

Figure 2. CST model of 12inch coaxial open circuit resonator

Figure 3(a) and 3(b) shows the line diagram and lumped element equivalent of open circuit resonator. Input impedance ($\boldsymbol{Z_{in}}$) of the open circuit stub is matched with the RF source impedance ($\boldsymbol{Z_o}$) using L-C matching circuit. This matching circuit contains 12inch short circuit stub (L_o) and 12inch coaxial capacitor (C_o) as shown in figure 2 and 3. The coaxial capacitor is connected in series arm and short circuit stub is connected in parallel arm. Both the circuit components are made tuneable so that they can be used in all three resonator configurations. The characteristics impedance ($\boldsymbol{Z_o}$) of Stub, TEE and reducer section is 50 Ω.

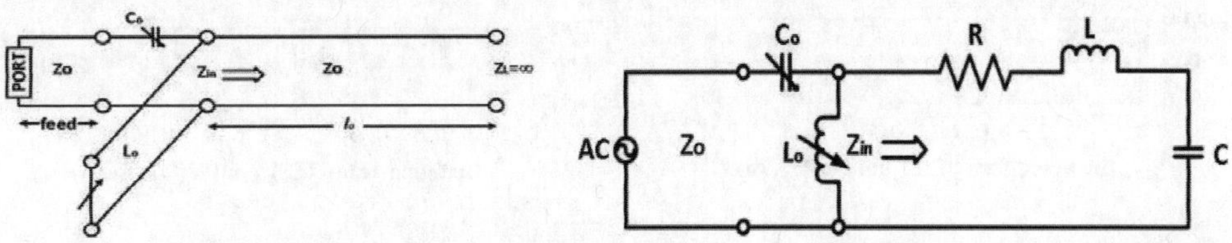

Figure 3. Coaxial open circuit resonator equivalent (a) line diagram (b) lumped circuit

Figure 4(a) and 4(b) shows smith chart and S-parameter simulation results for open circuit resonator in the quarter wave configuration. Input impedance ($\boldsymbol{Z_{in}}$) of open circuit quarter wave stub at 43.72 MHz (computed at port 3 of TEE1) is $0.87 + i98.99\,\boldsymbol{\Omega}$. Corresponding equivalent inductance is $L_{eq} = 360.54\,nH$ and required capacitance value to achieve resonance at 43.72 MHz is $C_o = 36.79\,pF$. The input return loss (S11) of resonator at source port at resonance frequency 43.72 MHz is – 28 dB as shown in figure 4(b).

IOP Conf. Series: Journal of Physics: Conf. Series **823** (2017) 012018 doi:10.1088/1742-6596/823/1/012018

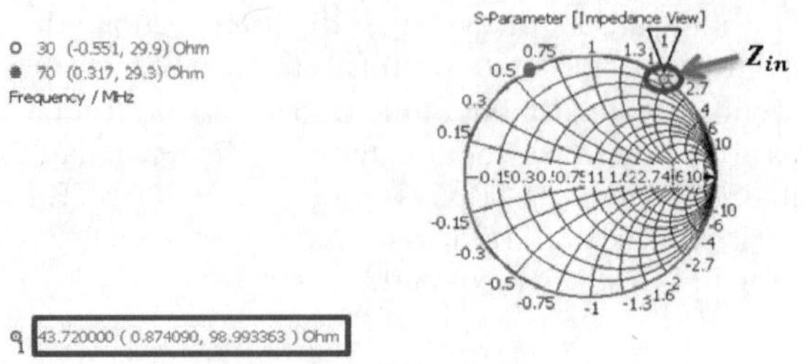

(a)

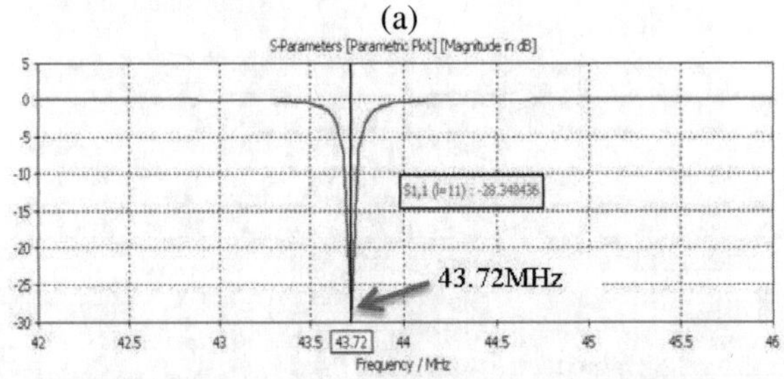

(b)

Figure 4. (a) Smith chart of open circuit stub (Z_{in}) (port 3 of TEE1) (b) Return loss (S11) at source port at resonance frequency 43.72 MHz

The resonator was simulated in quarter wave configuration for an input RF power of 20 kW. Peak voltage and peak current level inside the resonator were computed using inbuilt current and voltage probes. Distribution of the voltage and current along the length (***lo***) (see figure 2) of open circuit stub (taking centre of TEE1 as reference) is shown in figure 5(a) and 5(b) respectively. The computed max voltage and current level were ~ 32.1 kV at the position of open circuit and ~ 522.4 A near the TEE1 centre respectively.

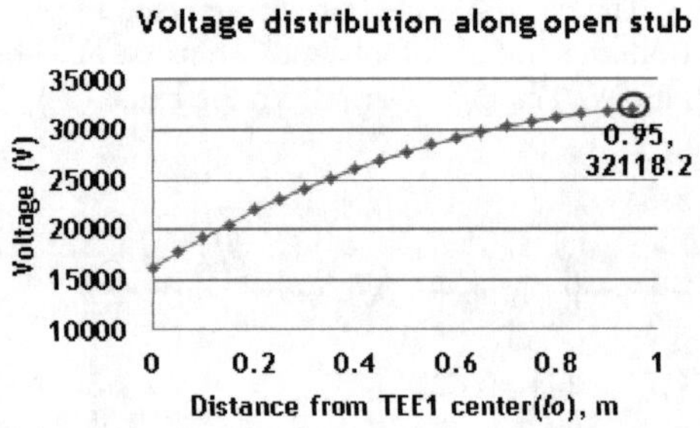

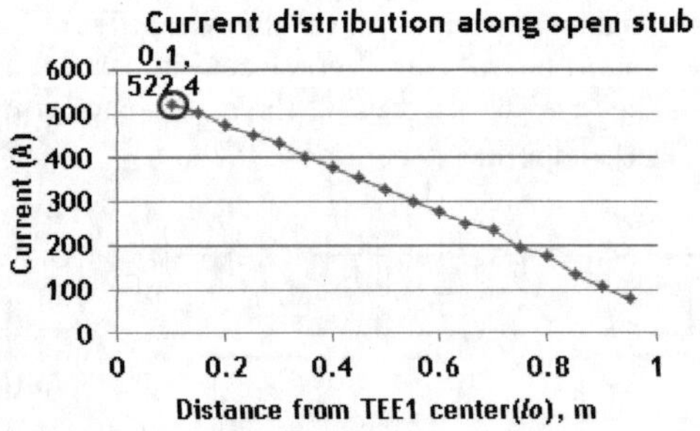

(a) (b)

Figure 5. Distribution along open stub length (*lo*) (λ/4 configuration) (a) Peak voltage value (b) Peak current value

With above defined method resonator was simulated for three quarter wave (3λ/4) configuration. Length (***lo***) of the open circuit stub was increased from ~ 1 m to ~ 4 m. The computed input impedance (Z_{in}) of stub is $0.48 + i86.19\ \Omega$. The equivalent inductance $L_{eq} = 279.88\ nH$ corresponding capacitance value to achieve resonance at 49.04 MHz is $C_o = 37.67pF$. The peak voltage and peak current level in the resonator reached up to a maximum value of ~ 32.7 kV (at 1.1 m from centre of TEE1) and 651.5 A (at 2.6 m from centre of TEE1) for an input RF power of ~ 20 kW.

Distribution of the voltage and current along the length (***lo***) of open circuit stub (taking centre of TEE1 as reference) is shown in figure 6(a) and 6(b) respectively.

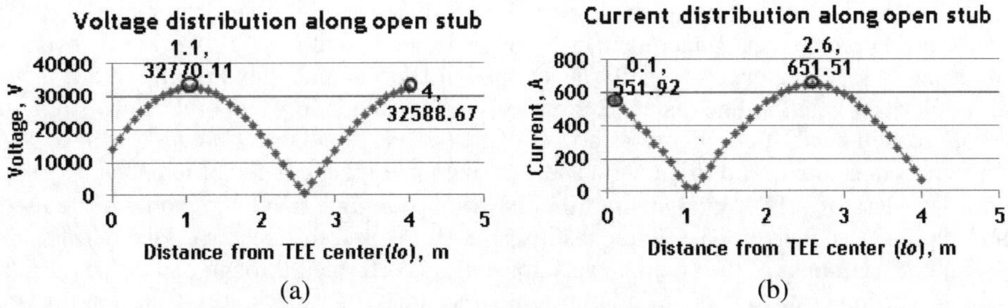

Figure 6. Distribution along open stub length (*lo*) (3λ/4 configuration) (a) Peak voltage value (b) Peak current value.

3.2. Half wave (λ/2) short circuit stub

In half wave configuration stub is shorted with a metal plunger at ~ 2.55m from TEE1 centre. The computed input impedance ($\boldsymbol{Z_{in}}$) of stub is $0.59 + i84.73\ \Omega$. The equivalent inductance $L_{eq} = 272.25\ nH$ corresponding capacitance value to achieve resonance at 49.56 MHz is $C_o = 37.91\ pF$. The peak voltage and peak current level in the resonator reached up to ~ 32.1 kV (at 1.1 m form centre of TEE1) and 646.3 A (near the short position) for an input RF power of ~ 20 kW. Distribution of the voltage and current along the length (***lo***) of short circuit stub (taking centre of TEE1 as reference) is shown in figure 7(a) and 7(b) respectively.

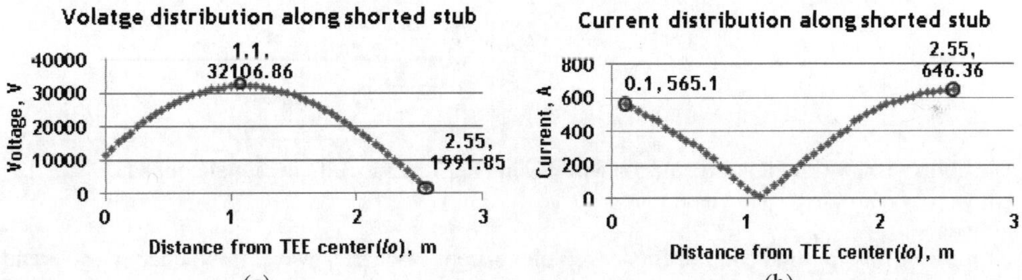

Figure 7. Distribution along short circuit stub length (*lo*) (λ/2 configuration) (a) Peak voltage value (b) Peak current value

In above three configurations the stub is critically coupled with the external L-C matching circuit at resonance frequency. Thus loaded Q is approximately half of the value of unloaded Q. The loaded Q of resonator (Q_L) computed from simulation is shown in table 1:

Table 1. $\boldsymbol{Q_L}$ of coaxial resonator

Q_L of resonator (critical coupling)[4]	
Configuration	Computed simulation value
λ/4 open circuit	6696
3λ/4 open circuit	7699
λ/2 short circuit	7170

4. Implementation as Transmission line test bed

Out of the three resonator configuration, quarter wave and three quarter wave resonators will be used depending on the length of the DUT's to be tested and the half wave shorted resonator configuration is useful to test shorting plunger and finger contacts in high current conditions. DUT's will be connected as a part of resonator section along the length (***lo***) of open or short circuit stub. Figure 8 shows 12 inch directional coupler connected in-line (as an example of DUT).Volume loss and breakdown strength of dielectric supports will be verified. Figure 8 shows position of view ports near the high voltage region for optical arc detection. Analytical calculation shows that copper inner conductor temperature rises up to 185 °C for heat loss of 1.17 kW. Dynamic flow of dry air through annulus region will be used to actively cool the inner conductor. The 4inch gas inserts will be used to inject and take out dry air as shown in Figure 8.The tuning of the coaxial capacitor will be performed through external motor drive attached to extended inner conductor. Minimum tuning accuracy of capacitor is 1mm for fine input matching. For example, a change in overlapping length of 1mm varies capacitance by ~ 420 pF which changes return loss from - 28 to - 19 dB at source port (in quarter wave configuration). The peak RF voltage level inside the resonator will be measured with calibrated 12inch loop directional coupler and voltage probes installed in the high voltage region shown in figure 8.

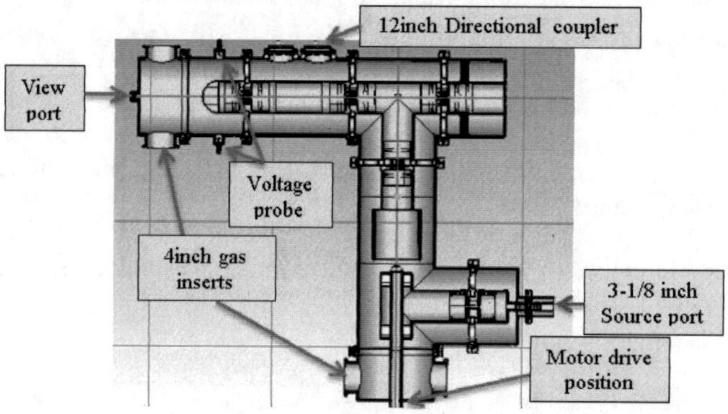

Figure 8. CST model of resonator with in-line DUT (12 inch Directional coupler).

5. Conclusion

Test bed for transmission line components based on concept of quarter wave, three quarter wave and half wave resonator is proposed. Detailed design, modelling and simulation conducted using CST MWS. A coaxial type tuneable capacitor is implemented to match the impedance of shorted/open stub with the RF source impedance. With an input power of ~ 20 kW, the peak voltage inside the resonators can be reached up to ~ 32 kV which is above the required 22.5 kV peak voltage level during operation at 2.5 MW with VSWR 2:1. The peak RF current at the short position in the half wave shorted stub is ~ 646 A which is high enough to test the present finger contacts. A detailed description for further implementation of the resonator configurations as a practical transmission line test bed is provided.

References

[1] ITER-Physics Expert group 1999 *Nuclear Fusion* **39** 2137
[2] Lamalle P U et al. 2013 *Fusion Eng. Des.* **88** 517
[3] Mukherjee A et al 2015 *Fusion Engineering and Design* **96–97** 542
[4] Pozar D M 2012 *Microwave Engineering* fourth edition (John Wily & Sons, Inc), chapter 2 pp 57-58, chapter 6 pp 272-282 pp 298-299
[5] Mutoh T et al. 1999 *Fusion Science & Technology* **35** 297
[6] BAE Y et al. 2002 *Journal of the Korean Nuclear Society* **34** 211
[7] Goulding R H et al. 2014 *AIP Conference Proceeding* **1580** 370

IOP Conf. Series: Journal of Physics: Conf. Series **823** (2017) 012019 doi:10.1088/1742-6596/823/1/012019

Solving the Capacitive Effect in the High-Frequency sweep for Langmuir Probe in SYMPLE

Pramila[1], J J Patel[1], R Rajpal[1], C J Hansalia[1], V P Anitha[1], K Sathyanarayana[1]

[1]Institute for Plasma Research, Near Indira Bridge, Bhat, Gandhinagar 382 428, India

E-mail: pramila@ipr.res.in

Abstract. Langmuir Probe based measurements need to be routinely carried out to measure various plasma parameters such as the electron density (n_e), the electron temperature (T_e), the floating potential (V_f), and the plasma potential (V_p). For this, the diagnostic electronics along with the biasing power supplies is installed in standard industrial racks with a 2KV isolation transformer. The Signal Conditioning Electronics (SCE) system is populated inside the 4U-chassis based system with the front –end electronics, designed using high common mode differential amplifiers which can measure small differential signal in presence of high common mode dc- bias or ac ramp voltage used for biasing the probes. DC-biasing of the probe is most common method for getting its I-V characteristic but method of biasing the probe with a sweep at high frequency encounters the problem of corruption of signal due to capacitive effect specially when the sweep period and the discharge time is very fast and die down in the order of μs or lesser. This paper presents and summarises the method of removing such effects encountered while measuring the probe current.

1. Introduction

The “SYMPLE”(System for Microwave Plasma Experiment) is an experimental system conceived few years back in IPR to investigate the physics of interaction of extremely intense electromagnetic waves in an over dense plasma. Theoretically, such waves are predicted to be absorbed in the plasma unlike their low amplitude counterparts that do not propagate in over-dense plasma. The experimental setup of the SYMPLE is shown in figure 1.In order to identify the right parametric regime where the plasma meets with the required pre-requisites conditions for density in axial and radial distance etc. Langmuir Probe based measurements need to be routinely carried out to measure various plasma parameters such as the electron density (n_e), the electron temperature (T_e), the floating potential (V_f), and the plasma potential (V_p).

Figure 1.View of SYMPLE system

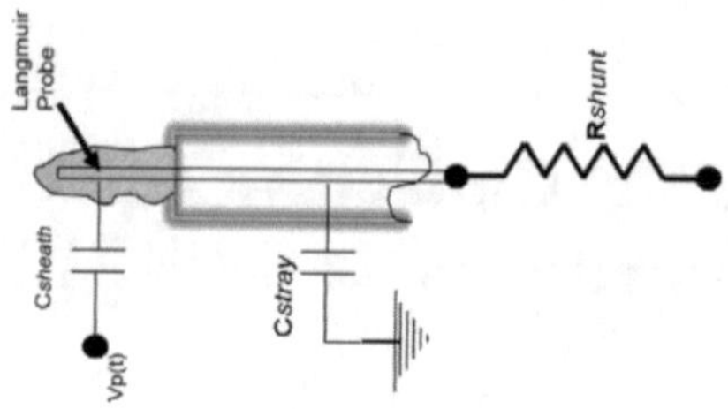

Figure 2. Langmuir Probe circuit immersed in plasma

The Langmuir probe in its simplest form is a small piece of metal inserted inside plasma as shown in figure 2. The I-V characteristic of the probe, as obtained by sweeping the probe potential from a negative to a positive value, contains information about the local parameters, namely, the electron temperature and the electron density. Thus, the Langmuir probes are known for their ability to provide the important measurements of local plasma at low cost with high speed and reliability.

Although a conventional Langmuir probe diagnostic method is very popular among plasma physicists for measuring various plasma parameters with constant density but it cannot be used to measure the plasma having transient characteristics. The current-to- voltage characteristic curve obtained in the latter case is present in every sweep period for which the probe is biased, thus giving variation in electron temperature with time by calculating every I-V curve acquired from every half sweep period [1]. The probe bias is swept at twice or higher frequency as compared to the plasma fluctuations from which we can derive the electron temperature by the equation (1) if the condition of $\ln(I_p) \sim V_p$ is met.

$$\text{Slope} = 1/\,T_{ev} \tag{1}$$

where T_{ev} is the electron temperature.

2. Electronics

The signal conditioning electronics for the Langmuir probe diagnostic system consists of two parts viz: sweep biasing circuit and the current sensing circuit which is shown as one complete unit in the figure 3. The optical fiber based system is being used to safely transmit an extremely low jitter trigger for simultaneous triggering of ignitron and the function generator. It is also employed to provide the complete isolation between the diagnostic and the signal conditioning electronics in order to avoid any ground lifting. The sweep biasing involves the triggering of the function generator through the optical link for generating the required sweep pulse and amplifying it using the power amplifier PA-85. The current sensing involves the current measurement circuitry along with the signal conditioning electronics having programmable gain amplifiers and programmable filter module.

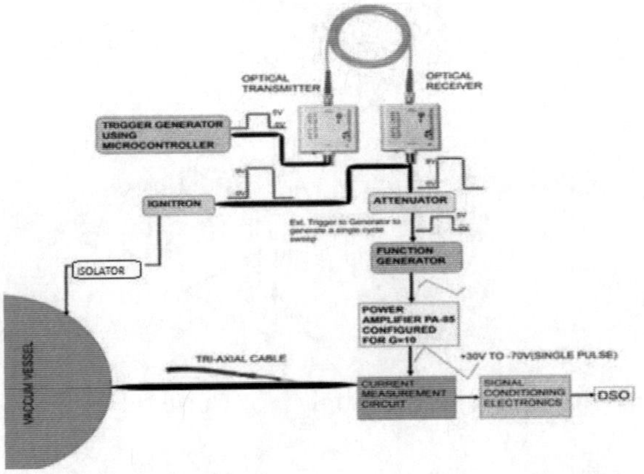

Figure 3. Complete system level diagram

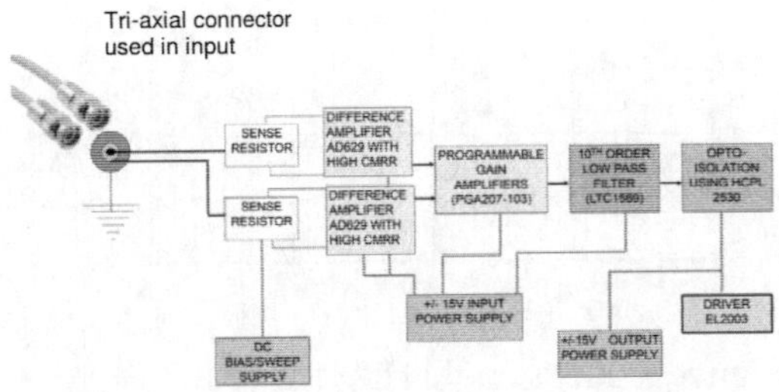

Figure 4. Current Sensing and Signal Conditioning Electronics

2.1. *Sweep biasing and the probe current sensing*

The electronics was well proven when the probe was dc-biased by looking at the probe characteristic received at the final output. But, when the probe is subjected with the sweep bias of high frequency of around 100 kHz with amplitude ranging between + 30 V to -70 V, the capacitive current becomes significant. Such high frequency presents severe limitation of differential operations for the front –end difference amplifier AD629 which possess poor ac CMRR of around 65 dB at these frequencies as shown in its datasheet in figure 5. Due to clear limitation of CMRR, the sweeping frequency is reduced to 10 kHz instead of the initial plan of 100 kHz by specially crafting the symmetry while retaining the same sweep duration of 10μs (duration of the discharge).

2.2. *Signal conditioning Electronics*

As the bias voltage is swept between the ion (~ - 70 V) and the electron (~ + 30 V) saturation regions , the probe collects the electrons of different energy which represents the electron energy distribution function determining the electron temperature. The probe current is pick-up by a sensing voltage over a small value shunt (200 Ω) resistor in the input of the difference amplifier AD629 followed by a signal conditioning electronics. The voltage signal corresponding to the current is fed to a programmable gain instrumentation amplifier PGA207 configured in a differential mode, which also receives the signal from another dummy difference amplifier AD629. This configuration not only remove any unwanted signal generated due to change in bias voltage or due to left-out ac-CMRR error of the difference amplifiers as shown in figure 8 ,but also imparts gain to the signal . The signal then passes through the filter-stage implemented using 8th order Bessel filter MAX296 configured as low pass filter for filtering-out the high frequency noise generated by the plasma itself or other non-plasma sources like vacuum pumps, magnetic coils etc. [2]. The bandwidth of the filter module is set at 10 kHz .The optical isolation is implemented using opto-isolator HCPL2530 and associated circuitry before data acquisition. The opto-isolator is configured to be operated with a current source of 3 mA for each channel to maintain the linearity in the ± 5 V range which the expected signal will never cross.

Thus, the signal obtained after passing through the signal conditioning electronics as shown in the figure 4 is less noisy and data processing may be easier with lesser data points.

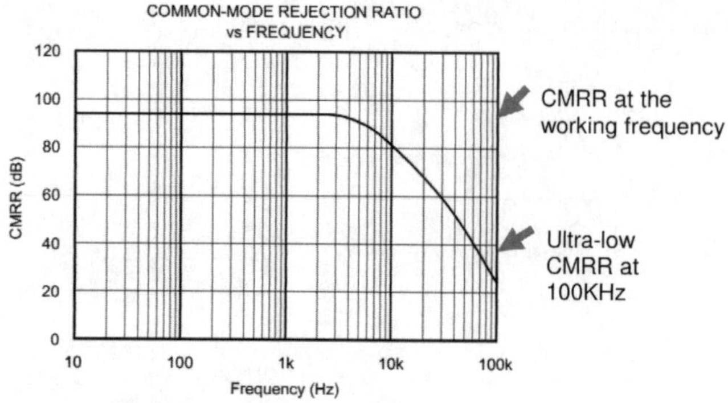

Figure 5. Plot showing CMRR of IC AD629 with frequency.

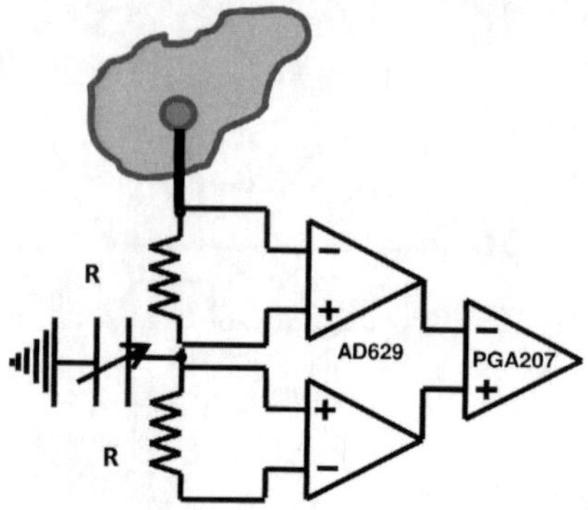

Figure 6. Front-end Electronics for Current Sensing.

2.3. *Capacitive Effect*

The swift variations of the sweep potential in large voltage range (+ 30 V to - 70 V) within short time-scale (~ 10 μs) leads to the milli-amperes of the unavoidable capacitive currents of the order of the ion saturation current thus, distorting the I-V characteristic. For a Langmuir probe, there are many capacitive sources but the one which is most important for us in high frequency sweep is the parasitic capacitance or stray capacitance, formed between a co-axial conductor's core and its shield or environment [3] in case of unshielded cables. Various types of cablings were tried out like shielded twisted pair, co-axial cable etc. but the double-shielded cables called as tri-axial cables are employed for the rapid sweep biasing of the probe. In tri-axial cable configuration, the inner shield and the central conductor are simultaneously biased with the same signal, while a shunt resistor placed in line with the central conductor measures the current. The outer shield is at the potential of the plasma chamber. Most of the capacitive effect was removed using the tri-axial cable connected in the configuration as depicted in the figure 7.Some left-out error as shown in figure 8 was removed

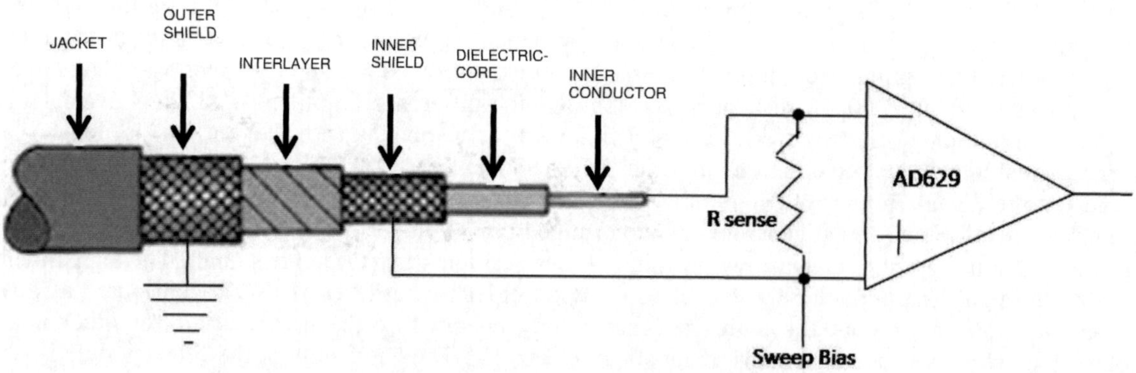

Figure 7. Configuration of the tri-axial connector with Front-end electronics.

using the differential configuration of instrumentation amplifier PGA207 after the current-sensing stage as shown in the figure 6. Figure 8 shows the mV noise that is still present after the tri-axial cable configuration while figure 9 shows the complete elimination of noise after the differential configuration implemented in PGA207.

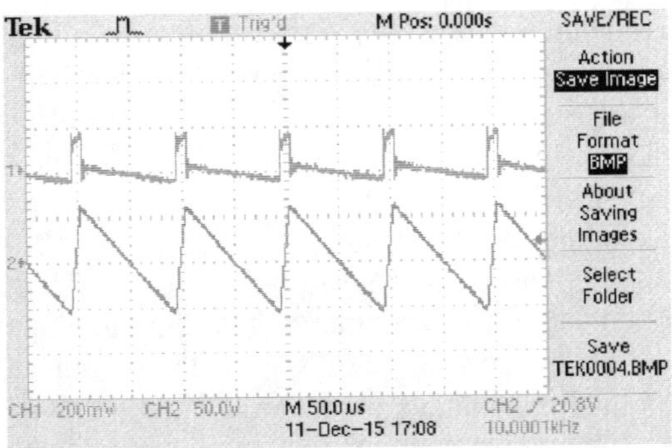

Figure 8. Reduced Capacitive effect implementation of tri-axial cable configuration.

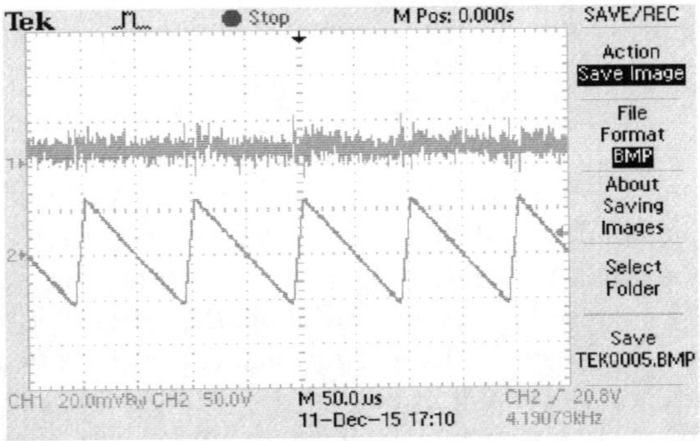

Figure 9. Complete noise elimination after differential configuration of INA.

The signal is now completely clean from the capacitive effect which is then further conditioned in the consecutive signal conditioning stages.

3. Experimental Setup and Testing

A total of 8-channels of the signal conditioning electronics mounted in a 4U- chassis will be used using the tri-axial cable configuration for sensing the current from the probe biased with the sweep. The whole of the electronics system is kept in the corridor, away and outside the exposure of the experimental setup. A microcontroller based circuit is used for generating a 5 V TTL pulse for triggering the optical fiber transmitter link which transmits a 9V pulse for simultaneous triggering of the ignitron as well as function generator(via an attenuator). The function generator then generates the desired sweep pulse of the required rise time for biasing the probe.

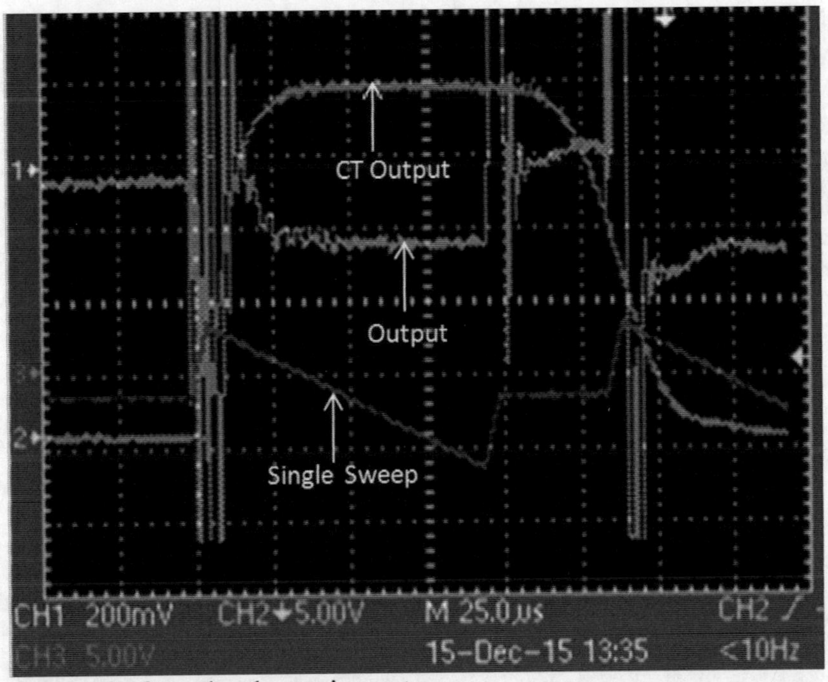

Figure 10. Testing results from the electronics system.

The probe current is sensed by a shunt resistor connected across the dedicated difference amplifier IC AD629 which works in the presence of high common mode voltage. The capacitive effect left-out from the tri-axial cable correction also could not be removed fully by the difference amplifier IC AD629 due to its low ac CMRR of about 80 dB at 10 kHz (figure 5), leaving some residual capacitive effect of amplitude of around 100 mV in its output. This inability to sufficiently suppress the common mode signal leads to a noise signal generated as a result of common mode signal which was comparable to the useful signal. A dummy combination of the same configuration using AD629 is used for feeding at one of the differential terminals of the programmable gain instrumentation amplifier IC PGA207. The residual capacitive effect is finally removed when seen at the output of the instrumentation amplifier as shown in figure 6. The final current-voltage characteristic curves as result of the testing is displayed in figure 10.

This figure has three plots containing the single sweep pulse, CT output and the final output. This depicts the temporal profile obtained for a discharge of 100 μs with an average voltage of 200 mV measured across a shunt of 200 Ω. The captured data (200 mV) in the Langmuir probe corresponds to the ion saturation currents (I_s) across a 200 Ω shunt resistance. Using the formula $I_s = n_e A v$, for a given probe area and the ion saturation current of 1 mA, the density of the plasma for this particular discharge is $2 \times 10^{17}/m^3$.

4. Conclusion

With the use of the tri-axial cable and connector configuration, the capacitive effect reduces drastically for the high –frequency sweep which is further removed by using one dummy difference amplifier in the electronics system. Thus, the effect of the change in the bias voltage and the capacitive effect are completely nullified at the output of the INA. The time dependent current-voltage characteristic is obtained using the Langmuir probe by sweeping the probe at a frequency of 10 kHz with 10 μs rise time.

Acknowledgement

I would like to extend my sincere thanks to Electronics group of Institute for Plasma Research (IPR) and all the team members of the SYMPLE experiment group for their technical and experimental support who enabled me to do the verification of the designed electronics in the real system.

References

[1] Tang E, Xiang S, Yang M ,Li L 2012, *Plasma Science and Technology,* **14** *No.8*
[2] Cheetham D, Davidson L, Jakobsen J, Lund T, and Rayner J P 1997, *Review of Scientific Instruments* **68** *3405*
[3] Lobbia R B and Gallimore A D, *Physics of Plasmas* ***17****, 073502 _2010*

IOP Conf. Series: Journal of Physics: Conf. Series **823** (2017) 012020 doi:10.1088/1742-6596/823/1/012020

The Determination of Plasma Radial Shafranov Shift (ΔR) and Vertical Shift (ΔZ) experimentally using Magnetic probe and Flux loop Method for SST-1 Tokamak

Subrata Jana[1], Jasraj Dhongde[1], Harish Masand[1], Subrata Pradhan[1], Sameer Kumar[1] and SST-1 Team[1]

[1]Institute for Plasma Research, HBNI, Bhat, Gandhinagar -382 428, Gujarat, India

E-mail: subrata.jana@ipr.res.in

Abstract. The radial Shafranov shift (ΔR) and vertical shift (ΔZ) has been calculated for steady state superconducting tokamak (SST-1) experimentally using magnetic probes and flux loops. The SST-1 plasma at the present phases of operations is circular in shape and leans against the limiters. The radial and vertical shift formulated from Shafranov equation have been used for computation. We have selected best possible pairs of flux loops and magnetic probes location for the measurements. The comparison of results obtained from these two methods also has been performed for numerous numbers of shots for repeatability and reliability purpose. We have observed that the results of both the methods are in good agreement for almost all SST-1 experiment campaigns. Since the control of plasma position plays an important role in plasma confinement and optimizing tokamak operations, the computed ΔR and ΔZ could later on be used as a control parameter for feedback control of plasma position in SST-1.

1. Introduction

Steady state superconducting tokamak (SST-1) [1, 2] has been commissioned since June 2013 and is in operations since then. SST-1 produces circular ohmic plasmas having maximum currents of ~ 100 kA in a central field of 1.5 T for a maximum duration of ~ 400 ms. Experimental efforts are presently underway for extending the duration of SST-1 plasma beyond 1 second. SST-1 being a highly complex and multivariate device, it necessarily requires inputs from several sensor signals that will be computed upon using suitable algorithms for generating feedback signals for actuators. The study carried out would be helpful in incorporating robust and precise feedback control system and in turn obtaining long duration confined plasma in SST-1.

In ohmically heated low beta ($\beta < 1$) tokamaks, the plasma equilibrium is achieved by the controlled vertical field for radial force balance of outward hoop force. In force balance problem, the opposite force may not be equal and therefore plasma intends to shift inward or outward and experiences a radial Shafranov shift (ΔR) and vertical shift (ΔZ). It is considered as one of the critical problem for tokamak plasma control and equilibrium studies. The Grad-Shafranov analytical solution and different experimental methods [3-7] are available to measure the plasma horizontal shifts. This paper describes calculation of radial Shafranov shift (ΔR) and vertical shift (ΔZ) in section 2 and 3 followed by result, analysis and conclusion.

IOP Conf. Series: Journal of Physics: Conf. Series **823** (2017) 012020 doi:10.1088/1742-6596/823/1/012020

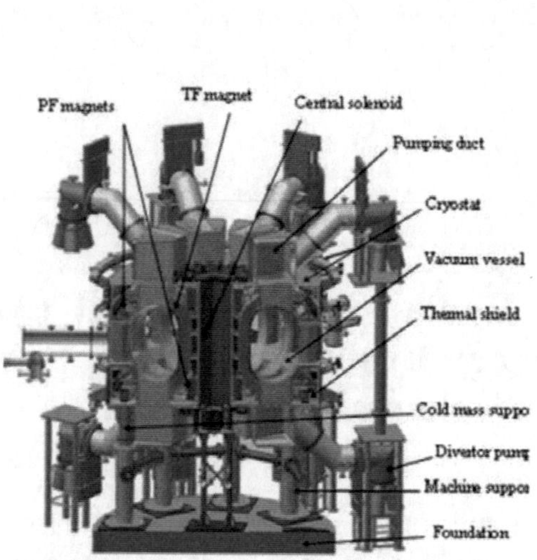

Figure 1. 3D cut view of SST-1 Machine.

Table 1. Parameters of SST-1 Tokamak.

Parameters	Values
Major radius (R_0)	1.1 m
Minor radius (a_0)	0.2 m
Toroidal Field	1.5 T
Plasma current (I_P)	~100 kA
Plasma Duration	~400 ms
surface area of VV	75 m^2
Plasma species	Hydrogen
Operating pressure range	5.0×10^{-6} mbar

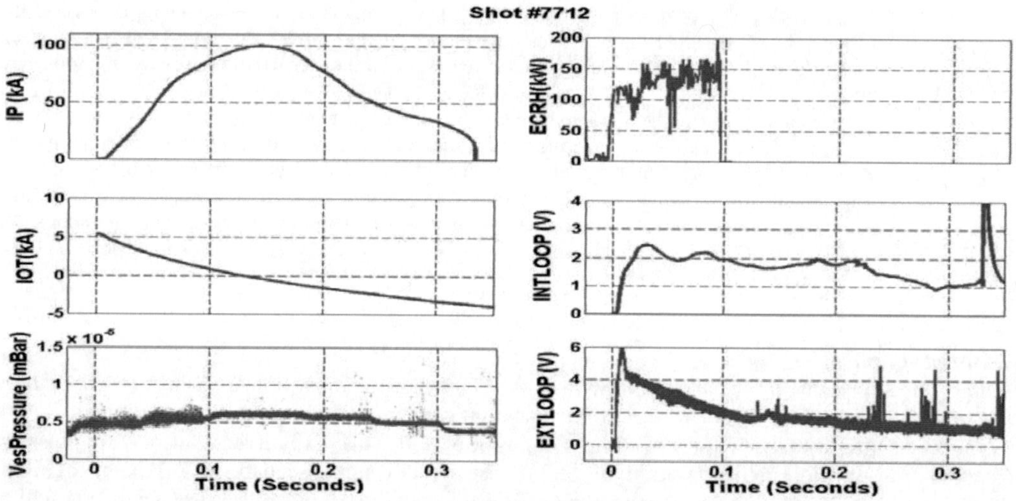

Figure 2. (LEFT) Plasma current (I_P), Central solenoid current (I_{OT} and Vessel pressure (RIGHT) ECRH Power, Internal loop voltage and External loop voltage for SST-1 shot # 7712

SST-1 plasma formation is dependent on synchronization of its different systems and sub systems. The primary magnetic configuration is provided by Super-Conducting Magnet Systems (SCMS) comprising of sixteen superconducting D-shaped Toroidal Field (TF) magnets and nine superconducting Poloidal Field (PF) magnets. An air core Ohmic Transformer (OT) with ECRH system is used for the pre-ionization, break-down and initial current start-up. A pair of resistive vertical field magnets outside the cryostat are placed symmetrically around the mid plane to provide the initial equilibrium. The particle environment within the plasma chamber is maintained using gas puffing. Figure.1 shows various components of SST-1 machine shell. In figure 2, plasma current (I_P), central solenoid current (I_{OT}), vessel pressure, ECRH Power, internal loop voltage and external loop voltage for specimen shot # 7712 has been shown.

2. Radial Shift and Vertical shift Methods

2.1. Radial Shafranov shifts Calculation

For determination of the radial plasma position, we have considered both installed magnetic flux loop and magnetic probes. The positions of the flux loops and probes [8] in SST-1 machine used for this method has been presented in figure 3 and figure 4 in a cross sectional view.

Radial shafranov shift (ΔR_p) can be written as equation (1) and have been computed from magnetic probes [6, 9].

$$\Delta R_p = \frac{a_0^2}{4R_0}\left\{\left(\frac{a^2}{a_0^2}-1\right) - 2\ln\frac{a_0}{a}\right\} + \frac{\pi a^2}{2\mu_0 I_P}\left\{\Delta B_\theta\left(1-\frac{a_0^2}{a^2}\right)+\Delta B_r\left(1+\frac{a_0^2}{a^2}\right)\right\} \quad (1)$$

where, a_o, a, R_0, ΔR_p are the plasma minor radius (0.2m) ,chamber minor radius(0.35m), major radius, shift using magnetic probes and $\Delta B_\theta = B_\theta(\theta = 0) - B_\theta(\theta = \pi)$ and $\Delta B_r = B_r\left(\theta = \frac{\pi}{2}\right) - B_r\left(\theta = \frac{3\pi}{2}\right)$.

So for this we have installed four magnetic probes as per Fig.4. The Tangential component B_θ have been measured using probe located at θ=0 and θ=π. The normal component (B_r) is measured using probes located at $\theta = \frac{\pi}{2}$ and $\theta = \frac{3\pi}{2}$.

Using flux loop [3, 4] method we have calculated plasma Radial shafranov shift (ΔR_l) as

$$\Delta R_l = \frac{a_0}{R_0 \mu_0 I_P \cos\theta}\Delta\psi \quad (2)$$

where, $\Delta\psi = \psi_{P_{out}} - \psi_{P_{in}}$ and $\psi_{P_{out}} = \psi_{out} + B_o \Delta S_o$ and $\psi_{P_{in}} = \psi_{in} + B_i \Delta S_i$

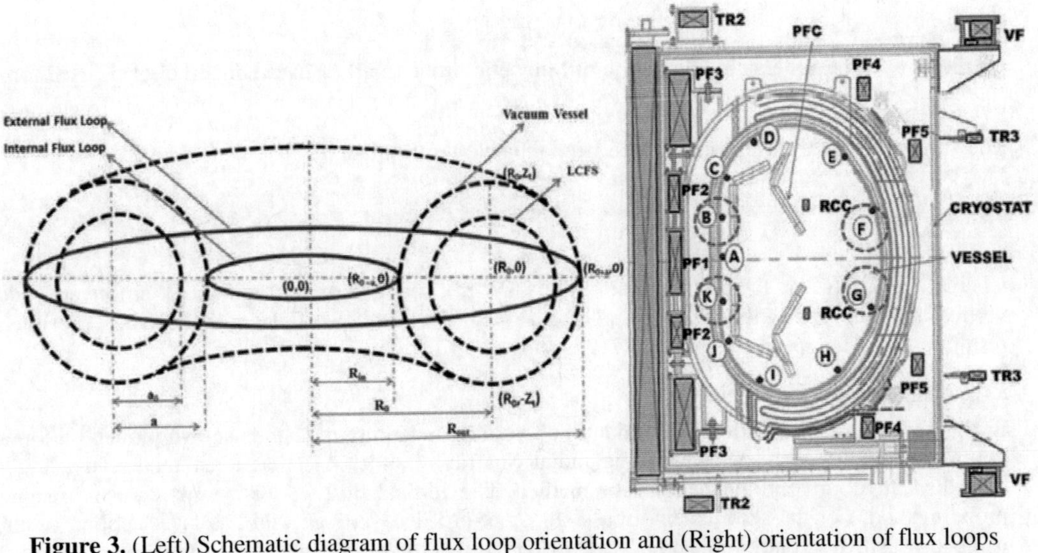

Figure 3. (Left) Schematic diagram of flux loop orientation and (Right) orientation of flux loops in SST-1 machine (marked in red dotted circle is used for measurement).

The ψ_{out} and ψ_{in} have been measured using outer and inner flux loop respectively. The B_o and B_i are the average magnetic field between outer and inner flux loops and plasma surface (LCFS) respectively obtained from magnetic probes, ΔS_i is the intervening area for internal loop and ΔS_o is the intervening area for external loop as per figure 3 (Left) and experimental result have been observed as per figure 5. We have used two set of flux loops (upper and lower) for radial shafranov shift measurement in SST-1. The lower flux loop (G, K) and upper flux loop (F, B) used are located as shown in figure 3 (Right) schematic diagram. According to Faraday's law, outputs of all magnetic diagnostics are proportional to the derivatives of the magnetic flux that pass through them. Therefore, the output of flux loops and magnetic probes signals are integrated after some compensation. A compensation technique is used to compensate central solenoid and VF current field contributions on probe and flux loop diagnostics signals. Necessarily, the compensation signals are the vacuum shot signals for the given ohmic and vertical field coil current profiles.

2.2. Vertical shifts Calculation

For determination of the vertical plasma position, we have used magnetic probes, their location and orientation is described in figure 4.

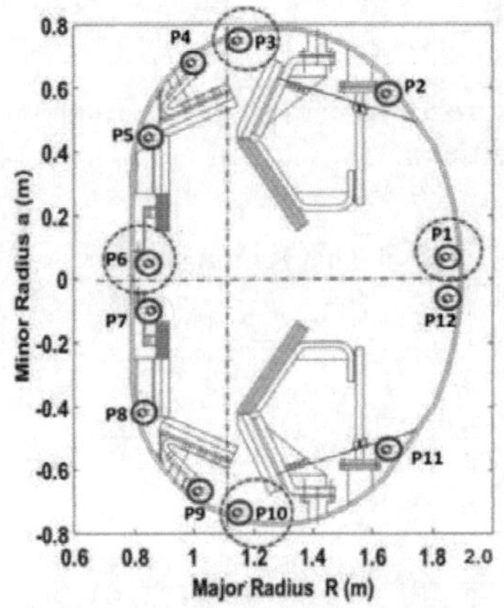

Figure 4. The Schematic orientation of Magnetic Probe (marked in red dotted circle is used for measurement).

The vertical shift (ΔZ) can be measured from the below formulation [6]

$$\Delta Z = \frac{\pi a^2}{\mu_0 I_P} \Delta B_{2\theta} \tag{3}$$

where a_o, a, R_0, ΔZ are the plasma minor radius (0.2 m), chamber minor radius (0.35 m), major radius, vertical shift using magnetic probes and $\Delta B_{2\theta} = B_\theta(\theta = 3\pi/2) - B_\theta(\theta = \pi/2)$. The experimental result observed for specimen shot no 7712 is shown in figure 5.

3. Results and Analysis

In figure 5, the plasma current (I_P) and applied vertical field current (I_{VF}) is shown plotted on top and bottom of the figure respectively. The radial Shafranov shift (ΔR) has been measured using the described flux loops and magnetic probe method. The vertical shift (ΔZ) is computed using magnetic probe method. For the shot under discussion, a pre-profiled vertical field has been applied (without using any feedback control method). It was observed that the plasma gets formed initially inboard side and maintains a stable position for some time before moving towards outboard side due to the inadequate availability of vertical field at the end. The pre-profiled VF (I_{VF}) reference signals take care of the vacuum vessel penetration time that is ~12 to 13 ms.

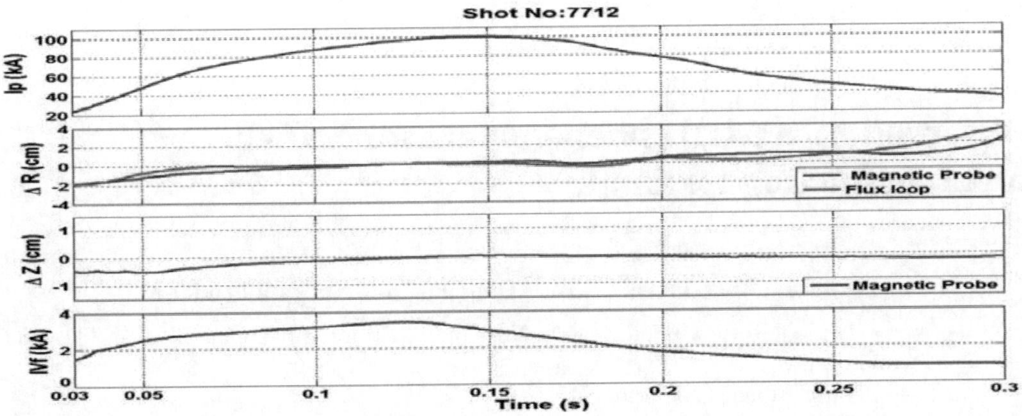

Figure 5. Plasma current (I_P), Plasma Radial shift (ΔR), Plasma Vertical shift (ΔZ) and Vertical Field current (I_{VF}) for SST-1 Shot No: 7712.

4. Observations and Conclusions

The Radial Shafranov shift measured using the flux loops method and magnetic probes methods are in good agreement, the results obtained are consistently repeatable in most of the shots and hence said to be reliable. To verify the results, the comparison and repeatability test with other diagnostics such as imaging was also carried out. The same trends of plasma movement were found between magnetic and imaging diagnostics, more details can be found in reference [10]. The studies carried out here on ΔR and ΔZ computations, later will be used as one of the vital input parameter in plasma position feedback control system to achieve long duration plasma discharges in SST-1.

Acknowledgement

We acknowledge the contribution of Mr. Kirit B Patel, Mr. Praveenlal Edappala and Dr. Santanu Banerjee for valuable suggestions related to EM diagnostics, electronic circuitry and fast imaging diagnostics in SST-1 machine. We also like to thank Mr. Debashis Ghosh, Mr. Hitesh Patel, Mr. Yuvakiran Paravastu and all the member of SST-1 operation control team for valuable suggestions and helps to carry out this work.

References

[1] Pradhan S and Team SST-1 mission 2010 *J. Fusion Res. Series* **9** 650
[2] Pradhan S et al. 2015 *Nuclear Fusion* **55** 104009
[3] Rahimirad A et al. 2010 *Journal of Fusion Energy* ***29(1)*** 73
[4] Salar Elahi A and Ghoranneviss M 2012 *Journal of Nuclear and Particle Physics* **2**(6) 142
[5] Shafranov V D 1963 *Journal of Nuclear energy* **5** 251
[6] Salar Elahi A and Ghoranneviss M 2010 *Brazilian journal of physics* **40**(3) 323
[7] Zheng S B,Wootton A J and Solano E R 1996 *Physics of plasmas* **3** 1176
[8] Raju D et al. 1997 *Fusion Engineering and design* **34-35** 717
[9] Ninomiya H and Suzuki N 1982 *Japanese journal of applied physics* **21**(9) 1323
[10] Jana S et al. 2016 *Accepted in Fusion Engineering and Design with title "Magnetic flux surfaces and radial Shafranov shifts(ΔR) in SST-1 Tokamak Plasma"*

IOP Conf. Series: Journal of Physics: Conf. Series **823** (2017) 012021 doi:10.1088/1742-6596/823/1/012021

Observation of MHD phenomenon for SST-1 superconducting tokamak

Manisha Bhandarkar[1], Jasraj Dhongde[1] and Subrata Pradhan[1]

[1]Institute for Plasma Research, Near Indira Bridge, Bhat, Gandhinagar 382 428, India

E-mail: manisha@ipr.res.in

Abstract. Steady State Superconducting Tokamak (SST-1) is a medium size Tokamak (major radius = 1.1 m, minor radius = 0.2 m) and is operational at the Institute for Plasma Research (IPR), India. In the last few experimental campaigns SST-1 has successfully achieved plasma current in order of 60-70kA and plasma duration in excess of ~ 500 ms at a central magnetic field of 1.5T. An attempt has made to study the behavior of the magneto-hydrodynamic (MHD) activity during different phases of plasma pulse which leads to major/minor disruptions, its present modes (poloidal/toroidal mode number i.e. m = 2, n = 1) impact on plasma confinement and signature of lock mode and its frequency in the SST-1 plasma using experimental data from Mirnov signals. Observed MHD phenomenon has also been correlated with other diagnostics (i.e. ECE, Density, Soft X-Ray etc.) and heating system (ECRH) for the recent campaigns of SST-1.

1. Introduction

Steady State Superconducting Tokamak (SST-1) is a medium size Tokamak (major radius R = 1.1 m, minor radius = 0.2 m, toroidal magnetic field B_t = ~1.5 T) in operation after successful refurbishment since June 2013. [1] During last few campaigns, experiments at SST-1 has successfully produced circular ohmic plasmas having current ~102 kA and plasma duration in excess of ~ 400 ms at the central toroidal magnetic field (TF) of 1.5 T. The plasma is generated by a resistive central solenoid (ohmic transformer) with the pre-ionization being provided by a 42 GHz Electron Cyclotron source.

Analysis of the experimental data from prior/post plasma facing components (PFC) campaigns shows that major disruptions in the plasmas are always connected with MHD phenomenon. MHD instability is limiting factor for improving plasma. Many tokamaks use Mirnov coils for the detection of plasma instability (e.g. mode locking activity, saw teeth activity etc.) during the discharges. SST-1 has Mirnov Diagnostic consists of an array of Mirnov coils to measure the fluctuations in magnetic field at the plasma edge. Structurally Mirnov coil arrays are distributed in both Toroidal and Poloidal directions to gather the information on mode structure, mode number and mode frequency during disruption. In SST-1, there are arrays of eight in-vessel poloidal Mirnov coils and four in-vessel toroidal Mirnov coils are installed to gather the information regarding MHD activities.

At SST-1, MHD activity has been observed during different phases (Ramp-Up, Flat-Top and Ramp-down) of plasma pulse which leads to major/minor disruptions. The analysis includes multiple Mirnov bursts with increasing amplitude and locked modes (phase locking between modes) which leads to degradation of confinement and discharge terminates permanently [5]. The analysis also includes to study its present mode (poloidal/toroidal mode number i.e. m = 2, n = 1) and its impact on plasma confinement, Fast Fourier Transform (FFT) for understanding dominant frequency component, Wavelet spectrogram for time resolved frequency for cross confirm with dominant mode frequency, Singular Value Decomposition (SVD) analysis for spatial structure of mode number, magnetic island width (W) and rate of island growth (dW/dt) during lock mode duration [3,8].

In this paper, the major disruption caused by mode lock instability has been studied. It has observed that during mode locking, the frequencies of the mode rotation gradually slow down while the MHD instabilities grow up, and then mode lock appears when the rotation stops suddenly. Due to the mode locking, as the plasma cannot flow through a large island at any significant rate, it is leads to a stopping of the plasma rotation at the resonant surface and an appreciable slowing down of the plasma as a whole due to viscous coupling [2,7]. The modes do not lock at every crash, but when they do, this sequence is extremely reproducible and obviously harmful to the plasma confinement. [5] Some times after a minor disruption the plasma pressure and rotation recovers, but MHD activity may reappear and slowing-down process starts again. This may sometimes leads to a series of minor disruptions. [6]

2. Lock mode observation at SST-1 plasma

Observation started with analyzing mode locking duration for different plasma pulses, found out its present mode (poloidal/toroidal mode number i.e. m = 2, n = 1) using SVD [8] and its co-relation with drift frequencies.

For ohmic heated plasmas, the rotation frequency of magnetic perturbation is about the electron drift frequency f_{De} [2, 4].

$$f \approx f_{De} \propto \frac{kT_{e0}}{eB_t a^2} \tag{1}$$

where k is the Boltzmann constant and T_{e0} is the central electron temperature.

At SST-1, mode lock instability has been observed at different phases of the plasma pulse (ramp-up, flat-top and ramp-down), which leads to degradation of confinement or majorly terminates the discharges abruptly. In particularly, it has observed that they more dominant to appear in low density discharges [2, 4].

Statistical observation on mode lock instabilities encourages studying the characteristics of field errors influencing the mode rotations and its co-relation with drift frequencies [2,3] for the plasma at SST-1. Rotational frequencies have been calculated before and during mode lock occurrence to find out its drift frequency.

To understand the conditions leading to locked mode instabilities, onsets of disruptions are also correlated with other SST-1 diagnostics. (e.g ECE Radiometer, Density from Homodyne diagnostic, Soft-Xray etc.).

IOP Conf. Series: Journal of Physics: Conf. Series **823** (2017) 012021 doi:10.1088/1742-6596/823/1/012021

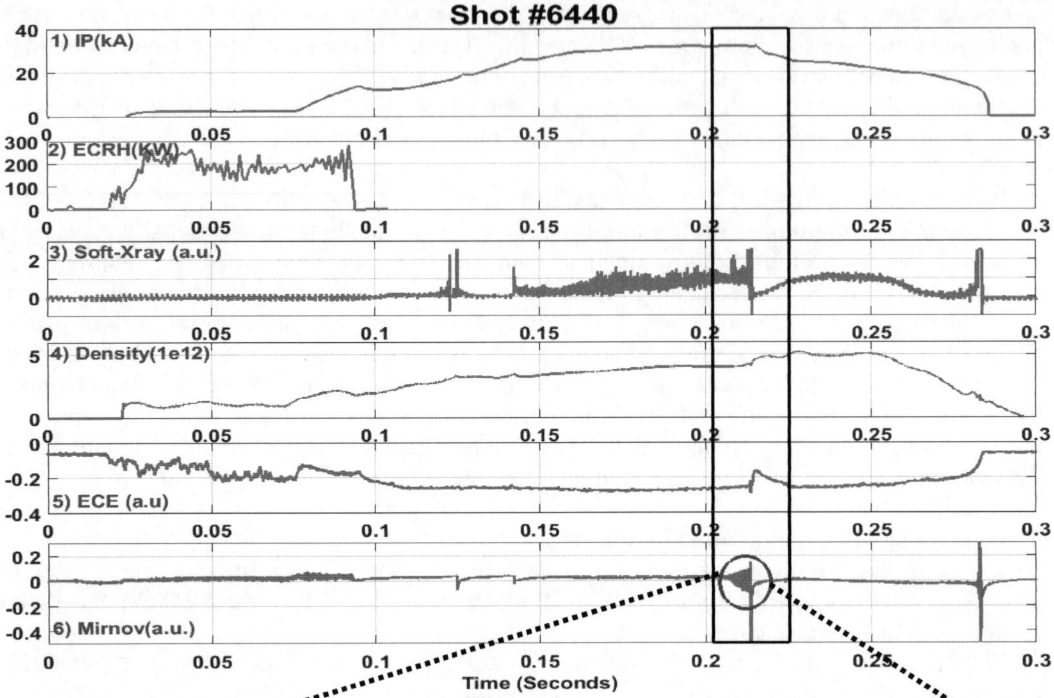

Figure-1

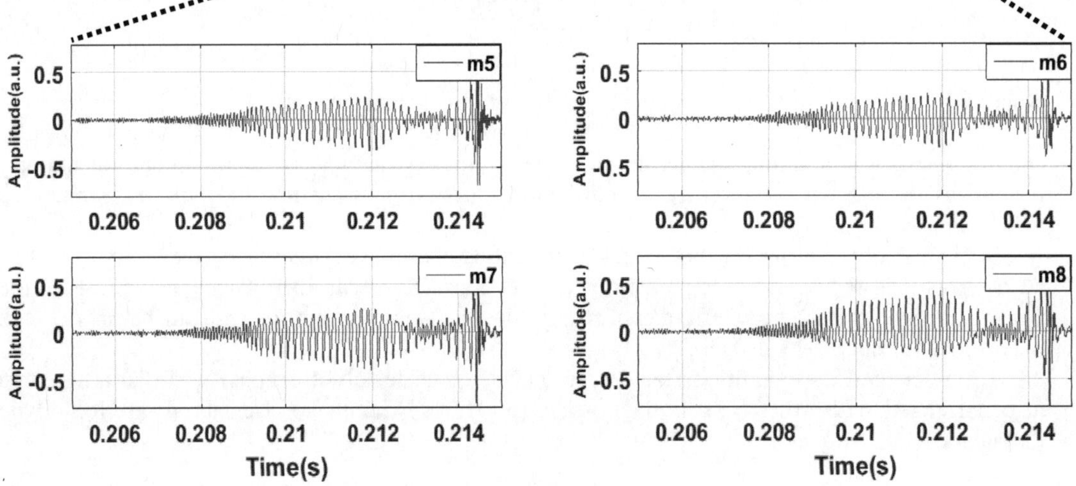

Figure-2

Figure 1. Shot#6440 (1) Plasma Current kA (Ip), (2) ECRH power (kW), (3) Soft X-ray signal (4) Density measurement from Homodyne(1e12) diagnostic (5) ECE Signal (6) Mirnov signal shows lock mode bursts during ramp-down phase of Plasma Current (Ip)

Figure 2. Zoomed time range of Mirnov signal during lock mode duration at ramp-down phase: Lock mode bursts during the 205-215ms duration for poloidal Mirnov coils

In figure 1, Mirnov signal observes the lock mode bursts at ramp-down phase during 205-215 ms. From many experimental data at SST-1, it has observed that sometimes minor dirsuption occurs at ramp-up phase but plasma pressure and rotation recovers and plasma current sustains. In this typical shot # 6440, major MHD activity has appears during ramp-down phase (during 205-215 ms), which have resulted in degration of confinement followed by major disruption.

It has observed that during major disruption, multiple Mirnov bursts are observed with increasing amplitude (figure 2), which leads to frequencies of the mode rotation gradually slows down and results in degration of confinment. As shown in figure 1, during mode locking duration (205-215 ms), major disruption occurs and it has co-related with the reduction/degration of temperature/density (from Homodyme diagnostics)/ECE signals with plasma current.

3. Analysis

In typical shot # 6440, from the lock mode duration (205-215 ms), found out its present mode using Singular Value Decomposition (SVD)[8] shows m = 2 (figure 4), predominantly mode (poloidal). As shown in figure 3, dominant frequency component is ~1-15 kHz; in the same figure Wavelet spectrogram also shows the same for time for the resolved frequency.

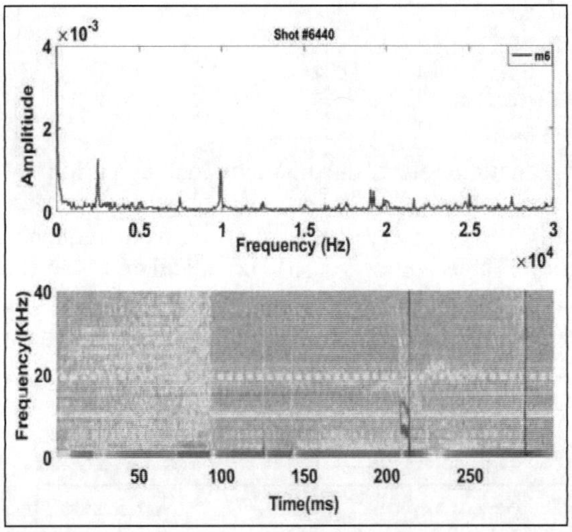

Figure 3. FFT & spectrogram of Mirnov (m6)

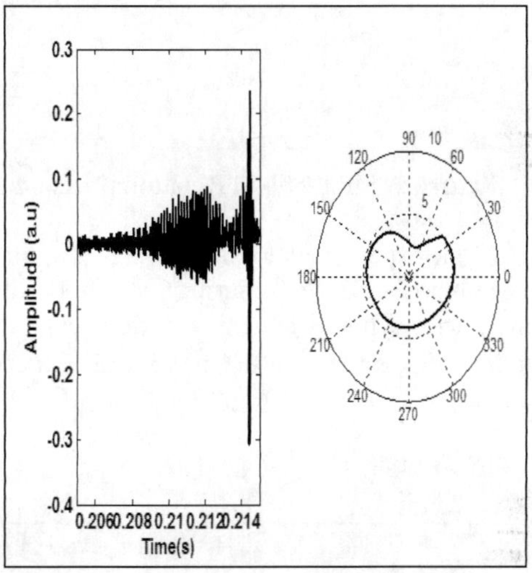

Figure 4. Shot # 6440-SVD shows m = 2

Figure 5 shows the rotational frequencies before and during the mode lock duration and found out its drift frequency. For the representative shot # 6440, it has observed that rotational frequency has reduced to ~ 37 % during lock mode duration (205-215 ms).

In figure 5, in 1st phase (Before lock mode), the Mirnov signal representing the magnetic perturbations are rather a harmonic, indicating a gradual slowing down of the plasma. In 2nd phase (or during island growth), the frequency drops an harmonic, indicating that the mode no longer rotates with constant angular velocity. This behaviour can be due to the combination of two different forces acting on a rotating magnetic perturbation (1) due to the interaction with the external error field or (2) due to the eddy currents induced in the resistive wall [7]. As shown in figure 5, reduction of rotational frequencies during lock mode event is ~ 37% for typical shot # 6440.

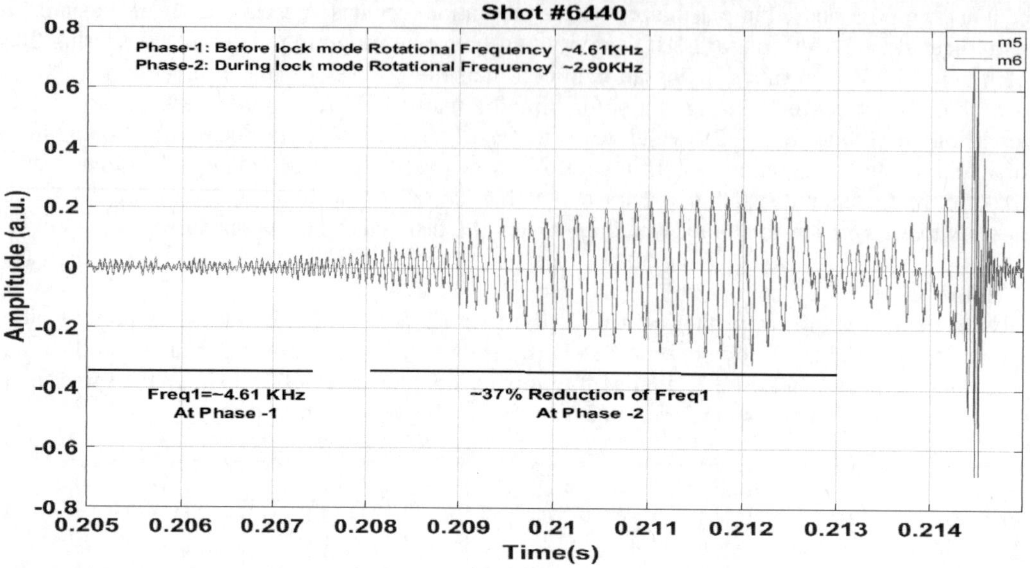

Figure 5. Shot # 6440 Rotational frequency reduction before/ during mode locking period

From experimental data of SST-1, statistically it has observed that the rotational frequency has reduced ~ 32-48 % during lock mode duration. Table 1 shows f_{de} (kHz) calculated is in the range with what had achieved during $f_{experimetnal}$ kHz) in many shots, shows the theoretic predications is very consistent with the experimental results, which are in range with other similar contemporary devices (e.g. COMPASS-C, DIII-D, HL-1M) [2,4].

Table1. SST-1's f_{de} calculated & $f_{experimetnal}$ is matching with other contemporary devices.

Tokamak	T_{e0} (eV)	B_t (T)	R_0 (m)	a (m)	f_{De} (kHz) Calculated	$f_{exprimental}$ (kHz)
COMPASS-C*	~ 600	1.1	0.56	0.18	~ 15	~ 13
DIII-D*	~ 920	1.3	1.67	0.67	~ 1.7	~ 1.6
HL-1M*	~ 700	2.1	1.02	0.26	~ 3.2	~ 3.1
SST-1#	~ 280	1.5	1.1	0.2	~ 4.7#	~ 4.61

*[2, 4]

for representative shot#6440, Campaign 12.

4. Summary and conclusion

The lock modes in the SST-1 have obvious similarities with other devices. It occurs at any phase (ramp-up, flat-top or ramp-down) during the discharge and lead to degrade the confinement. In major disruption it remains locked until termination. The analyzed results shows major disruptions were observed within a few milliseconds after the occurrences of the mode lock instabilities. Generally it appears with low density discharges. The rotational frequency has reduced to ~ 32-48 % during mode lock. The study is going on to suppress the mode locking instability, inherence error field and to increase the plasma density.

Acknowledgement
Authors would like to thanks SST-1 team & SST-1 diagnostics group for their necessary help and support.

References
[1] Pradhan S et al. 2015 *Nuclear fusion* **55** 104009 (10pp)
[2] Qingwei Y et al. 2002 *Brazilian Journal of Physics* **32** 160
[3] Asif M et al. 2005 *Physics Letters A* **342** 175
[4] La Haye R J, Fitzpatrick R et al. 1992 *Physics of Fluids B: Plasma Physics* **4** 2098
[5] Almagri A F, Assadi S, Prager S C et al. 1992 *Phys. Fluids B* **4** 4080
[6] Vries P C de et al. 1996 *Plasma Phys. Control. Fusion* **38** 467
[7] Zohm H 2014 *Magnetohydrodynamic Stability of Tokamaks* (Wiley-VCH: Garching, Garmany) p 146-152
[8] Dhongde J, Bhandarkar M, and Pradhan S 2016 *Fusion Engineering and Design* **108** 77

IOP Conf. Series: Journal of Physics: Conf. Series **823** (2017) 012022 doi:10.1088/1742-6596/823/1/012022

Thermal response of actively cooled tungsten monoblocks to inhomogeneous surface heat loading

Y Patil[1], S S Khirwadkar[1] and D Krishnan[1]

[1]Institute for Plasma Research, Bhat, Gandhinagar-382428, India

Email: ypatil@ipr.res.in

Abstract. Tungsten (W) monoblocks of the Plasma Facing Unit (PFU) are exposed to the inhomogeneous heat loads arises due to exponentially decaying power flux along Scrap of Layer (SOL) away from the separatrix as well as fabrication tolerances and misalignment of divertor targets in poloidal and toroidal directions. Thermal and structural analysis of a tungsten monoblock under such inhomogeneous heat loads are need to be investigated. The paper presents the results of Finite Element Analysis (FEA) performed on a tungsten monoblock loaded by inhomogeneous surface heat flux 10 MW/m^2 on the plasma wetted area of tungsten monoblock and 4 MW/m^2 on the shaded area with magnetically shadowed fraction of the monoblock length (L_{shad}) variation from 20% to 95% by a step of 15%. The other case is the total incident heat flux 20 MW/m^2 on the plasma wetted area of tungsten monoblock and no heat flux on the shaded area of a tungsten monoblock with L_{shad} variation from 20% to 95% by a step of 15%. FEA concluded that when L_{shad} is increased from 20% to 95% the total power to the monoblock decreases and the peak surface temperature decreases.

1. Introduction

Plasma Facing Unit (PFU) of the ITER Divertor vertical target consisting tungsten monoblocks bonded with the Copper-Chromium-Zirconium (CuCrZr) alloy coolant tube with Oxygen Free High Conductivity (OFHC) copper interlayer. Each vertical target of the ITER divertor consists of two sets of PFUs on separate support plates therefore neighboring monoblocks on each half of the target are separated by a 0.5 mm gap, with 3 mm gap between the two halves. Vertical target misalignment between neighboring PFUs will be source of leading edge for grazing incidence field line angles which will occur in the high heat flux strike point regions. These edges will receive a significant amount of inhomogeneous heat loads during steady state as well as transient scenarios of ITER operation. Inhomogeneous heat loads will be affected by tungsten monoblock due to varying magnetic shadowed length or fraction 'L_{shad}'. Within divertor cassette magnetically shadowed length of the tungsten monoblock (L_{shad}) can vary 25% in the case of perfect alignment monoblocks to 45% for worst misaligned monoblocks. The worst situation is faced by inter-cassette gap and where a much larger radial step might be introduced (~3.5mm) to protect leading edges. In that case, the full shadowing of the first monoblock and partial shadowing of the second monoblock can occur with L_{shad} up to 90% on the second monoblock. Surface area of tungsten monoblock exposed by total incident heat flux is called as 'plasma wetted area'. Surface area of tungsten monoblock exposed to magnetically shadowed fraction of the incident heat flux is called as 'shaded area'.

The ITER divertor will be operating in a partial detached plasma condition. The fraction of heat flux to the vertical target comes from radiation and Charge Exchange Neutral (CXN). In that case plasma, wetted area of a monoblock receives the total incoming heat flux, while a shaded area of tungsten monoblock receives only 40% of total heat flux. During slow transient negligible or zero amount of heat flux on the shaded area of tungsten monoblock while plasma wetted area of tungsten monoblock receives the total incoming heat flux [1]. Inhomogeneous surface heat flux loading of actively cooled tungsten monoblock with total incident heat flux 10 MW/m^2 on the plasma wetted area of tungsten monoblock and 4 MW/m^2 (40% of total incident heat flux) on the shaded area of a tungsten monoblock with L_{shad} variation from 20% to 95% by step 15% is referred as 'Inhomogeneous heat load study - I' in the further subsections of a research paper. Total incident heat flux 20 MW/m^2 on the plasma wetted area of tungsten monoblock and no heat flux on the shaded area of tungsten monoblock with L_{shad} variation from 20% to 95% by step 15% is referred as 'Inhomogeneous heat load study - II' in the further subsections of the research paper. These inhomogeneous heat load studies were simulated by Finite Element Analysis (FEA) using COMSOL Multiphysics software Heat Transfer and Thermal Stress module. Figure1 summarize details of inhomogeneous heat load study - I & II applied on the tungsten monoblock. Expected surface temperatures and thermal stresses during spatially inhomogeneous high heat flux loading were calculated using FEA and summarized in subsequent sections of present research paper. The main objective of this work is to study the thermal response of actively-cooled tungsten monoblocks subjected to spatially inhomogeneous heat loads using the High Heat Flux Test Facility (HHFTF) available at Institute for Plasma Research (IPR), India. Therefore, present FEA gives the initial guess for inhomogeneous high heat flux tests to be conducted using HHFTF at IPR. HHFTF, IPR is capable to simulate the steady state and transient heat loads. Previous high heat flux tests carried out on the straight as well curved tungsten monoblock type mock-ups using HHFTF, IPR is reported in the research publications [2-6]. The portable high temperature, high pressure water circulation system of HHFTF, IPR provides maximum water flow rate up to 85 liters per minute and maximum inlet temperature 90 °C.

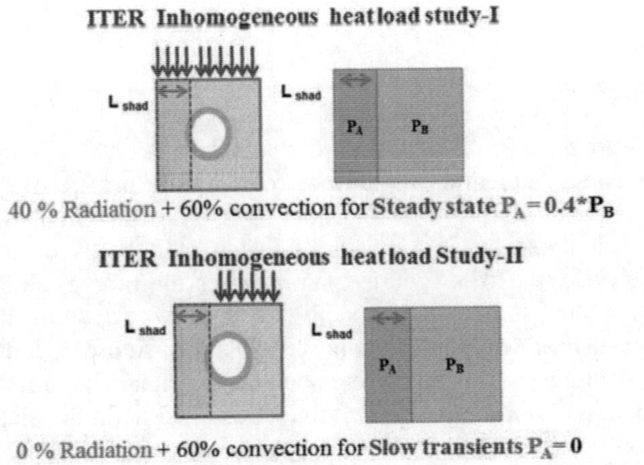

Figure 1. Schematic digram of inhomogeneous heat load study - I & II

2. Finite Element Analysis (FEA) model

Tungsten monoblock 28 × 28 × 12 mm^3 with OFHC copper layer thickness 1 mm, CuCrZr alloy tube with inner diameter 12 mm & outer diameter 15 mm was used for FEA analysis. Figure 2 gives material and dimensional details of the tungsten monoblock. Commercial available FEA software COMSOL Multiphysics Heat transfer & Thermal stress module was used for the thermal and structural analysis of a tungsten monoblock loaded with inhomogeneous heat load study - I & II. The 3D model of tungsten monoblock consisted approximately 41,220 tetrahedral mesh elements. Material properties of a pure tungsten, OFHC copper layer and CuCrZr alloy are used from available literature reference [8, 9]. The cooling conditions were applied to inner boundaries of a CuCrZr tube in the form heat transfer coefficient. Heat transfer coefficient is calculated using Sieder / Tate correlation for the inlet temperature 70 °C and flowrate 85 liters per minute without assuming enhancement tape inside coolant tube. Heat transfer coefficient used for FEA is 46000 $Wm^{-2} K^{-1}$. Spatial varying heat load in the form of a heat flux applied to the top surface of a tungsten monoblock by parametric swept variation of a L_{shad} from 20 % to 95 % by steps of 15 %. Thermal stress analysis was carried out by assuming fixed constraint at the bottom surface of the monoblock. FEA is carried out for transient study step for 10 s ON & 10 s OFF time.

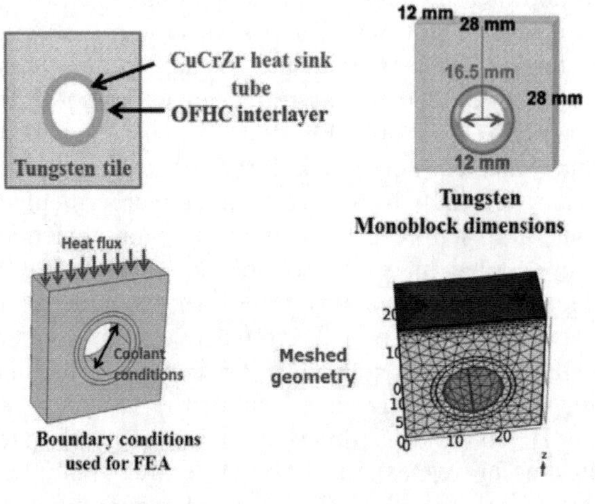

Figure 2. Geometric and mesh details of Tungsten monoblock used for FEA.

3. Results and discussion

3.1. Inhomogeneous heat load study - I

Figure 3 shows the calculated sequential 3D surface temperature profile of the tungsten monoblock during inhomogeneous heat load study - I for time 15 s. As illustrated in figure 3 peak surface temperature 1310 °C is calculated by FEA on the tungsten monoblock during inhomogeneous heat load study - I corresponds to L_{shad} 20%. Figure 4 summarizes the time evolution of averaged surface temperature on the edge of tungsten monoblock. Figure 5 shows stress on the tungsten monoblock during inhomogeneous heat load study - I for time15 s. As per figure 5, tensile or positive stresses followed by negative or compressive stresses were accumulated near the tungsten-copper interlayer in the direction of the heat flow. Maximum compressive stress 745 MPa is calculated by FEA for L_{shad} 20%. Ductile to brittle transition (DBTT) temperature range for polycrystalline pure tungsten is 200 °C - 300 °C [14] and it may be the reason of the compressive or negative stress found on tungsten monoblock near the copper interlayer.

3.2. Inhomogeneous heat load study - II

Figure 6 shows the calculated sequential 3D surface temperature profile of the tungsten monoblock during inhomogeneous heat load study - II for time 15 s. Figure 6 note that peak surface temperature 2669 °C, 2556 °C, 2382 °C, 2088 °C is calculated by FEA on tungsten monoblock corresponds to L_{shad} 20%, 35%, 50%, 65% respectively which is quite high and beyond the recrystallization temperature of the pure tungsten ~ 1500 °C. It concluded that the surface of the tungsten monoblock may undergo the recrystallization.

IOP Conf. Series: Journal of Physics: Conf. Series **823** (2017) 012022 doi:10.1088/1742-6596/823/1/012022

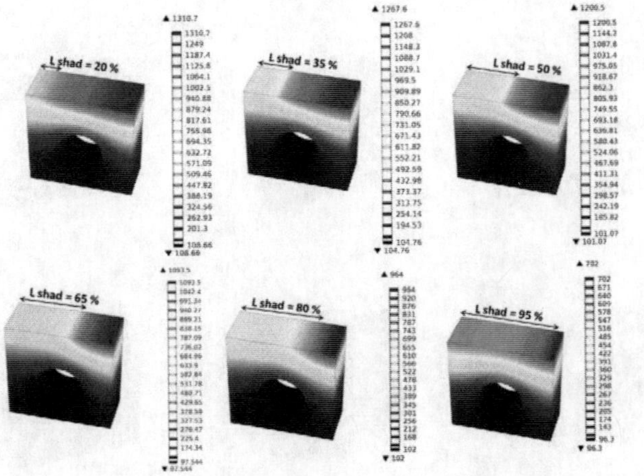

Figure 3. Sequential 3D surface Temperature profile of the tungsten monoblock for L_{shad} variation from 20% to 95% with a step of 15 % during Inhomogeneous heat load study - I.

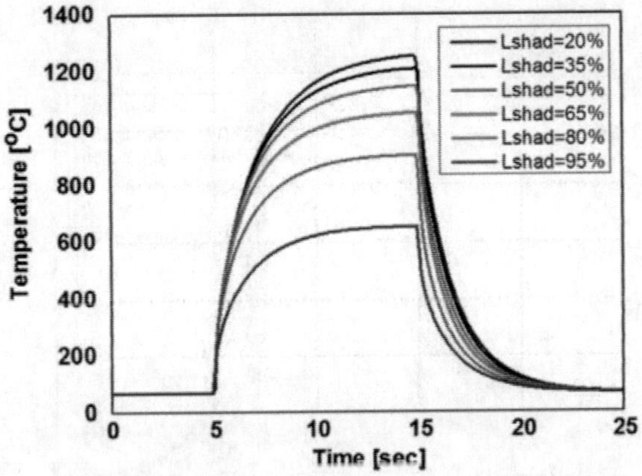

Figure 4. The time evolution of averaged surface temperature on the edge of tungsten monoblock for L_{shad} variation from 20% to 95% with a step of 15% during inhomogeneous heat load study - I.

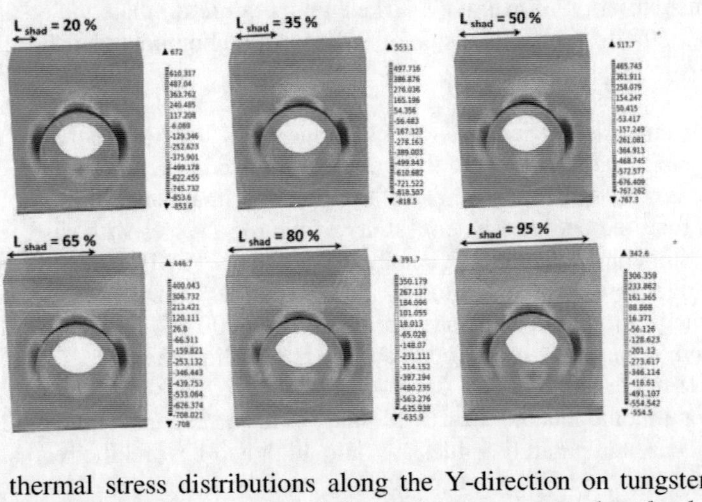

Figure 5. Equivalent thermal stress distributions along the Y-direction on tungsten monoblock for L_{shad} variation from 20% to 95% with a step of 15% during Inhomogeneous heat load study - I.

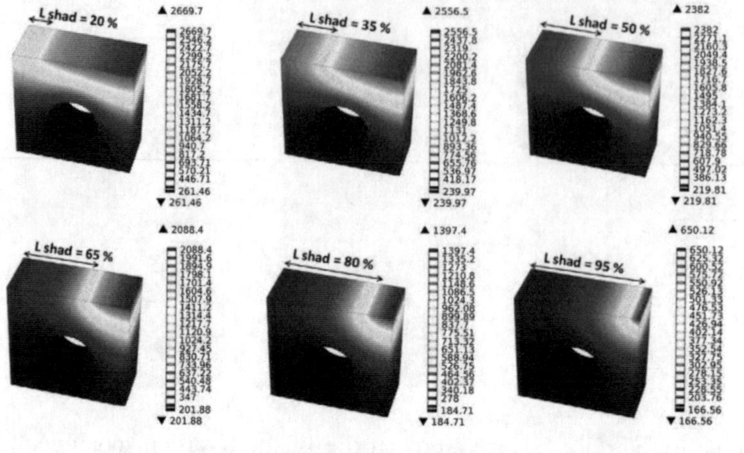

Figure 6. Sequential 3D surface Temperature profile of the tungsten monoblock for L_{shad} variation from 20% to 95% with a step of 15 % during Inhomogeneous heat load study - II.

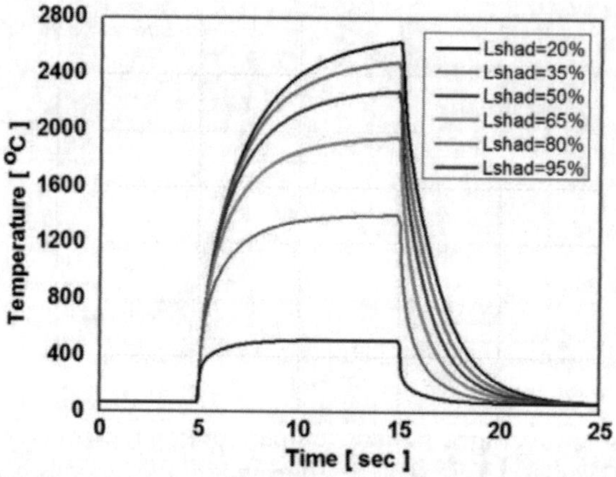

Figure 7. The time evolution of averaged surface temperature on the edge of tungsten monoblock for L_{shad} variation from 20% to 95% with a step of 15% during inhomogeneous heat load study - II.

4. Conclusions

1. Finite element analysis carried out for inhomogeneous heat load study - I & II concluded that when L_{shad} is increased from 20% to 95% the total power to the monoblock decreases and the peak surface temperature decreases. Therefore, during experiments calibration of surface measurement diagnostics is required because the emissivity of tungsten varies with surface temperature.
2. Peak surface temperature 1310 °C & compressive stress 745 MPa is found on tungsten monoblock during inhomogeneous heat load study - I for L_{shad} 20 %. Ductile to brittle transition (DBTT) temperature range for polycrystalline pure tungsten is 200 °C – 300 °C and may be the reason of the compressive or negative stress found around the cooling channel.
3. Peak surface temperature 2669 °C, 2556 °C, 2382 °C, 2088 °C is calculated by FEA on tungsten monoblock for inhomogeneous heat load study - II corresponds to L_{shad} 20%, 35%, 50%, 65% respectively. These temperature values are quite high and beyond the recrystallization temperature of the pure tungsten ~ 1500 °C. It concluded that the surface of the tungsten monoblock may undergo the recrystallization.

4. Inhomogeneous heat load study - I & II will be experimentally simulated using HHFTF at IPR by varying the scanning frequency electron beam as well as using actively cooled copper masks.

Acknowledgement
The authors would like to thank the Director of Institute for Plasma Research (IPR). The Authors would also like to acknowledge all the members of Divertor and First Wall Technology and Development Division at IPR for their support and help for various activities related to this work.

References

[1] Pitts R A, Carpentier S, Escourbiac F, Hirai T, Komarov V, Kukushkin A S, Lisgo, 2011 *J.Nucl. Mater.* **415** S957
[2] Patil Y, Khirwadkar S S, Belsare S M, Swamy R, Khan M S, Tripathi S, Bhope K, 2015 *NUKLEONIKA* **60** 285
[3] Patil Y, Khirwadkar S S, Belsare S M, Swamy R, Khan M S, Tripathi S, Bhope K, Krishnan D, Mokaria P, Patel N, Antwala I, Galodiya K, Mehta M, Patel T 2015 *Fusion Eng. Des.* **95** 84
[4] Patil Y, Khirwadkar S S, Belsare S, Swamy R, Tripathi S, Bhope K, Kanpara S 2015 *Phys. Scr.* **T167** 014020
[5] Patil Y, Khirwadkar S S, Krishnan D, Patel A, Tripathi S, Singh K P, Belsare S M, 2013 *J.Nucl. Mater.* **437** 326
[6] Sugandhi R, Swamy R, Khirwadkar S 2016 *Fusion Eng. Des.*
[7] Hirai T et al 2014 *Phys. Scr.T* **159** 014006
[8] Fukuda M, Nogami S, Guan W, Hasegawa A, Muroga T 2016 *Fusion Engineering and Design* **107** 44
[9] Muyuan Li , Jeong-Ha You 2015 *Fusion Engineering and Design* **101** 1
[10] Sieder E N, Tate G E 1936 *Industerial Engineering Chemistry* **28** 1429
[11] Antonella Li Puma, Richou M, Magaud P, Missirilian M, Visca E, Ridolfini V P 2013 *Fusion Eng. Des.* **88** 1836
[12] Crescenzi F, Roccella S, Visca E, Moriani A 2014 *Fusion Eng. Des.* **89** 985
[13] Missirlian M, Firdaouss M, Richou M, Languille P, Lecocq S, Lipa M 2013 *Fusion Engineering and Design* **88** 1793
[14] Erik Lassner Wolf-Dieter Schubert 1999 *Tungsten: Properties, Chemistry, Technology of the Element, Alloys, and Chemical Compounds* 20

IOP Conf. Series: Journal of Physics: Conf. Series **823** (2017) 012023 doi:10.1088/1742-6596/823/1/012023

Preparation of W/CuCrZr mono-block test mock-up using vacuum brazing technique

K Premjit Singh[1], S Khirwadkar[1], Kedar Bhope[1], Nikunj Patel[1] and Prakash Mokaria[1]

[1]Institute for Plasma Research, Bhat, Gandhinagar 382428, India

E-mail: kpsingh@ipr.res.in

Abstract. Development of the joining for W/CuCrZr mono-block PFC test mock-up is an interesting area in Fusion R&D. W/Cu bimetallic material has been prepared using OFHC Copper casting approach on the radial surface of W mono-block tile surface. The W/Cu bimetallic material has been joined with CuCrZr tube (heat sink) material with the vacuum brazing route. Vacuum brazing of W/Cu-CuCrZr has been performed @ 970°C for 10 min using NiCuMn-37 filler material under deep vacuum environment (10^{-6} mbar). Graphite fixture was used for OFHC Copper casting and vacuum brazing experiments. The joint integrity of W/Cu-CuCrZr mono-block mock-up of W/Cu and Cu-CuCrZr interface has been checked using ultrasonic immersion technique. The result of the experimental work is presented in the paper.

1. Introduction

Fabrication of the Plasma Facing Component (PFC) using Tungsten (W)-mono-block mock-up is a challenging area in Fusion R&D. The problem for joining of such dissimilar materials is due to thermal mismatches and metallurgical incompatibility. Various dissimilar materials joining have been discussed and attempted elsewhere for fabrication of the PFC of ITER like Tokamak application [1-4]. OFHC Copper material has been chosen as interlayer material in between the two dissimilar materials W and CuCrZr, has the advantages such as (i) minimizing the thermal stress, (ii) better wetting between W and Cu, (iii) better thermal conduction etc. Preparation of bimetallic W/Cu material have been made possible by several approaches such as (i) OFHC Copper casting in vacuum furnace / induction furnace under argon atmosphere, (ii) Melting and solidification using laser source (Laser cladding), (iii) Melting of solidification of OFHC Copper using Electron Beam heating source etc.

Vacuum brazing technique has been known for preparation of the neat, clean and sound joints with less alteration of the parent materials property as the brazing temperature has been carried out below the melting point of the base material. This technique (Vacuum brazing) has been acceptable approach in fusion research for the fabrication of the Plasma Facing component. In this paper, the methodology adopted for preparation of the PFC W mono block test mock-up using vacuum brazing technique will be primarily discussed.

2. Experimental Details

Pure Plansee W tiles (3 nos.) having dimensions of 30 mm (H) × 28 mm (W) × 10 mm (t) has been prepared from the W rod with wire-cut machining with the internal throughout hole of 17 mm (ID). The machined tiles are represented as W-Tile no.1, W-Tile no.2 and W-Tile no.3 respectively. The tiles were dimensionally checked, cleaned & inspected by contact probe ultrasonic testing (UT). No internal as well as surface defects were observed in the tiles. A list of materials used in the experiment is shown in table 1.

Table 1. List of materials.

Materials	Grade/Manufacture, Dimensions (mm)
Tungsten (W)	M/s. Plansee, 30 mm (H) × 28 mm (W) × 10 mm (t)
OFHC Copper	NFTDC, Hyderabad
CuCrZr	M/s. Carl Screiber, UNS C18150, 15 mm (OD), 12 mm (ID)
Graphite	M/s. Shunk Kohlenstofftechnik, grade FP479
NiCuMn-37 (AMS 4764)	M/s. Wesgo Metal, 50 micron thickness foil (Ni-9.5, Cu-52.5, Mn-38)

The CuCrZr tube having dimension of 15 mm OD and 12 mm ID with 81 mm length has been machined from CuCrZr block material. The dimensional tolerance of the CuCrZr tube plays an important role in vacuum brazing as a small non uniform gap can create pool of filler material and which further affects quality of the brazing joint due to lack of uniform filler flow. Moreover, the surface condition outer diameter (OD) of CuCrZr tube requires a good finishing as it will be going to join with WCu tile at inner surface radially. Figure 1 shows the flow chart for PFC preparation.

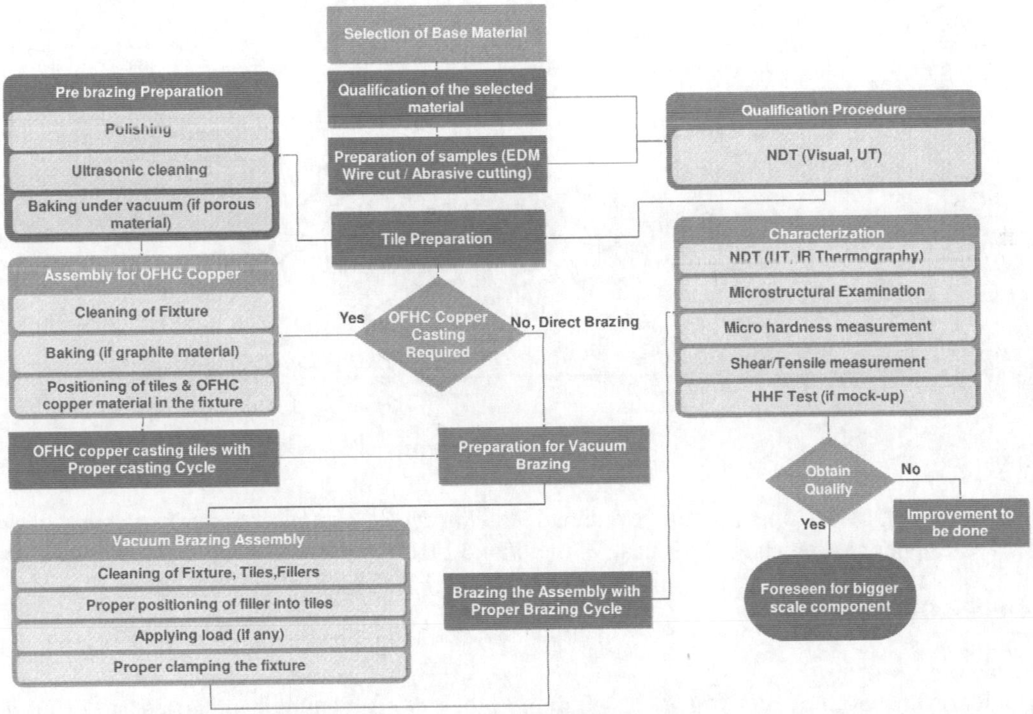

Figure 1. Flow chart for PFC preparation

Good quality and Defect-free graphite materials (grade FP479) manufactured by Schunk Kohlenstofftechnik was used for fabrication of the fixture for casting and fixture for brazing experiment. Baking of the Graphite fixture was carried at 350 °C for 2 h to remove the moisture and unwanted trapped gas. OFHC Copper casting in radial surface area of W tiles was carried out @ 1100

°C for 1 h under argon atmosphere using suitable fixture made by Graphite material. Figure 2 shows the OFHC Copper casting cycle and inset - assembly of the OFHC Copper casting on W Tiles and casted tile. After the OFHC Copper casting, the cast Copper surface has been machined to 1mm thickness radially on the W tiles. The W/Cu-Tiles with CuCrZr tube assembly is performed using custom made assembly fixture with maintaining gap of 0.5 mm in between the W tiles. Proper assembly of the W/Cu-tiles along with filler material is essential and important. The assembly fixture plays an important role to achieve the dimensional accuracy of the job by restricting the movement of the tiles under higher temperature. Vacuum brazing of the W/Cu-tiles assembly has been performed using brazing assembly fixture made of graphite material (high quality graphite has been chosen). The vacuum brazing is held for 970 °C for 10 min under vacuum environment of 10^{-6} mbar with NiCuMn-37 filler foil as brazing filler material. Placement of the component inside the hot zone area is also an important consideration taking into account of the distribution of the filler materials on the desired locations. The W-mono-block assembly has been kept vertically (figure 2 inset) inside the vacuum hot zone region. The brazing cycle used in W/Cu-CuCrZr brazing is shown in figure 2.

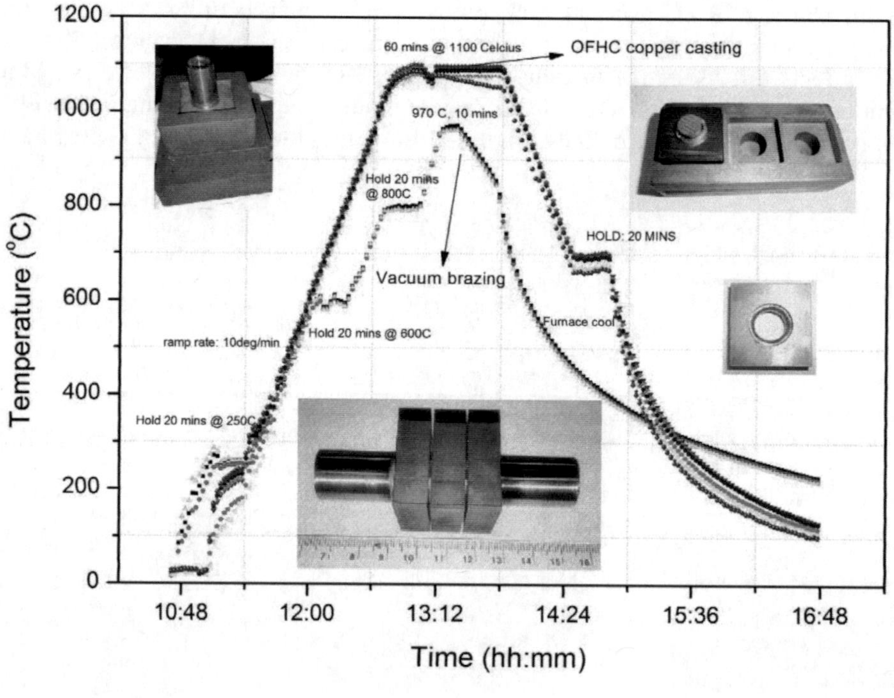

Figure 2. OFHC Copper casting cycle and Brazing cycle, inset pic – fixture assembly of the OFHC Copper casting and copper cast W tile, mono-block brazed mock-up and brazing assembly

3. Results & Discussion

3.1. Ultrasonic Testing

Ultrasonic Testing of three nos. of Cu Casted mono-block W tiles have been carried out using pulse-echo immersion technique with 20 MHz, 3.16 mm Dia. side looking probe. The C-scan images of W-Tile no.1, W-Tile no. 2 & W-Tile no.3 depicted no de-bond indication at OFHC Cu-W interface region are shown in figure 3. Also, shows the brazed mock-up image (left) and tiles marking for inspection scan. The brazed W/Cu-CuCrZr mono-block tiles mock-up has been inspected by Ultrasonic immersion probe (pulse echo) testing technique with static gain value set to 58 dB and water path was set to 4.7 mm [5].

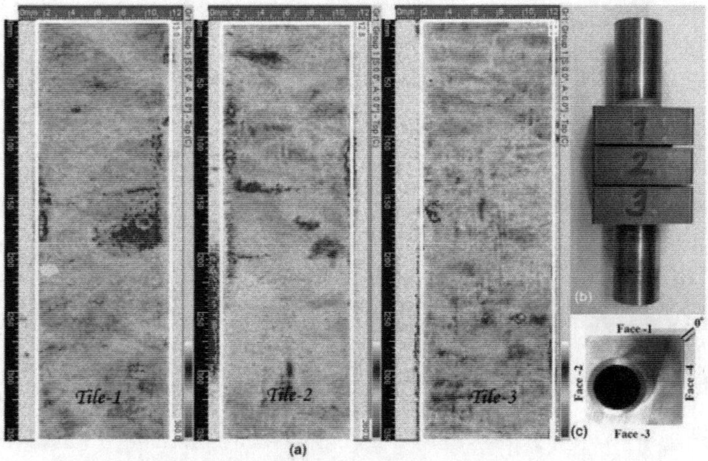

Figure 3. (a) C-Scan Image of WCu-Tiles, (b) Brazed W/Cu-CuCrZr mock-up, (c) Front face view of mock-up

From the above C-scan images of the W-Cu interface, the following points can be drawn.

(i) There is no signature of de-bonding at the interface region between OFHC Copper and W material surface.
(ii) No crack has been observed in interlayer copper material and W parent material

W/Cu-CuCrZr mono-block assembly has the encountered two interface regions as per the geometry, when seen from Inner tube diameter i.e. (i) 1st interface is CuCrZr tube to OFHC Copper, (ii) 2nd interface is OFHC Copper with W tiles. Conventional technique has been employed for the detection of de-bonding between OFHC Copper to CuCrZr material and Phase reversal technique is used to detect the de-bonding between Brazing Filler material and W tiles. A side looking 20 MHz frequency probe having 3.16 mm diameter and 12.5 mm focal length is selected to inspect the brazed geometry with the scanning resolution of 0.1 mm in step and 0.1 degree in rotational direction. Figure 4 shows the UT C-scanning of the WCu-CuCrZr brazed interface region where small dots of red circles are observed in Tile 1 and Tile 3. The blue and white areas are the signature of bonding region and small segregated red circles are the non-wetting regions. Figure 5 shows the UT C-scanning of the W-Cu interface region where still good intact of copper with Tungsten are seen.

From the C-scan images of the W/Cu-CuCrZr mono-block mock-up, following are the interpretations / observation points based on the ultrasonic testing results.

(i) The red small circle (s) observed in the C-scan image of W/Cu-CuCrZr mono-block mock-up represented the non-wetting region (de-bonding area) between W/Cu to CuCrZr tube.
(ii) Blue region (good bond) has been dominated over the red region (poor bond) in W/Cu-CuCrZr joining interfaces (figure 4).
(iii) It has been noticed that W-Tile 1 and W-Tile 3 contained the segregated isolated defects. The calculated de-bonding area of the tile no. 1 was 5.67 % and tile no.3 was 3.30 % which indicated the overall average de-bonding of the brazed mock-up was estimated to be less than 3 %.
(iv) No de-bonding was seen in W-Tile no.2 and was observed to be good bonding compared to W-Tile no. 1 & W- Tile no. 3.
(v) No phase reversal phenomena were observed for OFHC Copper with W bonding. No defect or de-bonding was observed in the interface of Cu-W after the vacuum brazing (figure 5).

IOP Conf. Series: Journal of Physics: Conf. Series **823** (2017) 012023 doi:10.1088/1742-6596/823/1/012023

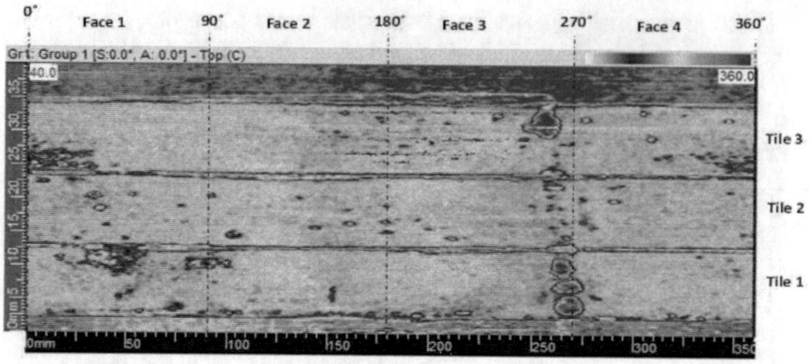

Figure 4. C-scan Image of W/Cu-CuCrZr mono-block mock-up

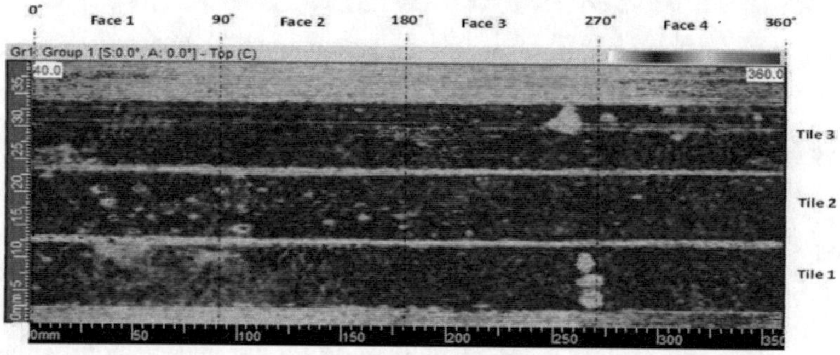

Figure 5. C-Scan image of OFHC Cu to W bonding from the brazed mock-up

3.2. Microstructural Examination

The microstructural examination of the polished bimetallic W/Cu tiles has been performed using SEM (Scanning Electron Microscope) at FCIPT-IPR. The sample was polished using different grade of SiC abrasive paper (180, 240, 400, 600, 800, 1000, 1200, 1500 and 2000 grit sizes), followed by final polishing using diamond paste using velvet cloth. Sample was cleaned ultrasonically using acetone after the final polishing. Figure 6 shows the SEM images of W/Cu interface, Cu-NiCuMn-CuCrZr interface and W/Cu-NiCuMn-CuCrZr interface. Spot analysis has been carried out on the W/Cu-CuCrZr brazed mono-block W tile using SEM- EDAX (see figure 6 right side). Twelve points have been taken across the specimen to identify the elemental distribution of the joints. From the microstructural image analysis, following observation points are the made.

(i) The OFHC Copper has been uniformly cast and no porosity in the copper is observed.
(ii) The proper diffusion of the brazing filler material (NiCuMn-37) is seen in both the OFHC Copper and CuCrZr tube.
(iii) No porosity or defect observed in the interface region of W/Cu and Cu-CuCrZr brazed region.

IOP Conf. Series: Journal of Physics: Conf. Series **823** (2017) 012023 doi:10.1088/1742-6596/823/1/012023

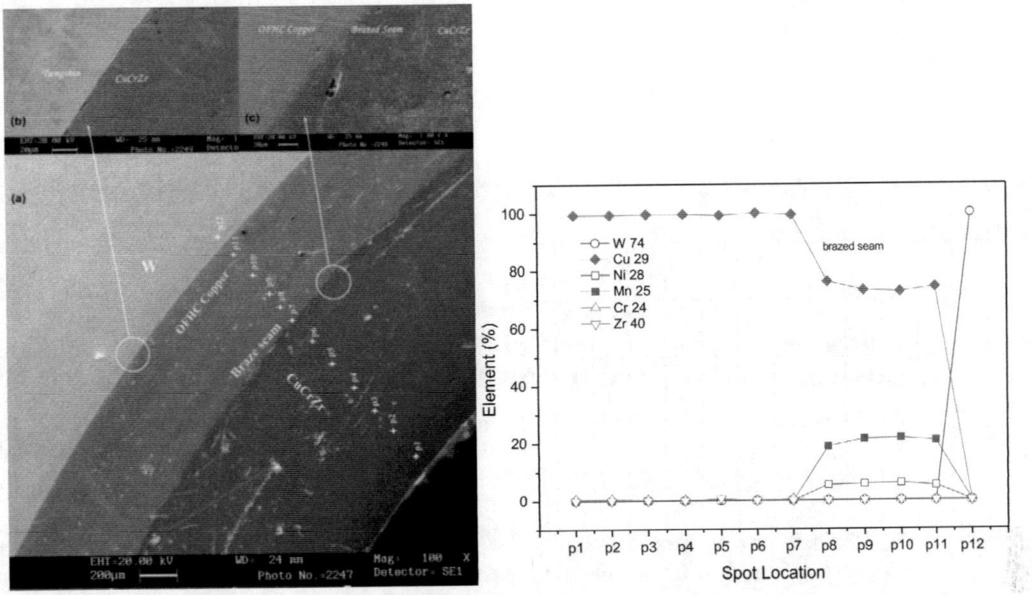

Figure 6. (a) SEM image of W-Cu-NiCuMn-CuCrZr of brazed mock-up, (b) W/Cu interface, (c) Cu-NiCuMn-CuCrZr interface, (right) –EDAX elemental analysis plot at brazed interface.

4. Conclusion

Straight geometry of W/Cu-CuCrZr mono-block mock-up of having three tiles has been successfully fabricated using vacuum brazing technique at 970 °C for 10 min using NiCuMn-37 filler materials. Custom made Graphite casting fixture has been fabricated to perform OFHC Copper Casting and Vacuum Brazing at high temperature. W/Cu bimetallic has been prepared by OFHC Copper casting technique. Ultrasonic testing of the brazed Cu-W mono-block assembly has been performed successfully using side looking probe. It is noted that tile no. 1 and 3 observed a minor de-bond signature of 5.67 % and 3.30 % areas at Cu alloy-Cu interface while no de-bond presented at Cu-W interface regions. The overall total brazed joint region was observed to be ~ 97 % area and the depicted total average defect(s) observed in the assembly was 2.99 % only which is acceptable as per ITER like PFC is concerned. From the microstructural image analysis, it was observed the good interaction of the filler material with parent's materials. No porosity or defect was observed in Tile No.2. Tile No.1 and Tile No. 3 has witnessed a small area of non-wetting region.

Acknowledgement

Author thanks to Mr. Naren Chauhan, FCIPT for helping in SEM/EDAX image analysis and Mr. Vinod Kaila for careful milling of inner copper cast surface. Thanks to all members of Divertor and Firstwall Technology Development Division for contributing in the work.

References

[1] Khirwadkar S S, Singh K P, Patil Y, Khan M S, Buch J J U, Patel A, Tripathi S, Jaman P M, Rangaraj L, Divakar C, 2011 *Fusion Eng. Des.* **86** pp. 1736-1740

[2] Appendino P, Ferraris M, Casalegno V, Salvo M, Merola M, Grattarola M, *J. Nucl. Mater.* ,2004 **329–333** pp. 1563-1566

[3] Casalegno V, Salvo M, Murdaca S, Ferraris M, 2009 *J. Nucl. Mater.* **393** pp. 300-305.

[4] Singh M, Shpargel T P, Morscher G N, Asthana R et.al, 2005 *Mater. Sci.Eng.* A, **412** pp. 123-128

[5] Bhope K, Mehta M, Khan M S, Khirwadkar S S *The e-Journal of Nondestructive Testing*, Vol.**19** No.02, ISSN 1435-4934

IOP Conf. Series: Journal of Physics: Conf. Series **823** (2017) 012024 doi:10.1088/1742-6596/823/1/012024

Design and performance of vacuum system for high heat flux test facility

Rajamannar Swamy Kidambi[1], Prakash Mokaria[1], Samir Khirwadkar[1], Sunil Belsare[1], M S Khan[1], Tushar Patel[1] and Deepu S Krishnan[1]

[1]Institute for Plasma Research, Near Indira Bridge, Bhat, Gandhinagar 382 428, India

E-mail: rajamannar@ipr.res.in

Abstract. High heat flux test facility (HHFTF) at IPR is used for testing thermal performance of plasma facing materials or components. It consists of various subsystems like vacuum system, high power electron beam system, diagnostic and calibration system, data acquisition and control system and high pressure high temperature water circulation system. Vacuum system consists of large D-shaped chamber, target handling system, pumping systems and support structure. The net volume of vacuum chamber is 5 m^3 was maintained at the base pressure of the order of 10^{-6} mbar for operation of electron gun with minimum beam diameter which is achieved with turbo-molecular pump (TMP) and cryo pump. A variable conductance gate valve is used for maintaining required vacuum in the chamber. Initial pumping of the chamber was carried out by using suitable rotary and root pumps. PXI and PLC based faster real time data acquisition and control system is implemented for performing the various operations like remote operation, online vacuum data measurements, display and status indication of all vacuum equipments. This paper describes in detail the design and implementation of various vacuum system for HHFTF.

1. Introduction

Plasma Facing Components (PFC) viz. Divertors and First Wall Components are expected to withstand severe heat load due to energetic plasma particles. Incident heat load on PFCs depends on a large variety of parameters related to plasma operation, plasma events, design of the components including their position with respect to. plasma etc. [1,2]. Heat load on these plasma facing components can be divided primarily in two categories viz. steady state heat load and transient heat load. In an ITER-like tokamak device, during steady state operation, plasma facing components are expected to withstand steady state heat load in the range 1 to10 MW/m^2. In case of transient events, the time scale of various types of events varies from fraction of a millisecond up to several seconds [3,4]. Since these time scales are less than or comparable to the typical thermal diffusion time scales in the plasma facing materials & components, it is more appropriate to define heat load in the form of total heat energy deposited during an event, frequency of the event and total area of heat deposition. Depending on the type of transient event in the plasma, incident transient heat energy on plasma facing components of an ITER-like device typically ranges from fraction of a Joule to several Joules.

In order to test the PFCs for their thermal performance under heat load conditions, various types of heat sources are used including electron beam, laser beam, neutral beam, plasma torch, plasma accelerator, etc. Each one of these sources has some advantages as well as some limitations. Amongst these sources, electron beam source is the most popular choice because of lots of advantages. It can be easily produced with high electric power efficiency, controlled, focused or defocused, rastered, steered

[5]. Its power output can be adjusted over a wide range easily by adjusting electron energy and/or electron current. Unlike other sources such as ion beam or neutral beam, electron beam have a major advantage as electrons do not chemically interact with the target materials.

High Heat Flux Test Facility (HHFTF) set-up at IPR [6] for testing thermal performance of plasma facing materials and components mainly consists of the following sub-systems like vacuum system, high power electron beam system, diagnostic and calibration system, data acquisition and control system and high pressure high temperature water circulation system. The major subsystem of HHFTF is vacuum system which integrates various subsystems. The design approach, implementation and testing are discussed in below sections.

2. Design approach

Vacuum chamber and target handling facility is designed for testing large size (1 m length × 1 m height) and heavy components weighing up to 1000 kg such as divertor and first wall components of ITER like device. D-Shaped vacuum chamber of 2.5 m diameter and 1.5 m height is used for mounting of the electron beam gun, diagnostic & monitoring equipment/sensors and vacuum pumping. Rectangular flange (2.5 m length × 1.5 m width) mounted vertically on a rail mechanism acts as support structure for assembly of the target. The D-chamber with target handling system is assembled on a support structure and both are clamped using hydraulic power pack system with hydraulic press plungers of 10 nos. Silicon O-ring is used as a gasket to ensure vacuum leak tight between chamber and target handling system.

Rail mechanism allows easy movement of rectangular flange away from the vacuum chamber so that targets can be assembled away from the vacuum chamber, connected with the coolant water supply lines as well as thermal & calorimetric diagnostics and pre-tested for any water leaks during high pressure high temperature water supply. Inside target handling system, CuCrZr plate and 2 nos. of CuCrZr blocks are mounted with cooling arrangements. In absence of any target, these thermal shields also act as beam dump.

Ports for diagnostic purposes are arranged in a radial pattern on semi-circular vertical wall of the vacuum chamber so that central region of rectangular flange where target is mounted can be easily viewed from any of the ports. Total number of ports and their usage is as follows:

a) For mounting electron beam gun: 3 ports of 250 CF;
b) For diagnostics: 29 Ports (4 ports of 40CF; 13ports of 63 CF; 12ports of 100CF);
c) For water coolant supply to vacuum chamber: 12 ports;
d) For water supply to component to be tested: 6 ports of 40 CF;

Electron gun is mounted on horizontal port opposite to the target so that XY deflection of the beam can be used to deposit heat flux on part or full surface of the target. Additional two inclined ports are also provided on the top wall of vacuum chamber that can be used either for mounting of electron gun or for diagnostic purposes. Vacuum pumping ports are provided on either side of the side walls of vacuum chamber. Ports provided on the rectangular flange allow supply of high pressure high temperature water to the targets. The design of D-shaped chamber is based on ASME section VIII division-1 using SS 304L material. The thickness of vacuum vessel was varying from 6 mm to 16 mm at various sections of vacuum chamber. Figure 1 shows the schematic of vacuum system with target handling system.

IOP Conf. Series: Journal of Physics: Conf. Series **823** (2017) 012024 doi:10.1088/1742-6596/823/1/012024

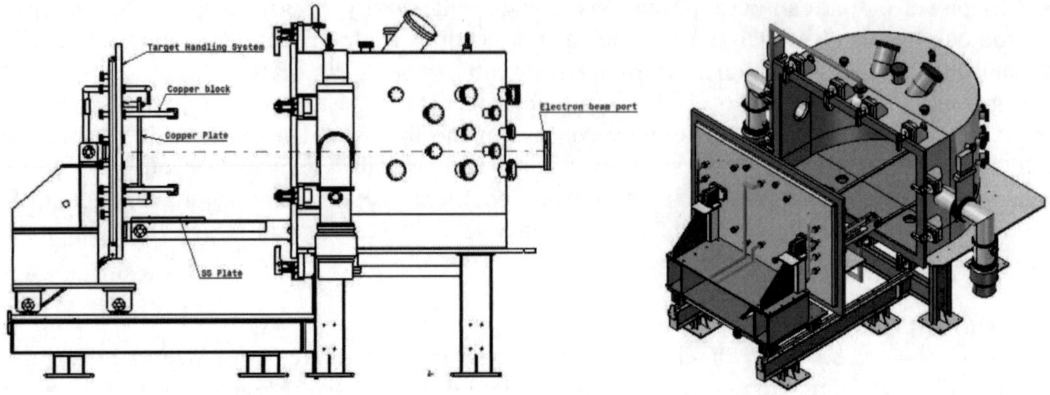

Figure 1. Vacuum system of HHFTF with target handling system

2.1. Thermal hydraulic analysis
The D-shaped chamber is double walled water cooled vacuum chamber with rectangular ribs arranged radially on top and bottom plates and longitudinally on the shell to remove the radiated heat from the target. The flow path at top and bottom plates is considered in radial direction. Flow of coolant (Chill Water at ~ 16 °C, pressure at 1.5 bar(g), flow rate at 3.3 kg/s) will enter from bottom plate and will flow towards radially outwards in channels between the both bottom plates (in between inner vessel and outer vessel). Exit coolant from bottom will enter in the vertical channels of domes and coolant exit from domes will enter radially inward on top plats channels. This will make a single flow path on all the three parts of double walled vessel. Thermal hydraulic analysis of vacuum chamber has been carried out to obtain process parameters for extraction of continuous 100 kW of power on the vacuum chamber. The rib size height is optimized to 30 mm and the pressure drop of 0.147 bar is obtained for the radial flow distribution of water flowing from bottom to top plates.

2.2. Steady state thermal analysis
Finite element analysis (FEA) of vacuum chamber is carried out using ANSYS 17.1 work bench considering various boundary conditions. The average heat transfer coefficient estimated through thermal hydraulic analysis for top and bottom plates is 505 W/m^2K and for dome is 416 W/m^2K is considered as an input parameter for steady state thermal analysis. Heat load of 200 kW is applied on the inner vessel as heat flux. The maximum temperature is observed at the internal dome area which is of 65 °C and the temperature developed at the outer shell region is within the room temperature range as shown in figure 2 and 3.

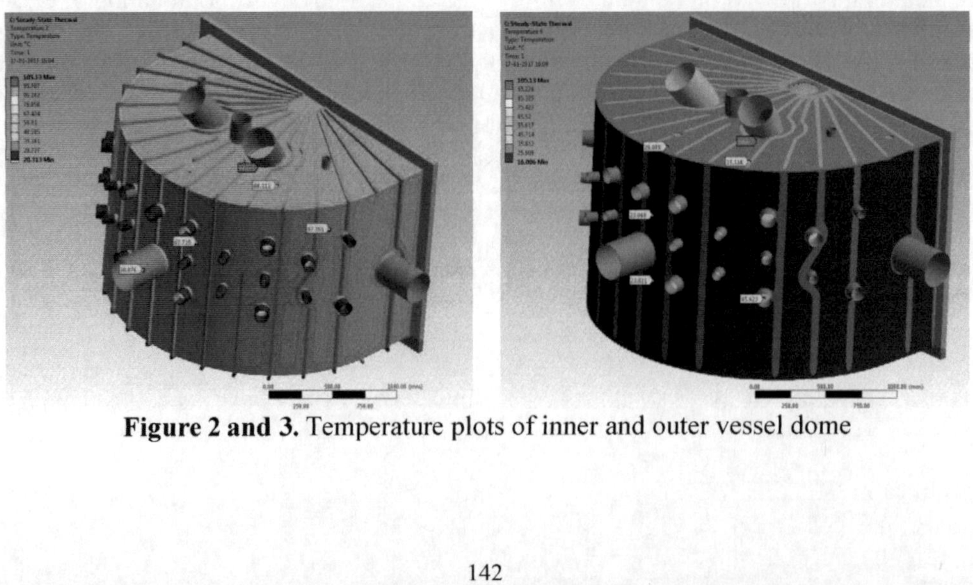

Figure 2 and 3. Temperature plots of inner and outer vessel dome

2.3. Thermal static structural coupled analysis
The results of thermal analysis are coupled with structural analysis considering various load conditions like fixed support at the bottom plate, water pressure of 1.5 bar(g) applied on inner and outer vessel of vacuum chamber, atmospheric pressure on the vacuum chamber, vacuum pressure inside the chamber of 10^{-6} mbar. The maximum displacement of 1.65 mm is observed at the top plate and all other parts has deformation less than the permissible value. The stresses developed on the shell are in the range of 60 MPa to 110 MPa which are less than the allowable stress value. Maximum stress of 236 MPa is observed at junction of collar to top plate and is localized at a single point. The stress and deformation plots are shown in figure 4 and 5. Parametric studies were carried out for optimization of thickness of the top and bottom plates of outer vessel from 12 mm to 24 mm in order to achieve the maximum deformation less than 2 mm and the rib height is optimized to 30 mm considering cross-section of 20 mm × 30 mm..

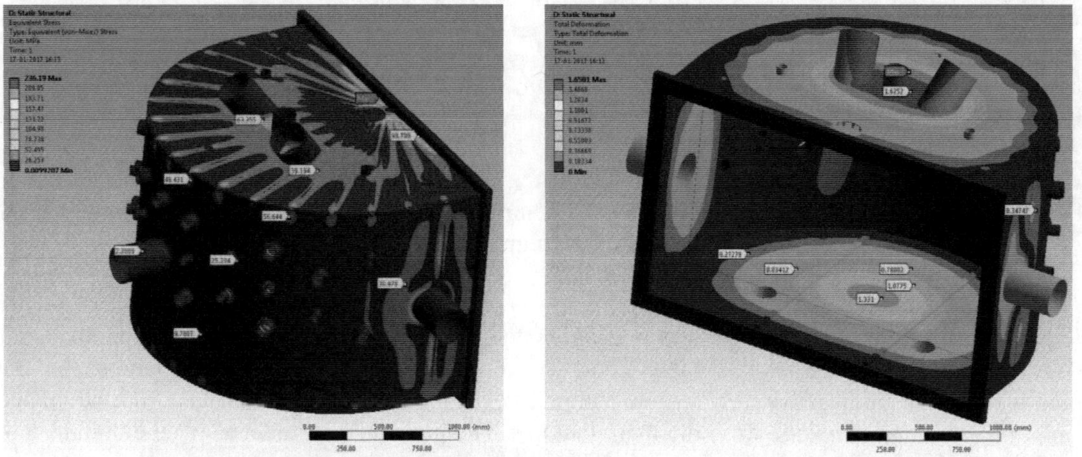

Figure 4 and 5. Equivalent Von-Mises plot and total deformation plot

3. Vacuum pumping system
Volume of vacuum chamber including ports is 5 m^3 and it needs to be evacuated to pressure of the order of 10^{-6} mbar for operation of electron gun with minimum beam diameter. All Con-Flat (CF) opening were enclosed using copper gaskets. During design of D-shaped chamber, SS 304L electro-polished material having out gassing rate of 6.65×10^{-8} mbar l/s/m^2, copper plate and copper block has out gassing rate of 6.6×10^{-8} mbar l/s/m^2, silicon O-ring has out gassing rate of 6.6×10^{-7} mbar l/s/m^2 were considered so that an ultimate vacuum of 1×10^{-6} mbar could be achieved inside vacuum chamber. Such base pressure is much adequate for constraining impurities and hence providing better environment for EB firing on target material.

Figure 6 shows the schematic drawing of pumping system including different vacuum measuring components. Vacuum chamber has TMP and Cryo pumping system mounted at the diagonally opposite radial ports and each of them is comprised of circular pumping lines of various lengths with elbows. D-Chamber has one roughing line with combination of root and rotary pump attached on port no P1 to achieve rough vacuum of 5×10^{-2} mbar from atmosphere. To achieve ultrahigh vacuum, TMP pumping system with splinter shield having an effective pumping speed of 2000 l/s (N_2 gas) and a baking rotary pump is mounted on 250CF port P2. Cryo pump of effective pumping speed of 4500 l/s also mounted on other 250 CF port P3. Each TMP and cryo-pump unit has an electro-pneumatic gate valve of respective size to isolate the pump in the case of either vacuum break of the D-shaped chamber or in case of the pump failure to protect each in either condition. A variable conductance gate valve is used for maintaining required vacuum in the chamber. The schematic diagram shown in figure 7 represents the various pumps mounted at the ports of vacuum chamber.

Various kinds of the vacuum gauges are installed at the several positions of D-shaped chamber ports and pumping lines to measure the vacuum with higher accuracy. A residual gas analyzer (RGA) is mounted on 65CF port P4 to measure partial pressure of different gases as well as the leak tightness of D-shaped chamber at the time of experiments. The exhaust of this pumping line was connected to main exhaust line fabricated using PVC pipe of 50 mm ID whose opening is outside the room in open air at height of 15 feet from ground.

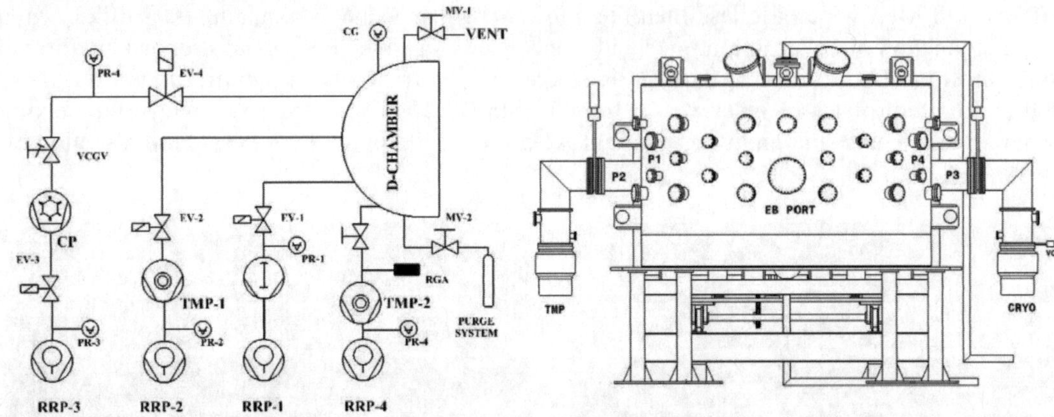

Figure 6 and 7. Schematic diagram of vacuum pumping system and various pumps mounted at the ports of vacuum chamber

3.1. Pump down time

The pump down time of vacuum system is simulated using Vac Tran software and compared with the results obtained during operation of the system. The complete vacuum system with pumps, gas loads, piping is modelled in the software. Various pumps used in the system are modelled using pump speed vs inlet pressure data obtained from the manufacturer. The gas loads considered for the system model are (a) outgassing load due to different materials used like stainless steel, copper, tungsten and silicone rubber (b) permeation due to silicone O-ring (c) gas load due to O-ring element.

The pumping line-1 is used for roughing of the vacuum chamber. It consists of rotary pump, root pump; solenoid operated gate valve connected to D-shaped vacuum chamber through pipes, bends and elbows. The pump down time required to achieve 1×10^{-3} mbar from atmosphere is approximately 40 min. The pumping line-2 has turbo molecular pump for evacuation of chamber in high vacuum range along with rotary pump and electro-pneumatic gate valve. This line is used once the vacuum reaches in the order of 10^{-2} mbar and targeted to reach vacuum in the order of 10^{-5} mbar. The pump down time required to achieve 1×10^{-5} mbar from 1×10^{-2} mbar is 1 min. The pumping line-3 has cryopump, rotary pump, variable conductance gate valve, electro-pneumatic valve and associated piping. First rotary pump will start and evacuate the cryopump in the order of 10^{-3} mbar through electro-pneumatic valve. Once the cryo temperature reaches 20 K, the electro pneumatic gate valve will be opened to evacuate the chamber in the order of 10^{-6} mbar. Before opening the gate valve the pressure inside the vacuum system shall be of order 10^{-2} mbar. The pump down time required by the cryo pump to reach pressure in the vacuum system of 10^{-6} mbar from 10^{-2} mbar is 1.5 min.

4. Control system of pumping lines

PXI and PLC based faster real time data acquisition and control system is implemented for performing the various operation like online vacuum data measurements, display and status indication of all vacuum equipment like electro pneumatic valve storing the data for further analysis and remote operation of D-chamber pumping system. This system is also used as safety protocol if any fault occurs during experiment.

Using PC based user interface the completed pumping system is controlled. A provision is adapted for switching over to manual mode in case of emergency. Both hardware and software interlocks are implemented for the safety of the equipment as well as for the protection the vacuum system in case a fault condition. The entire automation is developed using Lab view platform whose graphical users interface is shown in figure 8. Apart from the main vacuum system the various other parameters like temperature, flow, pressure and residual gas can also be seen here.

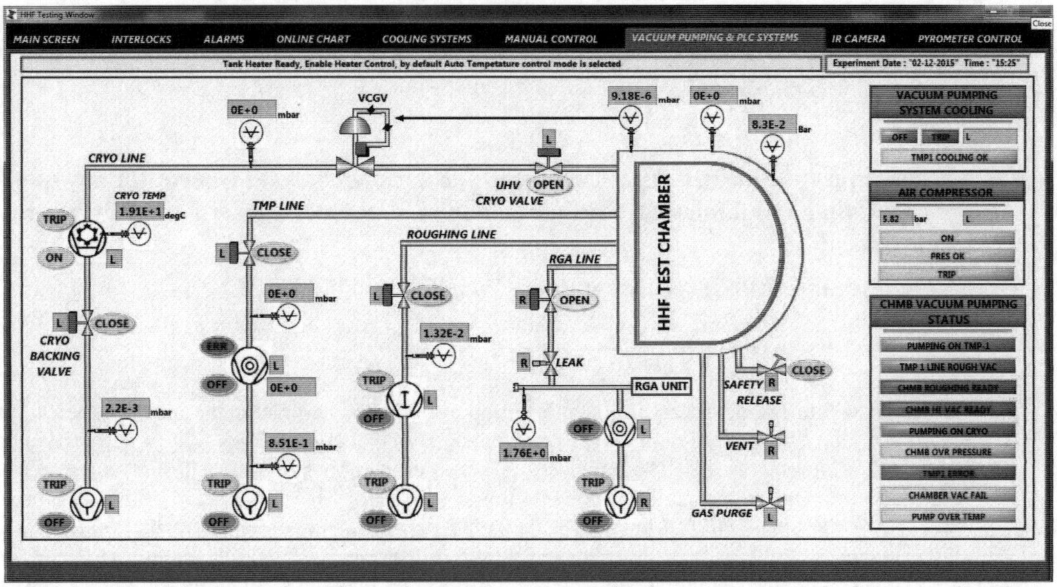

Figure 8. Graphical user interface of vacuum status and other parameters during pumping down operation of HHFTF

5. Conclusion

Design approach of vacuum system for high heat flux test facility has been discussed and validated with finite element analysis. The vacuum system was successfully commissioned and its functionality was tested as per the design requirements. Various subsystems of high heat flux test facility were integrated with vacuum system and their performances were tested. Hardwired interlocks were implemented for safe operation of vacuum system during high heat flux testing. PXI and PLC based faster real time data acquisition and control system is successfully implemented for performing various operations.

Acknowledgement

The authors would like to acknowledge all members of Divertor and first wall technology development division of IPR and M/s Vacuum Techniques Pvt. Ltd, Bangalore.

References

[1] Bolt H et al 2002, *J. Nucl. Mater.* **43** 307.
[2] Tivey R et al 1999, *Fusion Eng. Design.* **46** 207.
[3] Norajitra P et al 2008, *Fusion Eng. Design*, **83** 893.
[4] Linke J et al 2006, *Phys. Scr* **123** 45.
[5] Rodig M et al 2000, *Fusion. Eng. Design* **51** 715.
[6] Patil Y et al 2015, *Fusion.Eng. Design* **95** 84.

IOP Conf. Series: Journal of Physics: Conf. Series **823** (2017) 012025 doi:10.1088/1742-6596/823/1/012025

Design of a Prototype Positive Ion Source with Slit Aperture Type Extraction System

Sanjeev K Sharma[1], Prahlad Vattilli[1], Bhargav Choksi[1], Bharathi Punyapu[1], Rambabu Sidibomma[1], Sridhar Bonagiri[1], Deepak Aggrawal[1] and Ujjwal K Baruah[1]

[1]Institute for Plasma Research, Bhat, Gandhinagar – 382428

Email: sksharma@ipr.res.in

Abstract: The neutral beam injector group at IPR aims at developing an experimental positive ion source capable of delivering H^+ ion beam having energy of 30 – 40 keV and carrying an ion beam current of 5 A. The slit aperture based extraction system is chosen for extracting and accelerating the ions so as to achieve low divergence of the ion beam (< 0.5°). For producing H^+ ions a magnetic multi-pole bucket type plasma chamber is selected. We calculated the magnetic field due to cusp magnets and trajectories (orbits) of the primary electrons to investigate the two magnetic configurations i.e. line cusp and checker board. Numerical simulation is also carried out by using OPERA-3D to study the characteristic performance of the slit aperture type extraction-acceleration system. We report here the results of the studies carried out on various aspects of the design of the slit aperture type positive ion source.

1. Introduction

Neutral beam systems are demonstrated as one of the most effective techniques for heating, current drive and diagnostics in fusion related experiments. Recently, the NBI group at IPR has completed its test stand activities of the ion source PINI (plug-in neutral injector) [1] and is underway of transferring the injector system to the SST1 tokamak. The NBI group is now planning for indigenous development of a positive ion source capable of delivering H^+ ion beam having energy of 30 – 40 keV and carrying an ion current of 5 – 10 A. The ion source will be used for basic experimental studies on the extraction system. The neutral beam ion sources consist of an arc chamber (for production of ions) and a set of electrostatic grids (electrodes) for the purpose of extraction and acceleration of the ions. Large sized ion beams are produced by superposition of many beam-lets extracted through a set of apertures distributed uniformly on the surface of electrostatic grids. The resultant ion beam current I_b depends on the number of apertures, the current density J_b from each aperture and aperture size or radius r_a [2, 3].The arc chamber generates the required flux of H^+ ions. Ionisation in arc chamber is caused by primary electrons emitted by the hot cathode assembly. The magnetic multipole or bucket type plasma source is one of the most prominent sources considered for the application of neutral beam systems [4]. The ion source considered here is a cylindrical shaped plasma chamber with multi-cusp magnetic configuration. In designing these sources, it is essential to optimize magnetic configuration because it has strong influence on the performance of the ion source. Uniform magnetic field (< 20 G) in the range over the whole extraction area is necessary for keeping uniform plasma densities across the whole extraction area.

In order to optimize magnetic field configuration, a computer program is developed to calculate the value of magnetic field and the trajectories (orbits) of the primary electrons to investigate the role of various magnetic configurations on the confinement of these electrons. An optimization is carried out for the magnetic cusp geometry to produce field free region / uniform low magnetic field over the whole extraction area. Simulation of the trajectories of the primary electrons is performed to investigate the confinement behavior of these electrons in the magnetic multi-cusp geometry. The confinement of primary electrons and plasma uniformity for different magnetic configurations is analyzed using the spatial distribution of the ionisation collision points, which is analogous to the density distribution of the plasma [5,6]. The developed ion source is planned to have low angular divergence (< 0.5°). The study on the characteristic performance of the slit aperture based extraction system is carried out by the means of numerical simulation using OPERA-3D code [7].

2. Plasma Source and extraction-acceleration system

A magnetic multipole plasma source consists of an anode chamber equipped with several columns of permanent magnets on its outer surfaces and tungsten filaments kept at the potentials of the cathode. The energetic primary electrons emitted by the hot tungsten filaments are used to ionize the background gas to generate dense hydrogen plasma. The primary electrons are confined mainly in the magnetic field produced by the permanent magnets (SmCo, surface field ~ 4 kG). A numerical code (in MATLAB) was written to calculate the magnetic field and the particle trajectories of the primary electrons. The equation of motion for a primary electron with collision term (equation (1)) is solved to trace the particle trajectory and the collision points.

$$m dv/dt = q(E + v \times B) + F_{coll} \tag{1}$$

where m, q, v, t are mass, charge, velocity of the primary electron, and time, respectively. E and B are electric and magnetic fields respectively. F_{coll} is the ionisation collision term. For the simplicity other types of collisions are not considered. The initial energy of the primary electron is considered as 80 eV as the electron gains energy from the acceleration caused by the sheath potential at the filament surface, which is equal to the applied discharge voltage. An electron emitted from each filament is traced only until the electron causes its first ionisation or it is lost to the wall. The considered physical model assumes that the sheath thickness is thin enough and that a primary electron is traced only after the acceleration by the sheath field. In this simulation, the potential gradient in the plasma is also neglected for the simplicity. The magnetic field due to the current in filaments and due to permanent magnets is calculated by using Biot–Savart's law and the formulation as discussed by Y. Ohara et al [6], respectively. The magnetic field is first calculated for each magnet in their respective coordinate systems, and then transformed into the coordinate system of the ion source.

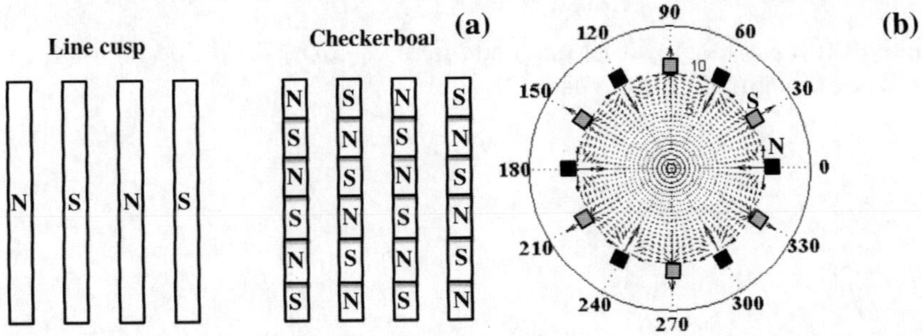

Figure 1(a). Schematic diagram of the two magnetic configurations: line cusp and checkerboard, (b) vector plot showing the magnetic cusps near the extraction plane. The size of a typical magnet is (9 mm × 13 mm × 40 mm) and the size of plasma chamber is 20 cm (diameter) and 20 cm (depth).

The considered extraction-acceleration system is a three grid system (triode type). The first grid is called the 'plasma grid' or extraction grid. The positive voltage, V_{pg}, required for accelerating the ions is applied to this grid. The second grid is called the 'suppressor grid'. A small *-ve* voltage is applied to this grid to prevent the back streaming of secondary electrons (produced due to interaction of ion beam with residual gas) from reaching the arc chamber. The second grid, which causes a small deceleration to the beam, is also known as the 'deceleration grid'. The third grid is at zero potential ($V_g = 0$) and therefore is known as ground grid. The considered width and length of the slits are 8 mm and 100 mm, respectively. The shapes/contours of the slit grids are considered as similar to those of PINI [1]. The gap between the plasma-deceleration grids and deceleration-ground grids are 8 mm and 2 mm respectively.

3. Optimization of magnetic geometry

The plasma chamber is equipped with an arrangement of permanent magnets on its outer surface to form the required magnetic configuration. There are two types of magnetic configurations commonly used in multi-cusp ion sources, namely, line cusp and checker board. They can be further classified into continuous and broken ones. The calculations are carried out for the continuous line-cusp and continuous checkerboard configurations. The two types of magnetic configuration are depicted in figure 1(a). In the continuous line-cusp configuration, columns of permanent magnets having north and south poles facing alternatively inward are placed on the outer surfaces of the ion source. The magnetic cusps are formed perpendicular to these magnetic columns as shown in figure 1(b). In the checkerboard configuration, each magnetic column has several magnets, facing their north and south poles alternatively inward in such a way that the magnetic cusps are formed along both parallel and perpendicular to the magnetic columns. In both the configurations, magnetic columns are parallel to the source axis i.e. along the longitudinal direction.

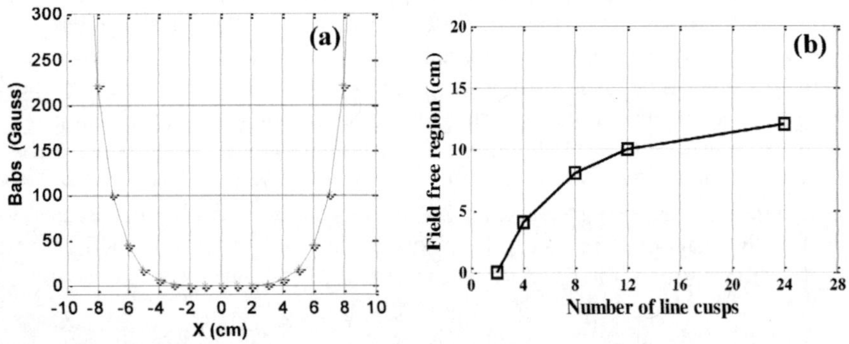

Figure 2(a). Radial profile of magnetic field, B_{abs}, (b) Field free region for various number of longitudinal line cusps, N.

IOP Conf. Series: Journal of Physics: Conf. Series **823** (2017) 012025 doi:10.1088/1742-6596/823/1/012025

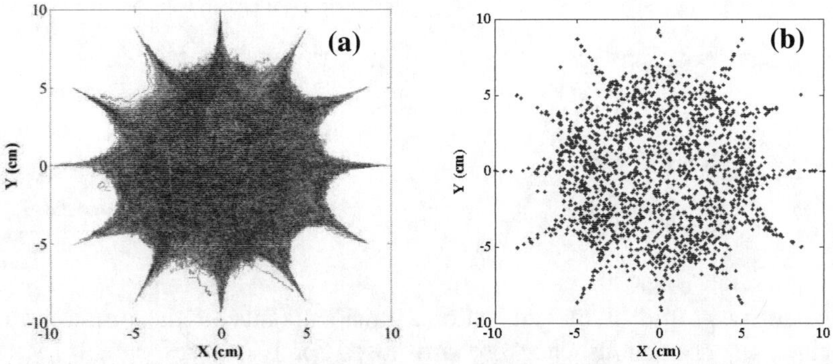

Figure 3(a). Trajectories of 1000 primary electrons (80eV) projected over the transverse plane, (b) distribution of ionisation collision points of the primary electrons. The calculations are obtained for the continuous line cusp.

A uniform plasma density over the whole extraction region is the basic necessity of the multi-cusp ion sources. Another important requirement is the achievement of maximum discharge efficiency. The uniform plasma is mainly produced in the field free region which has nearly uniform magnetic field (< 20 G). Figure 2(a) shows a magnetic field free region over ~ 10 cm for N = 12. N is the numbers of magnetic columns /cusps lines. In figure 2(b), the extent of field free regions is plotted as a function of number of cusps, N. It is evident that as the number of magnetic cusps increases, the field free region increases. More number of cusps causes strong magnetic field between the regions of the two cusps. When the number of cusps increases, the loss of the primary electrons also decreases due to the reduction in the cross field diffusion ($\propto 1/B$) of the plasma particles. The reduced loss of the primary electrons causes an increase in the discharge efficiency [8, 9]. It is also observed from figure 2(b) that the 24 magnetic cusps provide the maximum field free region of ~ 12 cm. It is however impractical to incorporate 24 magnetic columns on the periphery of the ion source having a diameter of 20 cm. The increment in the field free region from 12 cusps to 24 cusps is merely ~ 20 %. The numbers of magnetic cusps (or magnetic columns) are therefore chosen as N = 12. It is also observed that the field free region is increased by about 30 % for the checkerboard configuration. The magnetic field at the inner surface of the ion source is reduced by 50 %, which cause a rise in the cusp loss area i.e a decline in the discharge efficiency.

4. Particle orbit simulation

A three dimensional particle orbit simulation is carried out to analyse the particle confinement in the continuous line-cusp and checkerboard configurations. Figure 3(a) shows trajectories of about 1000 primary electrons (80 eV), projected over a transverse plane. The starting positions of the primary electrons are randomly chosen near the filaments (outside sheath). These trajectories are calculated for the continuous line cusp configuration. It can be seen that the electrons are well trapped by the cusp magnetic field produced by the permanent magnets.

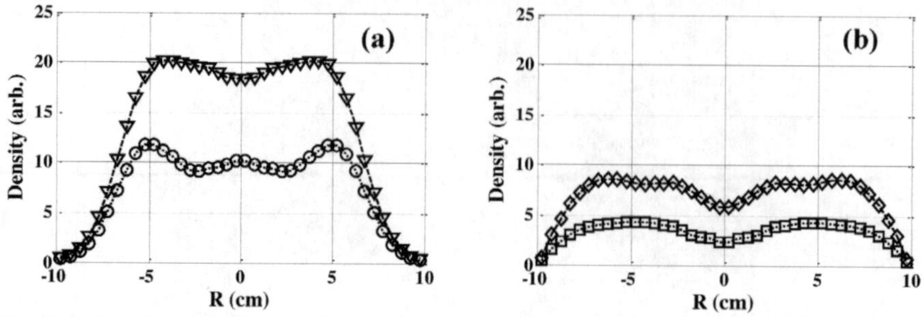

Figure 4. Radial density profile of the ionisation collision points of the primary electrons for (a) continuous line cusp, (b) continuous checkerboard cusps for two cases namely, with grid biasing (inverted triangle and diamond) and without grid biasing (circle and squares).

5. Spatial distribution of ionisation collision points

The spatial distribution of ionisation collision points (electron-neutral collision points) inside the plasma source is calculated using particle orbit simulation for continuous line cusp and continuous checkerboard configurations. The distribution of these collision points basically represents the pattern of ion birth locations. The primary electrons emitted from the filaments are traced inside the plasma source until they are lost due to the ionisation of hydrogen neutrals. The event of ionisation is decided by the completion of their ionisation mean free path, which is 38.2 cm for 5 mTorr hydrogen gas pressure and the primary electrons with 80 eV energy [8]. These positions are recorded for several electrons and used for calculating spatial distribution of the ionisation collision points. The distribution of ionisation collision points for the continuous line cusp configuration as shown in figure 3(b) suggests that the ionisation occurs mainly in the magnetic field free region. Figure 4(a) shows the radial density profile of these ionisation collision points. The curve with inverted triangles shows the density of these collision points for the case when plasma grid is biased at a fixed negative potential as similar to that of cathode i.e – 80 V. The curve with open circles shows this density for the unbiased plasma grid. It suggests that the density of the collision points ($\propto$ plasma density) would be higher for the biased plasma grid case. This is due to the increased confinement of the primary electrons caused by the reflection from the negatively biased plasma grid. Similarly the density of the ionisation collision points for the checkerboard configuration is calculated for both the biased and unbiased plasma grids as depicted in figure 4(b). It is seen that the ionisation occurs over a large region in case of checkerboard as compared to the line cusp. The density of the ionisation collision points is reduced significantly for the checkerboard configuration. Also an unbiased plasma grid further reduces the density of the ionisation collision points.

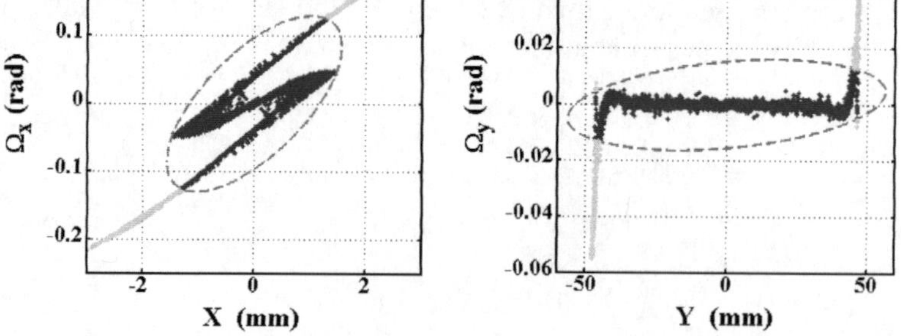

Figure 5. Emittance plots for the ion trajectories (a) along the width of slit (X) and (b) along the slit (Y). The emittance plots were obtained at a distance z = 29.2 mm from the surface of the emitter.

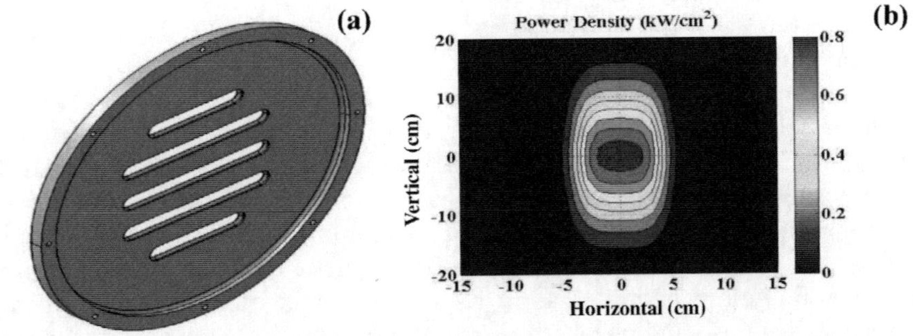

Figure 6. (a) 3-*D* model of the plasma slit grid, where width and length of slits are 8 mm and 100 mm (max) (b) beam power profile calculated with the bi-Gaussian beam extracted from the whole grid having angular beam divergences 1.8° and 0.2° (along width of the slit (vertical) and along the slits (horizontal) respectively) at 1 m distance from the extraction plane.

6. Simulation of ion beam trajectory

A multi aperture extraction-acceleration system can be described as superposition of several individual apertures. Hence the analysis of a single aperture of the extraction-acceleration system is adequate for understanding the performance of the complete extraction-acceleration system. Simulation of ion trajectories for the slit shaped apertures is performed by the OPERA code. The emittance calculations for the ion trajectories were performed at a distance z = 29.2 mm from the extraction plane. Slit aperture of widths 8 mm was considered for the simulation.

Figure 5(a) and (b) shows the emittance plots along the width of slit (8 mm) and along the slit. The abscissa represents the radial location (*X* or *Y*) of the individual ion from the beam axis (*Z*-axis).The parameter in the ordinate of the emittance plot denotes the transverse angles Ω_x and Ω_y subtended by the velocity of individual ion with beam axis in the *X-Z* and *Y-Z* planes respectively. Ellipses drawn over the emittance plots depict the spread of particles in phase space. The area of ellipse represents the geometrical emittance of the beam. The major and minor axes of the ellipses generally represent the spread in space and transverse angles. The rotation angle, θ of the ellipses, is estimated from $tan(2\theta) = 2\overline{X\Omega'}/(\overline{X^2} - \overline{\Omega'^2})$.This angle is *+ve* for convergent (*-ve* for divergent) beam. It may be noted that particles lying outside the ellipse have abnormally large transverse angles. The above behaviour can be explained by the fact that the trajectories lying outside the ellipse represent those particles that have originated from the edges of the slit aperture. As expected for slit aperture, the emittance in X-direction (along width of the slit) is entirely different from that of the *Y*-direction (along the slit). The small expansion of beam observed at the ends is due to the presence of large transverse electric fields at such locations. This expansion is more prominent in case of the slits with larger width. Transverse angles in the direction along the slit (i.e. along *Y*-axis) are nearly zero. Along the width of the slit (i.e. *X*-axis) the transverse angles vary from +/- 1° to +/- 1.6°. The minimum divergences along x-axis for 8 mm slit shaped apertures is 1.8° which correspond to ion emission current densities 160 mA/cm^2 respectively. The divergences along the slit (i.e. *Y*-axis) are very small (< 0.2°) and nearly constant for all the current densities.

Slits having widths of 8 mm are selected for developing the extraction system of the ion source. Figure 6(a) shows 3-*D* model of plasma slit grid. Figure 6 (b) shows the beam power profile calculated using bi-Gaussian beam profiles having angular beam divergences 1.8° and 0.2° (along width of the slit and along the slits, respectively) at 1 m distance from the extraction plane. The beam focussing in the extraction system is not considered because beam focussing is not found helpful in reducing the size of the beam for such smaller sized grids.

7. Summary and conclusion

The neutral beam injector group is engaged in the indigenous development of an experimental ion source capable of delivering H^+ ion beam having energies of 30 – 40 keV and carrying an ion beam current of 5 – 10 A. Studies have been carried out on various aspects of the design of the slit aperture type positive ion source. The three dimensional particle orbit simulation is carried out to analyse the confinement of primary electrons to understand the performance of the plasma generation in the magnetic multi-cusp geometry. First, the number of magnetic cusps (or magnet columns) to produce the uniform magnetic field is optimised as 12 for the ion source of the diameter 20 cm. Next the cusp geometries for the line cusp and checkerboard configurations are compared for the same number of magnetic columns. The densities of the electron-neutral (ionisation) collision points (i.e. the ion birth locations) calculated using the particle orbit simulation for continuous line cusp and continuous checkerboard configuration are compared.

For achieving the maximum density, the line cusp with the negatively biased plasma grid is found an optimum solution for higher discharge efficiency. However the checkerboard configuration with negatively biased plasma grid would be the optimum solution when a uniform density over a large extraction area would be required. Ion beam extraction and acceleration have also been studied by means of numerical simulation using OPERA-3D code. For Extraction system with slit apertures, the angular divergence along the slit is negligible (< 0.2°), while the divergence along the width of the slit is ~ 2°. Slit based extraction-acceleration system with size 8 mm (width) is considered for developing this ion source. The beam focussing in the extraction system is not considered.

References

[1] S.K.Sharma et al 2014 *Rev. Sci. Instrum.* **85** 113504
[2] Coupland J R, Green T S et al 1973 *Rev. Sci. Instrum.* **44** 9
[3] Green T S 1937 *Rep. Prog. Phys* **37** 1257
[4] Speth E, 1989 *Rep. Prog. Phys.* **52** 57
[5] Kashiwagi M and ShunjiIdo et al 1998 *Rev. Sci. Instrum.* **69** 1117
[6] Ohara Y, Akiba M, Horiike H et al 1987 *J. Appl. Phys.* **61** 1323
[7] *OPERA 3D Reference Manual* (England:Vector Fields Limited)
[8] Uhlemann R et al, 1995 *Phys. Plasma.* **2** 2141
[9] Stirling W L, Ryan P M et al 1979 *Rev. Sci. Instrum.* **50** 102

IOP Conf. Series: Journal of Physics: Conf. Series **823** (2017) 012026 doi:10.1088/1742-6596/823/1/012026

Sensitivity analysis of upstream plasma condition for SST-1 X-Divertor configuration with SOLPS

M. Himabindu[1], Anil. K. Tyagi[1], Deepti Sharma[1], Devendra Sharma[1], and R. Srinivasan[1]

[1]Institute for Plasma Research, Near Indira Bridge, Bhat, Gandhinagar-382428, India

E-mail: mbindu@ipr.res.in

Abstract: Extensive power exhausts and target heat loads are anticipated in reactor grade fusion devices. Prototyping of an X-Divertor based power exhaust scheme is being attempted by means of simulations of Scrape-off Layer plasma transport in the diverted plasma equilibria of SST-1 tokamak using SOLPS5.1. Evaluation of the relative advantages of an X-Divertor configuration involves simulating the SST-1 standard divertor scheme plasma transport for the reference and then achieving equivalent upstream plasma conditions in the X-divertor equilibrium to ensure equivalent core plasma in both the cases. The first optimization is to be achieved by simulating effects of an external gas puff in the SOL region for controlling separatrix density in the X-divertor configuration with visible modifications in the downstream plasma conditions. The present work analyzes sensitivity of the upstream SOL plasma conditions to the gas puff intensity and its effect on the plasma neutral transport in the divertor region

1. Introduction

In order to handle the huge amount of heat loads in fusion grade reactors with Standard divertor (SD) configuration that are employed in the current machines (ITER) might be problematic. The X-divertor (XD) [1] was proposed for reactors to improve power exhaust by increasing flux expansion and line length. The magnetic geometry of the X-divertor is based on creating an extra X-point near the divertor plate to increase the flaring of field lines in the divertor region and hence plasma-wetted area and it considerably reduces heat flux on the divertor plates. The XD configuration is expected to reduce the heat load at strike point by efficiently depositing heat on the downstream. The neutral front is expected to move downstream in XD as compared to SD and will reduce the energy loss in the plasma. Before going to experimental XD configuration in SST-1 tokamak it is essential to quantify these advantages by modeling the plasma transport in this magnetic geometry.

The free boundary magneto hydrodynamic (MHD) equilibrium code CORSICA [2] is used for SD and XD configurations. The quantitative measure of convergence of flux surfaces relative to the Standard Divertor (SD) is given by the SOL Divertor Index (DI). For SST-1, DI = 6.82 for XD means the flux surfaces are more flared than DI = 2.69 of SD. On these SD and XD configurations we performed transport simulations using SOLPS5.1 in order to see the qualitative performance of XD over SD. The SOLPS5.1 [3] couples the 2-D multi-fluid plasma code B2.5, handling the transport of electrons and multiple ionized states of ions, with neutral Monte-Carlo code EIRENE [4]. The computational mesh, produced by using CAREE [5] code available with the SOLPS5.1, is based on 2D magnetic field equilibrium and the real geometry of divertor and other boundaries in the device. Simulating

IOP Conf. Series: Journal of Physics: Conf. Series **823** (2017) 012026 doi:10.1088/1742-6596/823/1/012026

the effects of gas-puff in the SOL to ensure the core equivalence of both geometries and their effect at the divertor.

2. SOLPS5.1 Modeling

MHD equilibrium of SD and XD divertor geometry shown in figure 1(a) & 2(a) are predicted by using CORSICA with operational parameters of plasma current I_p = 220 kA and toroidal field $B_Ø$ = 3T. The computational meshes used for the modeling of SD and XD are produced by using Carre in the SOLPS5.1 code package are shown in figure 1(b) & 2(b) which has a resolution of 98 × 36. The computational modeling region is based on Double Null (DN) and covers the outer SOL, the divertor targets below the mid-plane plus a small segment of the region of closed flux surfaces and of the private flux region. The two coordinates in 2D modeling are one along the magnetic flux surface (poloidal or x-coordinate) and other across the flux surface (radial or y-coordinate).

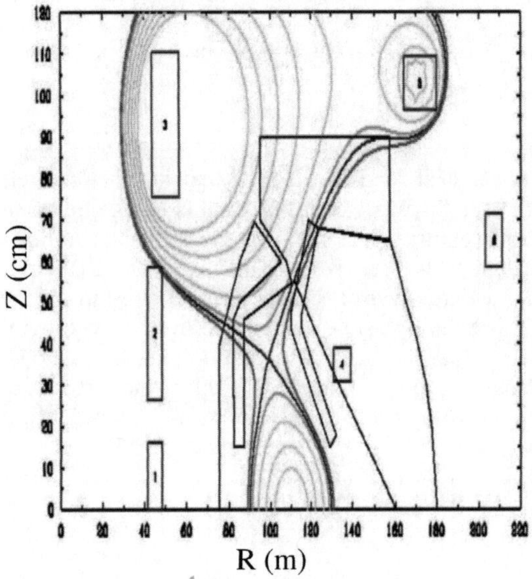

Figure 1(a). SD Plasma equilibrium

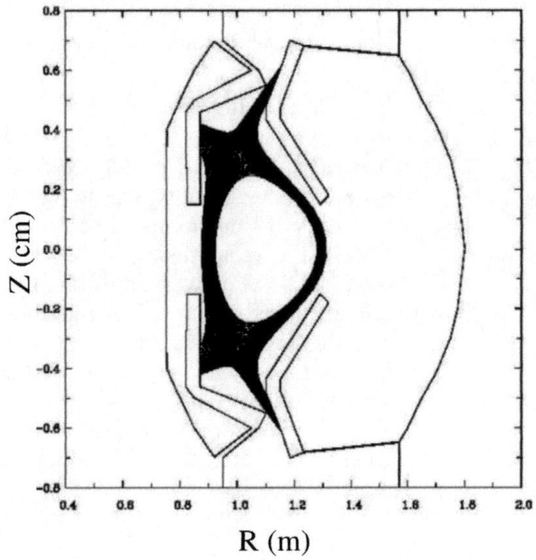

Figure 1(b). SOLPS computational grid 98×36

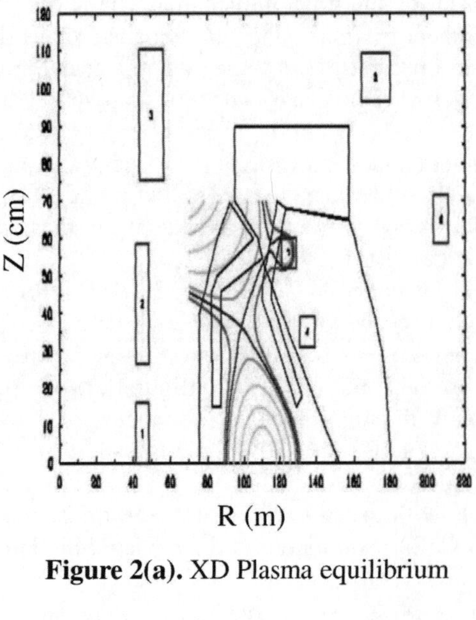

Figure 2(a). XD Plasma equilibrium

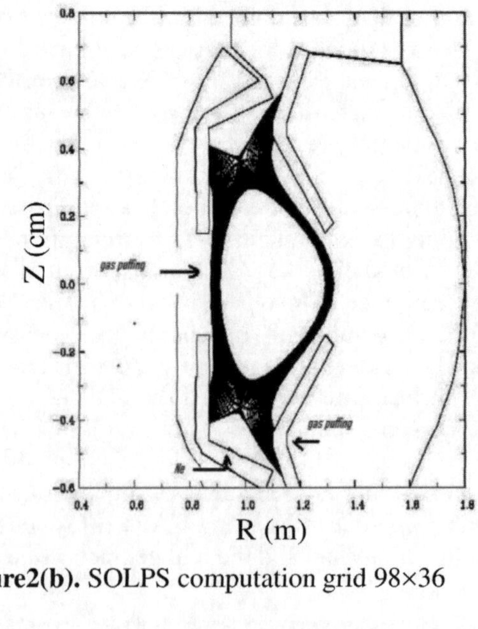

Figure2(b). SOLPS computation grid 98×36

The boundary conditions used in the SOLPS5.1 are at the core-edge interface(which is the inner most closed flux surface) is either plasma density n_e or fixed particle flux Γ_c, allowing one of them to be determined by the computations. The input power at the core-edge interface shared between the electron and ion equally and was taken as 1MW. The recycling coefficient at the walls and target surfaces are taken as unity. The out flow of all charged particles is at sound velocity. In the present modeling, the plasma species includes the neutral deuterium atom D^o, deuterium ion D+, and electron. The parallel heat and momentum transport are classical but flux limited. Perpendicular transport which is anomalous and their coefficients are kept as $D_\perp = 0.4\ m^2/s$ and $\chi_\perp = 0.6\ m^2/s$. In this study we try to check the advantages of XD over SD configuration through achieving the same upstream conditions to ensure the core equivalence of two configurations and see the effect of divertor geometry on the power deposition.

3. Optimized Modeling results of SD &XD

The equilibrium plasma configuration in SOL of SD and XD are determined by using input power P =1MW and fixed core-SOL interface density $n_e = 1 \times 10^{19}\ m^{-3}$. The radial distribution of plasma density and temperature at the inner and outer mid-planes of SD and XD are plotted in figure 3(a).The distribution of heat flux on divertors at the outboard side of XD is showing 25 % reduction as compared to SD is presented in figure 3(b). Since figure 3(a) is not showing same core conditions as expected it is difficult to say, XD geometry whose field lines are diverged as compared to SD has reduced the power deposition on the targets until unless we have the same upstream conditions.

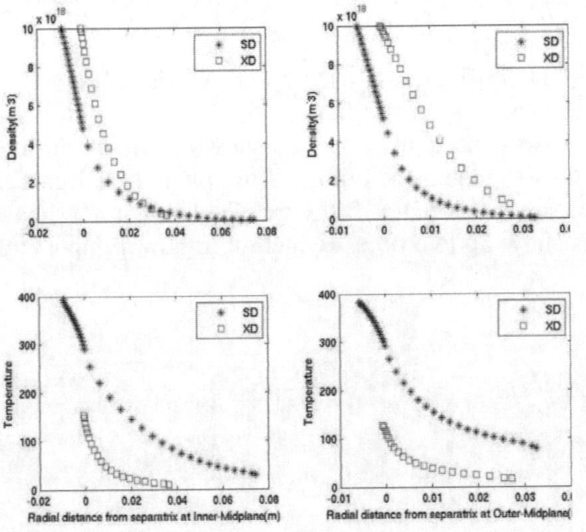

Figure 3(a). Plasma density n_e and temperature T_e at the inner and outer midplane for the input Power P = 1MW and $n_e = 1 \times 10^{19}\ m^{-3}$

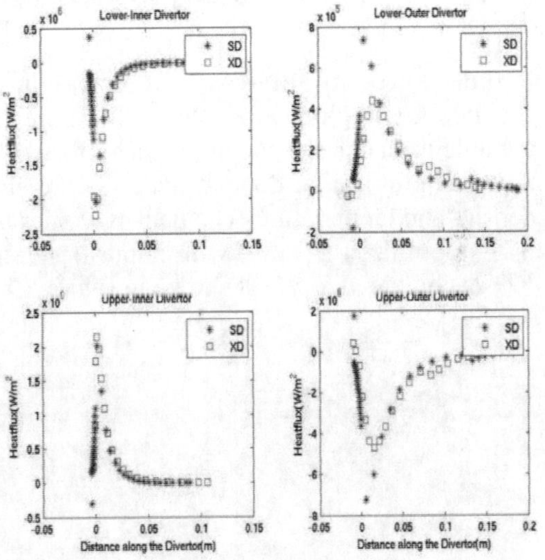

Figure 3(b). Distribution of Total heat flux as a function of distance from separatrix, at inner and outer divertors for input power P = 1 MW and fixed core-sol interface density $n_e = 1 \times 10^{19}\ m^{-3}$

To match the upstream conditions of SD and XD configurations, opted another boundary condition zero net particle flux across the core-SOL interface with gas-puff at inboard mid-plane, density values at upstream of SD and XD has 40% difference. The difference has come down when we optimized the gas-puff can be seen in figure 4(a). Optimized gas-puff condition is showing two times reduction is power deposition and shift in the peak value of heat flux as compared SD is shown in figure 4(b).

IOP Conf. Series: Journal of Physics: Conf. Series **823** (2017) 012026 doi:10.1088/1742-6596/823/1/012026

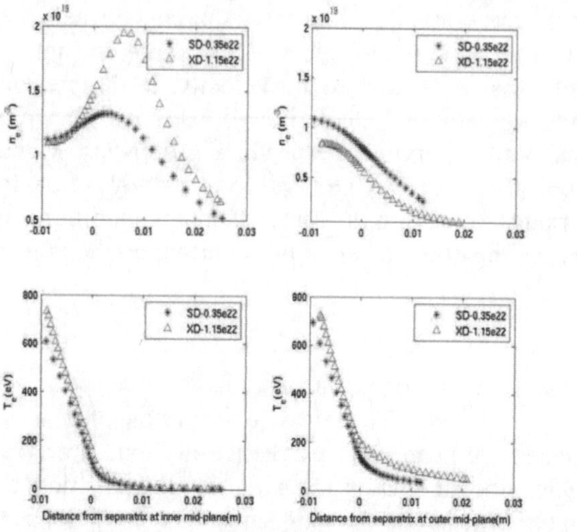

Figure 4(a). Radial distribution of Plasma density n_e and temperature T_e for input power P = 1MW with gas-puff intensities

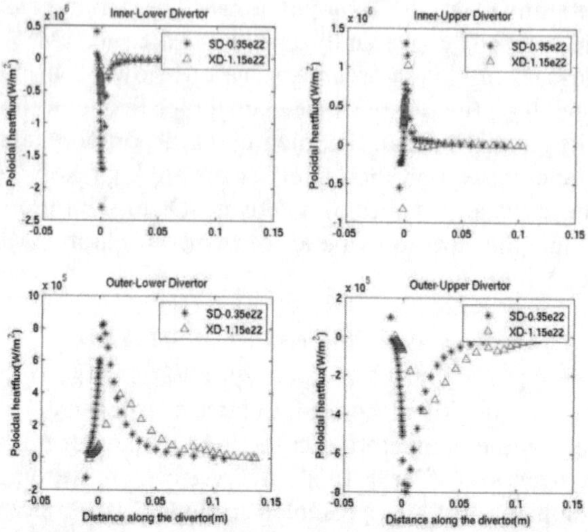

Figure 4(b). Distribution of Total heat flux on the inner and outer divertors for input power P = 1 MW with gas-puff intensities

In order to control the separatrix density in the XD configuration used Ne impurities (1×10^{20} s^{-1}) at Private Flux Region (PFR) shown in figure 2(b). When optimized the gas-puff at inner mid-plane, we are able to match the radial profiles of plasma density and temperature as shown in figure 5(a). Same kind of optimization done at the outboard side also to match the outboard midplane radial profiles of density and temperature. The density values at separatrix of SD and XD are matched but not the rest of SOL. Optimized gas-puff with impurities case is showing two times reduction in power deposition for XD as compared to SD is shown in figure 5(b).

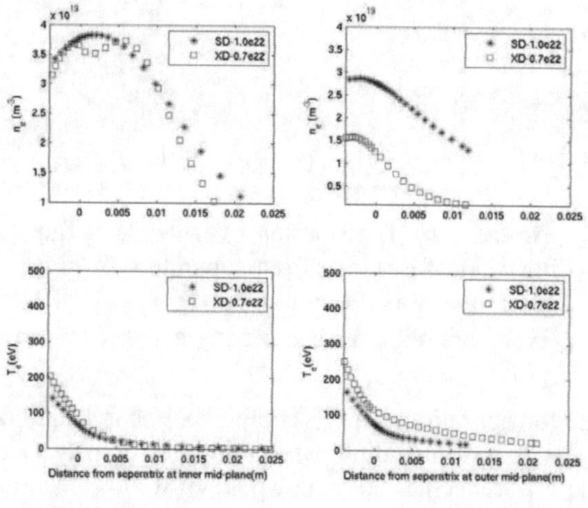

Figure 5 (a). Radial distribution of Plasma density n_e and temperature T_e for input power P = 1MW with gas-puff intensities

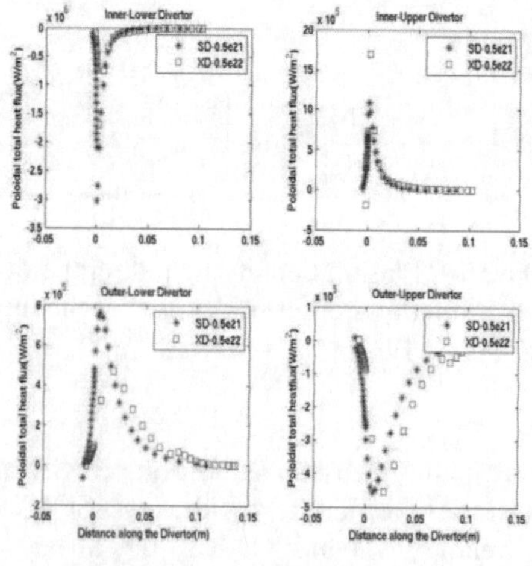

Figure 5(b). Distribution of Total heat flux on inner and outer divertors for input power P = 1 MW with gas-puff intensities

IOP Conf. Series: Journal of Physics: Conf. Series **823** (2017) 012026 doi:10.1088/1742-6596/823/1/012026

4. Summary and Conclusions

X-Divertor configuration for SST-1 tokamak analyzed computationally with SOLPS5.1 transport code to explore the relative advantages in power deposition over SD of SST-1. Optimized gas-puff values showed similar upstream conditions to ensure the core equivalence for both geometries. X-divertor configuration is showing reduction in power deposition as compared to SD. In the preliminary study, XD geometry option for SST-1 is showing very little effect on power deposition on the targets. We need further study to equivalence the core conditions and see the advantages of XD in reduction of power deposition.

References

[1] Kotschenrether M, Valanju P M, Mahajan S M, and Willey J C 2007 *Physics of Plasmas* **14** 072502

[2] Crotinger J ,LoDetro L, Pearlstein L, Tarditi A, Caeper T and Hooper E 1997 *LLNL Report UCRL-ID-126284.*

[3] Schneider R, Bonnin X, Borrass K, Coster D P, Kastelewicz H, Reiter D, RozhanskyV A and Braam B J 2006 *Contributions to Plasma Physics* **46** 3

[4] Reiter D *The EIRENE code user Manual version:11/2009*

[5] Marchand R, Dumberry M 1996 *Computer Physics Communications* **96** 232

IOP Conf. Series: Journal of Physics: Conf. Series **823** (2017) 012027 doi:10.1088/1742-6596/823/1/012027

Ponderomotive potential and second harmonic backward Raman scattering in dense quantum plasmas

Nisha Singh Rathore[1] and Punit Kumar [1]

[1]Department of Physics, University of Lucknow, Lucknow-226007, India.

E-mail: punitkumar@hotmail.com

Abstract. A theoretical model for Raman backscattering of the second harmonic of an intense laser beam propagating through quantum plasma in presence of an external magnetic field using the quantum hydrodynamical (QHD) model is presented. The effects associated with the Fermi pressure, the Bohm potential and the electron spin have been taken into account. The ponderomotive force imparts a longitudinal velocity to the electrons and the electrostatic upper hybrid Langmuir wave is excited. The second-order quantum perturbation theory shows that the second harmonic of the ponderomotive potential arises under the phase matching condition. The dispersion relation for upper hybrid Langmuir wave has been setup.

1. Introduction

Raman backward scattering (RBS) driven by intense laser wave through plasmas has long been an issue for inertial confinement fusion (ICF). It is a physical process that decreases the efficiency in ICF or particle acceleration experiments. It can also be used as diagnostic tool in the laser and plasma interaction experiments [1,2]. Raman scattering is a parametric instability in which an incident light wave decays resonantly into an electron plasma wave and a scattered light wave. Raman scattering process in the transversely magnetized plasma includes the decay of an electromagnetic pump wave into an upper hybrid wave and two scattered Stokes/anti-Stokes sidebands. The laser and the sidebands exert a ponderomotive force on electrons driving the upper hybrid wave [3,4]. RBS is significant for a number of reasons. As the Raman backward scattering mode grows to large amplitude, it can trap background plasma electrons, thus heating the plasma and creating a fast tail on the electron distribution.

Thomas et al. [5] have recently reported experimental results on Raman shifted second harmonic (RSSH) generation in the transverse direction. They have also reported a broadband emission, as a result of initial violent relativistic electron acceleration. Krushenick et al. [6] have also reported Raman shifted second harmonic radiation along with the second harmonics of Raman scattered (SHRS) light. Liu and Tripathi [7] have given a model to understand the second harmonic generation of Raman scattered light. The intense short pulse laser produces a Langmuir wave by Raman backscattering. In the plasma channel, formed in the region of wakefield, the Raman backscattered light exerts a ponderomotive force on electrons, which drives a nonlinear current at the second harmonic. Pathak and Tripathi [8] have reported the study of dispersion and growth of Raman shifted backscattered wave. All this work has been done for classical plasma.

Traditional plasma physics has mainly focused on high-temperature and low-density regimes, where quantum-mechanical effects play no role. Quantum effects become important when the particle approximation is not possible, i.e, whenever the de Broglie length does not vanish. A plasma can be regarded as quantum when the quantum nature of its particles significantly affects its macroscopic properties. Quantum plasma is a new emerging and rapidly growing subfield of plasma physics. The dense quantum plasma becomes increasingly important in the inertial confinement fusion, astrophysical plasmas, nanoscale plasmonic devices, quantum FEL, laser-plasma interaction, future generation plasma compression experiments, etc. [9-17]. The high-density, low-temperature quantum Fermi plasma is significantly different from the low-density, high-temperature "classical plasma" obeying Maxwell-Boltzmann distribution and understanding the quantum plasmas is challenging as physical processes deviate significantly from the classical prediction [18-21].

In the present paper, we have formulated the model for Raman backward scattering (RBS) in magnetized quantum plasma. The quantum mechanical excitation of the Langmuir wave in the RBS is analyzed, incorporating the electron density perturbation and quantum effects. A laser pulse propagating through highly dense plasma exerts a longitudinal ponderomotive force on electrons and imparts a longitudinal velocity to them. In the presence of transverse external static magnetic field, electrons also attain transverse velocity components and produce transverse current density components which may parametrically excite an electrostatic upper hybrid Langmuir wave and a backscattered Raman shifted second harmonic electromagnetic wave. The formulation has been built up using the quantum hydrodynamic (QHD) model [22-24]. The QHD model consists of a set of equations describing the transport of charge density, momentum (including the Bohm potential) and energy in a charged particle system interacting through a self consistent electrostatic potential [25,26]. The presentation of this paper as follows: In Section-2, the perturbed quantities have been obtained using the QHD model. Section-3 deals with the second harmonic Raman backward scattering generation and its dispersion relation has been established. Section-4 is devoted to discussion.

2. Mathematical model

We consider a laser beam, with electric field $\vec{E}_y = \hat{y}E_o \exp(k_o z - \omega_o t)$ and propagation constant $k_o = \omega_o(1-\omega_p^2/\omega_o^2)^{1/2}/c$, propagating in magnetized quantum plasma, where n_o is the initial density and $\vec{B} = \hat{y}B_s$ is the external static magnetic field. The plasma frequency being defined as $\omega_p = \left[n_o e^2/(m\varepsilon_o)\right]^{1/2}$, where e and m are electronic charge and rest mass respectively. The QHD equations for the laser and quantum plasma interaction are given as,

$$\frac{\partial \vec{v}}{\partial t} = -\frac{e}{m}\left[\vec{E} + \frac{1}{c}(\vec{v}\times\vec{B})\right] - \frac{v_F^2}{n_o^2}\frac{\vec{\nabla}n^3}{n} + \frac{\hbar^2}{2m^2}\vec{\nabla}\left(\frac{1}{\sqrt{n}}\vec{\nabla}^2\sqrt{n}\right) - \frac{2\mu_B}{m\hbar}\vec{S}.\nabla\vec{B}, \tag{1}$$

$$\left(\frac{\partial}{\partial t} + \vec{v}.\vec{\nabla}\right)\vec{S} = \left(\frac{2\mu_B}{\hbar}\right)\left(\vec{B}\times\vec{S}\right), \tag{2}$$

$$\frac{\partial n}{\partial t} + \vec{\nabla}.(n\vec{v}) = 0. \tag{3}$$

where, $\vec{v}$ is the velocity, $\hbar$ is the Planck's constant divided by 2π, v_F is the Fermi velocity and S is the spin angular momentum with $|S_o| = \hbar/2$ and $\mu = (-g/2)\mu_B$,with $g = 2.0023193$ and $\mu_B = e\hbar/2m$ being the Bohr magneton. The wave equation for the current source is.

$$\left(\vec{\nabla}^2 - \frac{1}{c^2}\frac{\partial^2}{\partial t^2}\right)\vec{E}(r,t) = \frac{4\pi}{c^2}\frac{\partial \vec{J}}{\partial t}. \tag{4}$$

where, $\vec{J}$ is the current density.

The first order perturbative expansion of equations (1)-(3) gives,

$$\frac{\partial \vec{v}^{(1)}}{\partial t} = -\frac{e}{m}\vec{E}^{(1)} - \frac{v_F^2}{n_o}\vec{\nabla}.n^{(1)} + \frac{\hbar^2}{4m^2 n_o}\vec{\nabla}(\vec{\nabla}^2 n^{(1)}) - \frac{2\mu_B}{m\hbar}\vec{S}_o.(\vec{\nabla}B^{(1)}), \tag{5}$$

$$\frac{\partial \vec{S}^{(1)}}{\partial t} = -\frac{2\mu_B}{\hbar}(\vec{B}^{(1)} \times S_o), \tag{6}$$

$$\frac{\partial n^{(1)}}{\partial t} + \left(n_o \vec{\nabla}.\vec{v}^{(1)} + \vec{\nabla}n_o.\vec{v}^{(1)}\right) = 0. \tag{7}$$

Simultaneous solution of the above equations gives the transverse and longitudinal quiver velocity acquired by the plasma electrons,

$$\vec{v}_\perp^{(1)} = \left[\begin{matrix} \dfrac{2\eta k_o}{n_o\omega_o}K_q + \dfrac{\mu_B S_o E_o k_o}{m\hbar\omega_o}\left\{\dfrac{1}{\omega_o} - \dfrac{\omega_c}{\omega_o^2-\omega_c^2}\right\} + \\ \dfrac{eE_o}{im\omega_o} + \dfrac{K_q\eta k_o(\omega_o-\omega_c)}{n_o(\omega_o^2-\omega_c^2)} \end{matrix}\right] \exp i(k_o z - \omega_o t) - \frac{\omega_c}{i\omega_o}v_z^{(1)}. \tag{8}$$

and

$$v_z^{(1)} = \left[\frac{K_q\eta k_o(\omega_o-\omega_c)}{n_o(\omega_o^2-\omega_c^2)} - \frac{\mu_B S_o\omega_c k_o E_o}{m\hbar(\omega_o^2-\omega_c^2)}\right]\exp i(k_o z - \omega_o t). \tag{9}$$

where, $\omega_c (= eB_S/mc)$ is the cyclotron frequency and $K_q = \frac{ik_o}{n_o\omega_o}\left\{\frac{k_o^2\hbar^2}{4m^2} + v_F^2\right\}$. Using the continuity equation the first ordered perturbed plasma electron density is obtained as,

$$n^{(1)} = \eta \exp i(k_o z - \omega_o t), \tag{10}$$

where,

$$\eta = \left[\frac{\dfrac{2\mu_B S_o n_o k_o^2 E_o}{\omega_o(\omega_o^2-\omega_c^2)}\left\{\dfrac{1}{\omega_o} - \dfrac{\omega_c}{\omega_o^2-\omega_c^2}\right\} + \dfrac{en_o k_o E_o}{im\omega_o^2} + \dfrac{\mu_B S_o k_o^2\omega_c^3 E_o}{im\hbar\omega_o^2(\omega_o^2-\omega_c^2)}}{1 - \dfrac{2K_q k_o^2}{\omega_o^2} - \dfrac{K_q k_o^2(\omega_o - i\omega_c)}{\omega_o(\omega_o^2-\omega_c^2)}\left(1 - \dfrac{\omega_c}{i\omega_o}\right)}\right].$$

The quiver velocity exerts an oscillatory nonlinear ponderomotive force on electrons, $\vec{F}_P^{(2)} = -e\vec{v}^{(1)} \times \vec{B}^{(1)}/2$, where $\vec{B}^{(1)} = k_o \times \vec{E}^{(1)}/\omega_o$. This ponderomotive force impart an oscillatory velocity to the electron, which couple with the external magnetic field to give transverse and longitudinal velocity components at frequency $\left(2\omega_{o,} 2\vec{k}_o\right)$as,

$$\vec{v}_x^{(2)} = \left[-\frac{\omega_c}{2i\omega_o}v_z^{(2)} - \frac{\omega_o k_o B_s S_o \mu_B^3 E_o}{(m\omega_o^4\hbar^3 + 2\mu m\hbar^2\omega_o^4(B_s)^2}\right]\exp 2i(k_o z - \omega_o t), \tag{11a}$$

$$\vec{v}_z^{(2)} = \left[\frac{\dfrac{e^2 E_o^2}{4cm^2\omega_o^2} + \dfrac{k_o K_q \eta E_o}{4in_o\omega_o^2} + \dfrac{2\omega_c i\omega_o k_o S_o B_s \mu_B^3 E_o}{(m\omega_o^3\hbar^3 + 2\mu m\omega_o\hbar^2 (B_s)^2}}{4\omega_o^2 - \omega_c^2 + \omega_p^2} \right] \exp 2i(k_o z - \omega_o t). \quad (11b)$$

The laser pump via current density $\vec{J}_2 = -en_o\vec{v}^{(2)} + J_s^{(2)}$, parametrically excites a Langmuir wave $\phi_\omega = A\exp i(kz - \omega t)$ and a backscattered electromagnetic wave with electric and magnetic fields $\vec{E}_1 = A_1 \exp i(k_1 z - \omega_1 t)$ and $\vec{B}_1 = \vec{\nabla}\times\vec{E}_1 / i\omega_1$ respectively, where $\omega = 2\omega_o + \omega_1$ and $\vec{k} = 2\vec{k}_o + k_1$.

The scattered wave and the external magnetic field impart oscillatory velocity to electrons and the scattered wave exerts a ponderomotive force on electrons at (ω, k). The components of ponderomotive force are,

$$(F_{P\omega})_x = \frac{e}{2c} v_z^{(2)} B_y^{(1)} + \frac{e\omega_c v_z^{(2)}\left(\dfrac{k_1}{2} + k_o\right)}{2i\omega_1\omega_o\left(1 - \dfrac{\omega_c^2}{\omega_1^2}\right)} \left[\left\{E_{1x} - \frac{\omega_c}{i\omega_1}E_{1z}\right\} - \frac{imK_{q1}k_1^2}{\omega_o\omega_1\delta_q}\left\{E_{1z} + \frac{\omega_c}{i\omega_1}E_{1x}\right\}\right], \quad (12)$$

$$(F_{P\omega})_z = \frac{e\omega_c}{4i\omega_o c} v_z^{(2)} B_y^{(1)} - \frac{ev_z^{(2)}\left(\dfrac{k_1}{2} + k_o\right)}{\omega_1\left(1 - \dfrac{\omega_c^2}{\omega_1^2}\right)} \left[\left\{E_{1z} + \frac{\omega_c}{i\omega_1}E_{1x}\right\} - \frac{mK_{q1}k_1^2}{\omega_o\omega_1\delta_q}\left\{E_{1z} + \frac{\omega_c}{i\omega_1}E_{1x}\right\}\right]. \quad (13)$$

where,

$$K_{q1} = \left(v_F^2 + \frac{\hbar^2 k_1^2}{4m^2}\right) \text{ and } \delta_q = 1 - \frac{K_{q1}k_1^2}{\omega_1}.$$

This force drives the Langmuir wave at ϕ_ω. The electron's response at (ω, k) is obtained, on solving the above equations of force which give the velocities as

$$\vec{v}_{\omega x} = -\frac{k_1 e v_z^{(2)} E_{1x}}{2cmi\omega\omega_1} + \frac{e\omega_c v_z^{(2)}\left(\dfrac{k_1}{2} + k_o\right)}{2m\omega_1\omega_o\omega\left(1 - \dfrac{\omega_c^2}{\omega_1^2}\right)} \left[\left\{E_{1x} - \frac{\omega_c}{i\omega_1}E_{1z}\right\} - \frac{\omega m K_{q1}k_1^2}{i\omega_c\omega_1\delta_q\left(1 - \dfrac{\omega_c^2}{\omega_1^2}\right)}\left\{E_{1z} + \frac{\omega_c}{i\omega_1}E_{1x}\right\}\right]$$
$$- \frac{\omega_c}{i\omega} v_{\omega z}, \quad (14)$$

and

$$\vec{v}_{\omega z} = -\frac{ek\phi_\omega\omega}{m\omega^2\left(1 - \dfrac{\omega_c^2}{\omega^2}\right)} + \left[\frac{\omega_c k_1}{2mc\omega_1\omega^2\left(1 - \dfrac{\omega_c^2}{\omega^2}\right)}\chi_1 - \omega_c e\alpha\chi_5 + eK_{q1}\chi_2\alpha\right] ev_z^{(2)} E_{1x}$$

$$-\left[\alpha\left\{K_{q1}\chi_3+\chi_4\right\}\right]ev_z^{(2)}E_{1z}, \tag{15}$$

where, $\alpha=\dfrac{1}{\omega^2\omega_1\left(1-\dfrac{\omega_c^2}{\omega^2}\right)\left(1-\dfrac{\omega_c^2}{\omega_1^2}\right)}\left(\dfrac{k_1}{2}+k_o\right)$, $\chi_1=\left(\dfrac{\omega}{2\omega_o}+1\right)$, $\chi_2=\left(\dfrac{\omega\omega_c k_1^2}{\delta_q\omega_1^2}+\dfrac{eB_s k_1^2\omega_c^2}{2\omega_o mc\delta_q\omega_1^2}\right)$,

$$\chi_3=\left(\frac{\omega}{\omega_1\delta_q}+\frac{eB_s\omega_c ik_1^2}{2\omega_o mc\delta_q\omega_1}\right),\ \chi_4=\left(i\omega+\frac{\omega_c^3}{2\omega_o\omega_1}\right),\ \chi_5=\left(\frac{\omega}{\omega_1}-\frac{\omega_c}{2i\omega_o}\right).$$

With the help of the Poisson's equation $n_\omega=\dfrac{-k^2\varepsilon_o\phi_\omega}{e}$, we get

$$\left(1-\frac{\omega_p^2}{\omega^2-\omega_c^2}\right)\phi_\omega=A\,E_{1x}+B\,E_{1Z}. \tag{16}$$

where,

$$A=-\omega_p^2\left\{\frac{\omega_c k_1}{2\omega_1\omega kc}\chi_1-\frac{\omega_c\alpha m}{k\omega}\chi_5+\frac{\alpha}{k\omega}K_q\chi_2\right\}v_z^2,$$

$$B=\omega_p^2\left\{\frac{\alpha m}{k\omega}\chi_4-\frac{\alpha}{k\omega}K_q\chi_3\right\}v_z^2,$$

and

$$C=\left(1-\frac{\omega_p^2}{\omega^2-\omega_c^2}\right).$$

3. Dispersion relation

The current density at (ω_1,k_1) represent current density due to free electrons, coupling of Langmuir wave with the ponderomotive force at $(2\omega_{o,}2k_o)$ and current due to spin magnetic moment. The current density of this source current is,

$$\vec{J}=\vec{J}_c+\vec{J}_s=-en\vec{v}+\frac{2\mu_B}{\hbar}\vec{\nabla}\times\left(n\vec{S}\right), \tag{17}$$

whose components are,

$$\vec{J}_{1x}=\vec{J}_{c1x}+\vec{J}_{s1x}$$

$$= \left[\frac{e^2 n_o}{mi\omega_1\left(1-\frac{\omega_c^2}{\omega_1^2}\right)} \left\{ -1 + \frac{m\omega_c K_{q1} k_1^2}{\omega_1 \delta_q} + \frac{2\mu_B \omega_c S_o k_1^2}{\delta_q e \omega_1^2 \hbar} \right\} - \frac{4S_o n_o k_1^2 \mu_B^2}{h^2 \omega_1^2} \right] E_{1x} +$$

$$\left[\frac{e^2 n_o}{m\omega_1^2\left(1-\frac{\omega_c^2}{\omega_1^2}\right)} \left\{ -\omega_c + \frac{mK_{q1} k_1^2}{\delta_q} + \frac{2\mu_B S_o k_1^2}{e\delta_q h} \right\} \right] E_{1z} - \left[\frac{\varepsilon_o B_s k_o S_o k^2 \mu_B^2}{m\hbar^2 \omega_o^3 (\hbar + 2\mu_b \omega_o^4 (B_s)^2)} \left(e^{2i(k_o z - \omega_o t)}\right)^* \right] \phi_\omega, \tag{18}$$

and

$$\vec{J}_{1z} = \vec{J}_{c1z} + \vec{J}_{s1x},$$

$$= \left[\frac{e^2 n_o \omega_c}{m\omega_1^2\left(1-\frac{\omega_c^2}{\omega_1^2}\right)} \left\{ 1 + \frac{mK_{q1} k_1^2}{e\delta_q \omega_1 i} + \frac{2\mu_B S_o k_1^2}{\delta_q e \omega_1 \hbar} \right\} - \frac{4S_o n_o k_1^2 \mu_B^2}{h^2 \omega_1^2} \right] E_{1x}$$

$$- \left[\frac{e^2 n_o}{m\omega_1\left(1-\frac{\omega_c^2}{\omega_1^2}\right)} \left\{ \frac{1}{i} + \frac{mK_{q1} k_1^2}{\omega_1 e \delta_q} + \frac{2\mu_B S_o k_1^2}{\delta_q e\hbar\omega_1} \right\} \right] E_{1z} - \left[\frac{\varepsilon_o k^2 (v_z^{(2)})^*}{2} \right] \phi_\omega. \tag{19}$$

Using Maxwell's equation and the current density component from equations (18) and (19), we get the governing wave equations for the Raman scattered wave,

$$[A_1]E_{1x} - [B_1]E_{1z} = -[C_1]\phi_\omega, \tag{20}$$

$$[A_2]E_{1x} + [B_2]E_{1z} = [C_2]\phi_\omega, \tag{21}$$

where,

$$A_1 = \left[k_1^2 - \frac{\omega_1^2}{c^2} - \frac{1}{c^2} \frac{\omega_P^2}{\left(1-\frac{\omega_c^2}{\omega_1^2}\right)} \left\{ 1 + \frac{m\omega_C K_{q1} k_1^2}{\omega_1 \delta_q} + \frac{2\mu_B S_o \omega_c k_1^2}{\delta_q e\hbar\omega_1^2} \right\} - \frac{4iS_o m\omega_p^2 k_1^2 \mu_B^2}{e^2 h^2 \omega_1} \right],$$

$$B_1 = \left[\frac{1}{c^2} \frac{\omega_P^2}{\left(1-\frac{\omega_c^2}{\omega_1^2}\right)} \left\{ -\frac{\omega_c}{\omega_1} + \frac{miK_{q1} k_1^2}{\omega_1 \delta_q} + \frac{2i\mu_B S_o \omega_p^2 k_1^2}{\delta_q h \omega_1} \right\} \right],$$

$$C_1 = \left[\frac{i\omega_1 S_o B_s k_o k^2 \mu_B^2}{mc^2\hbar^2\omega_o^3(\hbar + 2\mu_B\omega_o^4(B_s)^2)} \left(e^{2i(k_o z - \omega_o t)}\right)^* \right],$$

$$A_2 = \left[-\frac{1}{c^2}\frac{\omega_c\omega_p^2}{m\omega_1\left(1-\frac{\omega_c^2}{\omega_1^2}\right)}\left\{i + \frac{mK_{q1}k_1^2}{\omega_1\delta_q} + \frac{2\mu_B S_o k_1^2}{e\omega_1\hbar\delta_q}\right\} - \frac{4iS_o n_o \omega_P^2 k_1^2 \mu_B^2}{e^2h^2\omega_1} \right],$$

$$B_2 = \left[-\frac{\omega_1^2}{c^2} - \frac{1}{c^2}\frac{\omega_P^2}{\left(1-\frac{\omega_c^2}{\omega_1^2}\right)}\left\{1 + \frac{imK_{q1}k_1^2}{\omega_1 e\delta_q} + \frac{2i\mu_B S_o k_1^2}{\delta_q eh\omega_1}\right\} \right],$$

and

$$C_2 = \left[\frac{i\varepsilon_o\omega_1 k^2 (v_z^{(2)})^*}{2c^2} \right].$$

Combining the equations (16), (20) and (21), we get the nonlinear dispersion relation for the coupling of second harmonic pump and decay wave as,

$$\left[k_1^2 - \frac{\omega_1^2}{c^2} - \frac{1}{c^2}\frac{\omega_P^2}{\left(1-\frac{\omega_c^2}{\omega_1^2}\right)}\left\{1 + \frac{m\omega_C K_{q1}k_1^2}{\omega_1\delta_q} + \frac{2\mu_B S_o \omega_c k_1^2}{\delta_q e\hbar\omega_1^2} - \frac{4iS_o m\omega_p^2 k_1^2 \mu_B^2}{e^2h^2\omega_1}\right\} \right] = \frac{B_1 \times \{(C_2 \times A) - (A_2 \times C)\}}{R}, \tag{22}$$

where,

$$R = \left(\frac{\frac{i\omega_1\alpha\omega_P^2 k^2}{2k\omega c^2}(m\chi_4 - K_q\chi_3)(\omega^2 - \omega_c^2)|v_z^{(2)}|^2}{(\omega^2 - \omega_c^2 - \omega_P^2)} \right) + \frac{\omega_1^2}{c^2}\left(1 - \frac{\omega_P^2}{\omega_1^2 - \omega_c^2}\right)\left\{1 + \frac{iK_{q1}k_1^2}{e\omega_1\delta_q} + \frac{2i\mu_B S_o k_1^2}{\hbar\omega_1 e\delta_q}\right\}.$$

4. Disucssion

A model for excitation of Raman shifted harmonic electromagnetic waves in plasma has been formulated using the recently developed quantum hydrodynamic (QHD) model. The effects of quantum statistical pressure, the Bohm potential and the electron spin have been taken in to account. The applied static field is responsible for the transverse velocity attained by the electrons and for dispersion of the Raman shifted harmonic backscattered wave. The dispersion relation contains the term due to the interaction of electron quiver velocity with the constant magnetic field under the influence of electron's spin and ponderomotive force exerted by the radiation field on plasma electrons under the influence of electron spin and other quantum effects. The quantum terms contribute to diffraction and results in an increase in dispersion. The second harmonic backscattering can be employed as a diagnostic tool for the presence of transverse static magnetic field in laser produced quantum plasma and also to study the properties of excited plasma wave in various experiments

IOP Conf. Series: Journal of Physics: Conf. Series **823** (2017) 012027 doi:10.1088/1742-6596/823/1/012027

Acknowledgement
This work was performed under the financial assistance from the UGC India, under fellowship award letter no. F.4-1/2006(BSR)/2007 (BSR) dated 22nd Oct., 2013.

References
[1] Kim J, Lee H J, Suk H. and Koa I S 2003 *Phys. Lett. A* **314** 464
[2] Purohit G, Sharma P and Sharma R P 2010 *Phys. Lett. A* **374** 866
[3] Paknezhad A and Dorranians D 2011 *Laser part. Beam* **29** 373
[4] Paknezhad A 2013 *Phys. Plasmas* **20** 012110
[5] Thomos A G R, Mangles S P D, Najmudin Z, Kaluza M C, Murphy C D and Krushelnick K 2007 *Phys. Rev. Lett.* **98** 054802
[6] Krushenic K, Ting A, Burris H R, Fisher A, Manka C, and Esarey E 1995 *Phys. Rev. Lett.* **75** 3681
[7] Liu C S and Tripathi V K 1996 *Phys. Rev. E* **54** 4098
[8] Pathak B V and Tripathi V K 2008, *Phys. Plasmas* **15** 092104
[9] Tabak M, Hammer J, Glinsky M E, Kruerand W L , Wilks S C, Woodworth J, Campbell E M,. Perry M J, and Mason R J 1994 *Physics of Plasmas* **1** 1626
[10] Son S and Fisch N J 2005 *Phys. Rev. Lett.* **95** 225002
[11] Son S and Fisch N J 2004 *Phys. Lett. A* **329** 16
[12] Son S and Fisch N J 2006 *Phys. Lett. A* **356** 65
[13] Son S and Fisch N J 2006 *Phys. Lett. A* **356** 72
[14] Son S and Fisch N J 2005 *Phys. Lett. A* **337** 397
[15] Strickland D and Mourou 1985 *G Phys. Fluids* **55** 447
[16] Perry M D and Mourou 1994 *Science* **264** 917
[17] Brabec T and Krausz F 2000 *Rev. Mod. Phys.* **72** 545
[18] Son S and Ku S 2009 *Phys. Plasmas* **17** 010703
[19] Son S and Moon S J 2012 *Phys. Plasmas* **19** 063102
[20] Son S, Ku S, and Moon S J 2010 *Phys. Plasmas* **17** 114506
[21] Shukla P K and Eliasson B 2010 *Physics-Uspekhi* **53** 51
[22] Gardner C L and Ringhofer C 1996 *Phys. Rev. E* **53** 157
[23] Shukla P K *and* Eliasson B 2006 *Phys. Rev. Lett.* **96** 24500
[24] Cheng Z E and Shan W U 2012 *EPL* **100** 45004
[25] Misra A P and Shukla P K 2007 *Phys. Plasmas* **14** 082312.
[26] Misra A P and Shukla P K 2008 *Phys. Plasmas* **15** 052105

IOP Conf. Series: Journal of Physics: Conf. Series **823** (2017) 012028 doi:10.1088/1742-6596/823/1/012028

Behaviour of TSI in Magnetized Plasmas for different Plasma Oscillation to Debye Wavelength ratio

Jyoti[1] and Rakhee Malik[2]

[1]Department of Physics, Manipal University Jaipur, Jaipur 303007, India.
[2]Centre for Energy Studies, IIT Delhi, New Delhi – 110 016, India.

E-mail: iitd.jyoti@gmail.com

Abstract. The continuity and the momentum equation which take into account the constant rate of ionization are formulated for ions and the electrons. Using normal mode analysis along with linear approximation, potential is found from Poisson's equation neglecting higher order perturbed terms. From Potential equation, dispersion relation is generated which is solved numerically for obtaining the value of ω using typical plasma parameters. The behavior of growth rate with magnetic field and the propagation angle along with ionization constant has been studied with different plasma oscillation wavelength to Debye length ratio.

1. Introduction

Investigators have used different techniques to find out the unstable waves (instabilities) in different plasma models [1-10]. For example, Srivastava et al. [1] have applied the perturbation technique and used macroscopic equations to investigate the nonlinear development of two stream instability of magnetized finite temperature plasma having two electron beams. They found the enhanced growth rate due to the nonlinear effects. El-Labany [3] has used the multiple timescales perturbation theory to investigate the case of a weakly unstable two warm streams near a marginally stable state. Physical interpretation of the quantum two stream instability has been given by Haas et al. [4], where they talked about the concept of negative and positive energy waves by factorization of dispersion relation into different sectors. Plasma thermal effects on two stream instabilities in a hot beam warm plasma system have been studied by Hao et al. [5]. They found that the inclusion of higher order nonlinear terms leads to periodic, but not sinusoidal, solutions for the perturbed electric field. Karimov and Shcheglov [8] have considered the influence of an external ionization source on the characteristics of Langmuir oscillations in partially ionized plasma by neglecting the pressure gradient for the electrons and their friction on neutral particles and ions. But they did not include the effect of magnetic field and plasma inhomogeneity. This is clear that the researchers have investigated the effect of ion temperature, density gradient, magnetic field, etc. on different types of instabilities. Two stream instabilities have been studied by many researchers but only few studied them in magnetized plasmas [3-8]. Some have considered them in magnetized plasmas too [2, 10] but without considering the effect of ionization. Localized Solitary waves have also been studied in magnetized plasmas under the ionization effect without considering the evolution of unstable waves which can arise into such plasmas as instabilities [11-14]. In this article we have considered TSI taking into consideration of both the magnetic field as well as ionization.

2. Basic formulation

An inhomogeneous plasma having ions and electrons is considered to be immersed in a static magnetic field $\vec{B}_0$ applied in the z-direction. One fluid is taken to move at a faster rate than the other. Consistent with the case of Karimov and Shcheglov [8], we have considerd the simplest case, where the ionization takes place at a constant rate. This type of situation is possible when the medium is ionized by an external source, for example, by a particle beam or laser radiation. Hence, the following basic fluid equations are written, which govern the plasma response to the wave excitation at an angle θ with the magnetic field in the (x, z) plane

$$\frac{\partial n_j}{\partial t}+\nabla\cdot\left(n_j\vec{v}_j\right)-\gamma_j=0 \tag{1}$$

$$m_j n_j\frac{\partial \overrightarrow{v_j}}{\partial t}+m_j n_j\left(\overrightarrow{v_j}\cdot\vec{\nabla}\right)\overrightarrow{v_j}+n_j Z_j e\vec{\nabla}\varphi-n_j Z_j e\overrightarrow{v_j}\times\overrightarrow{B_0}+\vec{\nabla}p_j+m_j\gamma_j\overrightarrow{v_j}=0, \tag{2}$$

$$\varepsilon_0\nabla^2\varphi-\left(n_e-Z_p n_p\right)e=0. \tag{3}$$

Here j=e, p corresponds to the electron and ion. Here n_p (m_p) and n_e (m_e) represent the density (mass) of the ions and electrons, respectively, $\vec{v}(\vec{u})$ is the ion (electron) fluid velocity, Zp is the number of charge on the ion, γ_p and γ_e are the ionization rate constants for the ions and electrons, φ is the electrostatic potential.

We carried out normal mode analysis of Equations (1) - (3), for which the physical quantities $n_p, n_e, v_x, v_y, v_z, u_x, u_y, u_z, \varphi$ ($\equiv M$) were expanded as $M=M_0+M_1$, where M_0 is the unperturbed part and M_1 is the perturbed part of the relevant quantity. This method is easy to apply as it does not require any strong mathematics because it includes the linearization of higher order terms. But at the same time this method is quite cumbersome as it results into high order of the polynomial equations with complicated coefficients and it is not easy to get any analytical solution for the growth rates of the instabilities.

We obtained perturbed densities for both the electrons and ions. Using those perturbed densities we found perturbed potential from Poisson's equation in terms of unperturbed quantities.

$$\varphi_1=\frac{(Ai+B)}{(Ci+D)}\varphi_0 \tag{4}$$

Since the first order quantities cannot be explained explicitly in terms of unperturbed quantities, we have made R. H. S. indetermined by putting both numerator as well as denominator equal to zero separately. This way we got dispersion relation in the form of polynomial equations of order seven and eight. These equations are solved numerically using MATLAB for the values of growth rate (imaginary part of ω) by giving typical values of $n_{e0}, n_{p0}, u_{x0}, u_{y0}, u_{z0}, v_{x0}, v_{y0}, v_{z0}$, and Te. The negative imaginary part of the complex root gives the instability, whereas its positive part gives the damped wave. It means the oscillations are unstable when Im(ω) < 0.

IOP Conf. Series: Journal of Physics: Conf. Series **823** (2017) 012028 doi:10.1088/1742-6596/823/1/012028

3. Results and discussion

In our basic fluid equations we have considered the Poisson's equation whose general solution is dependent on the Debye length. We have considered perturbations in potential which tends to fall off with Debye length. Since we have not obtained any analytical solution for the potential due to its complicated expression in terms of unperturbed quantities we observed the instabilities behavior with respect to different values of plasma oscillation to Debye wavelength ratio and it has been observed that growth rates are sensitive to Debye length. We mainly focus on numerical solutions using typical parameters. We obtained two type of instability one of which is growing at a larger wavelength of oscillations and another one at smaller plasma oscillation wavelength similar to the one obtained by Lashmore-Davies [9]. We studied the normalized growth rate $\Gamma = \omega_{Im} / \omega_{pi}$ of instability under the effect of magnetic field, ionization rate constant and the obliqueness (the angle between the magnetic field and the wave vector $\vec{k}$) for small ($\lambda > \lambda_{De}$) and large ($\lambda >> \lambda_{De}$) wavelength of oscillations where λ_{De} is the Debye length.

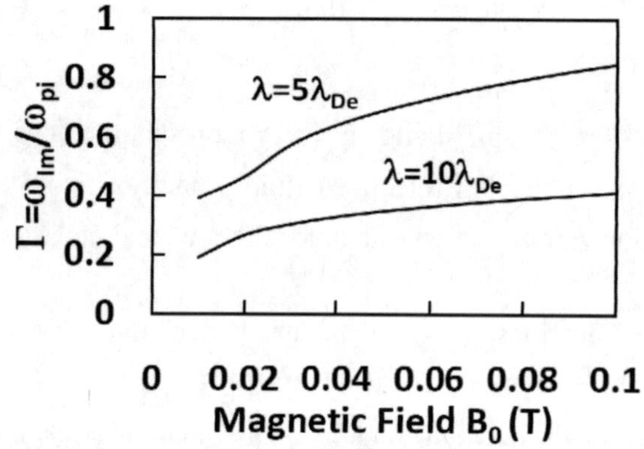

Figure 1. Variation of the growth rate of the instability growing at smaller oscillation wavelength w.r.t. magnetic field at two different values of plasma oscillation to Debye wavelength ratio when $n_{e0} = n_{p0} = 2\times10^{16}$ m^{-3}, $m_{p0} = 1.67\times10^{-27}$ kg, $m_{e0} = 9.1\times10^{-31}$ kg, $v_{x0} = v_{y0} = v_{z0} = 0.3$ Cs, $u_{x0} = u_{y0} = u_{z0} =$ 6Cs, $\gamma = 10^8$, $\theta = 60^0$, $T_e = 3eV$ and $T_p = 0.3eV$.

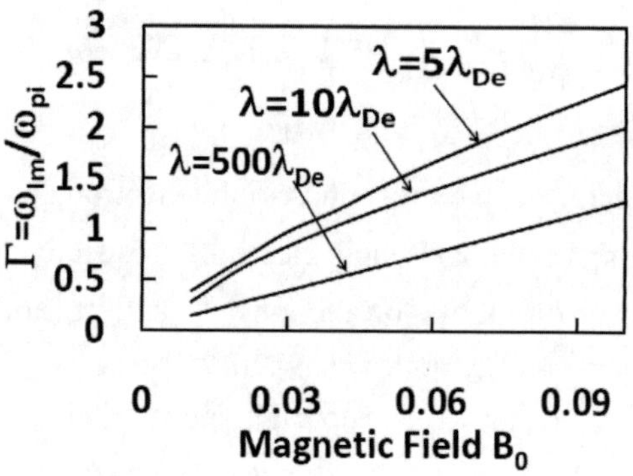

Figure 2. Variation of the growth rate of the instability growing at larger oscillation wavelength w.r.t. magnetic field at two different values of plasma oscillation to Debye wavelength ratio when $n_{e0} = n_{p0} = 2\times10^{16}$ m^{-3}, $m_{p0} = 1.67\times10^{-27}$ kg, $m_{e0} = 9.1\times10^{-31}$ kg, $v_{x0} = v_{y0} = v_{z0} = 0.3$ Cs, $u_{x0} = u_{y0} = u_{z0} =$ 6Cs , $\gamma = 10^8$, $\theta = 60^0$, $T_e = 3eV$ and $T_p = 0.3eV$.

We observed two instabilities as shown in figures 1 and 2. Both the instabilities have certain limit for the plasma oscillation wavelength for their excitation. One of the instability would occur at smaller oscillation wavelength while the other would occur at larger value of oscillation wavelength. We observed that the upper limit for the instability occurring at the smaller oscillation wavelength can be extended to even thousands of Debye wavelength with the increase in magnetic field while the instability occurring at the larger oscillation wavelength behaves opposite means it is seen growing at smaller oscillation of wavelength (in this case $\lambda = 5\lambda_{De}$ and $\lambda = 10\lambda_{De}$) even at magnetic field as strong as 30T and 20T while at larger oscillation wavelength ($\lambda = 500\lambda_{De}$) it diminishes at much smaller value of magnetic field ~ 0.2 T. Along with this we can see from the figures 1 and 2 that both the instabilities have higher magnitude of growth rate at smaller oscillation wavelength and it increases with the increase in the magnetic field . Higher growth rate with increase in magnetic field is also observed by Aggarwal [10] as shown in figures 1 and 2 but for the longer wavelength side we observe lower magnitude of growth rate which is opposite to Aggarwal [10].

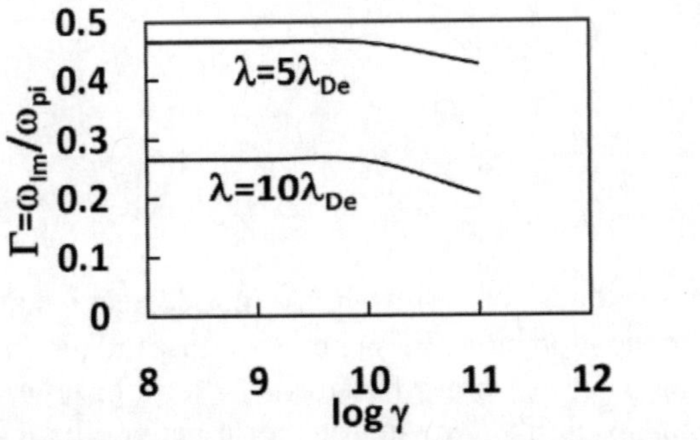

Figure 3. Variation of the growth rate of the instability growing at smaller oscillation wavelength w.r.t. ionization rate constant with B_0 = 0.02T, keeping rest parameters same as that of figure 1.

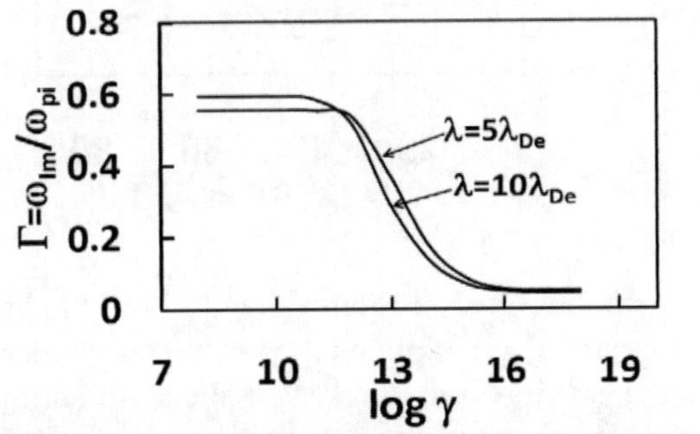

Figure 4. Variation of the growth rate of the instability growing at smaller oscillation wavelength w.r.t. ionization rate constant with B0 = 0.02 T, $n_{e0}=n_{p0}= 2\times10^{16}$ m^{-3}, $m_{p0}=1.67\times10^{-27}$ kg, $m_{e0}= 9.1\times10^{-31}$ kg, $v_{x0} = v_{y0} = v_{z0}$ = 0.3 Cs, $u_{x0} = u_{y0} = u_{z0}$ = 6 Cs, θ = 60^0, $T_e = 3eV$ and $T_p = 0.3eV$.

In figures 3 & 4, we observe higher growth rate of both the instability which suddenly start decreasing with further increase in the value of ionization rate and finally disappear or decreases to some almost constant smaller value of growth rate. This may be because when the ionization rate is smaller of the order of cyclotron frequency the wave particle interaction might have taken place and the energy of the particles is transferred to the wave and the instability starts growing at higher growth rate. Further when the ionization rate is increased to a much higher value than that of the cyclotron frequency then also wave particle interaction takes place because the phase velocity becomes the same order as that of the particle velocity but in this case the wave energy is transferred to the particles.

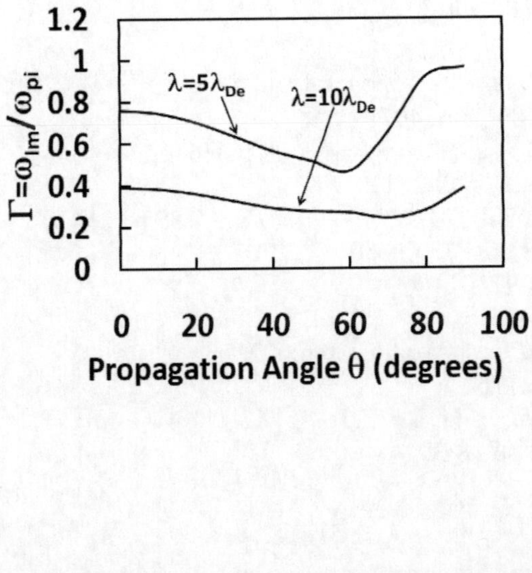

Figure 5. Variation of growth rate for ($\lambda = 5\lambda_{De}$) and ($\lambda = 10\lambda_{De}$) with θ (the angle between the magnetic field and the wave vector $\vec{k}$) when B_0 = 0.02T, $n_{e0}=n_{p0}=2\times10^{16}$m^{-3}, $m_{p0}=$ 1.67×10^{-27}kg, $m_{e0}= 9.1\times10^{31}$kg, $v_{x0} = v_{y0} = v_{z0}$ = 0.3Cs, $u_{x0} = u_{y0} = u_{z0}$ = 6Cs , γ=10^8, θ = 60^0, $T_e = 3eV$ and $T_p = 0.3eV$.

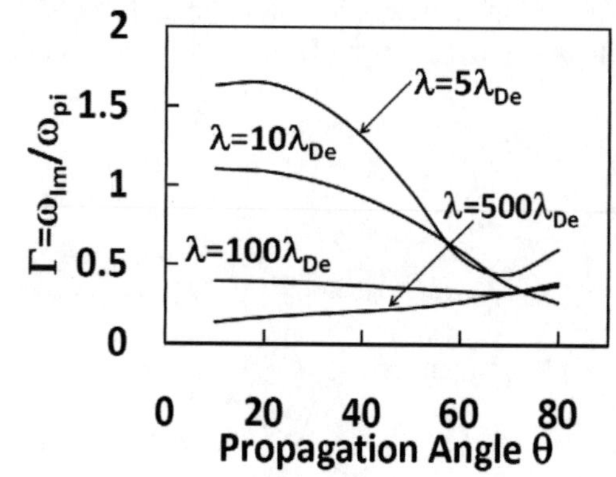

Figure 6. Variation of growth rate for different plasma to oscillation to Debye wavelength ratio with θ (the angle between the magnetic field and the wave vector $\vec{k}$) when B_0=0.02T, keeping rest parameters same as that of figure 1.

We observed in figures 5 and 6 that the growth rate for the both the instability first decreases and after attaining a dip increases further with increase in propagation angle θ. We can see that the dip for the instability growing at smaller oscillation wavelength shifts somewhat towards left for lower wavelength of plasma oscillations while for the second instability growing at the larger oscillation wavelength, as the wavelength is enhanced the growth rate behaves opposite (at $\lambda = 500\lambda_{De}$) and starts increasing after attaining almost the constant value at $\lambda = 100\lambda_{De}$ although the dip is attained at lower oscillation wavelength at smaller propagation angle. The behavior seen by Aggarwal [10] for the case of cold plasma is contrary to the behaviour shown in figures 5 and 6.

4. Conclusions

Two stream instability is obtained in a special case when both the species have same density and one stream is moving at a faster rate than the other. A more realistic inhomogeneous plasma under the effect of an external magnetic field support two types of instabilities whose growth rate changes significantly due to change the magnetic field. Both the instabilities are also dependent on the ionization constant whose behavior is shown in the figures above. The growth rate in both the instabilities has higher magnitude at smaller wave length of oscillations.

Acknowledgement

We thanks Centre for Energy studies, IIT Delhi for providing the technical support for writing this manuscript. Manipal University Jaipur is also gratefully acknowledged for rendering the financial assistance.

References

[1] Srivastava K M, Sudhanshu and Singh R 1982, *Phys. Plasmas* **25** p 57
[2] Lee H J 1986, *J. Korean Phys. Society* **19** p 273
[3] El-Labany S K 1990, *Phys. Scripta* **43** p 78
[4] Haas F, Bret A and Shukla P K 2009 *Phys. Rev. E* **80** p 066407
[5] Hao B, Ding W J, Sheng Z M, Ren C, and Zhang J 2009, *Phys. Rev. E* **80** p 066402
[6] Khrapak S A and Morfill G E 2010, *Phys. Plasmas* **17** p 062111
[7] Kompaneets R, Ivlev A V, Vladimirov S V and Morfill G E, 2012 *Phys. Rev. E* **85** p 026412
[8] Karimov A R and Shcheglov V A 1999, *Phys. Plasmas*, **7**, p 1050
[9] Lashmore-Davies C N 2007, *Phys. Plasmas*, **14** p 092101
[10] Aggarwal S S 1979, *Astrophys. Space Sci.* **66** p 341
[11] Jyoti, Kumar R, Malik H K and Dahiya R P 2013, *Phys. Plasmas* **20** p 092113
[12] Jyoti and Malik H K 2011, *Phys. Plasmas* **18** p 102116
[13] Malik H K, Jyoti and Kumar R, 2014 *J. Theor. Appl. Phys.* **8** p 123
[14] Jyoti, Malik H K and Dahiya R P 2014, *TOPPJ* **7** pp 18-30

IOP Conf. Series: Journal of Physics: Conf. Series **823** (2017) 012029 doi:10.1088/1742-6596/823/1/012029

Indigenous Manufacturing realization of TWIN Source

R. Pandey[1], M. Bandyopadhyay[2,3], D. Parmar[2], R. Yadav[2], H. Tyagi[2], J. Soni[1], H. Shishangiya[2], D. Sudhir Kumar[2], S. Shah[2], G. Bansal[1], K. Pandya[1], K. Parmar[1], M. Vuppugalla[1], A. Gahlaut[1] and A. Chakraborty[1]

[1]Institute for Plasma Research, Nr. Indira Bridge, Bhat, Gandhinagar-382428, India
[2]ITER-India, Institute for Plasma Research, Bhat-Motera Road, Ahmedabad - 380005, India
[3]Homi Bhabha National Institute (HBNI), Anushaktinagar, Mumbai- 400094, India

E-mail: rpandey@ipr.res.in

Abstract. TWIN source is two RF driver based negative ion source that has been planned to bridge the gap between single driver based ROBIN source (currently operational) and eight river based DNB source (to be operated under IN-TF test facility). TWIN source experiments have been planned at IPR keeping the objective of long term domestic fusion programme to gain operational experiences on vacuum immersed multi driver RF based negative ion source. High vacuum compatible components of twin source are designed at IPR keeping an aim on indigenous built in attempt. These components of TWIN source are mainly stainless steel and OFC-Cu. Being high heat flux receiving components, one of the major functional requirements is continuous heat removal via water as cooling medium. Hence for the purpose stainless steel parts are provided with externally milled cooling lines and that shall be covered with a layer of OFC-cu which would be on the receiving side of high heat flux. Manufacturability of twin source components requires joining of these dissimilar materials via process like electrode position, electron beam welding and vacuum brazing. Any of these manufacturing processes shall give a vacuum tight joint having proper joint strength at operating temperature and pressure. Taking the indigenous development effort vacuum brazing (in non-nuclear environment) has been opted for joining of dissimilar materials of twin source being one of the most reliable joining techniques and commercially feasible across the suppliers of country. Manufacturing design improvisation for the components has been done to suit the vacuum brazing process requirement and to ease some of the machining without comprising over the functional and operational requirements. This paper illustrates the details on the indigenous development effort, design improvisation to suits manufacturability, vacuum brazing basics and its procedures for twin source components.

1. Introduction

Large area multi driver inductively coupled RF negative ion sources have emerged as the baseline option for multi megawatt neutral beam injectors required for fusion applications [1]. RF based negative ion source couples RF power produced via RF generator for plasma production in driver region which expands in plasma box region and further negative ion beams are extracted via multi apertures grid segments. Such a RF based negative ion source consisting of two driver is defined here as TWIN source. The experimental objectives, conceptual design of TWIN source experiment have already been reported [3]. TWIN source would consist of two driver sub-assemblies and a plasma box

IOP Conf. Series: Journal of Physics: Conf. Series **823** (2017) 012029 doi:10.1088/1742-6596/823/1/012029

sub-assembly during plasma Phase of experiment and later on it would be supplemented by grid sub-assemblies during extraction phase of experiment. The two drivers of TWIN source would be coupled with a RF power of 180 kW. Source would be operated in continuous mode for 5 s and for 5 cycles with 5 Hz modulation for ~ 3 s ON 20 s OFF. As the components of various sub-assemblies of TWIN source would be receiving the heat flux, the engineering design of these components has been done accordingly to keep the thermo-mechanical stresses well within limit. Material selection of these components require high vacuum compatibility and strength at elevated temperature. SS 304L grade of stainless steel material are selected to provide structural integrity at elevated temperature and vacuum compatibility whereas OFC-Cu (UNS –C 10100) grade of copper has been selected for the plasma facing side being high on thermal conductivity value and with good brazing properties. These components are actively cooled via water as cooling medium. Major manufacturing requirements of these component are machining of complex cooling path over stainless steel material and covering these cooling paths with a copper layer. In the earlier manufacturing of similar kind of ion sources non-conventional processes like Copper electrode position has been opted. Taking the indigenous manufacturing attempt the readily available manufacturing process like Vacuum Brazing has been selected for joining of dissimilar materials. Thus the baseline engineering design has been revaluated and modified accordingly to accommodate the application of vacuum brazing process. In the following sections details on such design modification and glimpse of manufacturing activities have been provided.

2. Design and Engineering of TWIN source Components

Approach for engineering design on twin source mechanical components is three steps procedures as highlighted in figure 1. It includes pre-engineering conceptualization, detailed engineering design and design re-assessment steps. At pre- engineering steps several functional and operational parameters of twin source is generated. These parameters will be the design basis for detailed engineering of the twin source components that shall include overall dimensioning, thickness optimization as per applicable design code and standard engineering practices. A conceptual model is thus prepared using CAD that shall visualize these parameters, geometrical, space constraints and several interfaces with auxiliary components. At third step inputs from suppliers has been taken for the indigenous manufacturing realization of twin source. Based on the inputs few design improvisation done to accommodate the commercial availability of manufacturing techniques across country. These details are listed in following sub-sections.

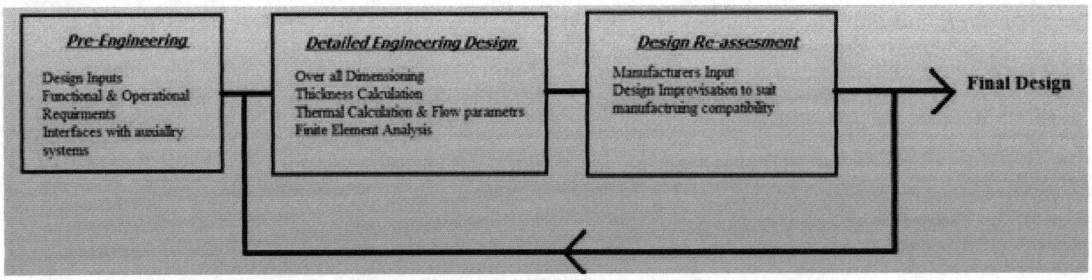

Figure 1. Design Route for TWIN source Components

2.1. *Detailed Engineering design with manufacturing assessment*

The source of weight ~ 2 tons is mounted in a cantilever position, placed between two vacuum vessels and supported by the vacuum vessels' flanges through a thick (~ 40 mm) stainless steel collar flange, as shown in figure 2. The source and the collar flange are interfaced through a thick vacuum compatible fiber reinforced polymer (FRP) ring structure to maintain the high voltage isolation between the source and the chamber as shown in figure 3.

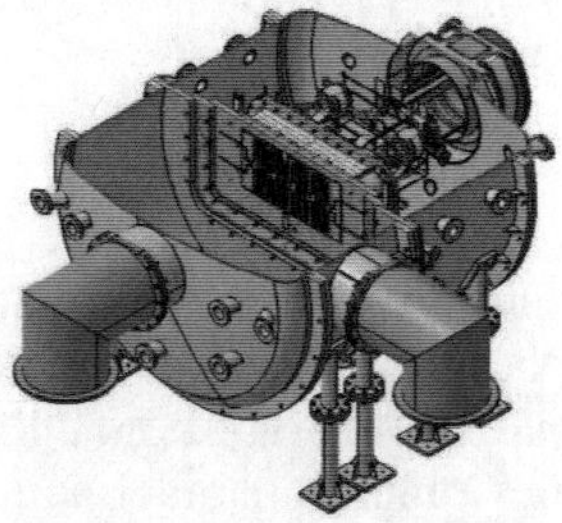

Figure 2. Sectional View TS vessel Assy

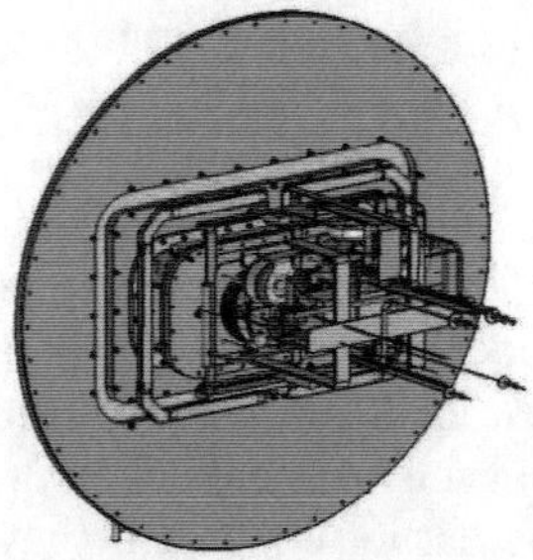

Figure 3. TWIN source assembly

Complete assembly of ion source basically consists of plasma box sub-assembly (figure 4) and driver sub- assembly (figure 5). Plasma box assembly holds the plasma box lateral wall (1000 × 500 × 200 mm^3) with end flanges holding source back plate. Whereas driver sub-assembly consist of faraday screen lateral wall, faraday screen back plate and alumina case housing the entire driver assembly. The two drivers are mounted over plasma box source plate through four number of supporting rods (for each driver). Drivers form a vacuum sealing over source back plate with O-ring mechanism and are electrically isolated with source back plate. Each drive assembly and plasma box assembly has separate cooling paths.

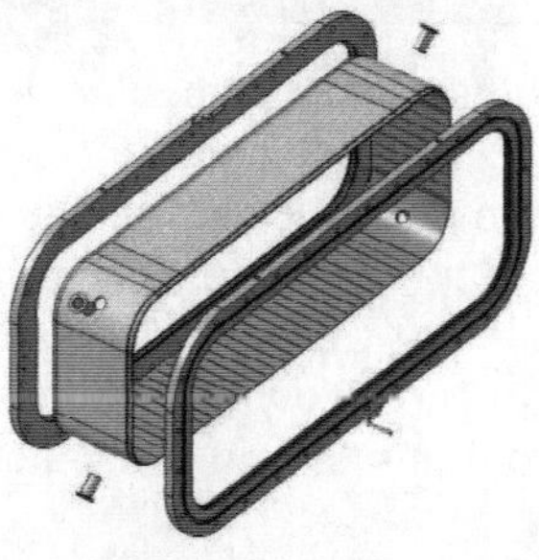

Figure 4. Plasma Box Assembly

Figure 5. Driver sub-Assembly

Plasma chamber lateral wall: It is a rectangular shape box structure as shown in figure 4, whose one side is covered by RF driver holding back plate and the other side will be covered by plasma grid (PG) – bias plate combination. The size of the structure is ~ 1000 (L) × 500 (B) × 200 (H) mm^3. The base material is SS 304 of thickness 6 mm. Milled cooling lines are provided in the lateral wall on the base material. The inner side walls are provided with OFC-Cu layer to distribute the heat load due to plasma loading uniformly over the whole surface. *Source back plate:* It holds both the RF drivers in horizontally demountable configuration. The size of the plate structure is ~ 1000 × 500 mm^2 of thickness 30 mm. The base material is SS 304L. Cooling lines are milled on the back side of SS plate on plasma facing side, which shall be covered with 5 mm thick copper layer.

Driver: Driver assembly comprises of many components, whose exploded view is shown in figure 5. The outer most component is a ceramic (Alumina) cylinder of grade A6 – A9, diameter 280 mm, length 150 mm, thickness 8 mm. The ceramic cylinder (CS) acts as a vacuum boundary for RF window which will house the Faraday shield (FS). RF helical coil antenna will be mounted on the CS. FS consists of few different parts. FS is placed inside the CS to protect the CS from plasma sputtering and is coated with 3 micron Molybdenum from inside. The whole OFC Cu cup is then mechanically coupled with a SS 304 plate (Driver magnet plate) which houses Samarium Cobalt magnets in the other side (rear side). Magnets are placed in chequer board configuration inside the channels milled on the rear side of the SS plate in matrix formation. To protect magnets from fly-off, all the magnets are mechanically covered with a SS thin plate. The whole driver back plate, few opening are provided to hold starter filament, windows for diagnostics, gas feed nozzle, Cesium injection nozzle etc. The vacuum tightness between the CS and the FS is ensured by an O-ring pressed by Driver cover plate, made of SS 304L of thickness 26 mm and of diameter 300 mm.

2.2. *Design re-assessment based on manufacturing input*

Based on feasibility study and inputs from various manufacturers, engineering design of few components were re-assessed and some modifications done to suit with the manufacturing requirement and ease of the process. *Plasma box lateral wall:* Plasma box lateral wall is split in to four segments and on each segments cooling lines are provided by external milling on the stainless steel part (figure 6). The earlier concept of having deep drilled cooling lines running parallel across length of lateral wall is converted in to rectangular slots with same hydraulic radius. Difficulties pointed by in deep drilling 150 mm straight path with conventional processes thus resolved by application of vacuum brazing process, by which the externally milled cooling slots would be covered with 1mm thin OFC-Cu sheet by surface to surface bonding.

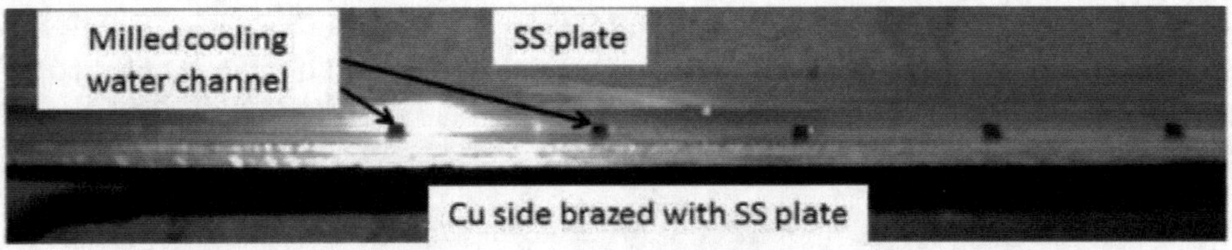

Figure 6. Plasma box lateral wall with milled rectangular slots externally

Faraday shield Lateral wall: Very significant modification done for faraday shield lateral wall, where the Z-slot configuration (figure 7) achieved with a proposal to split the design in two parts i.e. inner and outer cylinders. Inner cylinder will be manufactured with L-slot configuration on lateral surface and outer cylinder will have straight I-slot. Then these two parts would be vacuum brazed with surface to surface bonding by adjusting the L-slot and I-slot configuration in a manner that final assembly yields a z–cross section in the lateral wall.

Faraday shield back plate: Similarly for faraday shield back plate the requirement of a 10 mm thick copper plate (figure 8) having cooling channels embedded inside, is achieved by splitting the assembly into two parts. One having 6.5 mm thickness with 3.5 mm deep cooling lines milled externally on it, whereas the second part having 3 mm thickness is vacuum brazed to it.

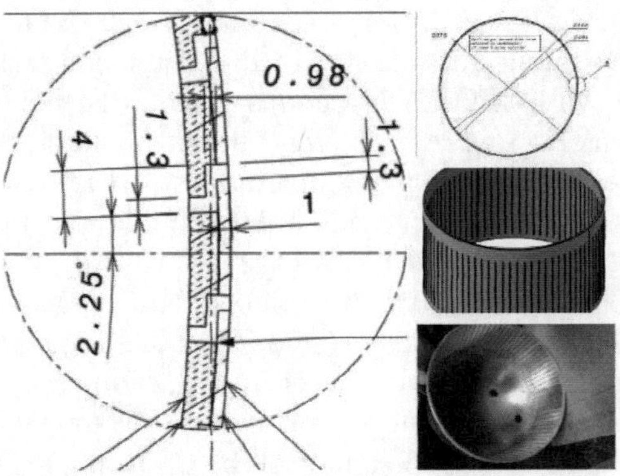

Figure 7. Z-Slot configuration of FS wall

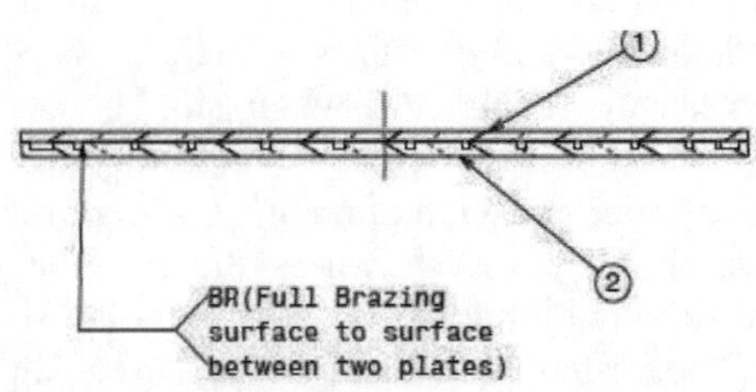

Figure 8. FS back plate assembly

2.3. *Finite element analysis for components*

Several finite element analyses have been carried out to estimate the structural integrity and thermal response of the individual sub-assemblies. Maximum stress is limited up to 115 MPa and deflection is limited to 1 mm only.

3. Manufacturing development of TWIN source experiments

3.1. Vacuum brazing for TWIN source components

Vacuum brazing being one of the most reliable techniques of joining dissimilar material for vacuum tight joint, the same has been opted for twin source components for joining Stainless steel and OFC-Cu. The major challenge in this was with the larger dimensions of components, the required clearance gap of 100 microns between the mating parts is never assured despite best machining practices. Hence, a lot of trials have been done with fixture design to assure this minimum gap clearance. The brazing filler alloy mostly used is Bv Ag-8 (Cu-Sil alloy), while in some places Bv Ag-29 is also tried to ensure greater fluidity based on geometrical consideration. SS part is Ni plated to increase the fluidity of molten braze alloy thus by giving a vacuum tight joint reasonably high on strength. Prior to brazing SS part was chemically cleaned in solution of 20 % sulphuric acid + 20 % hydrochloric acid + 60 % water followed by a dip in to 10 % nitric acid. OFC-Cu part was also chemically clean to remove all oxide and surface impurities in a solution of 25 % sulphuric acid + 10 % potassium dichromate followed by rinsing in hot water. The temperature for stabilizing is kept as 650 °C and that for brazing as 850 °C for a soak period of 10-20 min each. The heating rate was kept at 9 degree/min. The vacuum furnace was maintained at vacuum level of 10^{-5} mbar using a diffusion pumping system. Before the final run a sample run for Cu-Cu joint and SS-Cu joint was done under same condition. The test coupons then sent for tensile testing and macroscopic examination. Macroscopic results gives a fine eutectic mixture of brazing alloy where as in tensile testing failure occurs at parent material denoting strength more than 200 MPa.

Figure 9. Plasma Box Assembly

Figure 10. Faraday Shield Assembly

Manufacturing of plasma box assembly (figure 9) has been achieved with End flanges TIG welded to SS part of plasma box lateral wall. Manufacturing realization for the source back plate done by milling cooling lines over stainless steel part externally and then covering it with another copper sheet of 5 mm thickness through vacuum brazing process. 3 microns of Mo coating achieved by DC magnetron sputtering unit. The manufacturing realization of faraday shield lateral wall (figure 10) assembly achieved with combination of two sets of concentric cylinders made of OFC copper. The inner cylinder has the L-slot configuration and the outer cylinder has the straight I-slot configuration. Later on these two would be vacuum brazed keeping orientation of L-slot and I-slot in such a manner that the final assembly of FS lateral wall achieved the Z-slot configuration. Once the lateral wall assembly is achieved the faraday shield back plate assembly with SS cooling tubes is achieved with vacuum brazing option. Plasma grid segment (figure 11) is made of stainless steel whereas bias plate is made of OFHC copper. Both the components are required to have frame cooling. For both the components the cooling lines are attached with vacuum brazing post all the milling operations are finished on individual components. B Ni-3 brazing alloy is used for plasma grid segment (SS-SS joint) and Bv-Ag-8 braze alloy for bias plate assembly (Cu-Cu).

Figure 11. Plasma Grid Segment

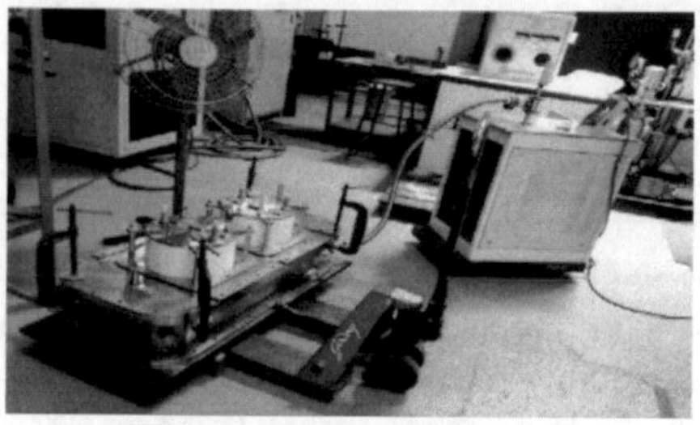

Figure 12. Leak Testing

3.2. Testing of brazed components

All the brazed components passed the leak tightness test up to level of ~ 10^{-9} mbar l/s. The leak testing was carried out by evacuating the cooling channels below vacuum level less than 10^{-3} mbar and then spraying helium gas all around the brazing joints (figure 12). The cooling lines were further pressurized to high pressure up to ~ 10 bar with demineralized water for 30 min. The voltage isolation test where performed over insulating parts using 2.5 kV AC supply and 5 kV DC supply. The IR resistance measured over more than 100 MΩ.

4. Conclusion

This indigenous manufacturing realization of twin source components has established the economical way of manufacturing such a large dimension ion source without compromising over its functionality. Learning experiences with the application of vacuum brazing process over such larger dimension components, issues with gap clearance of the mating parts and the need of design improvisation for components and fixtures to improve the capillarity of molten filler alloy were quite significant. Such lessons could be a part of standard vacuum brazing practices for future application.

Acknowledgement

Authors have acknowledged the contribution of M/S Design-Tech personnel for CAD modeling and M/S Hind High Vacuum, Bangalore for manufacturing the TWIN source components.

References

[1] Marcuzzia D, Dalla Palma M, Pavei M, Heinemann B, Kraus W, Riedl R 2009, *Fusion Engineering and Design* **84** 1253.

[2] Singh M J, Bandyopadhyay M, Bansal G, Gahlaut A, Soni J, Sunil Kumar, Pandya K, Parmar K G, Sonara J, Yadava R, Chakraborty A K (IPR, India), Kraus W, Heinemann B, Riedl R, Obermayer S, Martens C, Franzen P, Fantz U (IPP, Germany) 2012, *AIP Conf. Proc.* **1390** 604.

[3] Bandyopadhyay M, Pandey R, Shah S, Bansal G, Parmar D, Gahlaut A, Soni J,Yadav R K, Sudhir D, Tyagi H, Pandya K, Parmar K G, Mistri H S, Vuppugalla M and Chakraborty A K, 2014 *IEEE Transactions on Plasma Science* **42** 624.

IOP Conf. Series: Journal of Physics: Conf. Series **823** (2017) 012030 doi:10.1088/1742-6596/823/1/012030

Significance of ITER IWS Material Selection and Qualification

Bhoomi K Mehta[1], Jigar Raval[2], Abha Maheshwari[1], Rahul Laad[1], Gurlovleen Singh[1] and Haresh Pathak[1]

[1]ITER-India, Institute for Plasma Research, Gandhinagar 382025, India
[2]ITER Organization, Route de Vinon sur Verdon, 13115 Saint Paul Durance, France

E-mail: bhoomi.mehta@iter-india.org

Abstract. In-Wall Shielding (IWS) is one of the important components of ITER Vacuum Vessel (VV) which fills the space between double walls of VV with cooling water. Procurement Arrangement (PA) for IWS has been signed with Indian Domestic Agency (IN DA). Procurement of IWS materials, fabrication of IWS blocks and its delivery to respective Domestic Agency (DA) and ITER Organization (IO) are the main scope of this PA. Hence, INDIA is the only country which is contributing to VV IWS among all seven ITER partners. The main functions of the IWS are to provide Neutron Shielding with blanket, VV shells and water during plasma operations and to reduce ripple of the Toroidal Magnetic Field. To meet these functional requirements IWS blocks are made up of special materials (Borated Steels SS304 B4 & SS304 B7, Ferritic Steels SS 430, Austenitic Steel SS 316 L (N)-IG, XM-19 and Inconel-625) which are qualified, reliable and traceable for the design assessment. The choice of these materials has a significant influence on performance, maintainability, licensing, detailed design parameters and waste disposal. The main reasons for the materials selected for IWS are its high mechanical strength at operating temperatures, water chemistry properties, excellent fabrication characteristics and low cost relative to other similar materials. All the materials are qualified with respect to their respective codes (ASTM/EN standards with additional requirements as described in RCC-MR code 2007) and ITER requirements. Agreed Notified Body (ANB) has control conformity of materials certificates with approved material specification and traceability procedure for Safety Important Component (SIC). The procurement strategy for all the IWS materials has been developed in close collaboration with IO, ANB and Industries as per Product Procurement Specification (PPS). The R&D for sample, bulk material production, testing, inspection and handling as required are carried out by IN DA and IO. At present almost all IWS materials (~2500 Tons) has been procured by IN DA with spares to manufacture ~9000 IWS blocks. This paper summarizes IWS material selection, qualification and procurement processes in detail.

1. Introduction

ITER vacuum vessel is a double wall structure, comprising of inner shell and outer shell [1]. The space between two shells is filled up with In-Wall Shielding (IWS) blocks and cooling water. The main purpose of the IWS is, (i) to provide neutron shielding together with the blanket and VV shells and water during plasma operation and stop neutrons coming out of plasma and (ii) to reduce the ripple of the Toroidal magnetic Field (TF). Design of the IWS is almost completed, considering the assembly methods and the required tolerances. A detailed description is published in the DDD 1.5 Vacuum Vessel [6]. During the earlier design phase, a rigorous material R& D program in support of material section was carried out. The results of R & D program were summarized in the ITER Material Handbook [4].

The Procurement Arrangement (PA) for IWS was signed between IO and Indian Domestic Agency (IN DA) in September 2009. The IWS design is a build-to-print (BtP) item for the PAs that requires completion of engineering design by the ITER Organization (IO). Engineering design of the IWS has been performed with a model to verify structural integrity that was reviewed and approved by the IO and the ANB in order to start manufacturing design. The procurements for the IWS is "in-kind" and materials for these components are procured by IN DA. In accordance with the regulatory and quality requirements for operation, common material specifications were prepared by ITER Organization (IO) in collaboration with the DA and theirindustries.IN DA has placed the contract with M/s. Industeel for SS 304 B4, SS 430 and SS 316 L (N)-IG materials, M/s Carpenter Powder Product (CPP) for SS304 B7, M/s Valbruna for XM-19 and Alloy 625 and M/s Avasarala (ATL) for the IWS manufacturing. The IWS components will be delivered to the EU and KO DAs for completion of sectors as well as to the IOfor field joints of the torus.

This paper will describe the function oriented description, selection, qualification and procurement strategy for various IWS materials.

2. Materials for IWS

About 60 % volume between the vessel shells is filled up with IWS blocks made of 40 mm thick SS304B7 (UNS number S30467) in the inboard region and SS304B4 (UNS number S30464) in the outboard region while about 85 % volume between two shells near toroidal magnetic field (TF) coil locations is filled up with SS430 (UNS number S43000) to provide adequate neutron shielding and reduce the ripple in toroidal magnetic field [1]. The steel has low ductility (minimum total elongation of 6 %) and low fracture toughness and special attention should be made for the design and fabrication of these shielding blocks. In the design of these shields, numbers of plates are stacked with spacer in-between to allow required water flow. Actual numbers of plates in each structure vary from one poloidal location to other. This is because of the fact that the space between outer and inner shells is different at different poloidal locations. Shields are fixed to poloidal ribs between out/inner shells or to special support to withstand mechanical forces. Welding of SS 304 B4 and SS 304 B7 is not permitted as welding heat changes properties of these materials.

There are two types of IWS block, primary shielding block using material SS304B7 (with 2 % – 2.25 % boron) and SS304B4 (with 1–1.25 % boron) provides an effective neutron shielding while ferromagnetic shielding block made of material SS430provides neutron shielding and reduces the toroidal field ripple. A ferritic stainless steel type SS 430 is used under the TF coils in the out board area (12 o'clock to 5 o'clock except the equatorial port region) to reduce toroidal field ripple. This steel has a saturated magnetization at ~ 1.6 T. The borated and ferritic steels have high corrosion resistance in water and acceptable fabrication characteristics. IWS blocks contain several individual shielding plates with thickness 40 mm which are bolted together (minimum three to maximum eleven) by bolts M30. The length of shielding plates is normally less than 0.3 m poloidally and less than 0.15 m toroidally. The shielding blocks are supported by upper brackets and lower brackets attached to the ribs of VV and positioned with bolts M20 to with stand the mechanical forces. The connecting bolts, nuts, washers and spacers are made of XM19, while VV shells and ribs are made of SS 316L (N)-IG [1,2]. There are about ~ 9000 blocks distributed over nine vessel sectors (VSs) and Field Joints (FJs).

IWS components and materials are shown in figure1 and the detail design of one IWS block is shown in figure 2.

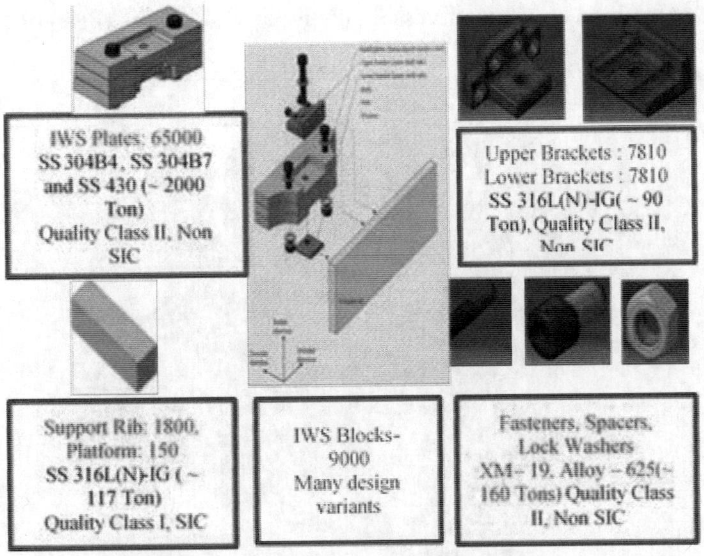

Figure 1. IWS components and materials

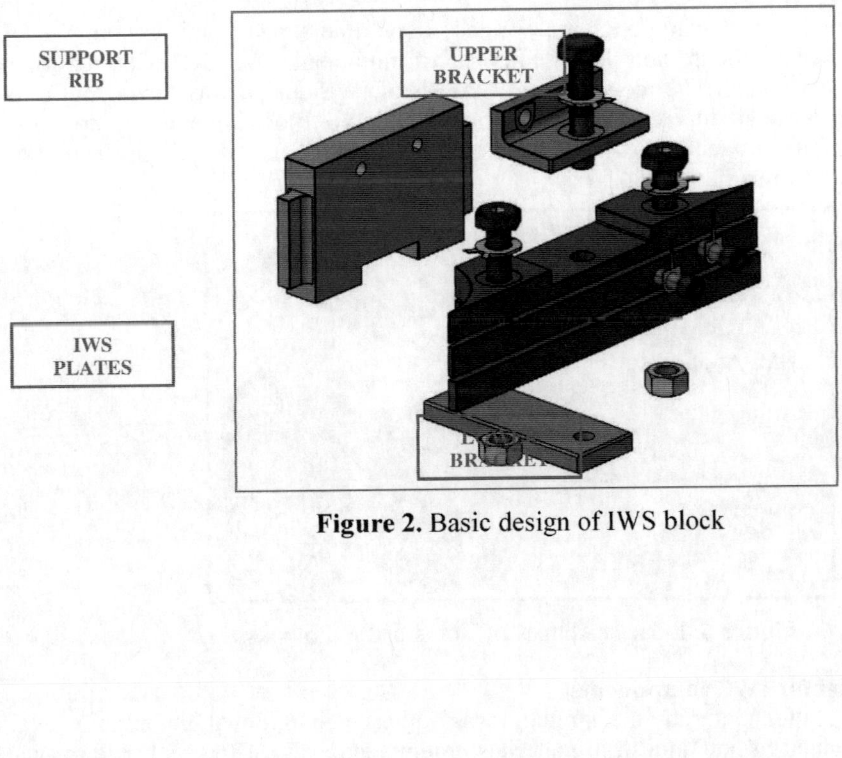

Figure 2. Basic design of IWS block

These blocks are assembled on vacuum vessel either on poloidal ribs (PRs) or support ribs (SRs) as shown in figure 3 as an example for VS6/PS2.

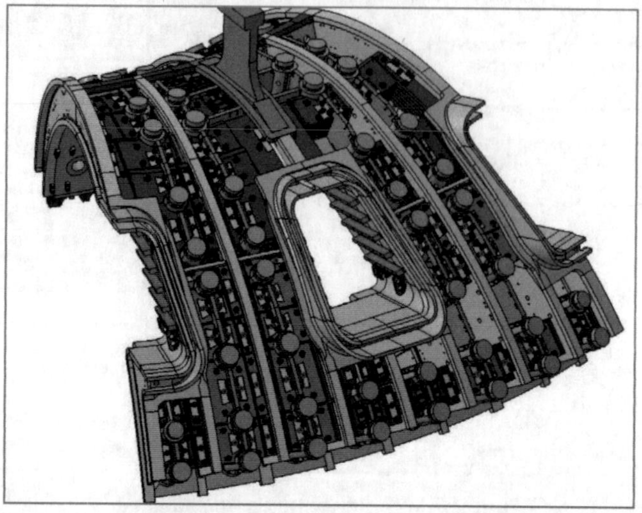

Figure 3. IWS blocks mounted on PRs and SRs of VS6/PS2

PRs are part of vacuum vessel while SRs are welded to flexible housing of vacuum vessel. All components of IWS, except SRs are non-Safety Important Component (non-SIC) with quality class - 2. Support rib is SIC component. Thermo-structural analysis of IWS design was carried out to ensure the structural integrity against different load conditions. As IWS block should fill up maximum volume between two shells of vacuum vessel, shapes of many IWS blocks are very intricate. Some of such intricate shapes are shown in figure 4.

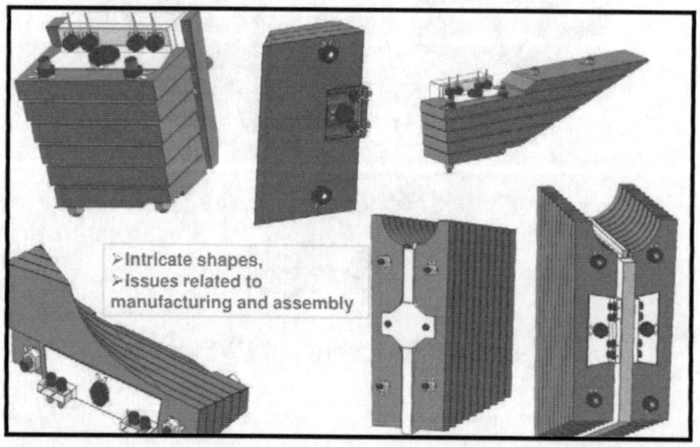

Figure 4. Intricate shapes of some of IWS blocks

3. Selection of Material for IWS components

During the selection of the material, the strategy was to use conventional industrially available materials. The development of new industrial materials requires extensive effort and time for material qualification, which was not acceptable for ITER project. However, the specific IWS materials (requires high neutron cross section, high Saturation Magnetization properties etc. to provide neutron shielding and reduce ripples) required some modification of the existing materials [3]. IWS materials selection was based on a general engineering approach, taking into account operational conditions, safety requirements, physical and mechanical properties, reliability, maintainability, corrosion performance, join-ability, etc. The main goal of the materials activity in IWS is to provide qualified, reliable and traceable recommendations needed for the design assessment done by design team. The required properties data are specified in the design codes (ASME, RCC-MR, etc.) used for the different ITER components.

From the analysis done by design team the expected damage dose is less than 0.02 dpa, when IWS materials inside Vacuum Vessel operate at 100–200 °C. The selected materials must provide reliable and safe operation for the planned 20 years of the ITER operation. The materials for the IWS components are summarized in figure 1.

All the information related to design, properties of Vacuum Vessel, In Vessel materials and statistical analysis to decide the parameters is available and documented by ITER team [4, 5].

To achieve the adequate radiation shielding efficiency, the effects of the fraction of the steel in IWS and the steel composition have been investigated by IO. Based on the nuclear analysis result, the steel fraction in the range from 60 % to 85 % including the VV shells provides the good shielding efficiency. The doping of boron in the steel of IWS was considered to improve the neutron shielding efficiency. The study shows that the TF coil (TFC) heating rate is reduced by a factor of 1.33 when 2 wt % of boron-doped stainless steel is used. The difference in TFC heating between 1% and 2% of boron content is 4 % [6]. Based on this radiation shielding study, borated steel SS304B7and SS304B4 were selected as primary IWS materials for VV inboard and outboard regions respectively. The shielding material under the TF coil in the outboard area (12 o'clock to 5 o'clock region) additionally has to provide the reduction of toroidal field ripple. This is achieved by using shielding plates made of ferromagnetic (FM) material. Type SS 430 has been selected because of its mechanical strength comparable to SS 304 (but with lower ductility), acceptable fabrication characteristics, corrosion resistance, availability and cost. High saturated magnetization is required for the ferromagnetic insert material to provide effective ripple reduction. SS316 L (N)-IG is selected as Support Rib, Bracket and Platform material considering its mechanical properties, weldability and thermal expansion. XM-19 and Alloy 625 are selected as fastener materials. In limited areas studs made of standard SS316 L are used.

4. Qualification of Material for IWS components

The ITER Vacuum Vessel is classified as an assembly of nuclear pressure equipment in accordance with French Order Dec 12 2005 concerning Nuclear Pressure Equipment [7]. IWS includes 2 types of components which are both part of Vacuum Vessel nuclear pressure equipment:

Support elements made of 316L (N) stainless steel, namely the connecting ribs, need the compliance with regulatory requirements (Nuclear Pressure Equipment Order) because these elements are permanently attached to pressure bearing parts. They are to be designed and manufactured according to RCC-MR 2007 Subsection C (Class 2 Equipment). ANB shall control conformity of materials certificates with approved material specification and traceability procedure.

Other elements (made of borated steels, ferritic steel and fasteners) and possible connecting ribs fixed by bolting are part of the pressure equipment, but they are non-pressure bearing parts and not permanently attached to pressure bearing parts in the sense of Nuclear Pressure Equipment Order and French Decree 99-1046. ANB may check the certificates of materials.

Since all the materials of IWS were specially developed, specific code and standards were followed to manufacture and qualify each IWS material. Description of these code and standard is given in Table 1.

Table.1: Code and Standard required for IWS Materials

Materials	*Code Standards requirements and other comments*
Alloy 625	ASTM B B446 (Non-pressure bearing)
Austenitic Steel XM-19 (B & R)	ASTM A 193 (Non-pressure bearing)
Ferritic steel 430 (UNS 43000)	ASTM A240/ A 240M with additional requirements
SS 316L(N)-IG (for connecting ribs)	Welded to pressure bearing parts –Steel from RCC-MR, Edition 2007 with additional requirements, not in EN harmonized standard
SS304B4 (UNS S30464)	ASTM A887 with additional requirements 1-1.25% B Grade B
SS304B7 (UNS S30467)	ASTM A887 with additional requirements, 1.75-2.25%B Grade B

5. Procurement of IWS Materials

IN DA procured all the materials as per the approved Product Procurement Specification (PPS) received from IO. The sample production has been completed based on the specification defined in the PPS and it was inspected physically by IN DA and IO/ANB representatives. An inspection certificate type 3.1 or 3.2 according to the EN 10204:2005 was submitted for the structural material welded to pressure bearing parts. For other materials certificates 3.1 (EN 10204:2005) were submitted.

Figure 5. IWS materials received and stored at ATL

The total required IWS materials to manufacture ~ 9000 IWS blocks have been successfully manufactured as per required specifications by Industries and procured after fulfilment of all the quality requirements as mentioned in previous sections. All procured materials have been arrived and stored at IWS manufacturer ATL's site as shown in figure 5. IWS blocks manufacturing is under progress at ATL.

Acknowledgment

Author would like to thank all members of IWS team of ITER-India, ITER IO and Industries for their continuous support in various activities reported in this paper.

ITER Disclaimer

The views and opinions expressed herein do not necessarily reflect those of the ITER Organization.

References

[1] Wang X et al. 2014 *Fusion Engineering and Design* **89** 1814
[2] Ioki K et al. 2012 *Fusion Engineering and Design* **87** 828
[3] Barabash V et al.2007 *Journal of Nuclear Materials* **367–370** 21
[4] ITER Materials Properties Handbook (MPH), ITER Doc. G 74 MA 16 04-05-07 R0.1, (internal project document distributed to the ITER Participants).
[5] ITER Materials Properties Database, ITER Doc. G 74 LS 102-03-22 W0.1, (internal project document distributed to the ITER Participants).
[6] System Design Document (DDD) 1.5 Vacuum Vessel, ITER Doc. 22FPWQ (2012) V 4.0 (internal project document distributed to the ITER Participants).
[7] Procurement Arrangement document for IWS, ITER Doc. 2V4NK4 (2009) V 1.3 (internal project document distributed to the ITER Participants).

Prototyping of radial plates for fusion relevant superconducting magnets

M Ghate[1], D Bhavasar[1], A Panchal [1], S Udgata[2] and S Pradhan[1]

[1]Institute for Plasma Research, Near Indira Bridge, Bhat, Gandhinagar 382 428, India
[2]I-DESIGN Engineering Solutions Ltd., Wagholi, Pune 412207, India

E-mail: mghate@ipr.res.in

Abstract. The fabrication trials for prototype radial plate to support its conceptual design and development for fusion relevant superconducting magnet have been discussed in this paper. The simulation approach with CAD has been presented for prototyping of radial plates. Extensive trials have been done on SS316LN plates to estimate and establish machining sequences, machine parameters, machining tools to achieve required tolerances. The critical machining operation and parameters has been discussed in this paper. Inspection procedures with articulated arm coordinate measuring machine for prototype radial plate has been conceptualized and verified.

1. Introduction

"Magnet technology Development Division" is responsible for the development and fabrication of fusion relevant superconducting magnet (FRSM) and associated technologies at Institute for Plasma Research (IPR) in India. FRSM has been designed for magnetic field of 5.5 T / 12 T with maximum current carrying capacity of 30 kA. It consists of five insulated double pancakes stacked together to form winding pack. FRSM has been conceptualized considering the use of Niobium-Titanium (NbTi) or Niobium-Tin (Nb_3Sn) based cable in conduit conductor (CICC) with designed operating parameters [1,2]. Radial plate is primarily structural component of superconducting magnet required to transmit large electromagnetic forces on coil case, to enhance electrical and mechanical reliability of insulation, to support large electromagnetic force on conductor and to provide structural integrity and stability. The critical design drivers for conceptualization and development of radial plates are CICC configuration and its shape, configuration of superconducting magnet, operational field and operating current, electromagnetic stresses, manufacturing requirements, and assembly tolerances. SS 316LN material is widely used for manufacturing of radial plates due its high yield stress, machinability, weldability and its compatibility with cryogenic temperature.

Radial plate and cover plate for FRSM has been conceptualized considering CICC of Ø 30 mm as shown in figure 1 with respect to design drivers mentioned earlier. An arrangement of one double pancake for FRSM winding pack with radial plate is shown in figure 1. It consists of CICC with insulation, radial plate, cover plate and ground insulation. The conceptual design of radial plate and cover plate has been carried out considering SS316LN as structural material. Further design refinement with assembly considerations and FEM simulations needs to be carried out for optimization of radial plate configurations. Before going for final design and optimization, it is decided to evaluate manufacturing feasibility by prototyping of radial plates to achieve required tolerances.

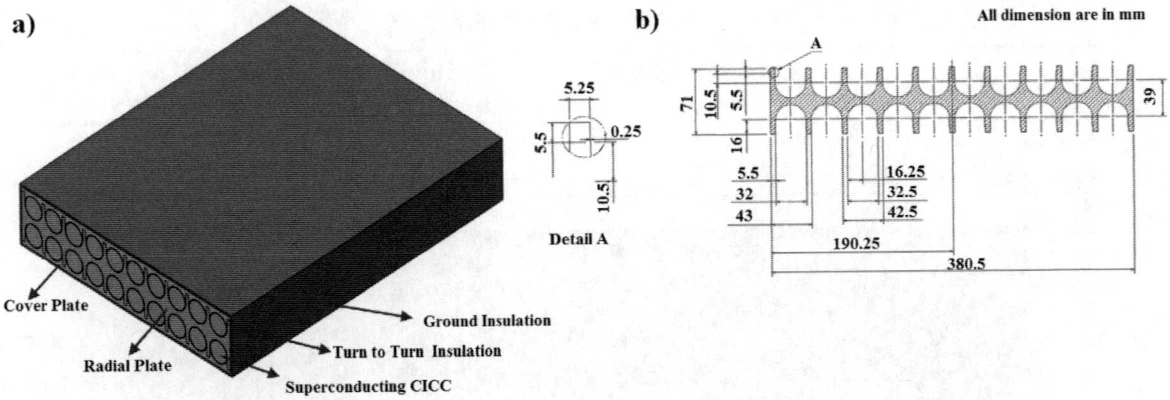

Figure 1. (a) Schematic for winding pack of FRSM, (b) conceptual design of radial plate

2. Prototyping of Radial Plate

The dimensions for prototype radial plate are shown in figure 2a. It is scaled down from actual size to demonstrate feasibility of manufacturing with available machines and tools. It consists of 10 cylindrical grooves and slots of 500 mm length with tolerances as shown in figure 2a. Some of the sectors of radial plate are curved as shown on figure 2b.

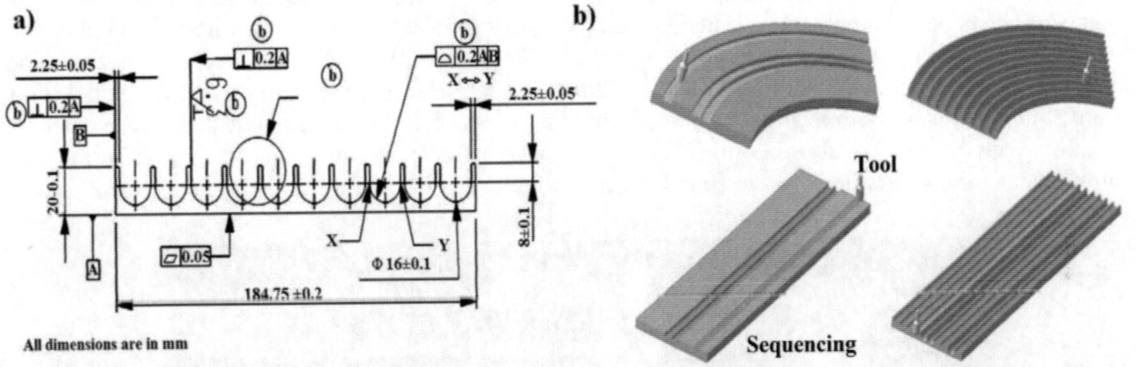

Figure 2. (a) Dimension of prototype for machining trials, (b) simulation of machining sequence

2.1. *Machining procedure & Parameters*

The important objective of these prototyping trials was selection of appropriate tools, machines, machining parameters and machining sequence. An extensive survey for appropriate tools and machine has been carried out. Ball end milling cutter (⌀10 mm with 3 numbers of carbide inserts for roughing and ⌀ 8 mm with 3 number of carbide inserts for finishing) and vertical machining center has been selected for this prototyping activities. The visual simulation for optimization of machining process to reduce deformation during machining has been carried out in commercially available CAD software as shown in figure 2b. Numbers of fabrication trials were performed for manufacturing of prototype radial plates on SS316LN. The machining sequences, procedures and parameters were optimized by number of trials as shown in Table 1. Typical machining processes include facing operation, slotting operation and milling operation. For facing operation of SS316LN plates, face milling cutter with carbide inserts of 45° entering angle has been used. Ball end milling cutter with 3 number of carbide inserts has been used for slotting operation. Typical sequence for slotting operation of prototype radial plates is shown in figure 3. The slotting operation was initiated from left side (1) of plate and alternatively from right side (2) of plate. The typical sequence as shown in figure 3 has been optimized and followed to reduce deformation during machining process.

Table 1. Final parameter for prototyping trials by SS316LN

Machining Pass	Cutter speed (rpm)	Feed rate (mm/min)	Depth of cut (mm)	Longitudinal tool travel (mm)
Roughing Pass	2000	1800	0.1 - 0.15	520
Finishing Pass	3200	1500	0.1	520

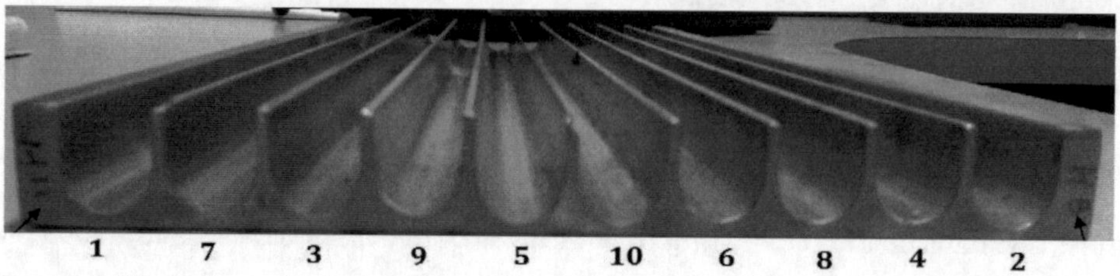

Figure 3. Typical machining sequence for slotting operation of prototype radial plates

2.2. Inspection and Validation

The number of prototypes for radial plate has been manufactured to optimize machining parameters and machining sequence. Dimensional metrology of these prototypes has been carried out with measurement equipment's, go – no go gauges (figure 4a) as well as with articulated arm coordinating machine (AACMM) (figure 4d). The suitability of AACMM and its probe for inspection of critical components has been demonstrated and reported [3]. During inspection, manufactured radial plate has been inspected with AACMM using contact probes. The wire frame model has been generated using collected data and analyzed with reference to datum surface of radial plate and its CAD model. Few initial samples during optimization of manufacturing process have been rejected due to dimensional variation with respect to the drawings. The subsequent prototypes after appropriate fixturing and optimize machining parameters were found to be satisfactory.

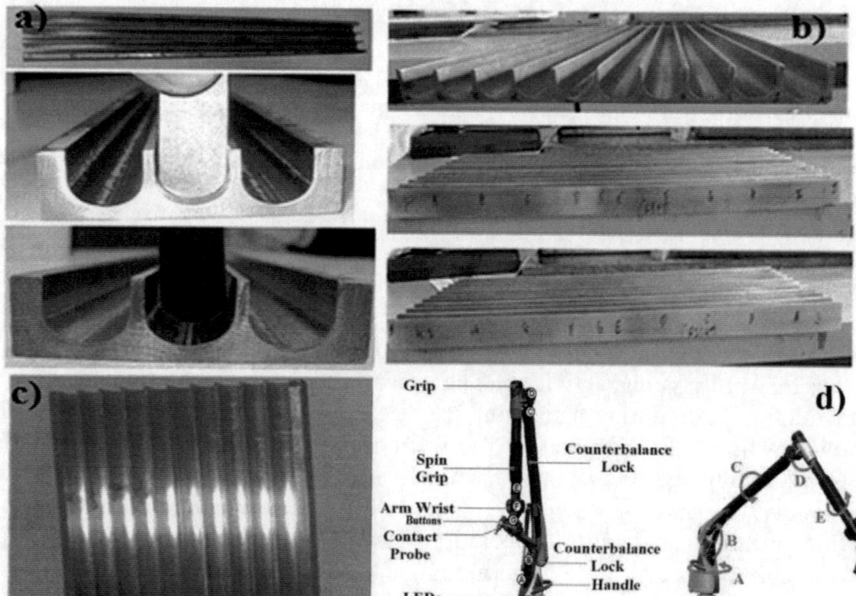

Figure 4. (a) Go- No go gauges, (b) Prototype radial plate-1, (c) Prototype radial plate-2 (d) AACMM

Detailed inspection of prototypes has been carried out by AACMM for its flatness, perpendicularity, linear variation, groove dimensions using contact probes as shown in figure 5. The average rib thickness was found to be 2.24 mm while average diameter of slots was found to be 16.11 mm. The perpendicularity of various surfaces was ranging from 0.07 mm – 0.18 mm. Three surfaces were found to be out of tolerance limit of flatness (0.2 mm).

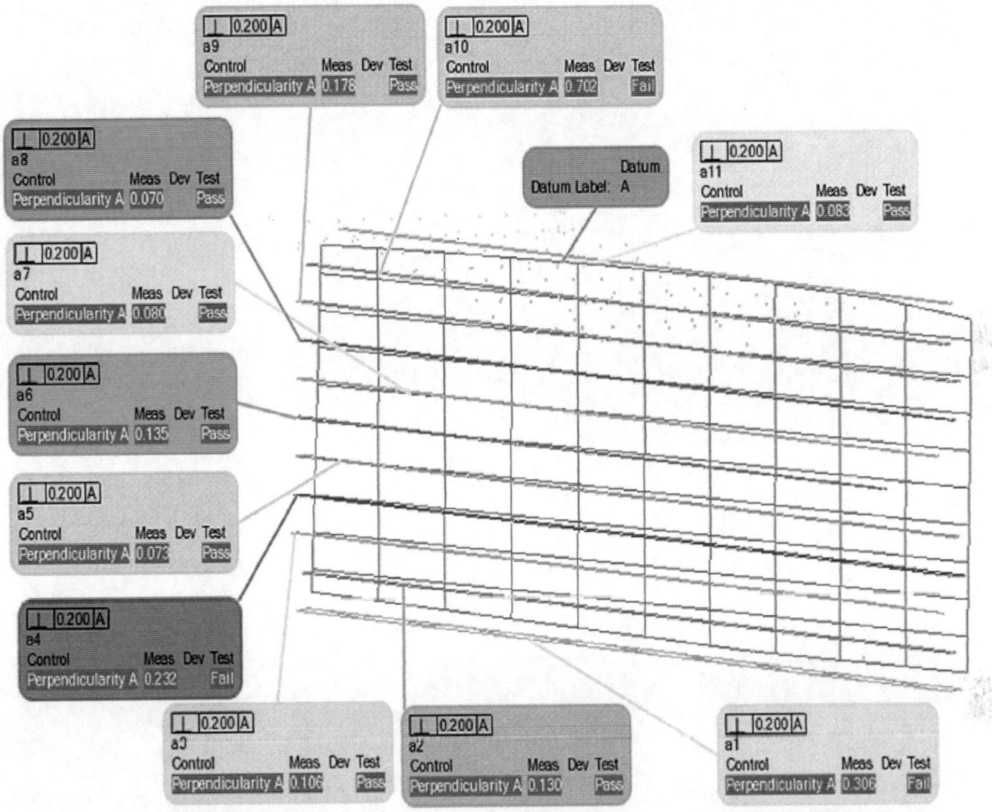

Figure 5. Dimensional accuracies of prototype radial plate by AACMM

3. Summary

- Survey for machining tools and cutter along with machining process sheet has been carried out. Ball end milling cutter mounted on vertical milling machine has been selected for this activity.
- Prototyping of radial plate on SS 316LN has been carried out for optimization of machining tools, procedure and machining parameters.
- Detail evaluation of prototype radial plate during trials has been carried out by gauges, measuring tool, and by AACMM.
- The prototyping trials for radial plate are found to be satisfactory as per all dimensional requirements. Manufacturing feasibility within stringent tolerance with prototypes has been demonstrated.

Acknowledgement

This work has been done under "Magnet Technology Development Programme". We are thankful to colleagues at MTDD, IPR for their assistance in this prototyping activity.

IOP Conf. Series: Journal of Physics: Conf. Series **823** (2017) 012031 doi:10.1088/1742-6596/823/1/012031

References

[1] Ghate M et al 2014 *Cryogenics* **63** 166

[2] Raj P , Ghate M et al., "Design and fabrication of indigenous 30 kA Nb3Sn CICC for fusion relevant superconducting magnet", March 2016, 26th International Conference on Cryogenic Engineering and Cryogenics Materials , New Delhi, India

[3] Ghate M , Bhavsar D et al, "Application of AACMM for Quality Control in Fabrication of Superconducting Magnet Components" , December 2013, *Proceedings of National Symposium on Plasma Science & Technology*, KIIT University , Bhubaneswar, India.

IOP Conf. Series: Journal of Physics: Conf. Series **823** (2017) 012032 doi:10.1088/1742-6596/823/1/012032

Assessment of delta ferrite in multipass TIG welds of 40 mm thick SS 316L: A comparative study of ferrite number (FN) prediction and measurements

Ramesh Kumar Buddu[1], P. M. Raole[1] and B. Sarkar[1]
[1]FRMDC Division, Institute for Plasma Research, Near Indira Bridge, Bhat, Gandhinagar 382 428, India.

E-mail: buddu@ipr.res.in; brkumar75@gmail.com

Abstract. Austenitic stainless steels are widely used in the fabrication of fusion reactor major systems like vacuum vessel, divertor, cryostat and other structural components development. Multipass welding is used for the development of thick plates for the structural components fabrication. Due to the repeated weld thermal cycles, the microstructure adversely alters owing to the presence of complex phases like austenite, ferrite and delta ferrite and subsequently influences the mechanical properties like tensile and impact toughness of joints. The present paper reports the detail analysis of delta ferrite phase in welded region of 40 mm thick SS316L plates welded by special design multipass narrow groove TIG welding process under three different heat input conditions. The correlation of delta ferrite microstructure of different type structures acicular and vermicular is observed. The chemical composition of weld samples was used to predict the Ferrite Number (FN), which is representative form of delta ferrite in welds, with Schaeffler's, WRC-1992 diagram and DeLong techniques by calculating the Creq and Nieq ratios and compared with experimental data of FN from Feritescope measurements. The low heat input conditions (1.67 kJ/mm) have produced higher FN (7.28), medium heat input (1.72 kJ/mm) shown FN (7.04) where as high heat input (1.87 kJ/mm) conditions has shown FN (6.68) decreasing trend and FN data is compared with the prediction methods.

1. Introduction

Austenitic stainless steels are the primary structural materials used in the fusion reactor fabrication by different joining techniques for different components [1,2]. The complex fusion reactor structural components have to sustain complex conditions like thermal, structural and neutron irradiation during plasma operation and have to meet the desired qualifications for materials as well welded joints. The final welded joint properties like tensile stress, impact toughness and fatigue properties are mainly depend on the microstructure of welds [3]. The welding process heat input conditions (weld current, voltage, weld travel speed, shielding gas pressure etc.) show prominent role in governing the final microstructure and mechanical properties of the welded joints [4]. In general, thicker stainless steel plates are joined by using multi pass Tungsten Inert Gas Welding (TIG) and Electron Beam welding for vacuum vessel and other components in fusion reactor [5,6]. The distortions and residual stresses have to be carefully controlled by choosing the appropriate process parameters [7]. The austenitic welds in general retain fraction of delta ferrite phase which is favourable for good toughness, retardation of stress corrosion cracking of weld joints [8]. The non-destructive method of measuring ferrite content is done by the magnetic measurement technique with using Feritescope. This technique is widely used to assess the ferrite (delta ferrite) in multipass welds. Further there are certain approaches to predict the Ferrite Number (FN) range in Austenitic welds by Schaeffler's, WRC-1992 & DeLong methods by using chemical elemental composition of welds with Cr_{eq} and Ni_{eq} measurements [9]. The present paper reports the weld microstructure evaluation carried on narrow groove TIG welds fabricated under different weld process parameters (three different heat inputs) and their FN prediction with analytical techniques and experimental data comparison.

2. Narrow groove TIG Weld samples development and microstructure characterization

AISI SS316L samples of 40 mm thick plates with narrow groove design are machined, joined with three sets of different TIG welding process parameters [10]. The heat input parameter data is calculated by using the weld current, voltage and weld speed for the three weld samples. AISI 40 mm thick SS 316LN samples of 150 mm × 150 mm are fabricated for the narrow groove TIG welding process with 2 mm root gap and 3 mm root height with bevel angle 20°. The weld samples are prepared with weld process parameters like high heat input (1.87 kJ/mm, Voltage: 16 VDC, Current: 175 A, weld speed: 100 mm/min), medium heat input (1.72 kJ/mm, Voltage: 16 VDC, Current:165 A, weld speed: 98 mm/min) and low heat input (1.67 kJ/mm, Voltage: 16 VDC, Current: 155 A, weld speed: 101 mm/min). The joining process was done by using ER316L filler wire with 1.6 mm dia and ~ 48 weld passes. The fabricated samples are cross sectioned, polished and etched for the macro and microstructure examination. The macro cross sections are shown in figure 1(a-c) for three welds. Further, the microstructure of welds has been characterized by optical microscope and scanning electron microscope (SEM) for base metal (BM), heat affected zone (HAZ) and weld zone (WZ). BM microstructure has shown austenite grain structure and HAZ shown coarse and finer grains and at interface combined structures which are due to the heat input, cooling rate and solidification mechanism associated. In WZ, the observed microstructure disclose the features of fine grains, dendrite, columnar and cellular within the austenite matrix. The observed optical microstructure for WZ of each sample is shown figure 2(a-c).

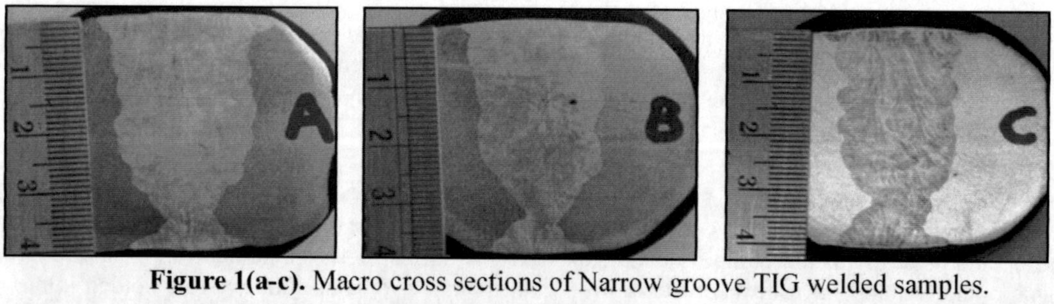

Figure 1(a-c). Macro cross sections of Narrow groove TIG welded samples.

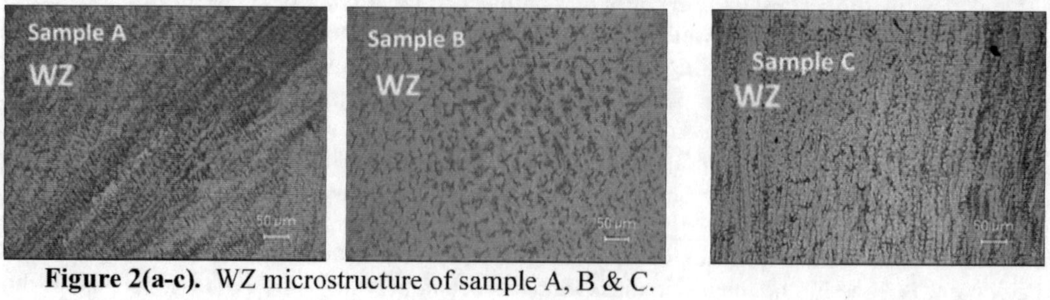

Figure 2(a-c). WZ microstructure of sample A, B & C.

The WZ of sample A has shown the complex behaviour of the combined features of dendrite, columnar and cellular structures with delta ferrite significant presence within the austenite matrix. Similarly for WZ of sample B and C are examined and the structures distribution has shown the similar observations as like sample A and the details are shown in figure 2 (b & c).

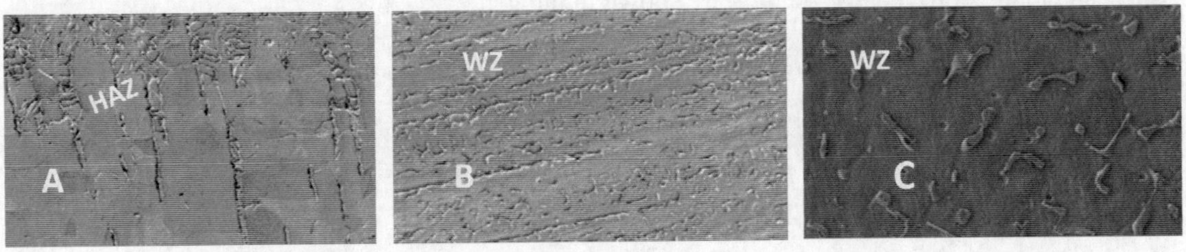

Figure 3(a-c). Delta ferrite distribution in three welded samples of A, B and C.

The WZ of three weld samples has been examined for the delta ferrite identification with suitable etching process (by using Villella's reagent with 1g picric acid, 100 ml Ethanol and 5 ml HCl). The observed delta ferrite in all the three different welds with low resolution (~ 20 μm) images are shown in figure 3(a, b & c). Figure 3(a) is shown with delta ferrite in HAZ region whereas figure 3(b) & (c) are for WZ of sample B & C. Scanning Electron Microscope with higher resolution (~ 1μm) is shown in figure 4(a-c) for three welds. The detailed examined delta ferrite features like acicular and lathy type ferrite distribution are attributed to the experienced heat input and cooling rates.

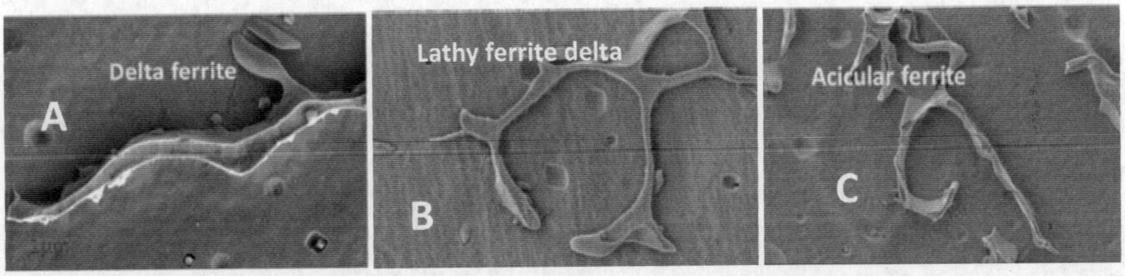

Figure 4(a-c). Delta ferrite distribution in welded samples of A, B & C.

3. Ferrite Number Prediction & Experimental methods

The chemical composition of the weld zone are examined with optical emission spectroscopy technique and the details are shown in table 1. The chemical elemental composition is used for the empherical estimation of the Ferrite Number range with the predefined calculation of Cr and Ni equivalent estimation with modified formulae.

Table 1. Chemical composition of the weld metal.

Weld Sample	Chemical composition (wt %)								
	C	Si	Mn	P	Ni	Cr	Mo	Nb	Ti
Base metal	0.023	0.230	1.820	0.041	10.30	16.62	2.060	0.017	0.001
A	0.013	0.350	1.780	0.027	11.15	18.70	2.020	0.013	0.002
B	0.013	0.360	1.760	0.026	11.23	18.73	2.030	0.012	0.002
C	0.013	0.360	1.770	0.026	11.21	18.69	2.040	0.013	0.002

IOP Conf. Series: Journal of Physics: Conf. Series **823** (2017) 012032 doi:10.1088/1742-6596/823/1/012032

3.1. Ferrite Number (FN) Prediction methods

Ferrite number prediction has been carried out with three techniques namely Schaeffler's diagram, WRC-1992 & De Long diagram for the desired calculation process. The formulae used for these calculations of Cr equivalent and Ni equivalent numbers with the chemical elements composition in weight % and their approximation details are shown in table 2.

Table 2. Cr and Ni equivalent estimation by different approaches used details

Equivalent	Schaeffler's diagram	WRC-1992 Diagram	De long Diagram
$Cr_{Equivalent}$	%Cr+%Mo+1.5%Si+ 0.5%Nb+2%Ti	% Cr + %Mo + 0.7%Nb	%Cr + %Mo + 1.5%Si + 0.5% Nb
$Ni_{Equivalent}$	%Ni+30%C+0.5%Mn	%Ni + 35%C +20% N + 0.25%Cu	%Ni + 30%C + 30% N + 0.5% Mn

By incorporating the weld metal chemical composition of the required elements in weight % data, the detailed Ferrite (in %) range is predicted for all the three different welds fabricated [11], software is used as per giving the inputs]. The detailed analyzed reports are shown in figure 5(a, b & c) for each sample with Schaeffler's diagram, figure 6(a, b &c) for the WRC-1992 diagram and figure 7(a, b & c) for the DeLong's diagram.

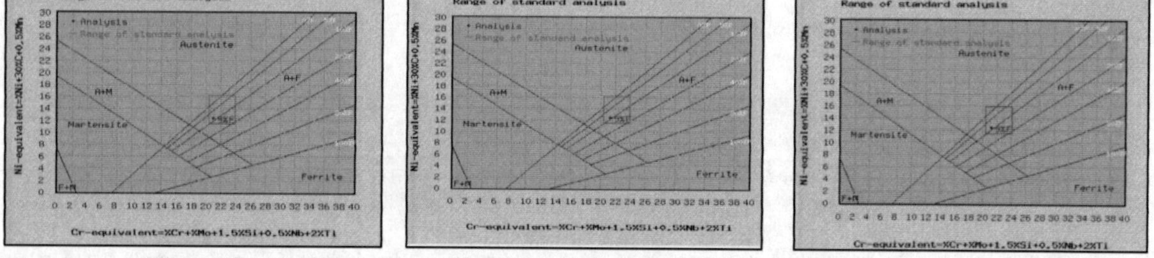

Figure 5(a-c). Schaeffler's diagram estimated FN for weld sample A, B & C.

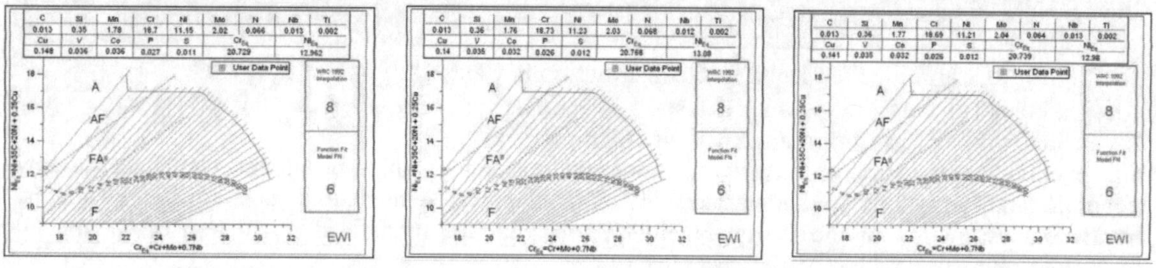

Figure 6(a-c). WRC-1992 diagram estimated FN for weld sample A, B & C.

IOP Conf. Series: Journal of Physics: Conf. Series **823** (2017) 012032 doi:10.1088/1742-6596/823/1/012032

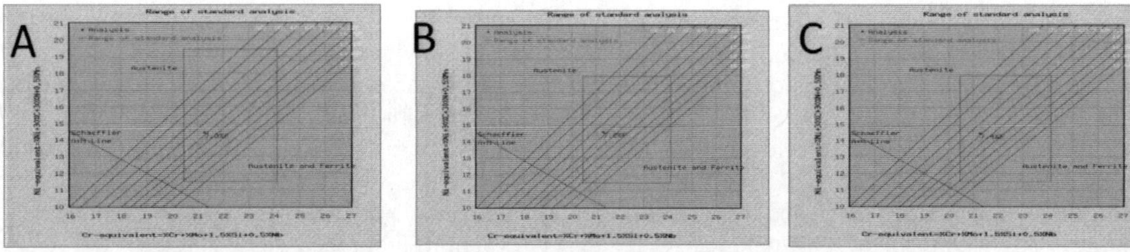

Figure 7(a-c). DeLong's diagram estimated FN for weld sample A, B & C.

3.2. Experimental measurement of FN number

Three welded samples (A, B & C) are polished and examined for the ferrite number (FN) data using the calibrated Ferrite scope (Fischer's make model FMP 30) for the magnetic delta ferrite phase determination. The each sample is scanned over (20 measurements) to estimate the average FN data for WZ and BM which are shown in figure 8(a) and (b).

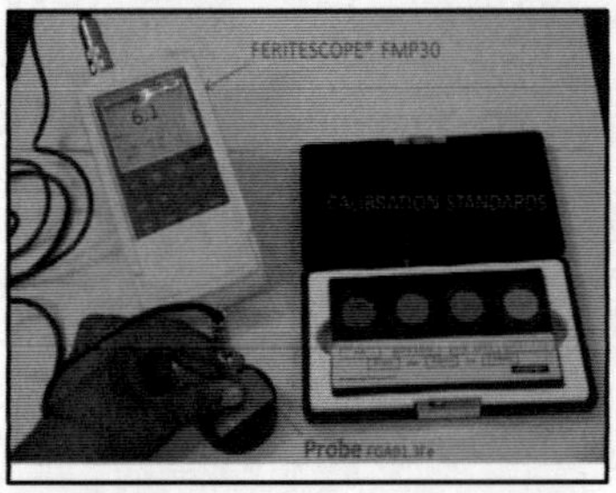

Figure 8(a). FMP 30 Ferritescope with calibration set-up.

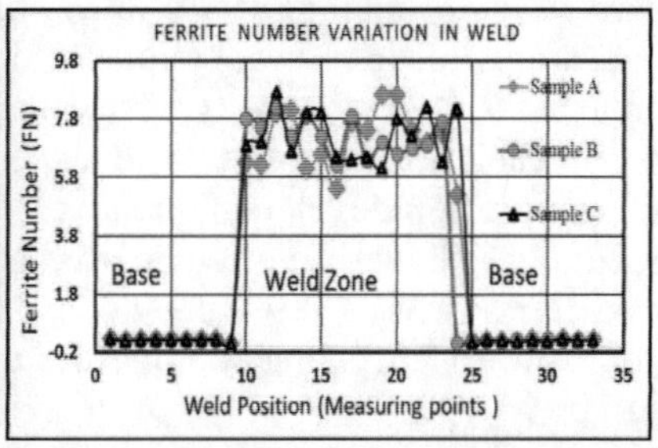

Figure 8(b). FN data for BM &WM of samples A, B & C.

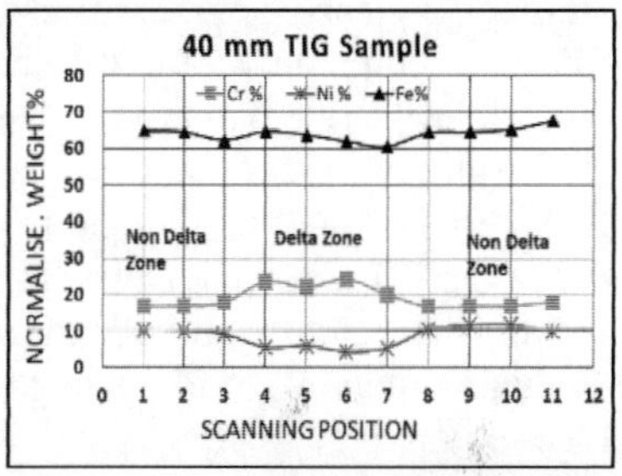

Figure 8(c). Mapped Cr, Ni & Fe in Delta ferrite zone.

As shown in figure 8(c), the delta ferrite zone is mapped for the elemental distribution analysis (Cr, Ni and Fe) to understand the variation. The enhancement in Cr (%) content in the mapped delta ferrite region shows the presence of ferrite phase (enhancement) and loss of Ni indicating the reduction in austenite phase in the delta ferrite.

4. Results and Discussion

The estimation of Ferrite percentage with different approaches is compared with the experimentally measured FN data of all the three welds and the details are shown in table 3.

Table: 3. Different FN prediction and experimental data.

Weld sample	Schaeffler's Diagram (% F)	WRC-1992 (% F)	De Long (% F)	Experimental Data (FN) ≈ (% F calibrated)
A(1.87 kJ/mm)	9%	8%	7.3%	6.68
B(1.72 kJ/mm)	9%	8%	7.2 %	7.04
C(1.67 kJ/mm)	9%	8%	7.4%	7.28

The three welds have shown the trend of high heat input used for weld sample A has exhibited lower FN (6.68), medium heat input used for samples B has shown intermediate FN (7.04) and lower heat input used sample C shown higher FN (7.28) data. Austenitic steel welds exhibited FN values with heat input trend in welded zone is similarly reported as high heat input show low FN comparatively which high heat input condition has shown lower FN and lower heat input has shown higher FN values [12]. However the predicted theoretical measurements (table 3) by using chemical composition data, DeLongs prediction method has shown close agreement with the measured data among others. It is always a tedious task to evaluate the delta ferrite with microstructure for multipass TIG welds. Feritescope measurements will give the idea of the state of the weld joints condition and the delta ferrite content estimation, hence this is a straight forward non destructive technique measurement on multipass austenitic welds (specimen cross section) with simplified procedure.

5. Conclusion

Multipass welds of 40 mm thick AISI SS 316L samples of have been fabricated by narrow groove TIG welding process and are characterized for the weld microstructure and ferrite number predictions using the standard techniques Schaeffler's, WRC-1992 and DeLong methods by Cr & Ni equivalent measurements. The experiments have shown the different high to low heat input conditions in austenitic stainless steel multipass welds (40 mm thick) exhibited the trend of low to high Ferrite Number in the welded joints. Macro and microstructure examinations have been carried out. Microstructure shows the different features of dendrites, cellular and columnar type structures along with delta ferrite within the austenite matrix. Different type structures like acicular and lathy delta ferrite are observed in weld zone of each weld sample. The experimental observation of Ferrite Number (FN) with Ferritescope is employed on developed three weld samples and compared with the predicted FN methods. DeLong method has shown the close approximation with measured FN data on welds when compared with the Schaffler and WRC-1992 methods. This estimation is quite useful to evaluate typical weld joints behaviour towards mechanical properties understanding and prediction.

Acknowledgement

We are sincerely thankful to Mr. Shamsuddin Shaikh for his valuable contribution during the course of this work.

References

[1] Onozuka M et al. 2001 *Fusion Eng. Des.* **55** 397
[2] Ramesh Kumar B, Gangradey R 2012 , *Journal of Physics: Conference Series* **390 (1)**, 012063
[3] Ramesh Kumar B, Chauhan N, Raole P M 2014 , *Fusion Eng. Des.* **89** 3194
[4] Rahul Unni Krishnan, Satish Idury K S N, Ismail T P, Bhadauri A et al. 2014 , *Materials Characterization* **93** 10
[5] Buddu R K, Chauhan N L, Raole P M 2014 , *Fusion Science & Technology* **65** 248.
[6] Guirao J, Rodríguez E, Bayón A, Cortizo J L, Jones L 2011 ,*Fusion Eng. Des* **86**, 2805
[7] Akella S, Ramesh Kumar B 2013 , *Advanced Materials Manufacturing & Characterization* **3** 199
[8] Elmer J W, Allen S M and Eager T W 1989, *Metall. Trans. A* **20A** 2117
[9] Kotecki D J, Siewert T A 1992 , *Weld. J.* **71** 171
[10] Buddu R K, Bhagvat S 2015 ,*Narrow groove TIG weld joints fabrication and mechanical properties characterization of 40 mm thick SS316L samples for fusion reactor components* **(Internal report : IPR/RR-766/2015)**
[11] http://www.migweld.de/english/service/welding-stainless-steels/schaeffler-diagram-for-standard-analysis/.
[12] Sejč P, Kubíček R, 2011 *Scientific Proceedings,* Faculty of Mechanical Engineering, STU in Bratislava **19** 8-14

IOP Conf. Series: Journal of Physics: Conf. Series **823** (2017) 012033 doi:10.1088/1742-6596/823/1/012033

ITER ECE Diagnostic: Design Progress of IN-DA and the diagnostic role for Physics

H K B Pandya[1], Ravinder Kumar[1], S Danani[1], Shrishail[1] P, Sajal Thomas[1], Vinay Kumar[1], G Taylor[2], A Khodak[2], W L Rowan[3], S Houshmandyar[3], V S Udintsev[4], N Casal[4] and M J Walsh[4]

[1]ITER-India, Institute for Plasma Research Bhat, Gandhinagar -382428, India
[2]Princeton Plasma Physics Laboratory, Princeton, NJ 08543, USA,
[3]Institute for Fusion Studies, The University of Texas at Austin, USA,
[4]ITER Organization, St Paul Lez Durance, France

E-mail: hitesh.pandya@iter-india.org

Abstract. The ECE Diagnostic system in ITER will be used for measuring the electron temperature profile evolution, electron temperature fluctuations, the runaway electron spectrum, and the radiated power in the electron cyclotron frequency range (70-1000 GHz), These measurements will be used for advanced real time plasma control (e.g. steering the electron cyclotron heating beams), and physics studies. The scope of the Indian Domestic Agency (IN-DA) is to design and develop the polarizer splitter units; the broadband (70 to 1000 GHz) transmission lines; a high temperature calibration source in the Diagnostics Hall; two Michelson Interferometers (70 to 1000 GHz) and a 122-230 GHz radiometer. The remainder of the ITER ECE diagnostic system is the responsibility of the US domestic agency and the ITER Organization (IO). The design needs to conform to the ITER Organization's strict requirements for reliability, availability, maintainability and inspect-ability. Progress in the design and development of various subsystems and components considering various engineering challenges and solutions will be discussed in this paper. This paper will also highlight how various ECE measurements can enhance understanding of plasma physics in ITER.

1. Introduction

The ECE diagnostic system on ITER [1] is used for the measurement of the core and edge electron temperature profile (T_e) with good spatial and temporal resolution. The diagnostic will also be used for measuring high frequency instability with frequencies of 100 Hz to 10 kHz, plasma energy, ECE radiated power, study of runaway electrons, ELMs and as L-H mode transition indicator.

There are two ECE radiation measurement lines foreseen for ITER. One is for the radial measurement and other is for the oblique ECE measurement [1]. The total ECE diagnostic system will be provided by two Domestic Agencies (DAs) and ITER Organization Central Team (IO-CT). The US DA will provide the front end quasi optical components, in-vessel high temperature calibration sources, 230 – 336 GHz radiometer and the required instrumentation and control system. The IN-DA will design and supply various sub-systems including four 40 meters long transmission lines (TL) with polarizer splitter units, a 122 – 230 GHz radiometer, two Fourier transform spectrometers (FTS) based on Michelson interferometers that detect 70 – 1000 GHz radiation, a high temperature black body calibration source and a RF stray radiation protection system. All the sub-systems are in preliminary design and prototype development phase. For the design of the sub-systems, necessary numerical simulation and analysis are performed.

2. Diagnostics Requirements & Challenges

Designing the diagnostics to meet measurement requirements for ITER plasma is a major challenge due to its size and parameters. The electron temperature profile for the ITER plasma will be measured with spatial resolution of *a*/30 (~ 6.7 cm), where "*a*" is the minor radius of the ITER plasma (~ 200 cm) and temporal resolution of 10ms. In the ECE diagnostic temperature measurement, the relativistic broadening might degrade the spatial resolution of the temperature measurement [2]. The high magnetic field (~ 5.3 T) and large size of the plasma cross-section gives wide bandwidth (> 100 GHz) of the ECE frequency range of each harmonic. This demands to develop wide band (122 – 230 GHz) radiometer with low noise, high sensitivity with minimum number of mixers. The ultra-wide band (70 – 1000 GHz) ECE radiation measurement for the estimation of the ECE radiated power loss will require the transmission line for the frequency range 70 to 1000 GHz with low attenuation over a long distance of 40 – 50 meters. This transmission line is needed to be evacuated to avoid water vapour absorption in this frequency range. This ultra-wide band will be measured with 10 – 20 ms time resolution. Therefore, a Fourier Transform Spectrometer with high throughput and fast (10 - 20 ms) continuous scanning mechanism for the moving mirror in vacuum is required. Assembly and alignment within ± 5 mm over a long (40 to 50 m) transmission line at ITER site will be required. There is high power RF source for the current drive and heating in ITER. This high power source will give RF stray radiation through ECE window in the ECE transmission line. ECE diagnostic systems need to be protected from this stray radiation.

3. Design progress of the ITER ECE diagnostic sub-system

This section describes about the design progress of the ITER ECE diagnostic system provided by IN-DA.

3.1. Transmission line

The ITER ECE diagnostic transmission line layout is shown in figure 1. This transmission line is rough vacuum compatible to avoid water vapor absorption within the ECE frequency range 70 to 1000 GHz. The transmission line includes two quasi optical polarizer selectors, smooth walled circular waveguide components and two quasi optical power splitter units. Progress on preliminary design of these components is described in this subsection.

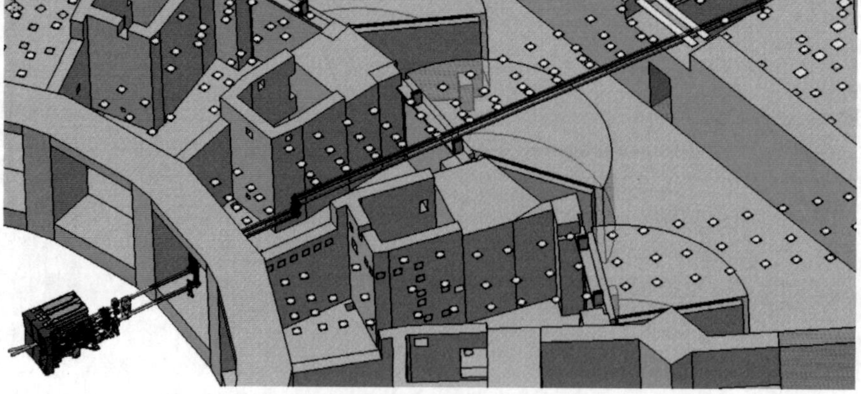

Figure 1. ECE transmission line layout at ITER

3.1.1. Preliminary design of the polarization splitter unit

The ECE radiation from the two measurement lines are collected by the front-end optics arrangement in the port plug and radiation beams are then routed out from the tokamak vessel through a double vacuum window represented by point D in figure 2. These beams contain both modes (i.e. O-mode and X-mode) of radiation. Therefore, it is necessary to split the polarization before transmitting through the waveguide towards the ECE diagnostic room to avoid polarization mode mixing. A polarizer splitter unit is used to meet this requirement.

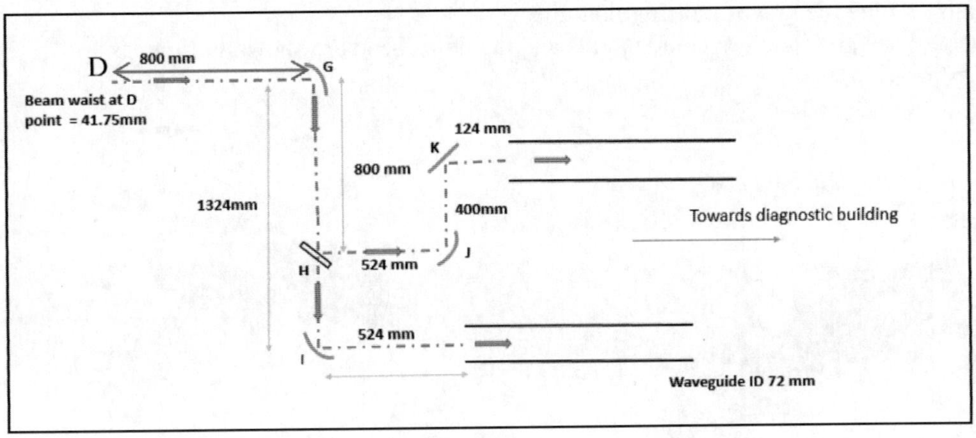

Figure 2. Optical layout of polarization splitter unit

The above schematic (figure 2) depicts two Gaussian beam telescopes and a wire grid polarizer inside the splitter unit. In this, mirror G is the ellipsoidal mirror with focal length 800 mm and other two ellipsoidal mirrors I & J have focal length of 524 mm. The wire grid polarizer is used to select the polarization. The telescope magnification is 0.65. This gives the splitter unit output beam size as 27.35 mm to match with waveguide input. The beam waist size at the circular waveguide mouth should be 0.76 times radius of waveguide for optimized coupling of 87 % between the waveguide and quasi-optical component of the splitter box [3].The radius of the waveguide is 36 mm and following equation [4] is used for calculation of the beam size at the mirror.

$$\omega\,(Z) = \omega_0 \left[1 + \left(\frac{\lambda Z}{\pi \omega_0^2}\right)^2\right]^{1/2} \qquad (1)$$

where ω_0, λ and Z are beam waist at Z=0, radiation wavelength and distance travelled by beam respectively.

The Gaussian Beam Mode (GBM) [5] software is used to analyse this design for estimation of spurious power in the cross-polarized and high order mode in the splitter unit. These are -30 to -34 dB respectively. The value of spurious power is acceptable. .

3.1.2. Design of Smooth walled Circular waveguide components

The waveguide components consist of straight waveguide section, miter bend to divert beam in 90°, pump out "T" for vacuum generation in the transmission line and vacuum compatible joints. The design of the all waveguide components are such that the transmission attenuation of the ECE radiation in frequency range 70 to 1000 GHz should be low. There are mainly three types of losses in the waveguide components; one is the Ohmic loss that depends on the dimensions of the waveguide and conductivity of the waveguide material. The other type of losses like higher order modes and cross polarization modes depend on dimension and alignment accuracy.

Aluminium alloy 6061 T6 is chosen for waveguide material with inner diameter of the waveguide as 72 mm with ±0.05 mm accuracy. The inner surface finish is 7.5 micron. The waveguide straightness of centreline, maximum gravitational bending and concentricity over a 2 meter length are 0.25 mm, 0.05 mm and < 0.5 mm respectively. Ellipticity of the waveguide tube inner surface and outer surface are < 0.1 mm and < 1 mm respectively. Waveguide components with above specified dimension tolerance are being fabricated for attenuation measurement during prototype testing with approximately 10 meter length of transmission line.

The proposed design of the waveguide joints and the miter bend are shown in figure 3.

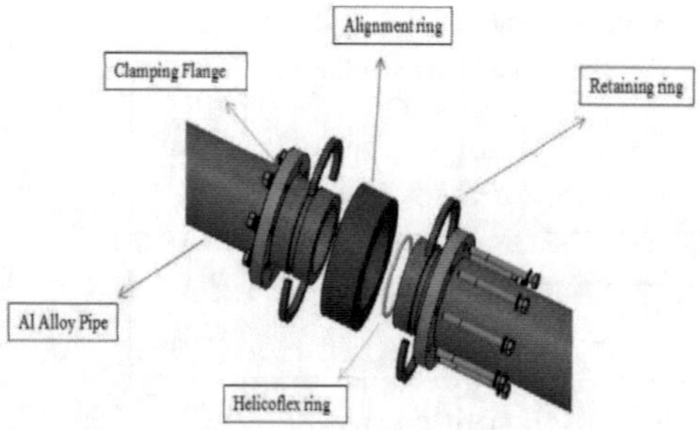

Figure 3a. Waveguide vacuum joints design

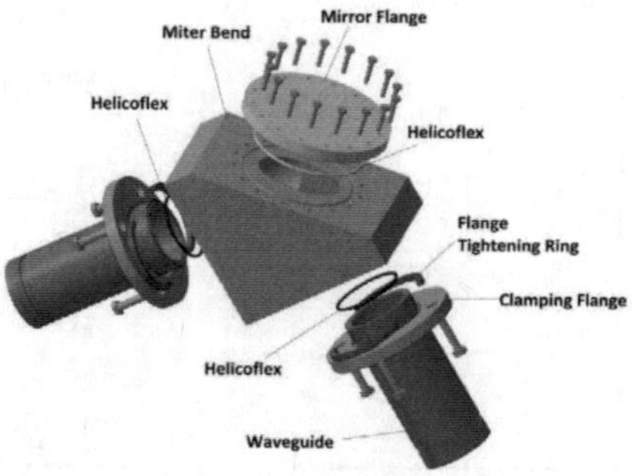

Figure 3b. Miter bends design

3.2. *Preliminary design of 122 – 230 GHz radiometer*
The Electron Cyclotron Emission in frequency range 122-230 GHz would be measured by making use of four Radiometers with intermediate frequency (IF) bandwidth between 16 to 30 GHz with frequency resolution of 1 to 2 GHz. The total numbers of channels are 58 that give good spatial resolution. The total noise temperature is 10 eV that is less than plasma noise. The design of radiometer has been optimized by considering the advanced technology, cost and maintenance.

3.3. *Design of Fourier Transform Spectrometer (FTS)*
The FTS based on polarizing Michelson interferometer (70 – 1000GHz) will be used to measure the power loss from the ITER plasma due to ECE radiation and to study of non-thermal electrons in the ITER plasma. This spectrometer is a polarizing Martin-Puplett type [6]. The spectrometer has been designed with frequency resolution of 3 - 5 GHz and temporal resolution of 10 – 20 ms. The schematic diagram of the spectrometer is shown in figure 4. The spectrometer consists of three Gaussian beam telescopes made of the six ellipsoidal mirrors, one fixed and one moving rooftop mirrors, two wire grid polarizers, one beam splitter, two attenuators and cryogenically cooled Indium Antimonide hot electron bolometer detectors.

There are two inputs to the spectrometer. Either one of the inputs can be used for the measurement of the radiation and other input is used as reference. The input polarizer selects the polarization of the input beam and the beam splitter divides the beam and sends it in two perpendicular directions. The roof top mirrors reflect the beam in the same direction and both the beams are combined at the beam splitter. The movable mirror creates the path difference between these two reflected beams and due to this path difference, an interferogram is generated. By taking Fast Fourier transforms of the interferogram, one gets the spectrum of the input radiation.

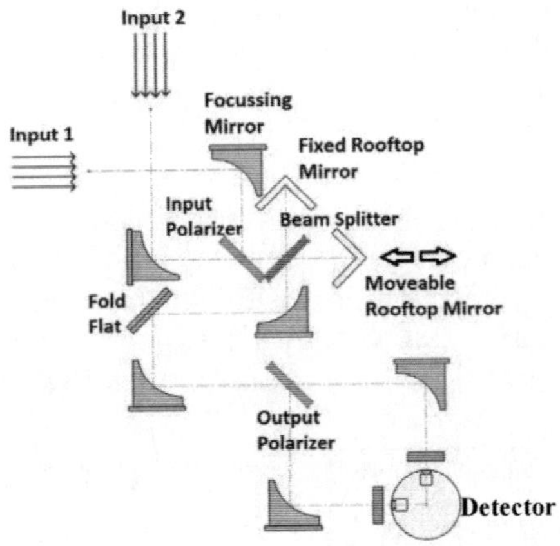

Figure 4. Schematic of the Fourier Transform spectrometer

3.4. Design of high temperature black body source

This black body source will be used to check the functionality of the ECE receiver systems (i.e. radiometers and Michelson interferometers) in ECE diagnostic room at ITER site. The radiation temperature of the source is 700 K with short term temperature stability of ± 2 K and surface temperature uniformity of ± 10%.

The high temperature black body source as shown in figure 5, consists of three main parts; an emissive surface, a heating element and surface temperature measurement arrangement. The emissive surface is fabricated separately. A commercially available high temperature hot plate is used as heating element with a provision for the surface temperature measurement inside this hot plate. A Silicon Carbide circular plate with an array of small pyramids at one side heated to 700 K is used as an emissive surface for the source. This surface is a black body surface for the millimetre wave frequency. The Silicon Carbide is chosen as the emitter because it has good thermal conductivity and high melting point. Therefore, the Silicon Carbide plate surface can achieve the required high temperature. The pyramidal contour on the surface of the Silicon Carbide plate creates multiple internal reflections of the incident radiation, resulting in increased emissivity. A prototype has been designed and developed and is now being tested. To check the temperature uniformity, IR camera is used. The preliminary test results are shown in figure 6. This source is tested up to 500 °C (i.e. 772 K) and temperature uniformity is observed to be ± 11%. Our requirement is ± 10%. This minor deviation in the temperature uniformity is due to the hot plate surface temperature non-uniformity. Therefore, an improvement is needed in the hot plate heating coil design to achieve required surface temperature uniformity.

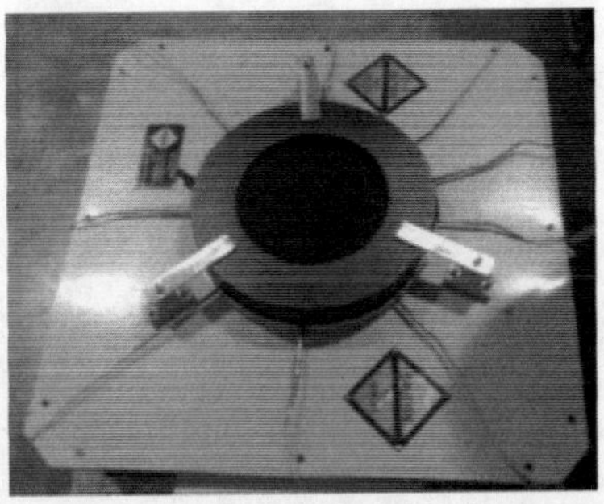

Figure 5. High temperature black body source

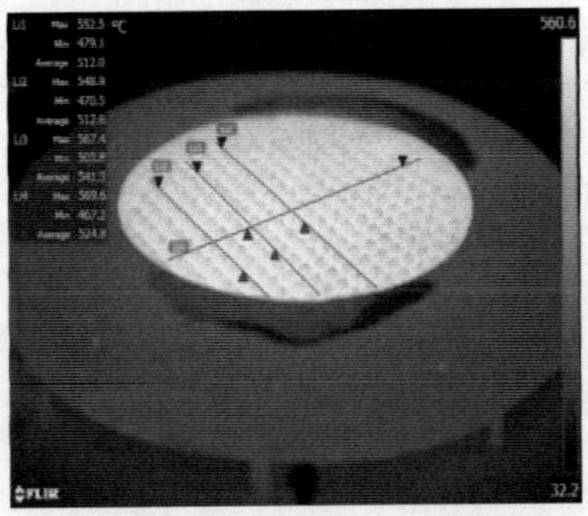

Figure 6. IR image of the Source

3.5. RF stray radiation protection system

High power RF sources will be used during ITER plasma operation for the heating and current drive at frequency of 170 GHz and for Collective Thomson Scattering (CTS) diagnostic at a frequency of 60 GHz. These high power sources will be a source of stray RF radiation during operation of ITER. This stray radiation can potentially damage the ECE diagnostic system. Therefore, a radiation protection system needs to be developed. As a first step, the power level of this stray radiation should be measured. For this, we designed and developed a detector (sniffer) that is polarization insensitive, multimode sensitive with wide angle reception, insensitive to standing waves and spatial interference patterns. The prototype sniffer is under test as shown in figure 7 and measured attenuation of the sniffer in D-band frequency range 110 to 170 GHz is shown in figure 8.

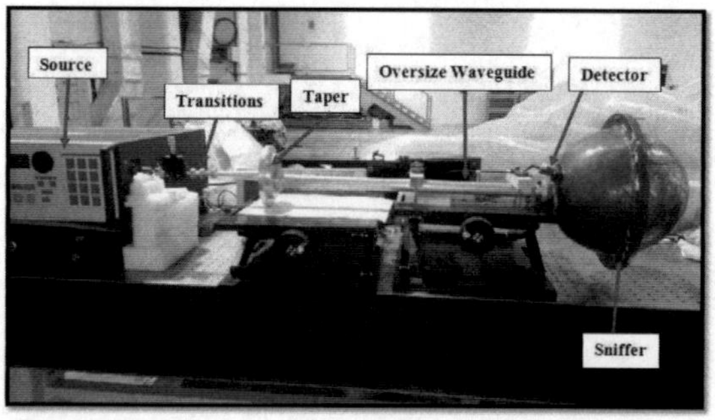

Figure 7. Testing set up for the prototype RF stray radiation sniffer

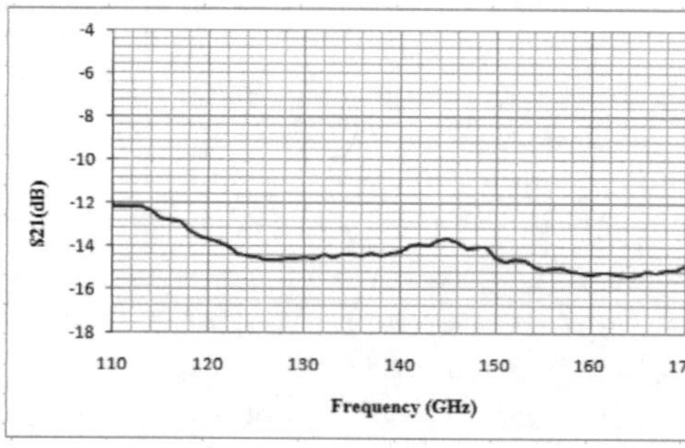

Figure 8. Attenuation (S_{21}) of the sniffer

4. Summary

This paper describes the progress of the preliminary design of the ITER ECE diagnostic system. Some of the subsystems components have been designed and prototypes are being tested. Among these, the FTS and waveguide components are under fabrication and the black body source and the RF sniffer are under testing as a part of prototype work. The measurement of surface temperature uniformity of the high temperature source was carried out and it is close to the required uniformity.

References

[1] Taylor G, et at. 2014 *18th Joint Workshop on ECE and ECRH*, Nara, Japan p 22
[2] Danani S, Pandya H K B, Vasu P and Austin M E 2011 *Fusion Sci. and Techno.* **59** 651
[3] Self S A 1983 *Applied Optics* **22** 658
[4] Goldsmith P F, Quasioptical system, *Gaussian beam Quasioptical propagation and Applications IEEE Press*, Institute of Electrical and Electronics Engineers Inc. New York
[5] Gaussian Beam Mode 3D Engine (running within PTC's Creo Parametric 2.0) Thomas Keating Ltd. www.terahertz.co.uk
[6] Martin D H and Puplett E 1970 *Infrared Physics* **10** 105

Preliminary Design of O-mode Radiometer for ITER ECE Diagnostic

S Danani[1], H K B Pandya[1], Ravinder Kumar[1], M E Austin[2], V S Udintsev[3] and Vinay Kumar[1]

[1]ITER-India, Institute for Plasma Research, Bhat, Gandhinagar, 382 428, India
[2]Fusion Research Center, the University of Texas at Austin, Austin, TX, USA
[3]ITER Organization, Route de Vinon sur Verdon, 13115 St Paul-Lez-Durance, France

Email: suman.danani@iter-india.org

Abstract: The ITER Electron Cyclotron Emission (ECE) diagnostic system will provide information about plasma electron temperature and its fluctuations and other important plasma parameters which are required for plasma control and physics studies. The temperature profile measurement in the first harmonic ECE frequency range from 122-230 GHz (for $B_T = 5.3$ T) is obtained by using an O-mode heterodyne radiometer. It is difficult to cover this wide frequency band by one radiometer, due to technological challenges in achieving wide bandwidth for the mixers. So, the present radiometer design has been optimized by considering four receivers, each of bandwidth ~ 30 GHz which can provide reliable temperature measurements. The splitting of frequency band into four receiver bands is efficiently achieved by considering a combination of quasi-optical and waveguide diplexers, optimizing power loss and cross-talk between the channels. The target spatial resolution of *a*/30, where "*a*" is the minor radius of the plasma cross section, is achieved by choosing Radiometer IF filter bandwidth of 1-2 GHz. Further, the radiometer is designed to achieve noise temperature < 10 eV. In this paper, the preliminary design and performance of O-mode Radiometer has been discussed.

1. Introduction

The Electron Cyclotron Emission (ECE) diagnostic system in ITER provides essential information for plasma control and for evaluating the plasma performance. It measures the electron temperature profile (edge/core), electron temperature fluctuations and radiated power in electron cyclotron frequency range from the plasma. These measurements yield information about important parameters such as NTM and TAE $\delta T/Te$, βp, ELM associated temperature perturbations and runaway electrons and are vital for understanding the evolution of plasma, thereby contributing significantly to transport studies and plasma confinement [1]. The ECE detection instruments include two Michelson interferometers (Ordinary, O-mode and eXtraordinary, X- mode) over a broad spectral range from 70 GHz to 1 THz, O-mode radiometer covering 122-230 GHz and X-mode radiometer covering 220-340 GHz. For full-field operation of ITER (5.3T), the O-mode radiometer will be used to measure the 1st harmonic O-mode ECE while the X-mode radiometer will measure the 2nd harmonic X-mode radiation. During ITER early operational stage, it is expected to run at half-field. In this case the O-mode radiometer will have to be used for 2nd harmonic X-mode measurements.

The primary role of O-mode Radiometer is core electron temperature radial profile measurement and NTM temperature fluctuation ($\delta T/T_e$) measurement in the frequency band 122-230 GHz. From the ITER measurement requirements, the electron temperature profile needs to be measured with a spatial

IOP Conf. Series: Journal of Physics: Conf. Series **823** (2017) 012034 doi:10.1088/1742-6596/823/1/012034

resolution of $a/30$ which is ~ 6.7 cm for plasma minor radius of a = 200 cm, temporal resolution of 10 ms and accuracy of 10% in the plasma core, for the temperature range 0.5 – 40 keV [2]. The principal measurement limitations of the system are restricted radial region of observation due to harmonic overlap and degraded spatial resolution due to the relativistic broadening [3,4].

The O-mode radiometer has been designed and optimized to meet the ITER requirements. The paper discusses the requirements, challenges and the design of O-mode Radiometer.

2. Design Requirements and Challenges of O-mode Radiometer

The following design requirements for O-mode radiometer are considered:

2.1. Frequency Bandwidth

For ITER operation at toroidal magnetic field $B_T(0)$ ~ 5.3 T, the 1st harmonic ECE frequencies extend from 122-230 GHz. Therefore the O-mode radiometer should provide measurement of plasma electron temperature profile by measuring 1st harmonic frequency range 122-230 GHz for the full B_T operation.

A set of four receivers with different frequency bandwidths is chosen to cover this wide bandwidth of ~110 GHz. A frequency splitter unit is needed to split the frequency range 122-230 GHz into four receiver bandwidths. The design requirement of the splitter unit is that it should have low insertion loss and a sharp frequency roll-off. As the frequency band is very wide, a hybrid approach of splitting has been adopted: first the power will be split optically and then waveguide diplexer will be used to get the desired four frequency bands.

2.2. Spatial Resolution

The spatial resolution of ECE measured Electron temperature depends on the ITER plasma parameters, the front-end radiation collection system and the instrument frequency resolution. However the radial resolution is restricted by ECE frequency broadening effects, primarily due to the relativistic broadening. In Ref [3], the effects of relativistic broadening were investigated for Scenario 2 plasma with a central electron temperature of 25 keV.

In order to achieve the spatial resolution of $a/30$, the radiometer is designed with frequency resolution of 1 and 2 GHz across the frequency band. For the low-field side edge region of the plasma, in and near the pedestal, the temperature gradient is sharp, so frequency resolution of 1 GHz is required for the lowest frequency band. Towards the core region, temperature gradient is flatter, so frequency resolution of 2 GHz is sufficient. This is achieved by proper selection of cavity filters in the IF section of the radiometer.

2.3. Temporal Resolution

For the core electron temperature measurement, the time resolution requirement is 10 ms (100 Hz) and for the high frequency instability measurement, it is 100 μs (10 kHz). For ECE electron temperature measurement by a radiometer, the time resolution depends on time response of the radiometer IF detectors, video amplifier bandwidth, signal to noise ratio, and sampling rate of digitizer. With the current technology, the time response of commercially available IF detectors is 1 μs and the video amplifier bandwidth can be adjusted between 100 Hz to 10 kHz according to measurement requirement.

By choosing the appropriate video bandwidth (~10 kHz), required temporal resolution is achieved.

2.4. Accuracy

The stated measurement accuracy for the electron temperature profile is typically 10%. The uncertainty in ECE measured electron temperature depends on the absolute calibration accuracy and stability of the radiometer.

To achieve good stability in the measurements, the RF and IF parts of the radiometer with active components shall be temperature-controlled to keep the temperature fixed within their instrumentation enclosures.

IOP Conf. Series: Journal of Physics: Conf. Series **823** (2017) 012034 doi:10.1088/1742-6596/823/1/012034

2.5. Protection from the High power ECRH stray radiation of frequency 170 GHz

The stray radiation level due to ECH is of the order of few kilowatts and it can damage the sensitive ECE Radiometer components. Two protection measures have been taken: (1) the splitting of the frequency band into four receiver bands has been done in such a way that 170 GHz radiation does not fall in the four frequency bands as mentioned in section 3. This is ensured by choosing appropriate band pass filters at the input of the mixers. (2) A quasi-optical or a waveguide band stop notch filter at 170 GHz may be used in case of requirement. This will be in addition to the protection provided by the fast shutter that is in front of the polarization splitter.

3. Preliminary Design of O-mode Radiometer

Based on the above design considerations, radiometer design is developed as shown in figure 1. The O-mode Radiometer consists of a set of four receivers with IF bandwidth between 16-30 GHz. as shown in figure 1:

a) Receiver 1: 122 - 138 GHz bandwidth with sixteen channels of 1 GHz bandwidth
b) Receiver 2: 141 - 168 GHz bandwidth with fourteen channels of 2 GHz bandwidth
c) Receiver 3: 172 - 200 GHz bandwidth with fifteen channels of 2 GHz bandwidth
d) Receiver 4: 205 - 230 GHz bandwidth with thirteen channels 2 GHz bandwidth
e) Frequency splitter unit which will split the ECE radiation 122-230 GHz into above four receiver bands

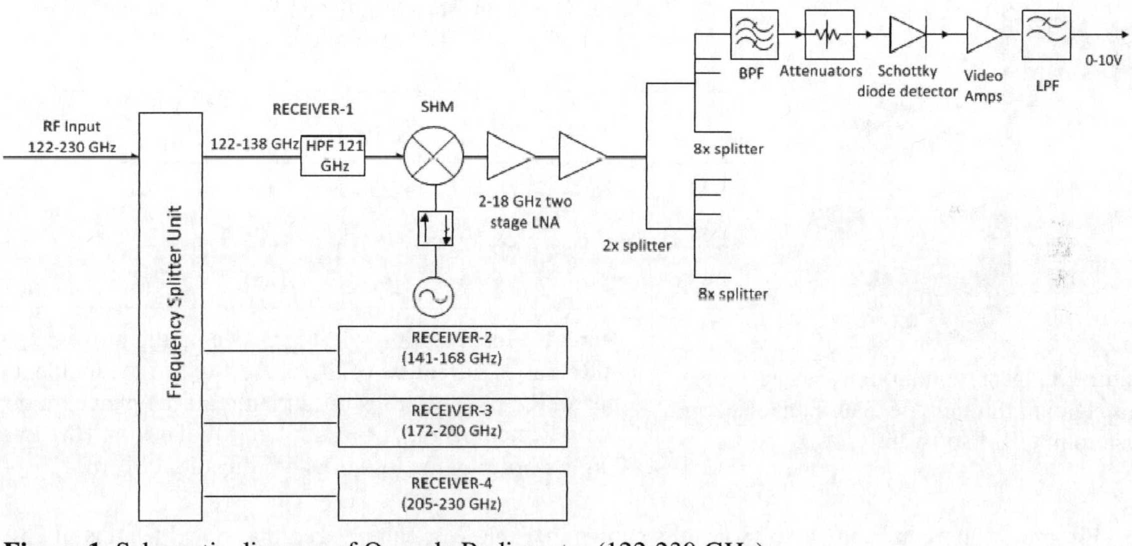

Figure 1. Schematic diagram of O-mode Radiometer (122-230 GHz)

3.1. Frequency splitter unit

The input frequency band 122-230 GHz needs to be split efficiently into four frequency bands corresponding to the four receiver bandwidths. Conventionally, directional couplers are used but the loss is high, and also the applicable frequency regime is limited due to the waveguide size. So, splitting needs to be done quasi-optically, but the roll-off is high in that case. A hybrid splitting mechanism is therefore adopted, making use of both quasi-optical splitter and waveguide diplexer as shown in figure 2.

IOP Conf. Series: Journal of Physics: Conf. Series **823** (2017) 012034 doi:10.1088/1742-6596/823/1/012034

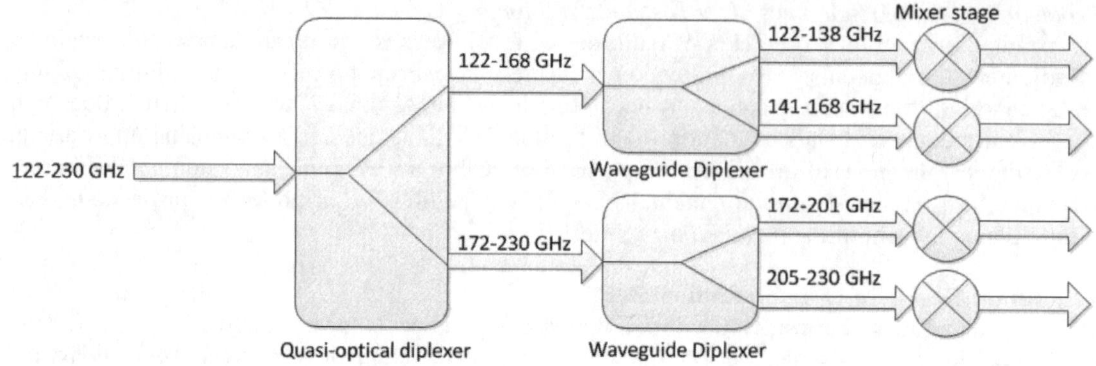

Figure 2. Frequency splitter unit for Radiometer

The quasi-optical diplexer unit consists of a Gaussian beam telescope and a dichroic filter (frequency selective surface). The analysis of the Gaussian beam telescope system has been done using 3D Gaussian Beam mode analysis software known as GBM CAD package [5,6] and summarized in Table 2. Gaussian beam propagation through the Gaussian Telescope System in GBM has been represented in Figure 3.

Figure 3. Representation of Gaussian beam propagation through the Gaussian telescope system in GBM software

Table 2. Gaussian Beam Analysis: Mirror parameters and Distortion levels

	f (mm)	w (mm)	R_i (mm)	R_e (mm)	XP-loss (dB)	HM-loss (dB)
EM1	250	28.19	-4271.3	265.5	25	28
EM2	250	28.19	-265.5	4271.3	25	28
EM3	250	28.19	-265.5	4271.3	25	28

Here, f = focal length, w = beam size on the mirror, $R_i = R_1$ = Radius of curvature of the phase front of the incident beam, $R_e = R_2$ = Radius of curvature of the phase front of the reflected beam, XP-loss = Cross Polar loss, HM-loss = Higher order mode loss, EM = Ellipsoidal Mirror

From the Gaussian beam analysis results shown in Table 2, it can be seen that focal lengths of 250 mm give acceptable distortion levels of the order of ~ -25 dB or lower. The splitter unit has a 3 dB power loss.

3.2. O-mode Receivers

For ITER ECE radiometer, the wide band 122-230 GHz is measured using four receivers, details of which are described in Table 3. A receiver consists basically of a down-converter, IF amplifier, power divider and detection system with band pass filters. The IF bandwidths of these four receivers are respectively 16, 27, 28 and 25 GHz. Broadband mixers with frequency bandwidths of around 30-35 GHz have already been developed and used in tokamaks such as Alcator C-mod [7]. Also the noise temperature of < 10 eV is achievable with such broadband mixers. To determine, the noise temperature of the radiometer, the overall noise figure is estimated using the Friis' formulae [8] as given below

IOP Conf. Series: Journal of Physics: Conf. Series **823** (2017) 012034 doi:10.1088/1742-6596/823/1/012034

$$T_r \sim (T_m + L_c T_{if}) \tag{1}$$

where T_m = mixer noise temperature, L_c = mixer conversion loss, T_{if} = Noise temperature of IF amplifier.

Assuming $T_m \sim 4000$ K, $L_c \sim 9$ dB, $T_{if} = 1000$ K, results in $T_r = 12000$ K

Table 3. Main specifications of the Receivers

Parameters	Receiver-1	Receiver-2	Receiver-3	Receiver-4
Frequency covered (GHz)	122-138	141- 168	172-200	205-230
Bandwidth (GHz)	16	27	28	25
No. of mixer	1	1	1	1
Frequency resolution (GHz)	1	2	2	2
IF band pass filter bandwidth	0.5 GHz	1 GHz	1 GHz	1 GHz
No. of channels	16	14	15	13
Noise Temp	Less than plasma noise ~ 10 eV [7]			
Remarks	Total no. of channels = 58, Total no. of mixers/down converters = 4			

3.2.1. Channel details for Receivers:
This section describes the channel details of each of the four receivers. A 2-way and an 8-way power divider are used after the mixer and IF amplifier to obtain the required number of channels in each receiver. This is then followed by cavity filters. The channel separation of the cavity filters decides the frequency resolution and hence the spatial resolution in the plasma. The frequency resolution will be different in different receivers in order to achieve the spatial resolution requirement of ~6.7 cm. The number of channels will depend on the frequency resolution. For Receiver-1, the IF frequency bandwidth is 16 GHz and frequency resolution is 1 GHz, so number of channels would be 16. For receivers 2, 3 & 4, the IF frequency bandwidths are 27 GHz, 28 GHz and 25 GHz respectively and frequency resolution is 2 GHz. Thus number of channels is 14, 15 and 13 respectively.

3.2.2. I & C requirement for the Radiometer:
In order to meet the temporal resolution requirement for temperature measurement and Plasma control system for the radiometer having four receiver systems with a total of 58 channels, it is planned to use video amplifier with 3 dB bandwidth of 1 MHz and sampling at 100 Hz (slow sampling) and at 500 kHz (fast sampling). The data acquisition system will be based on 16 bit channel ADC modules.

4. Conclusion
The preliminary design of O-mode Radiometer for ITER ECE Diagnostic has been described in this paper. The Radiometer is designed considering the measurement requirements and present days' technical feasibility. The design has been optimized by considering four receivers, each of bandwidth ~ 30 GHz which can provide reliable temperature measurements. Splitting of the frequency band into four receiver bands is efficiently achieved by considering hybrid splitting mechanisms. The target spatial and temporal resolutions have been achieved by choosing appropriate IF and video bandwidths.

IOP Conf. Series: Journal of Physics: Conf. Series **823** (2017) 012034 doi:10.1088/1742-6596/823/1/012034

References

[1] Udintsev V S et al. 2012 *Proc. FEC* (San Diego) IAEA-CN-197/ITR/P5-41

[2] Taylor G, Austin M E, Beno J H, Danani S, Ellis R F et al. 2015 *EPJ Web of Conferences* **87** 03002

[3] Austin M E, Beno J H, Ellis R F et al. 2008 *Proc. of the 15th ECE/ECRH Workshop*, Yosemite, California, USA

[4] Danani S, Pandya H K B, Vasu P and Austin M E 2011 *Fusion Science & Technology* **59** 651

[5] Gaussian Beam Mode 3D Engine (running within PTC's Creo Parametric 2.0) *Thomas Keating Ltd.* www.terahertz.co.uk

[6] Sozzi C et al. *23rd Symposium on Fusion Technology*, 20 - 24 September 2004 - Fondazione Cini, Venice, Italy

[7] Chatterjee R, Phillips P E, Heard J et al. 2001 *Fusion Engineering and Design* **53** 113

[8] Hartfuss H J, Geist T and Hirsch M 1997 *Plasma Phys. Contr. Fusion* **39** 1693

IOP Conf. Series: Journal of Physics: Conf. Series **823** (2017) 012035 doi:10.1088/1742-6596/823/1/012035

Development of control system for multi-converter High voltage Power supply using programmable SoC

Rasesh Dave[1], Jagruti Dharangutti[2], N P Singh[1], Aruna Thakar[1], Hitesh Dhola[1], Sandip Gajjar[1], Darshan Parmar[1], Tanish Zaveri[2] and Ujjwal Baruah[1]

[1]ITER-India, Institute for Plasma Research, Bhat, Gandhinagar -382 428, Gujarat, India

[2]Institute for Technology, Nirma University, S G Highway, Ahmedabad-382 481, Gujarat, India

E-mail: rasesh.dave@iter-india.org

Abstract. Multi-converter based High Voltage Power Supplies (HVPSs) find application in multi-megawatt accelerators, RF systems. Control system for HVPS must be a combination of superior parallel processing, real time performance, fast computation and versatile connectivity. The hardware platform is expected to be robust, easily scalable for future developments with minimal overheads. This paper describes development of control system on Zynq All Programmable SoC (System on Chip) for HVPS. Typical HVPS control mechanism involves communication, generation of precise control signals/pulses for few hundred numbers of chopper and closed loop control in microsecond range for regulated output. Such kind of requirements can be met with Zynq All Programmable SoC, which is a combination of Dual core ARM Cortex A-9 Processing System (PS) and Xilinx 7 series FPGA based Programmable Logic (PL). Deterministic functions of power supply control system such as generation of control signals with precise inter-channel delay of nanosecond range and communication with individual chopper at 100kbps can be implemented on PL. PS should implement corrective tasks based on field feedback received from individual chopper, user interface and OS management that allows to take full advantage of system capabilities. PS and PL are connected with on-chip AXI-4 interface with low latency and higher bandwidth through 9 AXI ports. Typically PS boots first, this ensures secure booting and prevents external environment from tampering PL.

1. Introduction

High voltage power supplies (HVPS) are utilized for Multi-megawatt accelerator applications and RF system. HVPS with multi-level inverters have typical configuration as shown in Figure 1. Multisecondary transformers provide multiple isolated feeds, converted to DC voltage by chopper units where individual choppers are connected in series to generate high voltage. Precise chopper control is essential for fine resolution and high accuracy of output voltage. Chopper coverts fixed DC voltage to variable DC voltage. The main controlling element is IGBT switch. Each chopper has its own on-board controller for state transition & protection.

Control system for multi-converter High voltage Power supply should control multiple chopper units. Typical control system for HVPS with multiple choppers must be a combination of superior parallel processing, real time performance, fast computation and versatile connectivity. Same way the development platform should be robust, easily scalable for future developments with minimal cost overhead. Typical HVPS control mechanism involves communication, generation of precise control signals for few hundred numbers of chopper and closed loop control in microsecond range for regulated output.

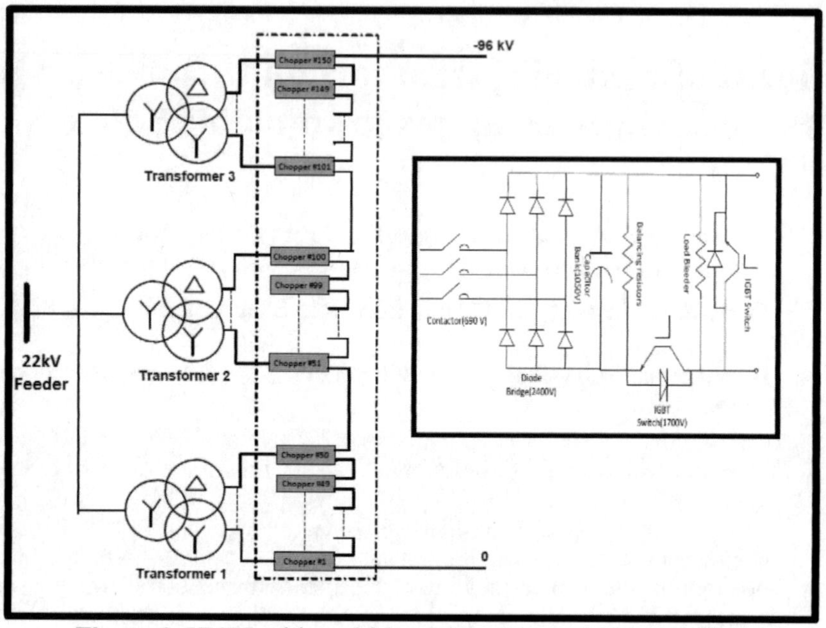

Figure 1. HVPS with multi-level inverters and Chopper Unit

2. Logic Blocks of HVPS Controller & Zynq SoC Architecture

HVPS controller needs to have a continuous communication link with each chopper module for state transition and health monitoring. For this purpose, serial communication is implemented with Universal Receiver Transmitter (UART). The maximum operating and response frequency in complete system is ~ hundreds of us. So, UART baud rate should be ~ 100 kbps. Independent UART link for number of choppers in HVPS configuration leads to custom UART creation. Multiple UART port cannot be available with any processor or controller so, the option is to use reconfigurable hardware platform: FPGA. The control pulses for choppers are generated as PWM pulses with predefined precise inter-channel delay. The duty cycle of PWM pulse is calculated depending on total voltage of chopper modules and set voltage given by user. Generation of PWM pulse is implemented on FPGA. GUI is also required for taking user input and monitoring of system parameters.

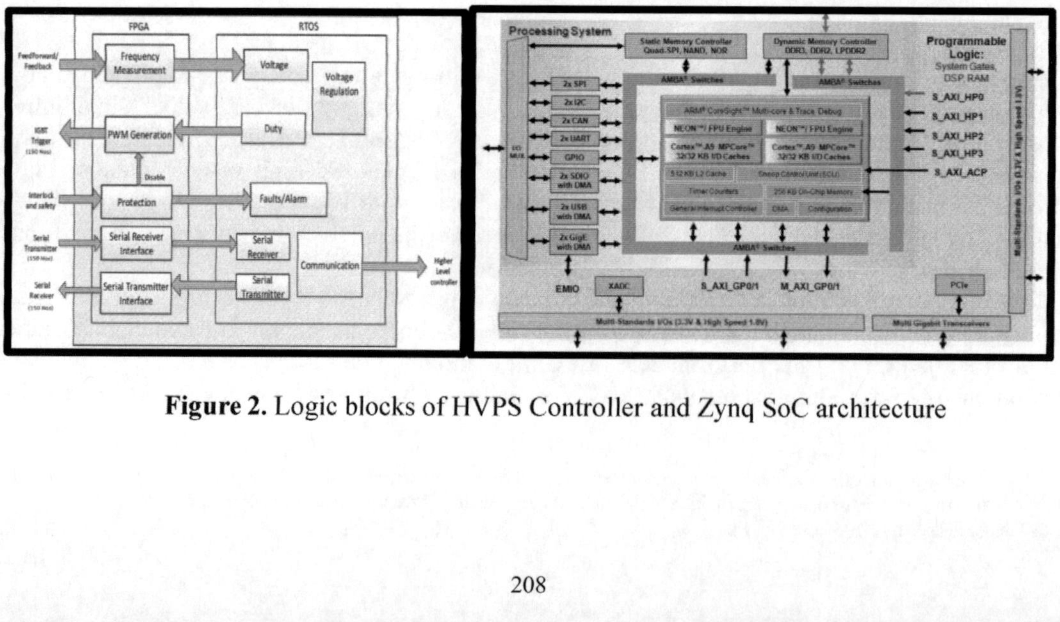

Figure 2. Logic blocks of HVPS Controller and Zynq SoC architecture

Control mechanism for duty calculation, voltage regulation and communication with Host PC/ Higher level controller should be implemented on processor.

Standard SoC can have both FPGA (Programmable Logic (PL)) and processor (Processing system (PS)) e.g. Zynq SoC from Xilinx [1]. Zynq is combination of a dual core processor and a FPGA on a single chip. The processor is ARMs Cortex A9 processor which is an application specific processor capable of running full OS like Linux, while FPGA is 7-series FPGA from Xilinx. The processor and FPGA are internally interfaced with AXI point to point interconnect.

3. Software suite for Controller Development

For synthesis and analysis of HDL designs, Vivado Design Suite by Xilinx is used. It also supports additional features for system on a chip development and high-level synthesis. The Vivado IP Integrator allows quickly integration and configuration of IP from the large Xilinx IP library. The Xilinx Software Development Kit (XSDK) is the Integrated Design Environment for creating embedded applications on any of Xilinx' for Zynq-7000 All Programmable SoCs. The SDK application IDE can deliver true homogenous and heterogeneous multi-processor design and debug. The key of communication between PS and PL is the highly sophisticated AXI interconnect. There are three variants of AXI that are AXI Full, AXI Lite and AXI Stream [2]. In this development while generating custom IP AXI Lite standard is used.

As shown in figure 3, Vivado block diagram is created with standard IPs of Zynq PS, AXI Interconnect, AXI Timer and Reset and clock. Application specific custom IPs [3] of PWM pulse generation, UART TX, UART RX and Frequency Counter are added to Vivado block diagram for HVPS controller requirements.

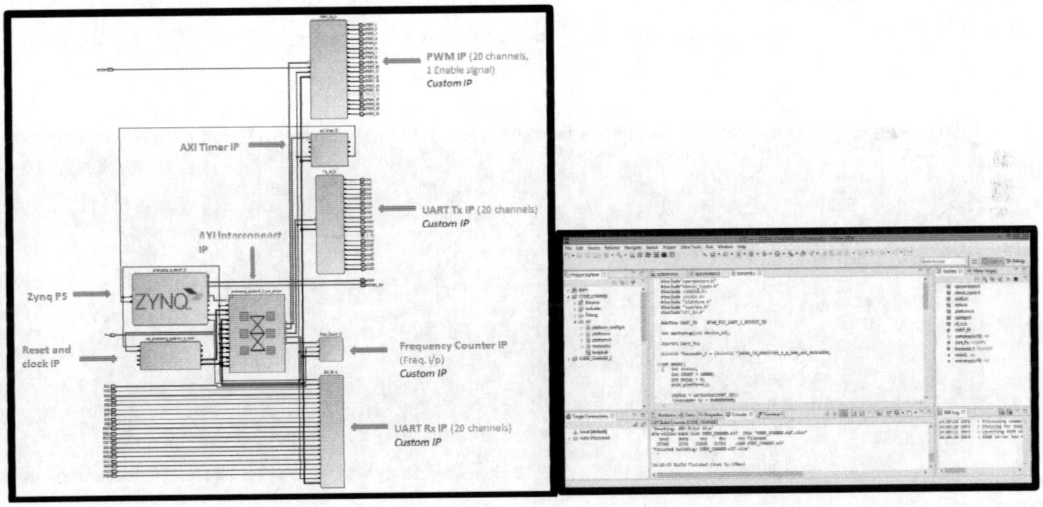

Figure 3. Vivado Block Diagram with Custom IP Blocks and XSDK Environment

Xilinx Software Development Kit (XSDK) is use to develop embedded application for taking user input via USB-UART interface, threshold setting & state command transmission from UART TX, state & error monitoring of each chopper module from UART RX data, calculation of feedforward voltage and PWM duty correction based on output voltage. LabVIEW based GUI is developed for ease of configuration parameter setting state transition prompting and output voltage monitoring. GUI can be used for error reading from each chopper module for fault finding

IOP Conf. Series: Journal of Physics: Conf. Series **823** (2017) 012035 doi:10.1088/1742-6596/823/1/012035

4. Test Setup and Results

HVPS Controller development is performed on ZC706 Evaluation board from Xilinx containing Zynq -7045 device. To validate controller functionality, test bench of 30V chopper unit with microcontroller is used as a test jig. These modules have MOSFET as a control element. The test setup consists of 20 chopper modules connected in series, which can produce maximum output voltage of 600 V.

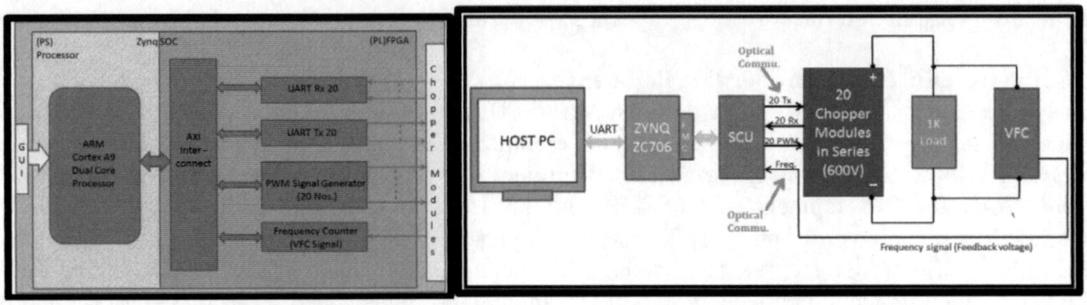

Figure 4. Block Diagram of Test Setup for HVPS Controller using Zynq SoC

Figure 4 shows the block diagram of test setup. All serial inputs and output are mapped on DIO (Differential Input Output) pins of LPC (Low Pin Count) connector of Zynq board. The communication between chopper module and Zynq board is optical for providing isolation. A Signal Conditioning Unit (SCU) is used for electrical to optical and optical to electrical conversion of input and output signals. Each chopper module has one UART receiver link, one UART transmitter link and one PWM link. Final voltage is taken by VFC based Voltage Monitoring Unit (VMU) for closed loop correction.

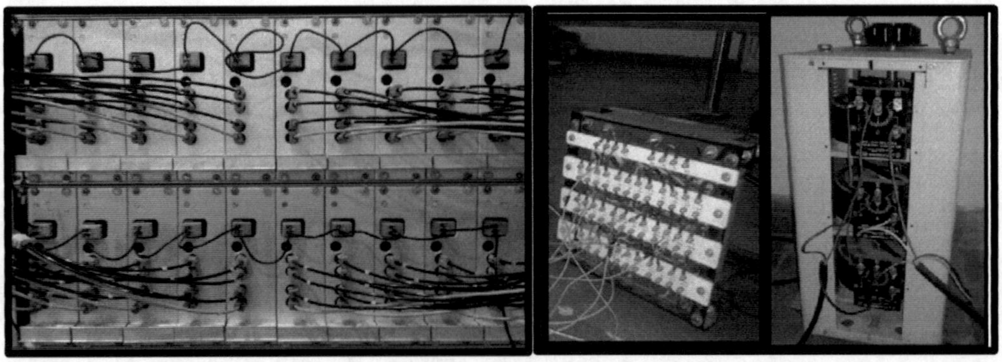

Figure 5. 20 Nos of cascaded chopper module with multisecondary transformer and Variac

Figure 5 shows Test setup of HVPS controller consisting of 20 Nos of chopper unit connected in series and multisecondary transformer for 22V AC Input. Three phase Variac is also used for simulation of under voltage, over voltage faults and feedforward correction in calculation of duty cycle. Figure 6 shows Zynq based ZC706 Evaluation board, FMC to FMC cable for connecting daughter board which is compatible with SCUs for taking Input/outputs. Control PC with Vivado Design Suite and running XSDK application program is also shown in figure 6.

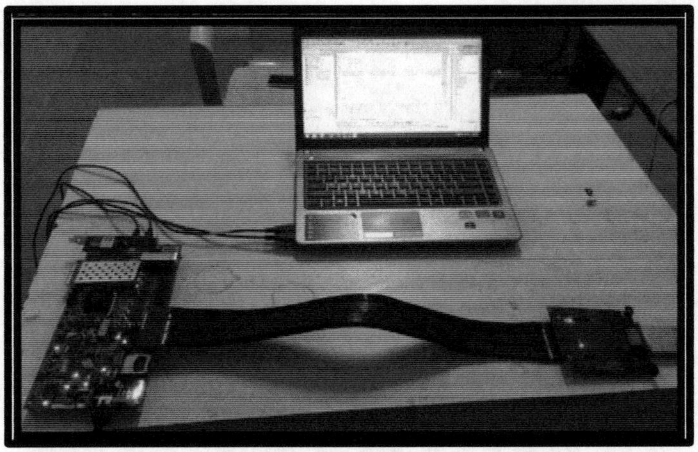

Figure 6. ZC706 Evaluation board with control PC

LabVIEW based GUI is developed for threshold parameter configuration, state transitions prompting & monitoring and error reading for each chopper module. Figure 7 shows screenshot of GUI in SPS Parameter Tab and State Transition Tab. User can configure chopper parameter i.e. PWM frequency, Temperature limit, Under Voltage, Over voltage, Over Current etc. Different buttons for state transitions, total DC link voltage, status and health of each chopper module are available on State Transition Tab

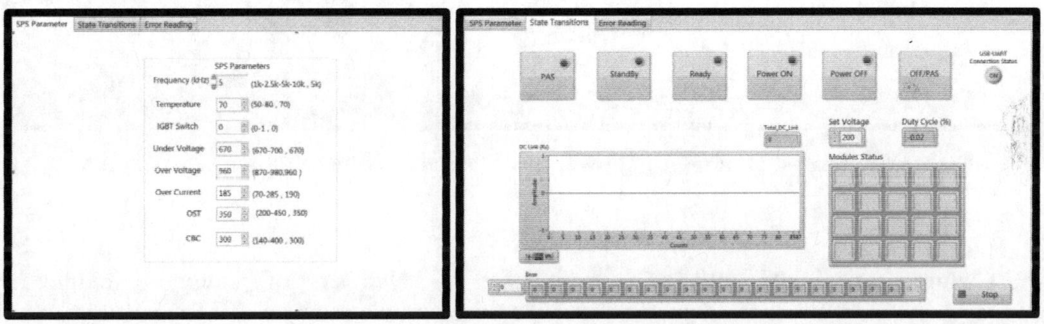

Figure 7. Screenshot of GUI for Parameter Configuration and State Transition (PAS State)

Figure 8 shows UART TX and UART RX channel's output on MDO (Mixed Domain Oscilloscope) in Parameter passing mode.14 packets of configuration parameter at 100kbps baud rate are transmitted by controller and receives response from chopper module. Parameter passing is a sequential process. Other state requires parallel transmission of commands to each chopper modules and processing of received data for DC link and error monitoring. Processing System (PS) calculates required duty cycle based on available DC link and set voltage provided by user. Based on variation in DC link voltage due to line voltage, correction in PWM pulse duty cycle is also applied for getting regulated output voltage. Figure 9 shows screen shot of GUI in Power ON state showing status of each chopper module by red colour in "Module Status" array. Final out voltage for set voltage of 200V is monitor on MDO.

IOP Conf. Series: Journal of Physics: Conf. Series **823** (2017) 012035 doi:10.1088/1742-6596/823/1/012035

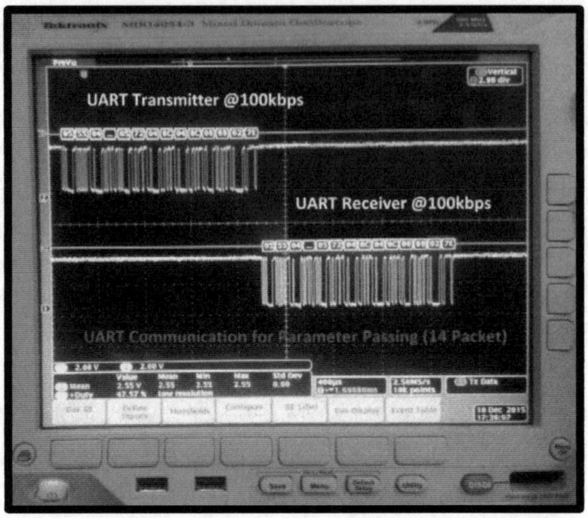

Figure 8. Screenshot of MDO for Parameter Configuration

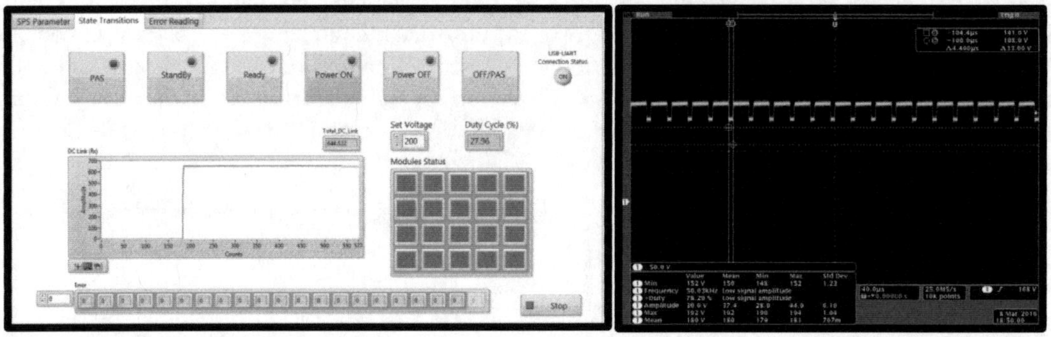

Figure 9. Screenshot of GUI in Power ON State and Voltage Output (200V) on MDO

5. Conclusion

UART Communication and PWM Generation for 20 Numbers of Chopper Modules were demonstrated by ZC706 board. Zynq SoC can be virtuous alternate of COTS HVPS Controller. Developed controller using programmable SoC can be expanded to cater for hundreds of PWM signals with predefined phase lag for HVPS.

References

[1] Architecture Matters: Choosing the Right SoC FPGA for Your Application,white paper, Altera Corporation, [Accessed: 2015, Nov 10]

[2] Crockett L H, Elliot R A, Enderwitz M A and Stewart R W "The Zynq Book", 1st Edition

[3] Creating a custom IP block in Vivado, [Online]. Available: http ://www:fpgadeveloper.com/creating-a-custom-ip-block-in-vivado.html, [Accesseed: 2015, June 30]

Integration & Validation of LCU with Different Sub-systems for Diacrode based amplifier

Kumar Rajnish[1], Sriprakash Verma[1], Dipal Soni[1], Hriday Patel[1], Gajendra Suthar[1], Hrushikesh Dalicha[1], Hitesh Dhola[1], Amit Patel[1], Dishang Upadhayay[1], Akhil Jha[1], Manoj Patel[1], Rajesh Trivedi[1], Harsha Machchhar[1], Raghuraj Singh[1] and Aparajita Mukherjee[1]

[1]ITER-India, Institute for Plasma Research, Near Indira Bridge, Bhat, Gandhinagar 382 428, India

E-mail: krajnish@iier-india.org

Abstract. ITER-India is responsible to deliver nine (8+1spare) ICH & CD Power Sources to ITER. Each power source is capable to deliver 2.5 MW at 35 to 65 MHz frequency range with a load condition up to VSWR 2:1. For remote operation of different subsystems, Local Control Unit (LCU) is developed. LCU is developed using PXI hardware and Schneider PLC with LabVIEW-RT developmental environment. All the protection function of the amplifier is running on PXI 7841R module that ensures hard wired protection logic. There are three level of protection function- first by power supply itself that detects overcurrent/overvoltage and trips itself and generate trip signal for further action. There are some direct hardwired signal interfaces between power supplies to protect the amplifier. Second level of protection is generated through integrated controller of amplifier i.e. Command Control Embedded (CCE) against arc and Anode over current. Third level of Protection is through LCU where different fault signals are received and processed to generate off command for different sub-systems. Before connecting different subsystem with High power RF amplifiers (Driver & Final stage), each subsystem is individually tested through LCU. All protection functions are tested before hooking up the subsystems with main amplifier and initiating RF operation.

1. Introduction

ITER is an International Thermonuclear Experimental Reactor constructed at Cadrache, France. Seven participant countries (Europe, Japan, Korea, Russia, USA, China & India) are involved in this project and contributing through in-kind subsystem [1]. One of the contributions from IN-DA is eight ICH & CD source which has to deliver 24 MW power to couple 20 MW power in plasma [2]. To fulfil the ITER requirement of ICRH source, ITER-India has started R&D activity with third party contractor to qualify final stage high power tube. To test the amplifier chain and check the performance of amplifier, Local control Unit (LCU) is developed and integrated with different sub system. It is validated by amplifier manufacture to ensure the safe and reliable operation of system.

ICH&CD source, shown in figure 1, consist of 2-chain of amplifier, each will generate 1.5 MW and combined using combiner to generate 2.5MW output power. In R&D phase only one chain of amplifier is developed and will be tested for their performance [2] i.e. power of 1.5 MW at 2:1 VSWR condition for 2000 s, operational frequency 35-65 MHz and for their bandwidth of 2 MHz. Each chain of amplifier is consisted of low power RF section, different auxiliary power supplies, Main high voltage power supply [2], different protection component like arc detectors, Flow switch, pressure switch & temperature switch and power measurement at output of each stage of amplifier. Main functions of LCU are to co-ordinate the operation of different sub systems and ensure protection of amplifier.

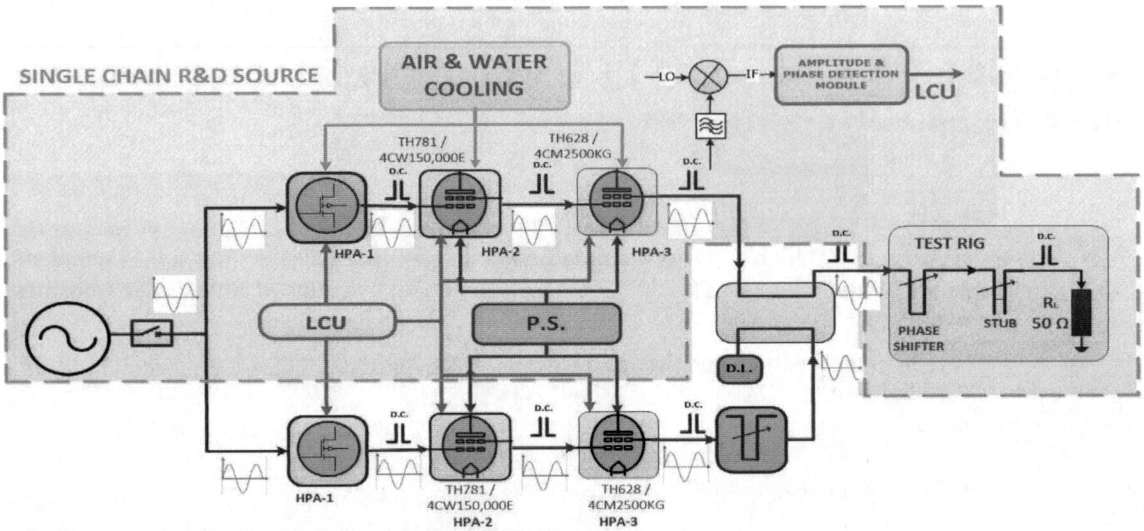

Figure 1. Block diagram of Proposed ICH&CD Source

2. Validation against Requirement of LCU

2.1. Sequential Biasing

LCU has to ensure sequential biasing of different electrode of amplifier i.e. Filament, Control Grid (CG), Anode & Screen Grid (SG). For this purpose Schneider PLC is used where different state of operations are defined. It ensures smooth transition from one state to another after achieving the safe condition. Figure 2 shows the sequential offing sequence of biasing voltage.

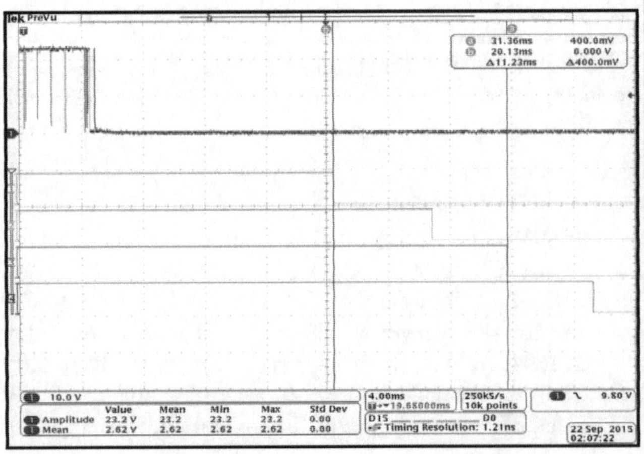

Analog Channel:
Channel-1: Air cooling Fault
Digital channel:
Ch-1: SGPS off
Ch-2: HVPS off
Ch-3: CGPS off
Ch-4: Filament PS off
Conclusion:
All power supply is switched off sequentially within ~ 32 ms through PLC once fault is detected.
De-bounce filter of 10ms for digital i/p line is implemented in PLC programming and verified.

Figure 2. Sequential switching off of different Power supply against Fault

2.2. Fast Protection

Different fault signal from power supply like Over current & trip acknowledgment, Fault from water & air distribution, Fault generated by CCE are considered as fault and in any case RF, Anode voltage and

SG will be switched off and then depending on the fault, CG & filament will be commanded. These protection functions are implemented on PXI 7841R, FPGA module that gives response time of <500ns. However each digital i/p of FPGA is feed through de-bounce filter of 10 μs. Logic diagram is described in figure 3. Response of power supply is validated and shown in figure 4.

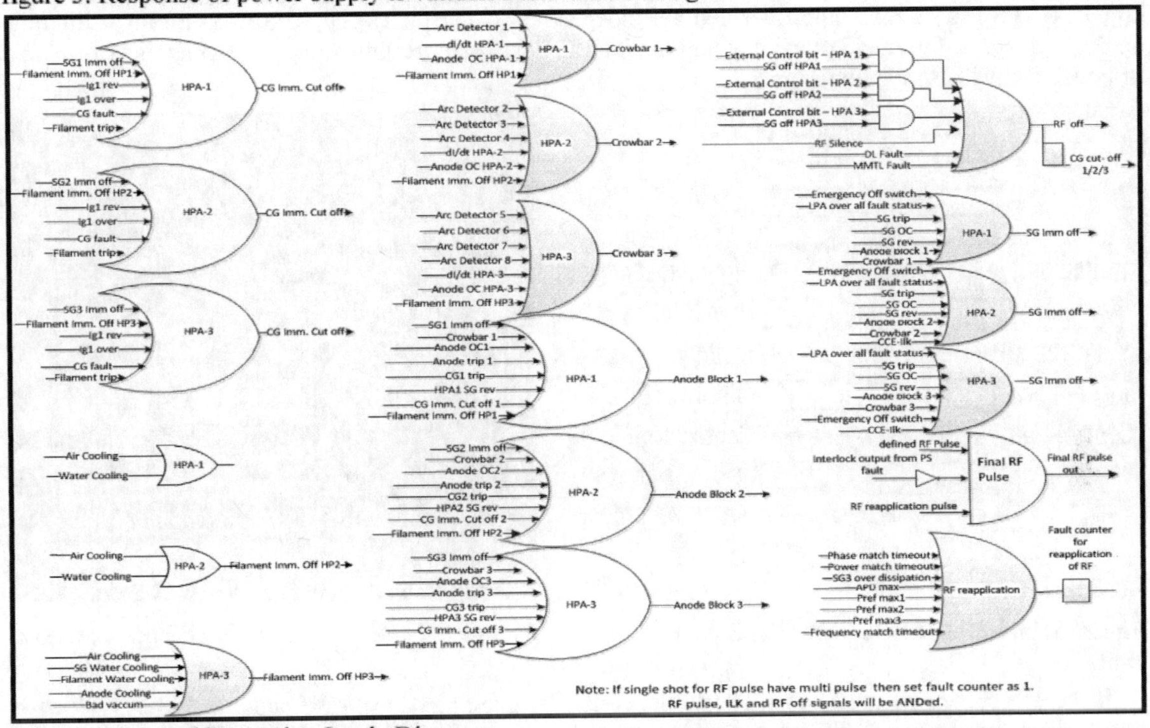

Figure 3. Internal Protection Logic Diagram

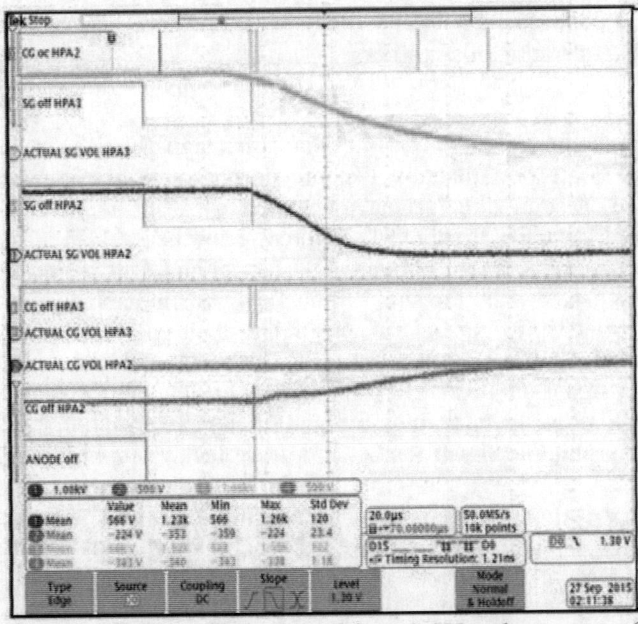

Analog channel:
Channel-1: Actual SG voltage HPA2
Channel-2: Actual CG voltage HPA2
Channel-3: Actual SG voltage HPA3
Channel-4: Actual CG voltage HPA3
Digital channel:
Ch 0: CG OC HPA2 (simulated fault)
Ch1: SGPS HPA3 on/off command
Ch2: Anode on/off command
Ch3: CGPS HPA3 on/off command
Ch5: SGPS HPA2 on/off command
Ch6: CGPS HPA2 on/off command
Conclusion:
Actual auxiliary power supply response time w.r.t. immediate off command generated from LCU is:

1. HPA2 SG: ~70 μs
2. HPA3 SG: ~100 μs
3. HPA2 CG: ~100 μs

As fault is simulated in HPA2 CG OC, HPA3 CG is not switched off.

Figure 4. Switch off Command from LCU and response of Power supply against Fault

IOP Conf. Series: Journal of Physics: Conf. Series **823** (2017) 012036 doi:10.1088/1742-6596/823/1/012036

Integration with HVPS:

Single High Voltage Power supply is feeding to driver stage and final stage of amplifier as it has dual output. HVPS has its own controller and act independently and interfaced with LCU through different hard wired signal. Interfacing topology and the interfacing signal are shown here in figure 5 and response of Power supply is shown in figure 6.

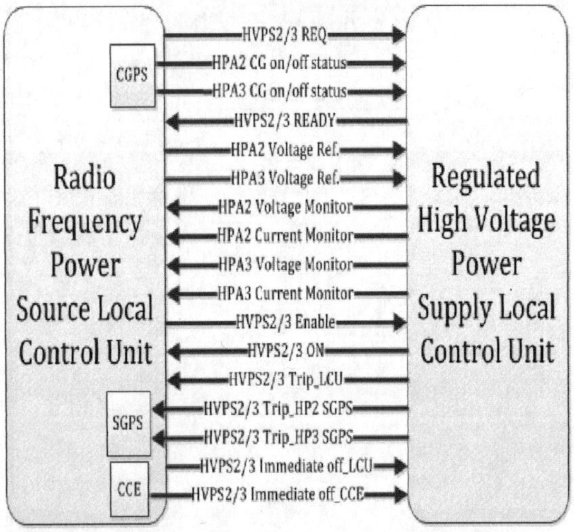

Figure 5. Interface of LCU with HVPS control Unit

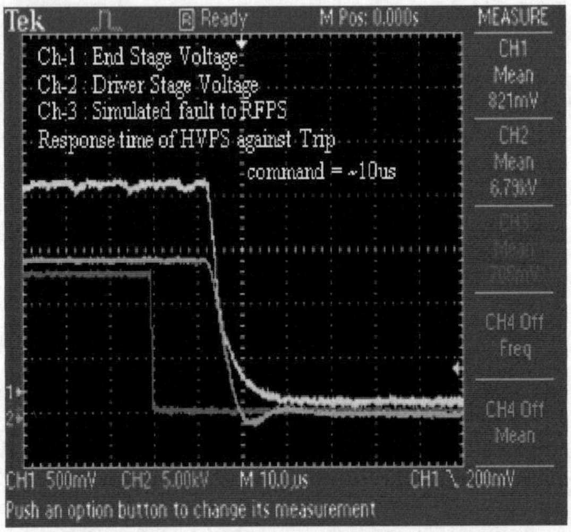

Figure 6. Fault response of HVPS power supply

If there is some internal fault in HVPS, its controller generates SGPS off command signal which is directly interfaced with SGPS power supply through optical fiber link. Similarly HVPS is interlocked with CGPS on/off status i.e. HVPS controller generates Ready signal only when it receives ON status of CGPS and it will be enabled only when LCU controller get HVPS ready status. These hard wired interlocks are checked and validated before actual operation of amplifier.

Integration with CCE:

Command Control Embedded (CCE) is integral controller of Amplifier that is managing the water/air distribution system, tuning motors and critical protection function like arc detector and anode over current. CCE act independently and controlled through direct signal generated by LCU application program. Once Start cooling request is received by CCE from LCU, different blowers get started in delayed fashion to reduce high in-rush current requirement. Air cooling & water cooling distribution is managed by CCE independently. If flow is going below the threshold limit, flow switch activates and make series combination of status of all flow switches to generate water cooling fault and air cooling fault. The interface and protection signal handled by CCE is shown in figure 7.

For tuning CCE will get Auto tuning enable signal only when RF amplifier is at filament ON-state. Once Auto tuning enable signal is available, CCE allow operator to go for auto tuning and once tuning is performed successfully, it will generate tuning system OK signal for LCU to move the system at higher state and make system ready for RF operation.

During initiation of the system, each morning system is tested for arc detector fault. LCU will generate ARC detector test request, against which CCE will generate signal to make test LED to glow as all the arc detector location is equipped with test LED and once all arc detector detect the light, CCE generate ARC detector OK signal for LCU to make auxiliary state OK and allow to go higher state. All these functionalities are checked during validation.

IOP Conf. Series: Journal of Physics: Conf. Series **823** (2017) 012036 doi:10.1088/1742-6596/823/1/012036

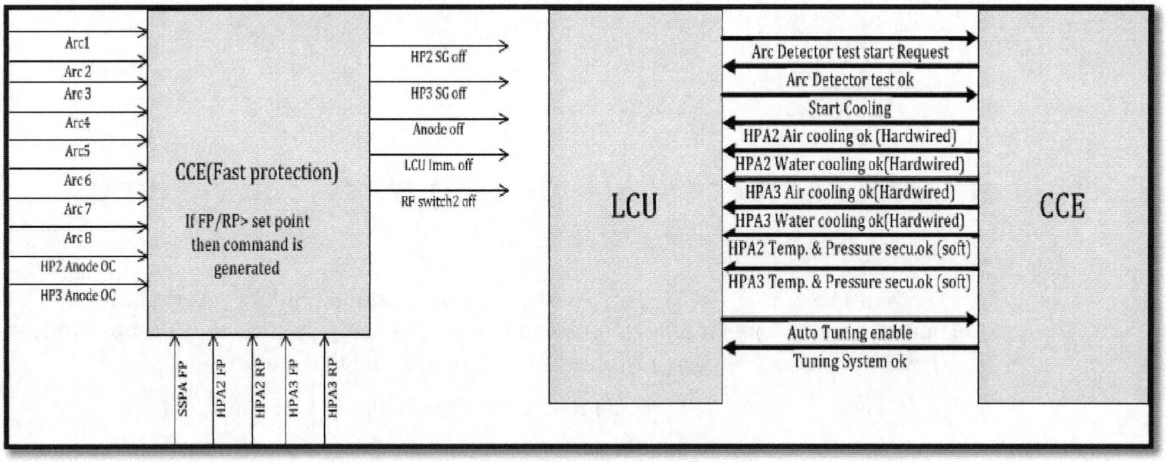

Figure 7. Protection function handled by CCE and Interfacing signal of LCU & CCE

3. Conclusion

LCU is developed using NI-make PXI hardware and Schneider make PLC hardware. LabVIEW tools are used to develop application program. After successful testing with simulated signals, LCU is integrated and validated with different sub system. Site acceptance testing of 1.5 MW amplifier system is initiated. Driver stage amplifier is tested successfully using this LCU and one of the typical RF shot is shown here in figure 8 where shot is for more than 3600 s and operating power level is ~130 kW. Further testing of integrated amplifier system is continued.

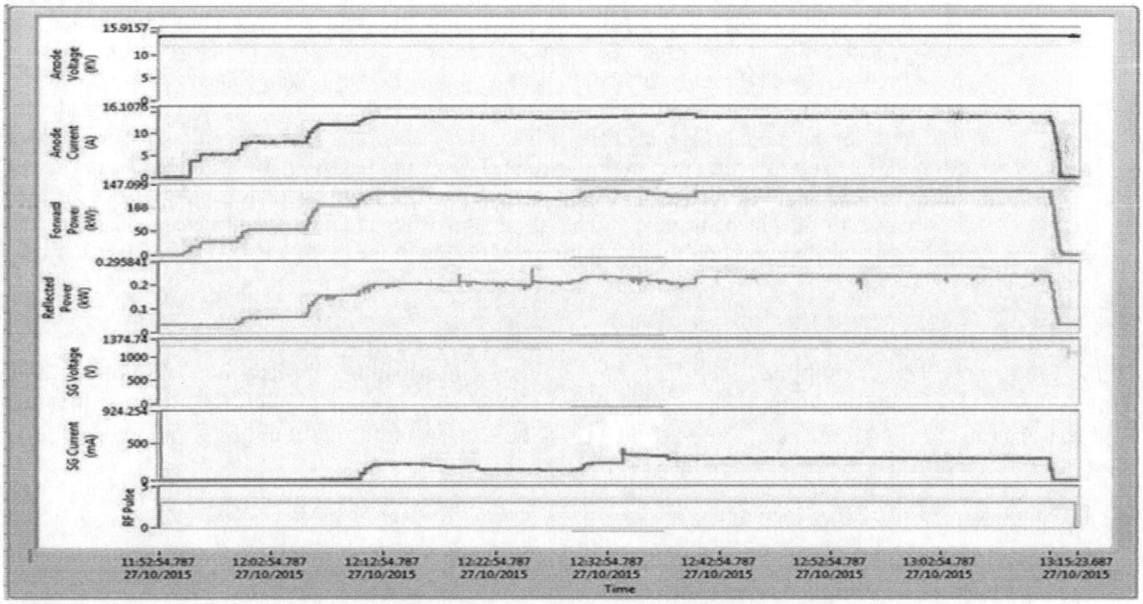

Figure 8. RF shot of more than 3600sec showing different operational parameter

References

[1] Ikeda K 2007 *Nuclear Fusion* **47** 1
[2] Mukherjee A et al. 2013 *Fusion Science & Technology* **65** 120

Design of 1 MHz Solid State High Frequency Power Supply

Darshan Parmar[1], N P Singh[1], Sandip Gajjar[1] Aruna Thakar[1], Amit Patel[1], Bhavin Raval[1], Hitesh Dhola[1], Rasesh Dave[1], Dishang Upadhay[1], Vikrant Gupta[1], Niranjan Goswami[1], Kush Mehta[1] and Ujjwal Baruah[1]

[1]ITER-India, Institute for Plasma Research, Gandhinagar, Gujarat, India

E-mail: darshan.parmar@iter-india.org

Abstract. High Frequency Power supply (HFPS) is used for various applications like AM Transmitters, metallurgical applications, Wireless Power Transfer, RF Ion Sources etc. The Ion Source for a Neutral beam Injector at ITER-India uses inductively coupled power source at High Frequency (~1 MHz). Switching converter based topology used to generate 1 MHz sinusoidal output is expected to have advantages on efficiency and reliability as compared to traditional RF Tetrode tubes based oscillators. In terms of Power Electronics, thermal and power coupling issues are major challenges at such a high frequency. A conceptual design for a 200 kW, 1 MHz power supply and a prototype design for a 600 W source been done. The prototype design is attempted with Class-E amplifier topology where a MOSFET is switched resonantly. The prototype uses two low power modules and a ferrite combiner to add the voltage and power at the output. Subsequently solution with Class-D H-Bridge configuration have been evaluated through simulation where module design is stable as switching device do not participate in resonance, further switching device voltage rating is substantially reduced. The rating of the modules is essentially driven by the maximum power handling capacity of the MOSFETs and ferrites in the combiner circuit. The output passive network including resonance tuned network and impedance matching network caters for soft switching and matches the load impedance to 50ohm respectively. This paper describes the conceptual design of a 200 kW high frequency power supply and experimental results of the prototype 600 W, 1 MHz source.

1. Introduction

High Frequency Power supplies (HFPS) are utilized in various applications including AM Transmitters [1], metallurgical applications [2], Wireless Power Transfer [3], RF Ion Sources [4], etc. The Ion Source for a Neutral beam Injector at ITER-India uses inductively coupled power source at High Frequency (~1 MHz), typical scheme is represented in figure 1.

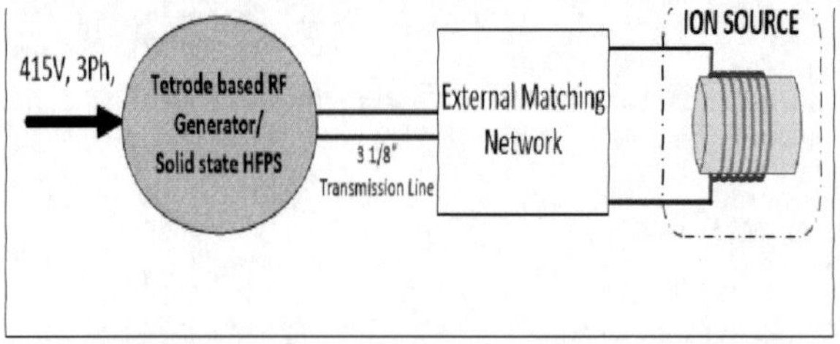

Figure 1. Scheme of ion source for plasma production

IOP Conf. Series: Journal of Physics: Conf. Series **823** (2017) 012037 doi:10.1088/1742-6596/823/1/012037

RF power is coupled in the external matching network through transmission line with characteristic impedance of 50 ohm. An external matching circuit would match the impedance of HFPS terminated at 50 ohm load to RF driver coils of the Ion Source for maximum power transfer.

Typical specifications of 200kW HFPS for RF Ion Source to be utilized for ITER neutral beam injector are shown in table 1.

Table 1. Major specifications of 200 kW HFPS for ITER ion source.

Parameters	Values
Active Power to the Load	200 kW, 10 – 100 % in increments of 1 kW
Accuracy of power control	1%
Mode of operation of RF power	RF output Power Control
Working frequency	(1 ± 0.1) MHz
Mode of operation of RF frequency	Internal tuning in real time to guarantee the matching conditions
Frequency accuracy	± 10 kHz
Output impedance of RF Gen	50 Ω
Possible Load Variation	25 to 75 Ω
Rise/Decay Time	< 0.3 ms
Efficiency	> 50 %
Type of duty	max pulse length 3600 s with pulsed duty

Traditionally, HFPS is realized through tetrode tubes based oscillator design where operational efficiency is < 60 %. This paper illustrates novel switching converter based topology used to generate 1 MHz sinusoidal output. A solid state system is expected to have advantages on efficiency (> 80 %) and reliability as compared to traditional RF Tetrode tubes based oscillators. Further, it works at lesser operational voltages hence provides better safety and works with smaller clearances between components for a compact design. Depending upon the availability of the switching devices at 1 MHz frequency and associated thermal management, modular architecture is proposed for the HFPS. The modular design would allow operation of HFPS even in case of failure in few HF modules. The combined output is fed to output passive network to have sine wave output terminated at 50 ohm resistive load. The necessary tuning required for frequency and load change may be addressed through the passive network. The typical block diagram representation of a high power HFPS is shown in figure 2.

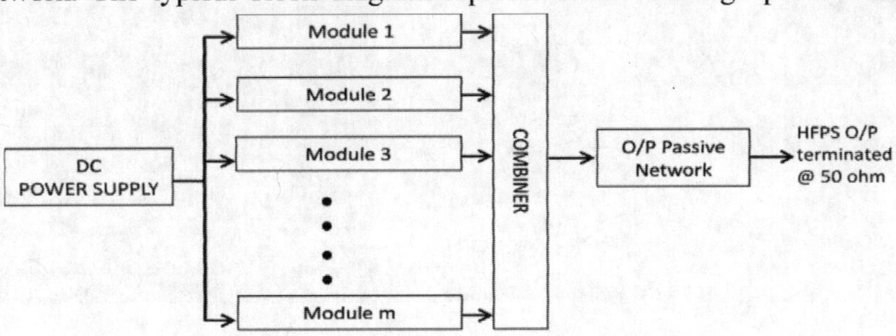

Figure 2. Solid State HVHF PS Block Diagram Representation

2. 600W Prototype Development

Class-D & Class-E switching amplifier topology are popular and widely been used in various HF applications [5]. These are switching amplifiers where output power no longer depends on the gate drive power but the DC supply voltage. The prototype design is attempted with Class-E amplifier [6] topology where a MOSFET is switched resonantly as shown in figure 3. As the switch opens and close, current flows in the switch and load network alternatively. The resonant load network shapes the voltage and current waveforms to prevent simultaneous high voltage and high current in the

MOSFET implying ZVS/ZCS (Zero Voltage/Current Switching); that minimizes power dissipation, especially during the switching transitions. In case of ZVS, the voltage across MOSFET falls to zero before the MOSFET is turned ON eliminating any overlap between voltage and current and minimizing losses.

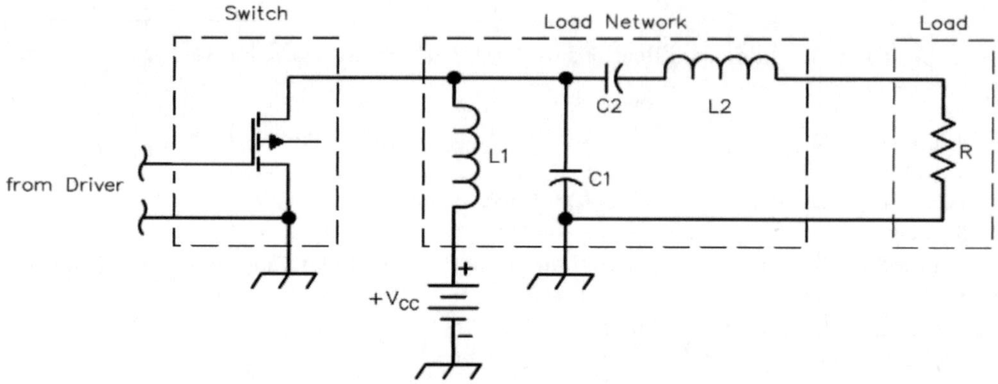

Figure 3. Schematic representation of Class-E topology

The main advantage of Class-E topology lies in a simple circuit with a single switch where isolated drivers are not essential. The impedance matching circuit is used to match the impedance of the Class-E output with 50 ohm load for maximum power transfer. The prototype development used handy items such as axial flow fan with Aluminum heatsink for cooling the MOSFET and PVC pipes for former of air core inductor. The nominal rating of the planned prototype module is 300 W module at 50 ohm termination, the output voltage is 122 Vrms and output current is ~ 2.4 Arms. The developed 300 W HF module and results demonstrating ~ 310 W on 50 ohm RF dummy load are shown in figure 4 and figure 5 respectively.

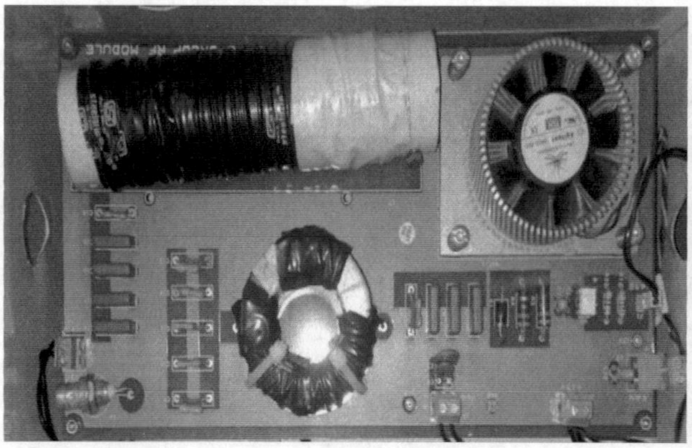

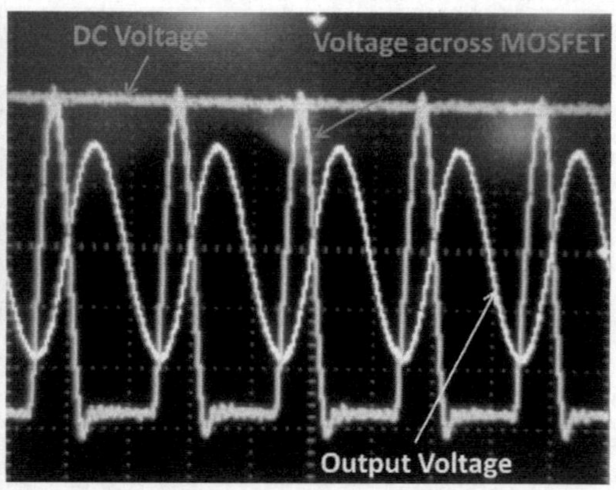

Figure 4. View of 300 W HF module

Figure 5. Results of 300 W module

Two such low power modules (300 W) are combined through a ferrite transformer and loaded on to RF dummy load. The setup with two modules combined through ferrite transformer is shown in figure 6 while results showing output of both modules and combined voltage is depicted in figure 7. Additionally, the setup is also demonstrated for plasma formation over electrodeless lamp through inductive coupling as shown in figure 8.

Figure 6. Snap of 600 W Setup

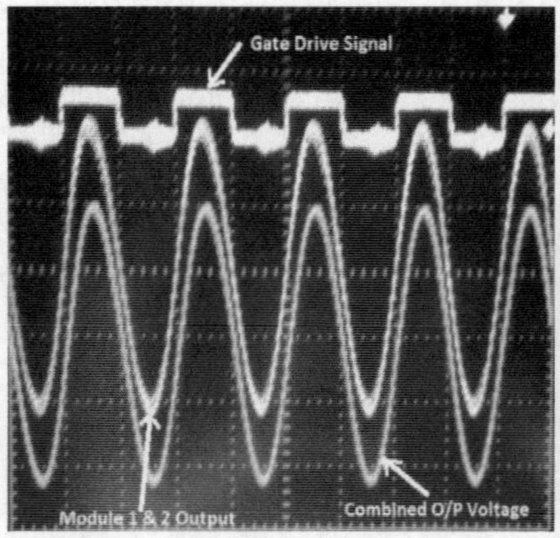

Figure 7. Results of 600 W setup

Figure 8. Inductive Plasma formation

The prototype development implies that thermal, layout and power coupling issues are major challenges at such high frequency. While being simpler, the major disadvantage of Class-E topology is the high voltage stress across the switch, which may reach a value as high as four times the input DC supply voltage. Also, the switching device is part of resonance circuit, hence frequency tuning would require module level alteration that may not be practical for higher power application where multiple modules are required.

3. 200 kW High Frequency Power Supply

Higher power and bandwidth are the essential requirements for 200 kW HFPS at such a high frequency. For higher power, the rating of the modules is essentially driven by the maximum power handling capacity of the MOSFETs and ferrites in the combiner circuit. The rated power handling capacity of the HF modules must be high enough to limit number of modules and associated interfaces. The HF module should be able to switch resonantly for the given frequency range moreover, variable passive components shall be used in impedance matching circuit for achieving the ± 100 kHz bandwidth criteria. The design started with the exploration of HF module topology, availability of high frequency MOSFET and ferrite combiner at given power.

3.1. 3 kW HF Module Design & Simulations

The design of the HVHF modules is considered on H-Bridge Class-D switching amplifier topology for the two major reasons; first it supports stable design for un-tuned events at output i.e. module itself do not contribute to resonance hence no module level alternation needed for frequency tuning. Secondly the switching device voltage rating is substantially reduced. The schematic representation of Class-D amplifier is shown in figure 9. A full bridge Class-D topology always has two of the four MOSFETs on at a time, it gives the load a positive voltage when switches Q2 and Q3 are on at the same time, a negative voltage when Q1 and Q4 are ON. In this way, the switching bridge generates a square waveform to excite tuned output passive network generating a resonant sinusoidal current. A Class-D topology employ zero voltage switching (ZVS) to eliminate turn-on switching losses. When using an H-bridge, it is extremely important to make sure that MOSFET on the same arm do not conduct at the same time because this would short the rails (shoot-through) and damage the converter. All soft switched ZVS bridges operate with the optimized dead time during the transitions between the low and high-side MOSFETs.

The simulation results of ideal 3kW HF module on PSIM simulation tool [7] is illustrated in figure 10 where DC voltage is considered as 250 V, essentially the voltage across the MOSFET during ON period. The output voltage of the bridge swings from positive rail voltage to the negative rail voltage while the MOSFET current is half sinusoidal (12 Arms) as an effect of resonant switching. Sinusoidal

IOP Conf. Series: Journal of Physics: Conf. Series **823** (2017) 012037 doi:10.1088/1742-6596/823/1/012037

output voltage is referred in the simulation results terminating at 50 ohm load terminating resistor through impedance matching network with output power resulting in to 3 kW.

In conjunction with the simulation results, the devices need to be realized that can work for such a power at 1 MHz frequency range. With ZVS in Class-D topology where the turn ON losses are nullified leaving only turn OFF losses, the conduction losses dominate, followed by the Coss-related losses. As a result, important parameters to look for in the MOSFET are $R_{DS(ON)}$, Q_{rr} & C_{oss} once satisfy the minimum requirement of drain voltage and drain current while C_{iss} is important from gate drive power perspective. Table 2 suggests the few MOSFETs that are candidate for the 3kW HF module design along with comparison of relevant parameters.

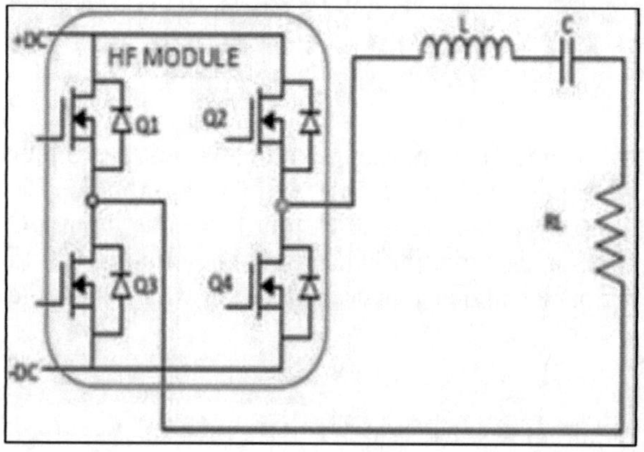

Figure 9. Schematic for Class-D HF module

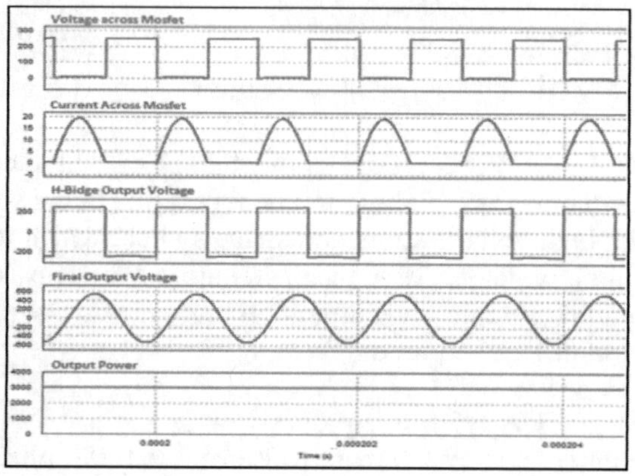

Figure 10. Simulation results of 3 kW HF module

Table 2. MOSFET parameters and relevant parameter comparison.

Parameters	Switching Devices		
	SCH2080KE (ROHM)	DE475-501N44 (IXYS)	IXFK55N50F (IXYS)
VDS (V)	1200	500	500
ID (A)	40	48	55
Ciss (pF)	1850	5100	6700
Coss (pF)	175	300	1250
RDSon (mohm)	80	130	85
Tf(ns)	28	8	9.6

3.2. Ferrite combiner

The combiner is typically a multi-primary single secondary ferrite transformer that adds up the output of the modules. Typical schematic representation of the ferrite combiner is shown in figure 11 below.

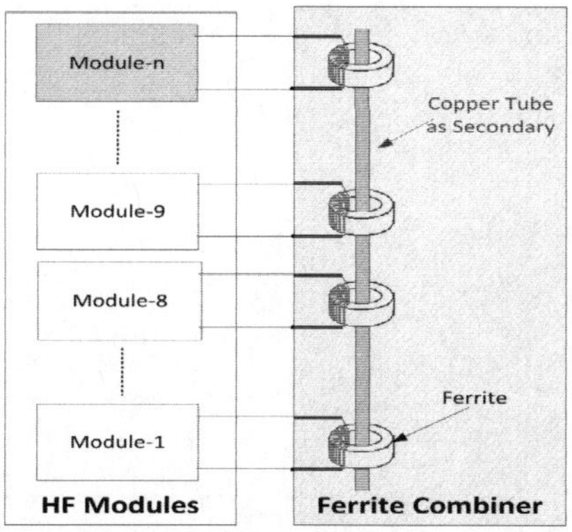

Figure 11. Typical representation of Ferrite combiner

Following transformer equation would apply to this multi-primary single secondary ferrite combiner.

$$V_s = \frac{m}{n} V_p \quad (1)$$

$$I_s = n\, I_p \quad (2)$$

where V_s and I_s are the secondary voltage and secondary current while V_p and I_p are the primary voltage and primary current, m is number of modules and n is turns ratio of the transformer.

Ferrites for the combiner can be realized with a ferrite imposing low core loss, higher Bsat at operational frequency of ~1 MHz, Epcos Toroid N49/N59 Material (MnZn) is one of the candidate. At such as high frequency, core losses in the ferrites are dominating, the losses for ferrite transformers are estimated to be ~ 2 % in accordance to the power loss density value in the core datasheet.

4. 200 kW HFPS Design & Simulations

The conceptual design of the 200 kW HFPS assumes 80 HF modules each having the rated capacity of 3 kW. The efficiency of the HFPS is targeted to be ~ 80%, selecting 80 modules of 3 kW rating is targeted with 20 % margin inclusive of ~ 8 % losses from passive network and ferrite transformers. Remaining 12 % margin is considered for redundancy factor with 4 redundant modules.

The ferrite combiner couples the output from all HF modules to achieve final 200 kW output power. The output of combiner connects to the passive network including filter, impedance matching, resonance circuit etc. as necessary to terminate the sinusoidal output at 50 ohm load. Regulated DC Power Supply should be a high accuracy low ripple power supply as it directly impact the output RF power. The output RF power control from 10 % to 100 % is achieved through change in DC voltage from regulated DC power supply feeding all the HF modules. Typical schematic representation of 200 kW HF Power supply is depicted in figure 12.

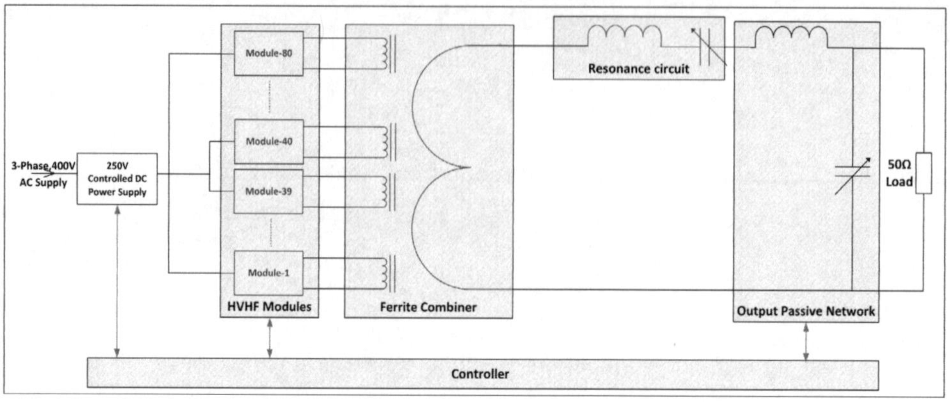

Figure 12. Typical schematic representation of 200 kW HFPS

The simulations results with above configuration for 200 kW power at 1 MHz frequency is shown in figure 13 where MOSFET is modelled including appropriate datasheet parameters while ferrite is modelled as saturable core. The design assumes 10:1 turns ratio for the ferrite transformer and the operational flux density is ~14 mT. The sag in the simulation results in output of HF module and ferrite combiner is due to the fact that simulation has included $R_{DS(ON)}$ of the MOSFET, where sinusoidal current through MOSFET due to resonance introduces corresponding voltage drops across $R_{DS(ON)}$. Further various simulations have been performed to validate the performance of solid state design for bandwidth & load variation ~ 2 nF vacuum variable capacitor may be required to cope up to the bandwidth requirement.

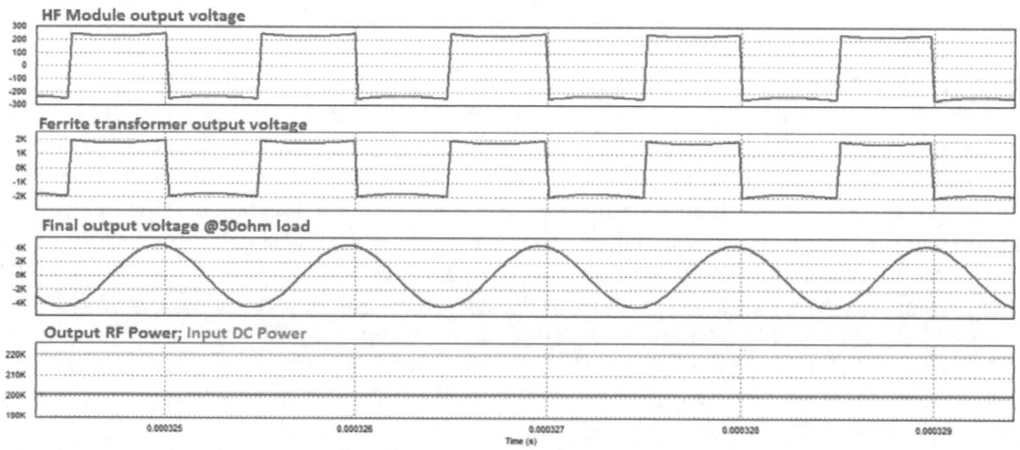

Figure 13. Simulation results of 200 kW HFPS

5. Conclusion

Conceptual design, analysis & component realisation of 200 kW HFPS along with 600 W prototype works established the manufacturing feasibility of such a solid state HFPS. The stated work including helped considerably conversing with prominent domestic suppliers involved in HF converters with limited frequency range within 300 kHz. The first target is to develop 40 kW HFPS based on solid state topology, the same is to be tested with ROBIN ion source at Indian Test facility. The contract on 40 kW solid state HFPS is been awarded to the domestic supplier where 3 kW modules are tested satisfactorily while delivery at site is expected by March 2017. The outcome of the 40 kW HFPS on ROBIN ion source would allow stepping in to 200 kW development on similar lines.

IOP Conf. Series: Journal of Physics: Conf. Series **823** (2017) 012037 doi:10.1088/1742-6596/823/1/012037

References

[1] Swanson H 1989 *IEEE Transactions on Broadcasting* **35** 131
[2] Tsujino J 1995 *Ultrasonics Symposium* **2** 1353
[3] Bani S 2014 *Proc. IEEJ Int. Power Electronics Conference* 2794
[4] Singh M J 2011 *Proc. AIP Conf.* **1390** 604
[5] Krauss Herbert L 1992 *Solid State Radio Engineering* (John Wiley & Sons Incorporate) 1047151410
[6] Sokal N O 2001 *Class-E RF power amplifiers QEX* **204** 9
[7] *PSIM user manual*, version 8.0 (by Powersim Inc)

IOP Conf. Series: Journal of Physics: Conf. Series **823** (2017) 012038 doi:10.1088/1742-6596/823/1/012038

Alternate Design of ITER Cryostat Skirt Support System

Manish Kumar Pandey[1], Saroj Kumar Jha[1], Girish Kumar Gupta[1], Avik Bhattacharya[1], Gaurav Jogi[1] and Anil Kumar Bhardwaj[1]

[1]ITER-India, Institute for Plasma Research, Block A, Sangath Skyz, Bhat-Motera Road, Koteshwar, Ahmedabad - 380005, Gujarat, India

E-mail: manish.pandey@iter-india.org

Abstract. The skirt support of ITER cryostat is a support system which takes all the load of cryostat cylinder and dome during normal and operational condition. The present design of skirt support has full penetration weld joints at the bottom (shell to horizontal plate joint). To fulfil the requirements of tolerances and control the welding distortions, we have proposed to change the full penetration weld into fillet weld. A detail calculation is done to check the feasibility and structural impact due to proposed design. The calculations provide the size requirements of fillet weld. To verify the structural integrity during most severe load case, finite element analysis (FEA) has been done in line with ASME section VIII division 2 [1]. By FEA 'Plastic Collapse' and 'Local Failure' modes has been assessed. 5° sector of skirt clamp has been modelled in CATIA V5 R21 and used in FEA. Fillet weld at shell to horizontal plate joint has been modelled and symmetry boundary condition at ± 2.5° applied. 'Elastic Plastic Analysis' has been performed for the most severe loading case i.e. Category IV loading. The alternate design of Cryostat Skirt support system has been found safe by analysis against Plastic collapse and Local Failure Modes with load proportionality factor 2.3. Alternate design of Cryostat skirt support system has been done and validated by FEA. As per alternate design, the proposal of fillet weld has been implemented in manufacturing.

1. Introduction

The Skirt support system of ITER cryostat is a support system which takes which takes all the load of cryostat cylinder and dome during normal and operational condition. Dimension of skirt consist of internal diameter as 28940 mm and internal diameter as 28040 mm respectively with 50 mm thickness. There are sliding systems installed on the skirt. In the accidental case when helium leakage will takes place the whole cryostat will try to shrink inside.

Cryostat skirt support is part of Base section and it is a support structure which provides support in vertical direction and allows displacement in radial direction. The present design of skirt contains full penetration joint for all components of the assembly (vertical cylinder, bottom horizontal plate and ribs). There is a stringent requirement of tolerance, to achieve the new requirement with full penetration is very difficult and fillet weld is alternate solution to have less deformation by welding. The calculation has been done to check the feasibility and structural impact due to proposed design. To verify the structural integrity and interface load during most severe load case, finite element analysis (FEA) has been done in line with ASME section VIII division 2.

IOP Conf. Series: Journal of Physics: Conf. Series **823** (2017) 012038 doi:10.1088/1742-6596/823/1/012038

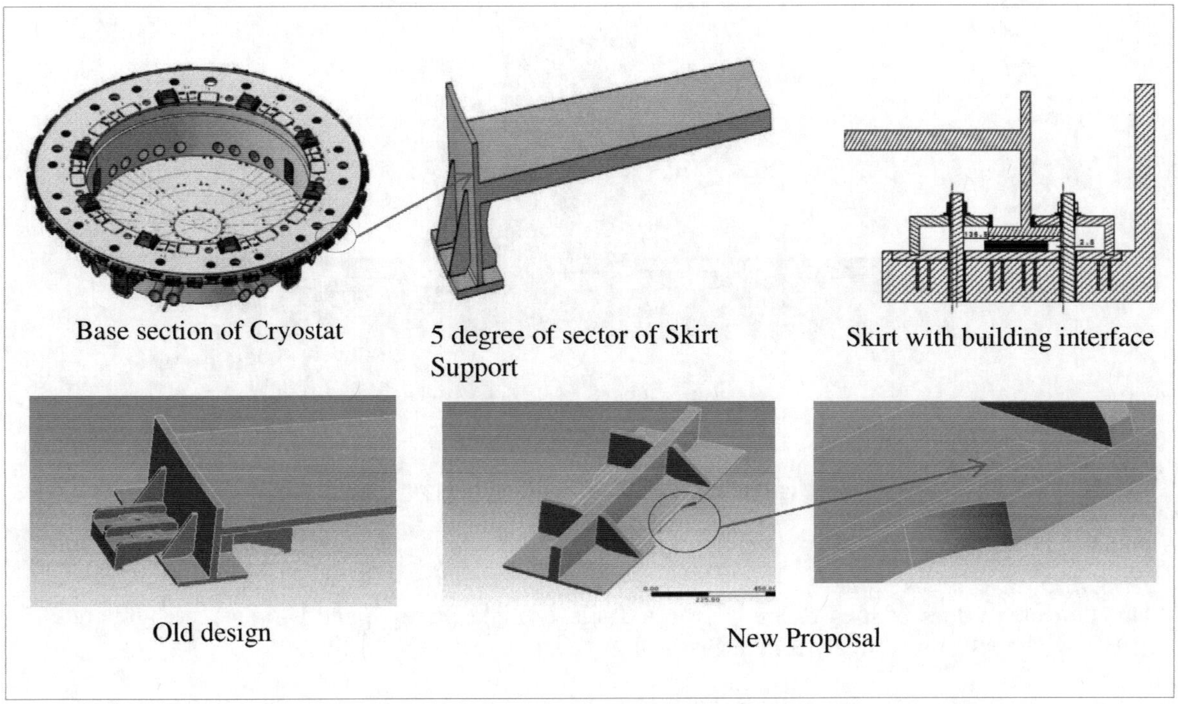

Figure 1. Old Design and New Design

2. Verification of Proposed Design by Calculation
Configuration of Proposed design has been decided by theoretical calculation.

2.1. Theoretical calculation
Fillet has been proposed instead of full penetration weld joint at skirt-shell location. Calculation of Fillet weld configuration has been done.

2.1.1. Design Loads and considerations
Material mechanical properties are considered as per approved Cryostat material (Dual Mark SS-304 / 30L). 20 degree sector of vertical cylinder and horizontal skirt plate has been considered for calculation, which consists of group of 8 clamps. Ribs are not considered for conservative approach. Efficiency of weld considered 85 % as per code compliance. Weld will be done as per ASME Section VIII Division-2 for category of E weld joint, which is applicable as per the code.

2.1.2.Geometry and section
The cross section of skirt support considered for theoretical design calculation is shown in figure 2. The reinforce ribs are not considered as conservative approach. As per ASME Section VIII Division 2, minimum leg length required for fillet weld is minimum 0.25 of thickness of web i.e. 13 mm.

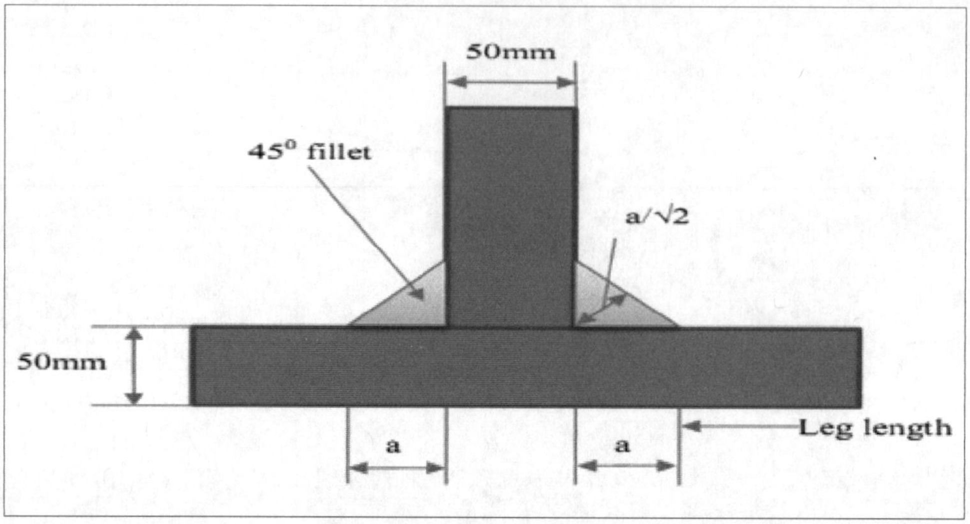

Figure 2. Fillet Weld joint configuration

2.1.3. Allowable Stress Values
The allowable values of stresses are shown in Table 1. Yield Strength and Ultimate strength are 207 MPa and 517 MPa respectively for the Cryostat material.

Table 1. Allowable Stress Values

Mechanical Properties	For Category II		For Category III	
	For cryostat Design	For this design validation considering 85% weld efficiency	For cryostat Design	For this design validation considering 85% weld efficiency
Allowable Tensile Stress	138 MPa	117.3 MPa	165.6 MPa	140.8 MPa
Allowable Shear Stress	76 MPa	58.65 MPa	82.8 MPa	70.8 MPa

2.1.4. Maximum Loads.
Loads for considered categories are shown in Table 2.

Table 2. Load Cases.

Load	**Cat II- Load Case-**	**Cat III- Load Case-2**
Vertical Load	7200 kN	10000 kN
Radial load	1440 kN	2000 kN
Toroidal Load	1440 kN	2000 kN

IOP Conf. Series: Journal of Physics: Conf. Series **823** (2017) 012038 doi:10.1088/1742-6596/823/1/012038

Theoretical calculation for worst load case (Load Case 2)

Radial Shearing for Load Case-2:

Maximum shear Force acting in radial direction (Fsr):
Fsr = Fver cos45 + Frad cos45 = (10000 + 2000) × 0.707 = 8485.3 kN =8.4853 MN

Allowable shear stress (σs) = Load (Fsr)/ Area (A)
70.4 = 8.4853 / (7.41 × *a*)
Minimum Leg length (*a*) = 0.00163 m = 16.3 mm

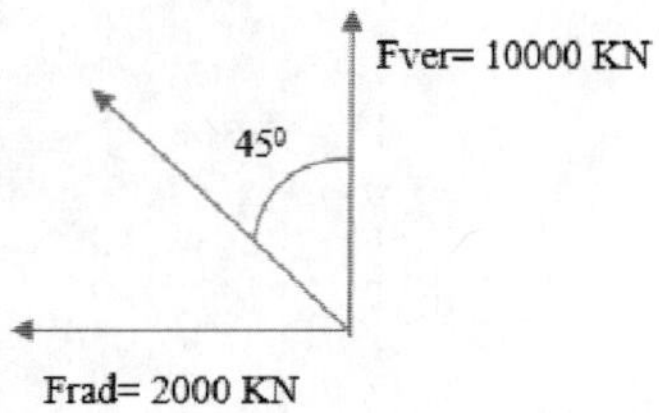

Toroidal Shearing for Load Case-2:

Maximum shear force in Toroidal direction
Fst = Ft = 2000 kN = 2.0 MN
Allowable shear stress (σs) = Load (Fst) / Area (A)
70.4 = 2 / (7.41 × *a*)
Minimum Leg length (*a*) = 0.0038 = 3.8 mm

Considering the above Cat-III Load case-2 (worst load case) conditions we have found that minimum leg length required is 16.3 mm. So it is recommended for the fillet weld the minimum leg length should be 16.3 mm.

3. Verification of Proposed Design by FE Analysis

Plastic collapse failure mode has been assessed by elastoplastic analysis in line with ASME section VIII division 2 part 5, design by analysis method. FE analysis has been done on ANSYS WORKBENCH 15.0.

3.1. FE Model.

5° Sector of skirt system has been considered for FE analysis due to cyclic symmetry. Fillet joint at shell and horizontal plate joint location has been modelled as solid. Bonded contact has been defined for the connections. Clamp and skirt plate interface has been defined as elastic support by defining the stiffness of clamp.

Meshing has been done with SOLID 186 element with element nominal size of 25mm. Material Non-linearity has been defined by incorporating true stress strain curve from ASME section II Part D, 2010 Edition for Cryostat material.

3.2. Loads

Load cases for FE analysis are same as that for theoretical calculation, shown in table 2. Load factors for elastoplastic analysis are considered as per [1].

3.3. Elastoplastic Analysis.

Elastoplastic analysis has been done in line with [1] and Plastic collapse failure mode has been assessed. In addition to this reaction forces on building have been also assessed.

IOP Conf. Series: Journal of Physics: Conf. Series **823** (2017) 012038 doi:10.1088/1742-6596/823/1/012038

4. Results and Discussions

4.1. Results of FE Analysis

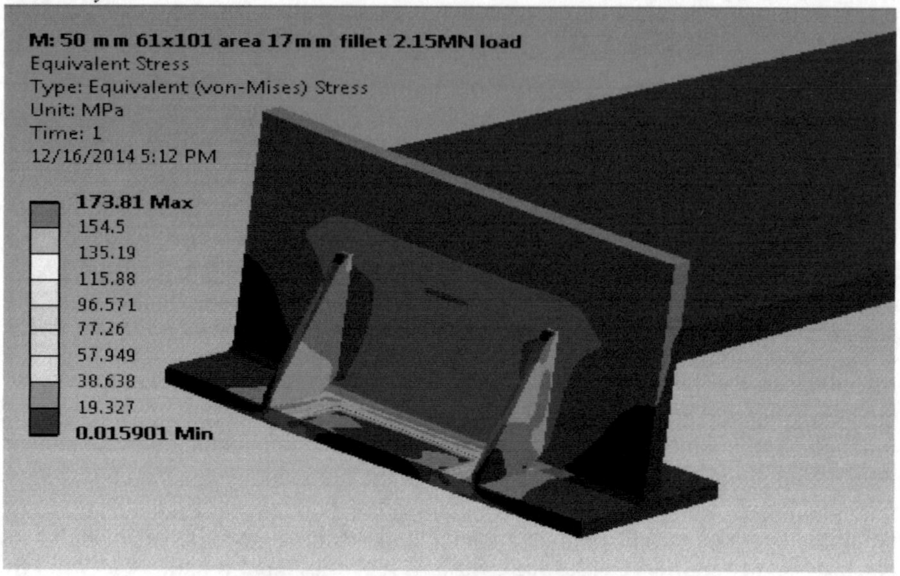

Figure 3a. Von Mises Stress

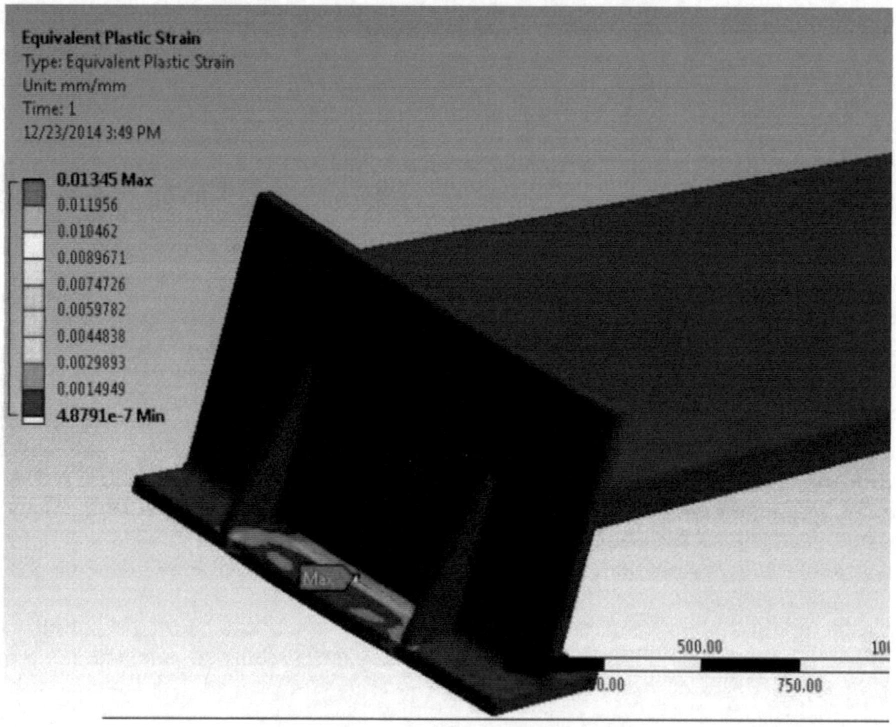

Figure 3b. Equivalent Plastic Strain

Plastic Failure mode has been assessed based on the convergence criteria as per [1]. Local Failure has been asses as per the equation (5.6), provided by [1] for local failure assessment by elastoplastic analysis.

4.2. Assessment of Reaction Forces

Table 3. Reaction Forces.

Design	Old Design		New Design	
	Inner Clamp	Outer Clamp	Inner Clamp	Outer Clamp
Reaction on clamp at contact region with skirt(Upward)	0.33MN	0.80MN	0.43MN	0.7MN
Reaction on bolts(downward)	0.45MN	1.03MN	0.60MN	0.75MN
Reaction on clamp support on building(upward)	0.12MN	0.23MN	0.17MN	0.05MN
Net reaction on the clamps (downward)	0.33MN	0.80MN	0.43MN	0.70MN
Total load transmitted to building through bolts by both clamps(upward)	1.48MN		1.35MN	
Total net load transmitted to building by both clamps(upward)	1.13MN		1.13MN	
Input Upward load applied on Cryostat Cylinder	1.13MN		1.13MN	

Minimum required leg length for the fillet weld is 13 mm and 16.3 mm for Category II and bed III event respectively and 13 mm is specified value in figure 4.2.2 of ASME Section-VIII Div-2 ($a \geq 0.25$ of thickness). These are very much conservative as described in design by loads. So it is recommended to adopt the change of weld joint from full penetration to fillet weld with 16.3 mm leg length at skirt cylinder to bottom horizontal plate welding along rib joint welding. After considering manufacturing feasibility the fillet weld configuration has been decided as leg length as 17 mm.

5. Conclusion

Total load transmitted to building through bolts is less in new proposal in comparison to old model for both the cases. Net load transmitted to building is same for new model and old model and it is equal to the upward load applied by the cryostat cylinder.

Acknowledgement
This study was supported by Cryostat & VVPSS Group, ITER-India.

References

[1] ASME, "Rules for the Construction of Pressure Vessels," *ASME Boiler and Pressure Vessel Code, Section VIII, Division 2, Alternative Rules*, (2010).

IOP Conf. Series: Journal of Physics: Conf. Series **823** (2017) 012039 doi:10.1088/1742-6596/823/1/012039

Development of electromagnetic welding facility of flat plates for nuclear industry

Rajesh Kumar[1], Subhanarayan Sahoo[1], Biswanath Sarkar[1], Anurag Shyam[1]

[1] Institute for Plasma Research, Near Indira Bridge, Bhat, Gandhinagar 382 428, India

Email: rkumar@ipr.res.in

Abstract. Electromagnetic pulse welding (EMPW) process, one of high speed welding process uses electromagnetic force from discharged current through working coil, which develops a repulsive force between the induced current flowing parallel and in opposite direction. For achieving the successful weldment using this process the design of working coil is the most important factor due to high magnetic field on surface of work piece. In case of high quality flat plate welding factors such as impact velocity, angle of impact standoff distance, thickness of flyer and overlap length have to be chosen carefully. EMPW has wide applications in nuclear industry, automotive industry, aerospace, electrical industries. However formability and weldability still remain major issues. Due to ease in controlling the magnetic field enveloped inside tubes, the EMPW has been widely used for tube welding. In case of flat components control of magnetic field is difficult. Hence the application of EMPW gets restricted. The present work attempts to make a novel contribution by investigating the effect of process parameters on welding quality of flat plates. The work emphasizes the approaches and engineering calculations required to effectively use of actuator in EMPW of flat components.

1. Introduction

Electromagnetic pulse welding (EMPW) is a technique for welding of metals by means of plastic deformation generated by a repulsive force on account of the interaction between the magnetic field of the coil and the current induced in the workpiece. Being a high-strain rate forming method, the EMF process has all the advantages of high-velocity forming process, viz. increased formability, reduced springback and reduced wrinkling of formed parts [1-2]. In case of flat components controlling the magnetic field is difficult. Hence, the application of EM technique is restricted. The research on electromagnetic impact welding of flat sheets has not been extensively documented, and the literature prominently shows the activity of only few researchers in this field [3-4].This article discusses the theoretical aspects involved in the design of EMPW process. The behaviour of electromagnetic parameters such as magnetic field, voltage, current induced and Lorentz force is explained on the basis of Maxwell's equations. For effectiveness of EMPW process, it is desirable that more magnetic field is screened by the job piece and less diffuses out. Thus the discussion about diffusion process is important. The Lorentz force is to be coupled with electrodynamic and structural equations to get the information about deformation and the kinetic energy generated in the job piece. The article concludes with the detailed discussion on the theoretical aspects work with these the design procedure for welding coil. Recent studies and advances in the welding fraternity have been made in order to develop a newer process which enables us to achieve better quality, increase productivity and is environment friendly. EMPW is a step forward into the future of welding. In the past few decades the development of a joining technology to weld multimaterial combinations and efficiently join lightweight structures has been an area of significant and growing interest. In the specific case of multi-metallic assembly, the MPW process is known as a potential joining solution [5]. EMPW has wide applications in nuclear industry, automotive industry, aerospace, electrical industries [6].

Nuclear industry responsible for the discovery and the initial development of EMPW on the production of end closures and nuclear fuel rods. The need for a reliable method for joining T91 and ODS steels is central to the fabrication of fuel pins using these materials. Nuclear clad to tube welding, stainless steel joints, stainless steel telemanipulator cables crimping, Aluminium container covers crimping, Welding of closing caps, Metal canisters welding etc. are some examples related to nuclear industry where EMPW technique can be applied for better performance. Owing to the increasing demand in nuclear industry, there is a need to develop joining process [7-9].

The fusion and fission reactors are faced with important challenges, notably in the field of materials. A class of materials, susceptible to resist the anticipated severe environments, is the ODS (oxide dispersion strengthened) family, which are metallic alloys strengthened by a very fine oxide dispersion. Oxide particles, which have excellent high temperature stability and fine dispersoid distributions, provide the ODS steels with excellent resistance to neutron irradiation-induced swelling, hydrogen embrittlement over austenitic steels and high-temperature creep.

A barrier of these materials for application is their weldability. It has been found that conventional fusion welding methods can disturb the dispersion of the fine oxide particles in the alloy. Hence, fusion welding methods are not applicable and should be substituted by better welding methods. Oxide dispersion strengthened (ODS) steels have the microstructure and mechanical properties required for advanced nuclear reactor applications. These alloys have a very high strength at room temperature, but the strength diminishes slowly at higher temperatures. Fe-based oxide dispersion strengthened (ODS) alloys contain fine and uniformly dispersed yttrium oxide particles in a ferritic matrix. The high temperature stability of the nano-oxide dispersoids act as pinning points to inhibit dislocation movement and provide excellent creep strength and irradiation resistance. Therefore, these materials are of potential use in advanced energy generation such as hydrogen fuel cells and fission/fusion nuclear reactors. For ODS alloys, agglomeration of fine oxide particles would result in loss of strength in the ODS joints [10-11].

Friction stir welding (FSW), electron beam (EB) welding, Laser beam welding, Pressure Resistance Welding (PRW) are some solid state joining process tried on ODS steel by different researches for successful welding but each process have some limitations to qualify for successful welding of ODS due to affecting particle dispersion consistency near welding zone.

2. Basic Principle

Figure 1(a) shows a schematic diagram of a discharge circuit. The circuit for the present method consists of a capacitor for a supply of electrical energy, a discharge gap switch and an E-shaped one-turn flat coil. The two plates are placed above the coil, with a little space between them. The plate nearest the coil is termed the ''flyer plate'' and the plate above it, which is fixed firmly in place, is referred to as the ''parent plate''. When the capacitor is charged and the discharge gap switch is closed, a discharge pulse is released to the coil. The electrical energy stored in the capacitor is called ''discharge energy'', the welding conditions are mainly controlled by this parameter. Figure 2(b) shows a close-up around middle section of the coil. When a discharge pulse passes through the coil from the capacitor, it induces a high-density magnetic flux around the coil.

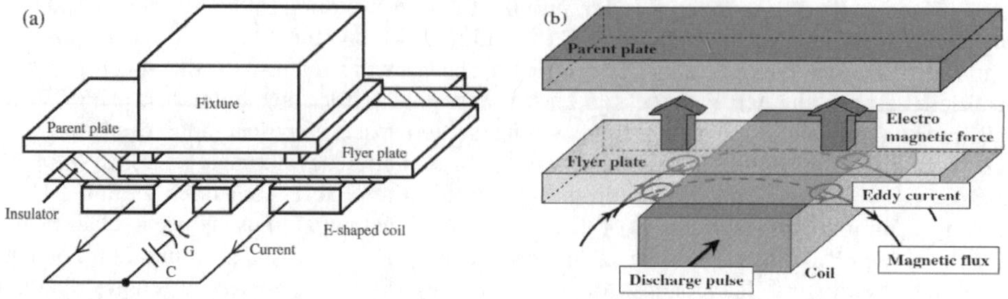

Figure 1. Schematic illustrations of welding process of magnetic pulse welding. (a) Set-up of the lapped plates over the E-shaped one-turn coil. (b) Principle of the magnetic pulse welding.

The generated magnetic flux lines intersect with the flyer plate, and in accordance with Lentz's law, eddy currents are excited in the surface of the flyer plate adjacent to the coil. In accordance to Fleming's left-hand rule, the eddy current and the magnetic flux generated around the coil induce an electromagnetic force upward. The electromagnetic force drives the flyer plate toward the parent plate at a high velocity and the flyer plate is welded to the parent plate [12].

3. Experimental Apparatus

Figure 2. Electromagnetic Pulse Welding facility consists of power supply, HMI control, Capacitor banks, Trigger generators, Spark gaps and EMPW actuators

Figure 2 shows the general outlines of the magnetic pulse welding apparatus, which consists of a capacitor bank (C) and a spark gap switch (G) with a one-layer, E shaped flat coil. It was attempted to make a low-inductance discharge circuit that can generate a high-density magnetic flux around the coil area. The capacitor bank that drives the discharge system of the MPW device consists of 16 capacitors of 4 μF/75 kV in parallel.it is connected to the gap switch and one-turn coil by a low-inductance transmission line. The circuit is designed to keep the inductance as low as possible to carry out fast welding. The flat, E-shaped, one-turn coil was made of Cu .The coil thickness is 20 mm and the inductance of the coil is 0.02 nH. When the gap switch is closed, an impulse discharge current from the capacitor bank (C) passes through the coil and the MPW process begins. The aluminum work sheet were 25 mm long, 25 mm wide and 1 mm thick whereas stainless steelwork sheet was 35 mm long, 35 mm wide and 2 mm thick. Distance between flyer plate & parent plate before start of welding is 1 mm. The contact surface between two samples was polished and cleaned .The 0.1~0.3-

IOP Conf. Series: Journal of Physics: Conf. Series **823** (2017) 012039 doi:10.1088/1742-6596/823/1/012039

mm-thick insulating sheets were loaded between the coil surface and the overlapped ends of the workpiece sheet.

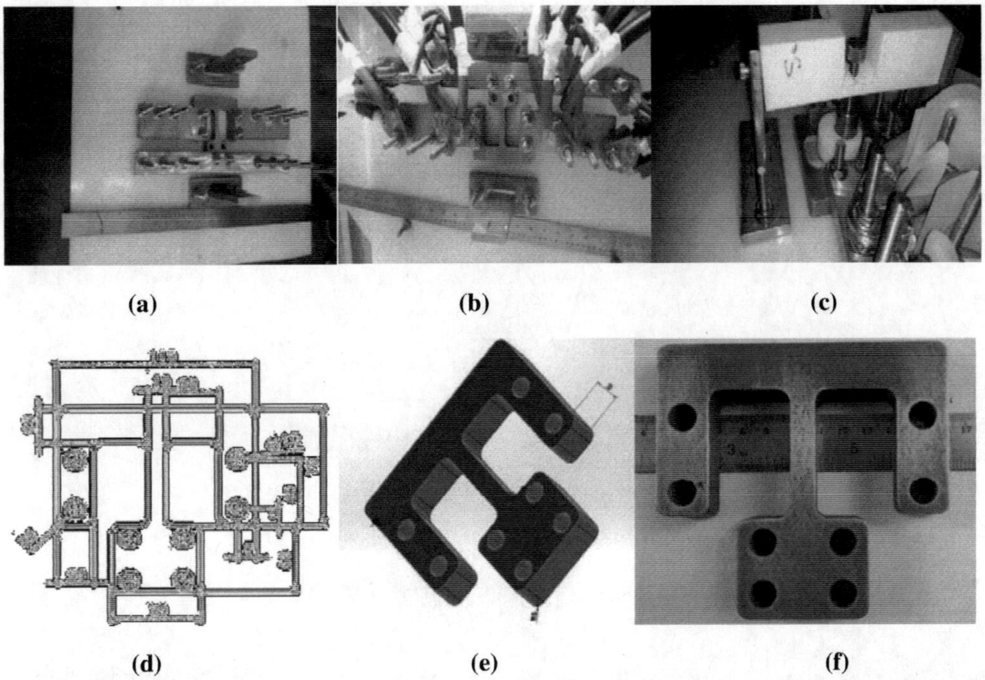

Figure 3. Photograph of three-dimensional flat coil for electromagnetic impact welding. (a) Bus bars connected to high voltage and ground terminal (b) High voltage and ground cables connected to coil by bus bar (c) Sample placed and tightened properly on coil web to withstand high pressure,(d) 2D model of coil, (e) 3D model of the coil, (f) Coil processed by water jet cutter for EMPW application

A three-dimensional, 'E'-shaped, flat one turn copper coil with its set-up as shown in Figure 3 was designed and fabricated for the electromagnetic impact welding of 1mm thick aluminum sheet to stainless steel sheet . The web of 10mm wide and 50mm long at the center leads to concentration of the current, thereby increasing the current density **J** [3]. An increase in current density gives a corresponding increase in the resultant Lorentz force that generates stronger impact between the sheets to be welded. The Lorentz force is given by [13].

$$\mathbf{F} = \mathbf{J} \times \mathbf{B}$$

Where, **F** is the Lorentz force in N, **B** the magnetic flux generated by the coil in Tesla and **J** is in Am^{-2}.

4. Result

A typical current waveform is shown in figure 4. This current signal was obtained at 9 kJ discharge by using a magnetic probe. The current signal shows that a damping and oscillating current flows through a one-turn coil for the duration of about 50 µs. The maximum current was measured at about 280 kA. If the discharge current flows uniformly on the surfaces of the middle portions of the coil, The aluminum work sheet were 25 mm long, 25 mm wide and 1 mm thick whereas stainless steelwork sheet was 35 mm long, 35 mm wide and 2 mm thick. Figure 5 (a) shows the photograph of 5 representative samples of Al sheets welded to SS sheets and cupper sheets at same energy values.

The bank capacity used for all the experiments was 9 kJ at 15 kV. Copper coil having inductance of 0.2 nH was used. Electromagnetic force generated to drives the flyer plate toward the parent plate is14.36 kN. Velocity of the flyer plate achieved is 92 m/s.The web of the coil was supported with a specially designed and fabricated nested type nylon fixture to avoid its bending in the process of welding of the sheets. Figure 5 (b) show the cross-section view of Al-Cu sample with welded zone at center and non-welded zone.

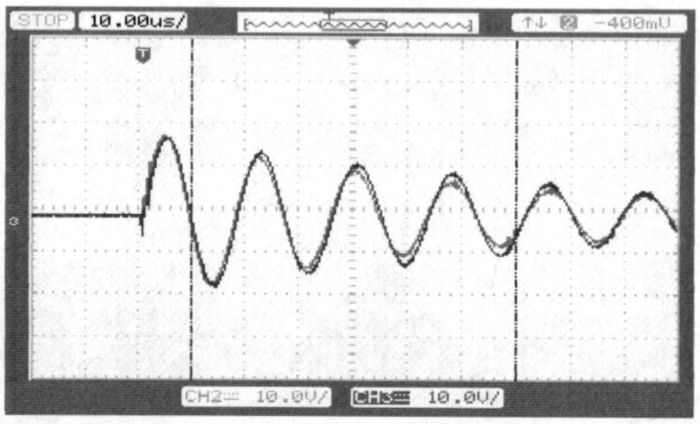

Figure 4. Current waveform at 9 kJ discharge energy

(a) **(b)**

Figure 5(a) Al-to-SS and Al to Cu electromagnetic impact welded samples. **(b)** Al to Cu electromagnetically welded sample showing weld and no-weld zones.

5. Conclusion

In this work, a process of electromagnetic impact welding of inferior electrical conductive stainless steel sheet to aluminum and copper sheets have been successfully established. The feasibility of electromagnetic impact welding of aluminum sheets has been established. Electromagnetic welding technique overcomes problems in welding Cu-to-Al, Al-to-SS sheets that could be encountered in the conventional techniques. It proves to be a potential welding technique for various industrial components requiring dissimilar sheet metal joints. In the future, the feasibility of electromagnetic welding process can be checked for the difficult-to-weld materials like magnesium, galvanized steel, etc. The effect of process parameters on the strength and metallographic structures of the EM welds can be studied in detail for various dissimilar metal combinations Numerical simulations of the process can lead to better understanding and prediction of the EM weld characteristics. EMF method

IOP Conf. Series: Journal of Physics: Conf. Series **823** (2017) 012039 doi:10.1088/1742-6596/823/1/012039

of joining dissimilar metals specially, which is difficult to weld by conventional method, has shown promising results. Although, the trial was basically a mechanical joint, this technique has potential for diffusion bonding between dissimilar metals. This has possibility of an alternative to the explosive welding in laboratory scale. Magnetic Pulse Welding has been growing since the 70's and there is an increasing interest in the welding process in several industrial sectors. It is a cold welding process in which two metal surfaces in placed in close contact by the effect of a high speed electromagnetic force produced in a short pulse capacitor in the range of few tens of millisecond. Due to this no heat developed in the interfaces.

EMPW is a technology with proven repeatability, with widespread applications in mass-production processes. The solid state joint produced does not adversely affect the particle dispersion consistency of ODS steel. Problems associated with a heat-affected zone, like grain growth, are eliminated. EMPW can potentially make joining of these materials simple, reliable and repeatable. It involves no physical contact with the cladding at the joint, which could even enable the joining of cladding materials with dedicated coatings.

References

[1] Padmanabhan, M. 1997 *Wrinkling and Springback in Electromagnetic Sheet Metal Forming and Electromagnetic Ring Compression*. MS Thesis. The Ohio State University

[2] Seth M, Vohnout V J, Daehn G S 2005 *J. Mater. Process. Technol.* **168** 391

[3] Aizawa T, Okogawa K, Yoshizawa M, Henmi N 2001 *Impact Eng. Appl.*, 827.

[4] Aizawa T, 2004 *Weld. Int.* **18** (11) 868

[5] Raoelison R N, Buiron N, Rachik M, Haye D, Franz G and Habak M 2013 *Journal of Materials Processing Technology* **213** 1348

[6] Ji-Yeon Shim, Bong-Yong Kang 2013 *Materials Sciences and Applications* **4** 856

[7] Kore S D, J. Imbert J, Worswick M J and Zhou Y 2009 *Science and Technology of Welding and Joining* **14** 549

[8] Kore S D, Date P P, S.V. Kulkarni S V 2007 *International Journal of Impact Engineering* **34** 1327

[9] Manoharan P, Manogaran A P, Priem D, Racineux G, Marya S, 2013 Experimental investigation of magnetic pulse welding of ODS alloys,I²FG Workshop on Impulse Forming ,Gent

[10] John M 2009 European Commission, JRC-ITU Karlsruhe, Germany, Electromagnetic pulse technology as a means of joining generation IV cladding materials, *Proceedings of the 17th International Conference on Nuclear Engineering ICONE17 July 12-16,* Brussels, Belgium

[11] Prabu M, Manogaran A P, Priem D, Marya S, Racineux G, 2013 *Weld World* **57** Issue 6 pp 867

[12] Watanabe M and Kumai S 2009 *Materials Transactions* **50** No. 2 286

[13] Moon F C 1984 *Magneto-Solid Mechanics* (John Wiley and Sons Inc.)

IOP Conf. Series: Journal of Physics: Conf. Series **823** (2017) 012040 doi:10.1088/1742-6596/823/1/012040

Electron Beam Welding: study of process capability and limitations towards development of nuclear components

Gautam R Vadolia[1] **and K Premjit Singh**[1]

[1]Institute for Plasma Research, Near Indira Bridge, Bhat, Gandhinagar 382 428, India

E-mail: gautamv@ipr.res.in

Abstract. Electron Beam Welding (EBW) technology is an established and widely adopted technique in nuclear research and development area. Electron beam welding was thought of as a candidate process for ITER Vacuum Vessel Fabrication. Dhruva Reactor at BARC, Mumbai and Niobium superconducting accelerator cavity at BARC has adopted the EB welding technique as a fabrication route. Study of process capability and limitations based on available literature is consolidated in this short review paper.

1. Introduction

EBW (Electron Beam Welding) equipment covers a range of power from 2 to 200 kW. Using vacuum chambers ranging from few litres to several hundred cubic meters have ensured a production rate of few parts per day (aircraft structures, heavy industry and the petrochemical industry) to more than a thousand parts per hour (automobile) [2]. The application of EB welding for the strategic use is reported. In France, in the 1980s, a big machine of 264 m^3 (with an internal gun of 45 kW) was used for welding internal core barrels for the nuclear industry. Later, even bigger chamber of 800 m^3 was realized by Techméta in France for welding submarine components [2]. EBW machine at NAMRC (Nuclear Advance Manufacturing Research Centre) has the vacuum chamber volume over 200 m^3 and beam power 30 kW (at 60 kV), which is capable to make full penetration single-sided welds of 100 mm thickness in steel. A few examples of large EB welding installations for other purposes include a machine at Nates (France) having chamber size of 860 m^3 volume (diameter 10 m and length is 11 m) [1] and a machine at Kazan Aircraft Production Association in Russia having total length of the chamber 38 m and the total volume of the vacuum chamber 1160 m^3 with beam power 120 kW at 120 kV [2]. Internationally as per 1979 estimate there are more than 3000 installations of EBW machines [3]. A survey of EBW and its capabilities is compiled in this short review paper.

Figure 1 shows current state of experience and information regarding material combinations and their welding compatibility for EBW.

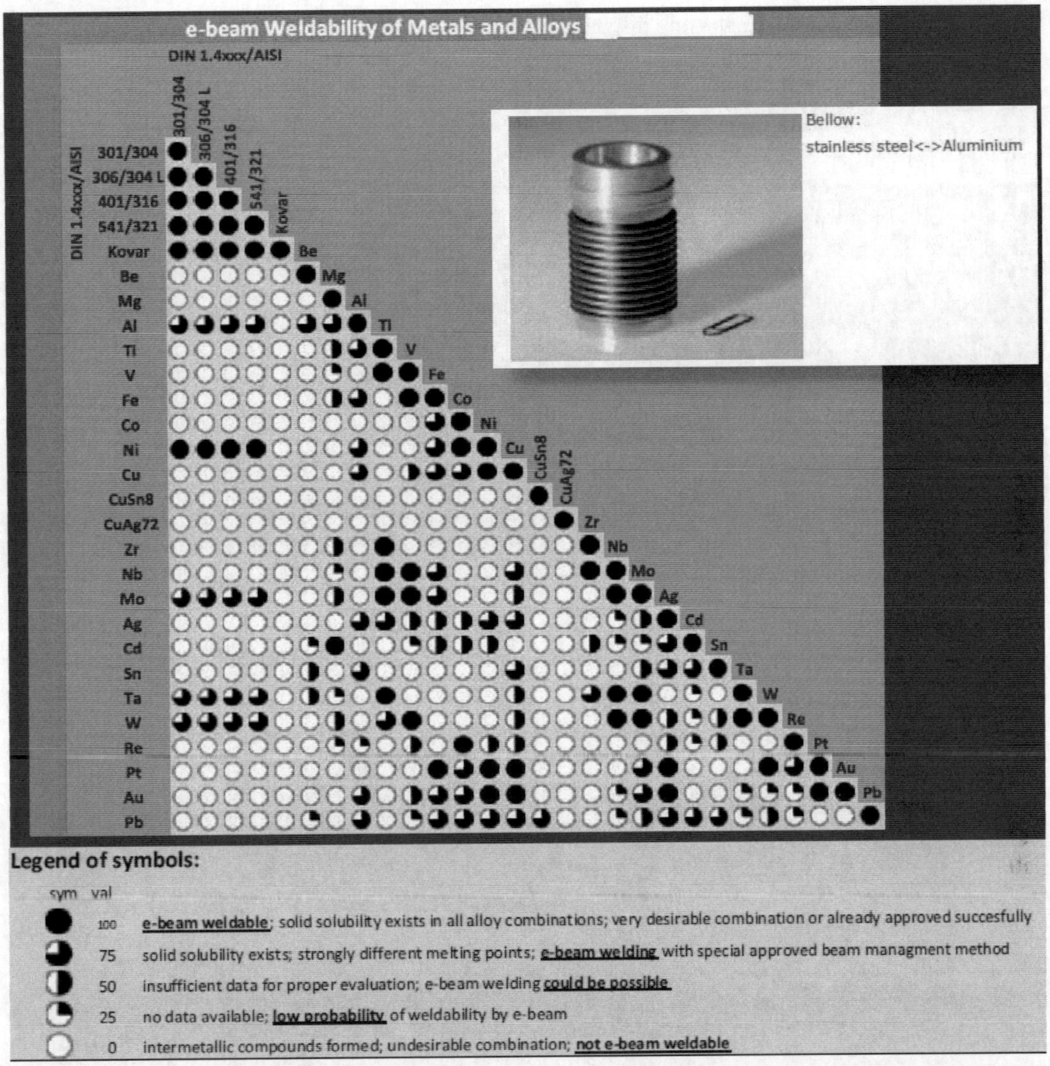

Figure 1. Current state of experience and information regarding material combinations and their welding compatibility [4].

2. EB Process Parameter

The output parameter of EBW, such as the depth of penetration, thickness of the thermal affected area, productivity, working time, surface precision etc. depend on input parameter [9]. A few important EB welding parameters are listed below.

- Vacuum parameter: vacuum in EB gun, vacuum in working chamber.
- Work piece material parameter: melting temperature and thermal conductivity of work piece.
- Electron beam parameter: electron beam diameter, electron beam power, beam current, beam current intensity, acceleration voltage.
- Beam focus parameter: form, amplitude and frequency of oscillation, focalisation current, focal point and beam oscillation, focal distance.

- Other parameters like welding distance, welding speed.

Some of the EB gun configurations commercially available (or were available) are shown in figure 2.

Figure 2. Partial list of commercially available electron beam gun configurations (left – moving EB gun [6], middle –EB gun for out of vacuum [5], right – EB gun for welding tubes in to heat exchanger using local vacuum [2].

3. EB capabilities

A few examples of various types of welded joints produced by EBW are shown in figure 3 and figure 4.

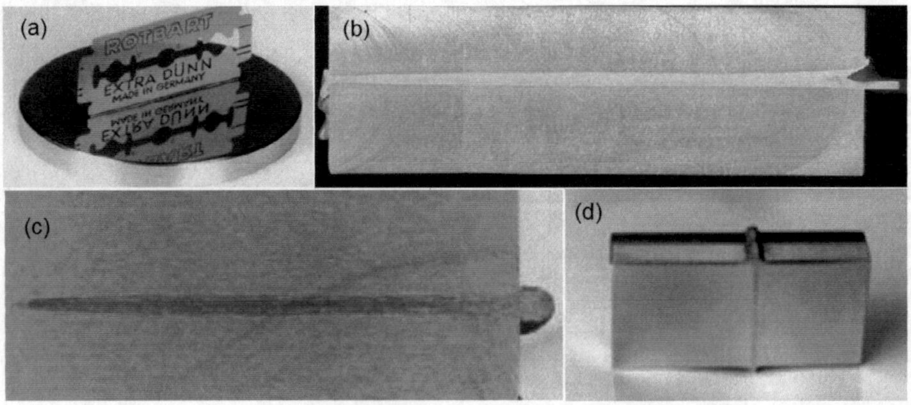

Figure 3. Examples of components produced by EBW, (a) Thick to Thin Welding [2, 10], and (b) Horizontally weld aluminum, penetration depth: 150 mm [12], (c) 45 mm deep weld seam in SS [10], (d) Bronze/Steel material welding of 30 mm [10].

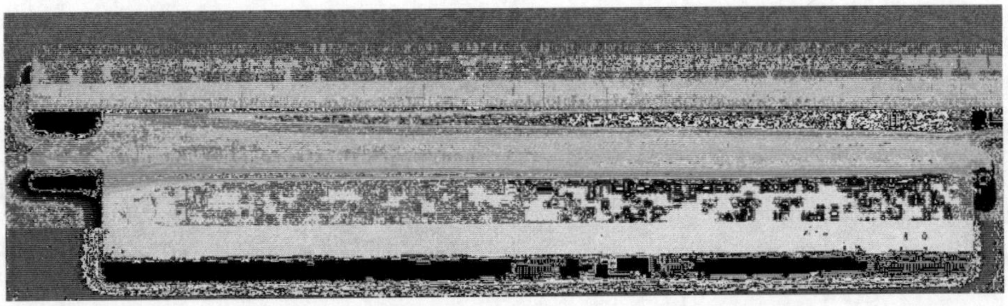

Figure 4. Single pass fully penetration EB welding in 280 mm thick C-Mn steel [8].

Electron beam welding experience (of a job shop) according to material can be summarized as below [10].

- In practice, most of steel materials are weldable by means of an Electron Beam without any special requirement such as pre-heating or using filler wire. 0.1 mm to 200 mm structural steel can be welded.
- Low alloyed steels as e.g. fine grained steels, case hardening or nitriding steels (for welding the case has to be removed or not applied in the welding zone), as well as steels for high temperature service, construction steels and most heat treatable steels are readily EB weldable.
- Most of the stainless and acid resistant steels, such as austenitic, ferritic or duplex steels are readily EB weldable. 0. 5 mm to 45 mm thick components can be welded.
- Wrought aluminium alloys, either forged or extruded are usually readily weldable. 50 mm to 200 mm thick Aluminum can be welded.
- Magnesium Alloys containing 6 % to 9 % of aluminium and up to 1 % of zinc, are readily weldable.
- SE, SF and OFHC copper are readily weldable. There are more than 1000 EB welded clamps of Cu Be 2 in the ASDEX fusion reactor.
- Pure nickel and pure cobalt as well as many of their alloys like Ni Mn 2, Ni Cr 80 20, Ni Cu 30 Fe are readily EB weldable.
- Reactive materials like titanium, zirconium, tantalum, vanadium or niobium exhibit a strong affinity to oxygen and hydrogen, especially at elevated temperatures. EB welding under vacuum provide the ideal environment for these materials.
- In principle, it is possible to EB weld any dissimilar metals to each other, even if their melting points are far apart, those metals, which alloy easily with each other are EB weldable without problems. Welding austenitic to martensitic steel, steel to bronze and many practical examples are mentioned in [10].

All electron- beam-welded wing carry-through structure of the F-14 Tomcat (very advanced fighter plane) was arguably one of the most sophisticated weldments ever put into production [11]. TWI has also reported the full penetration, 60 mm thickness, 1300 mm long seam weld 2350 mm diameter S355 steel using mobile electron beam welding concept [5].

4. Limitations of the technology

The introduction of a new technology almost invariably leads to exaggerated claims about its capabilities from over-zealous promoters of its virtues. The multi-functional electron beam welding process that has evolved over the past 50 years has definitely proved to be an extremely powerful welding tool when applied within the confines set by normal physical metallurgy and metallurgical thermodynamic limitations [2].

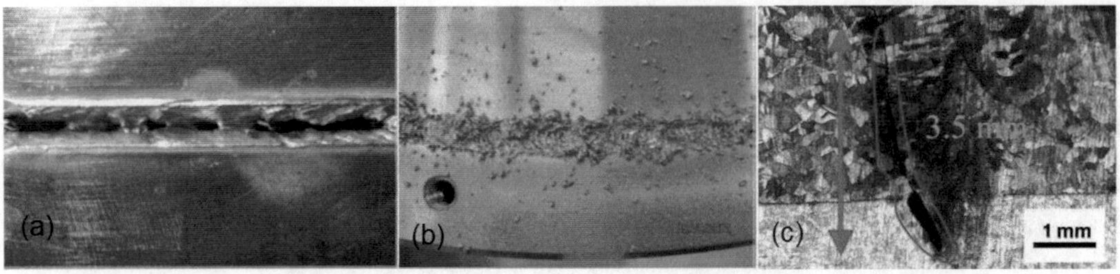

Figure 5. Few examples of potential weld defects in EBW. (a) Incomplete fusion due to burn through observed in EBW, (b) Spatter in EBW of copper [1], (c) Root spiking defect in copper EB weld [7].

Electron beam welds are not immune to the same type of defects as shown in figure 5 (e.g. porosity, spatter, shrinkage, cold shuts, cracking, etc.) [2].

5. Conclusion

From the survey, it is realized that EBW is a useful technology for joining of similar or dissimilar materials with a variety of material thicknesses. It has a wide range of application in industrial as well as in research field.

The knowledge gained at various Institutes in India like BARC (Bhabha Atomic Research Centre), RRCAT (Raja Ramanna Centre for Advanced Technology), IUAC (Inter University Accelerator Centre), ISRO (Indian Space Research Organisation) and BHEL (Bharat Heavy Electricals Limited) in the area of EB welding needs to be consolidated for a national level facility for nuclear / strategic application. Suitability of out of vacuum process needs to be explored in detail for large sized nuclear components.

References

[1] Salonen T, *State of the art of the welding method for sealing spent nuclear fuel canister made of copper Part 2 - EBW* (POSIVA-WR--14-21), Finland 2014.

[2] Dobeneck D 2007 *An International History of Electron Beam Welding* (Germany:Pro-beam AG) 1st ed.

[3] Schultz H 1993 *Electron beam welding* (Cambridge: Abington Publishing)

[4] Weglowski M, Blacha S and Phillips A 2016 *Vacuum* **130** 72.

[5] Sanderson A 2007 *Welding in the World* ***51*** Issue 1 37.

[6] Węglowski M, Dworak J, and Błacha S, *Institute of Welding Bulletin No.4/2014,* ISSN 2300-1674.

[7] Pitkänen J, Salonen T, Sandlin S, Ronneteg U 2007 *Proceedings of the Sixth International Conference on NDE in relation to structural integrity for nuclear and pressurised components,* Budapest, Hungary.

[8] Pike S, Allen C, Punshon C, Threadgill P, Gallegillo M, Holmes B and Nicholas J 2010 *Critical review of welding technology for canisters for disposal of spent fuel and high level waste*. Nagra Technical Report **NTB 09-05** Nagra, Wettingen, Switzerland, ISSN 1015-2636.

[9] Adriana M, Gheorghe N, *Proceedings of the 1st International Conference on Manufacturing Engineering, Quality and Production Systems* 2009 **Volume II,** 484

[10] Dobeneck D 2005 *Electron Beam Welding: Examples of 30 Years Job-Shop Experience (Germany:* pro-beam AG & Co.)

[11] Robert W Messler Jr. 2007 AWS *welding journal* **86** 41

[12] DVS Forschung, Research Association for Welding and Allied Processes of DVS 2014 *The electron beam as a tool for joining technology*, Düsseldorf, DVS Report Volume 299e p18

Conceptual & Engineering Design of Plug-in Cryostat Cylinder for Super-Conducting Central Solenoid of SST-1

Prabal Biswas[1], Prosenjit Santra[1], Kirit Vasava[1], Snehal Jayswal[1], Tejas Parekh[1], Pradeep Chauhan[1], Hitesh Patel[1] and Subrata Pradhan[1]

[1]Institute for Plasma Research, Bhat, Gandhinagar – 382 428, Gujarat State, India

Email: prosenjit@ipr.res.in

Abstract. SST-1, country's first indigenously built steady state super-conducting tokamak is planned to be equipped with an Nb_3Sn based superconducting central solenoid, which will replace the existing copper conductor TR1 coil for the purpose of Ohmic breakdown. This central solenoid (CS) of four layers with each layer having 144 turns with an OD of 573 mm, ID of 423 mm length of 2483 mm will be housed inside a high vacuum, CRYO compatible plug-in cryostat thin shell having formed from SS 304L plate duly rolled and welded to form cylinder along with necessary accessories like LN_2 bubble panel, current lead chamber, coil and cylinder support structure etc. This paper will present the design drivers, material selection, advantages and constraints of the plug-in cryostat concept, sub-systems of plug-in cryostat, its conceptual and engineering design, CAD models, finite element analysis using ANSYS, safety issues and diagnostics, on-going works about fabrication, quality assurance/control and assembly/integration aspects with in the existing SST-1 machine bore.

1. Introduction

SST-1 [1] country's first indigenously built steady state super-conducting tokamak is planned to be equipped with a Nb_3Sn based super-conducting central solenoid which will replace the existing copper conductor TR1 coil for the purpose of Ohmic breakdown. The main advantage of the plug-in cryostat is that it acts as an assembled unit containing the CS coil and accessories, which can be integrated with in the SST-1 machine bore of 690 mm without disturbing any sub-systems of machine.

2. Components of Plug-in Cryostat

The plug-in cryostat system figure 1 and figure 2 [2] consists of (a) Cryostat cylinder (CCY), (b) LN_2 shields and copper radiation guards for cryostat cylinder, (c) Current lead chamber (CLC), (d) LN_2 shields and copper radiation guards for current lead chamber, (e) CS Magnet support, (f) Base spacers and (g) CS Vacuum Line Pumping Duct line components.

(a) Cryostat cylinder: CCY of central solenoid is a high vacuum compatible ($\leq 1 \times 10^{-5}$ mbar) CRYO compatible (5 K) thin cylinder component having a wall thickness of 6mm, made of SS 304L material. It acts as housing to super-conducting central solenoid and supports CS coil and accessories dead weights. Its dimensions should have the assembly compatibility with SST1 in available cryostat bore space of 690 mm with a radial clearance of 20 mm.

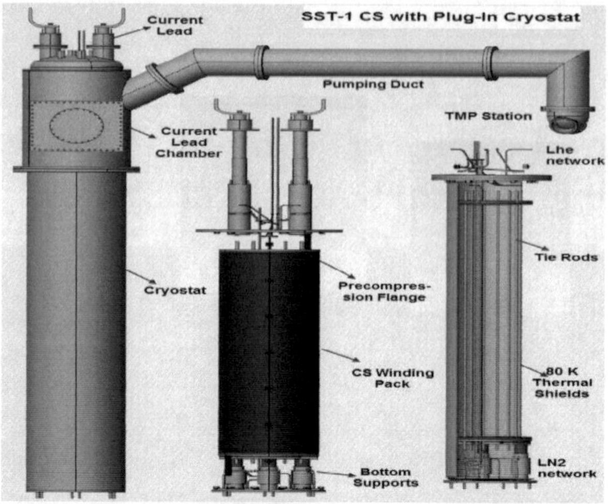

Figure 1. Plug-in Cryostat – General Assembly

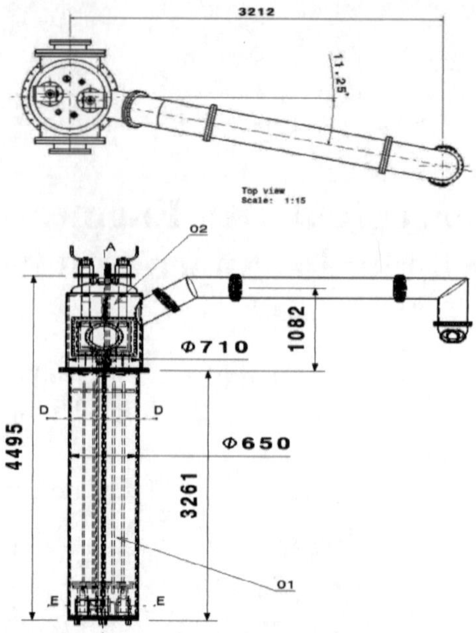

Figure 2. 2D View of Plug-in Cryostat

(b) LN_2 shield for CCY: LN_2 shields for CCY are made in two halves (diametrically) in order to minimize the eddy current formations. The material of construction is SS 316L, are of one-sided bubble type with a sheet thickness of 1.6 mm as per ASME Sec VIII – Div –1 appendix – 17 with respect to space restrictions. The advantage of bubble type geometry is the uniformity in temperature distribution along the entire surface area of the thermal shield. The LN_2 shield at 80 K acts as a thermal barrier between 5 K surface of CS coil and ambient surface of CCY so that the coil does not view the cylinder directly in order to reduce the thermal heat load on the coil. The gaps are covered by copper radiation guards, which are passively cooled due to conduction from the shields. The shields are mounted with insulated supports from the bore of the CCY. The radial clearance between shields and CCY is 10 mm, so that there is no interference between them during assembly. There are also three shields mounted on cylinder base in order to reduce the heat load on the coil supports

(c) *Current Lead Chamber:* CLC of central solenoid is a high vacuum compatible ($\leq 1 \times 10^{-5}$ mbar) CRYO compatible (5 K) thin cylinder component having a wall thickness of 6 mm, made of SS304 L material. At the top end it also has a 2:1 semi elliptical head of 6 mm thickness. It acts as housing and exit points of the two current leads of super-conducting central solenoid, man-hole ports for hydraulic connections, diagnostics requirements and opening of the vacuum line pump duct required for evacuation of the plug-in cylinder.

(d) *LN_2 shield for CLC:* LN_2 shields for CLC are made in two halves (diametrically) in order to minimize the eddy current formations and also with demountable type at port openings. The material of construction is SS316L, are of double-sided bubble type with a sheet thickness of 1.6 mm as per ASME Sec VIII – Div –1 appendix – 17. The advantage of bubble type geometry is the uniformity in temperature distribution along the entire surface area of the thermal shield. The LN_2 shield at 80 K acts as a thermal barrier between 5 K surface of CS coil current leads and ambient surface of CCY so that the current leads does not view the cylinder directly in order to reduce the thermal heat load on the current leads. The gaps are covered by copper radiation guards, which are passively cooled due to conduction from the shields. The shields are mounted with insulated supports from the bore of the CCY. The radial clearance between shields and CLC is 12 mm, so that there is no interference between them during assembly.

(e) *CS Magnet support:* CS coil is supported at the base with three equally angular spaced SS 304L/G10 supports raised from the cylinder base. SS 304L pipe are welded onto cylinder base and molded G10 pipes inserted on SS 304L pipe to reduce the heat conduction from cylinder base to coil surface. There are SS 304L and G10 pads mounted on the G10 pipe onto which the coil is supported. The SS 304L pipes are circumscribed by LN_2 bath to reduce the heat load on the coils due to conduction from ambient surfaces of cylinder base. The annular spaces between supports are used for interlayer joint connections of CS coil and also for hydraulic connections/joints for cryogenics fluids.
(f) *Base spacer:* The assembled base spacer is situated outside the vacuum environment consists of SS 304L ribbed structure in two halves tightened together with a sandwiched G10 pad in between to reduce the effect of eddy current formation. The cylinder with all its content sits on the base spacer having a G10 pad in between and is tightened with base spacer with nuts and bolts. The base spacer is mounted and assembled with the SST-1 machine frame structure.
(g) *CS Vacuum Line Pumping Duct:* It is an approximately 5 m long, 274 mm OD, 6mm thick SS 304L pipe connected at the current lead chamber opening, used for evacuating the plug-in cryostat assembly to vacuum level of $\leq 10^{-5}$ mbar. It is a height of 4.35 m from cylinder base and is toroidally oriented between radial port 3 and 4.

3. Assembly plans and sequence for CS coil and its cryostat in SST-1

Assembly sequence of plug-in cryostat is very important; as the system has to be inserted in the SST-1 cryostat bore of 690 mm figure 3 and figure 4 without disturbing any sub-systems. The design, dimensions, tolerances and radial clearances are finalized keeping in mind smooth and interference free insertion of the system in SST-1 machine bore. A tentative assembly sequence is presented here for reference purpose. A detailed assembly sequence/procedure will be worked out before actual integration of plug-in cryostat with SST-1.

- CS Coil mounted with support bases and hydraulic connection.
- LN_2 shields in two halves is mounted inside the cylinder bore.
- The cylinder with pre mounted LN_2 shields will be lifted to enclose the coil.
- This compound assembly of coil &cylinder will be inserted in SST-1 CS bore upon the base spacer and bolt tightened with the same.
- CLC semi elliptical head with lead connections is welded with the current lead chamber having pre mounted LN_2 shields.
- CLC chamber is lifted, mounted with already inserted CCY in proper toroidal orientation
- Hydraulic connections for Helium and LN_2 transfer lines and leak testing.
- Finally the vacuum line pump duct is lifted and mounted with respective flange in CLC so that the toroidal orientation of the duct line is between radial port 3 and 4.
- Closing & leak testing of the plug-in assembled system, thermal and electrical isolation checks .

4. Safety and Diagnostics

In case of a pressure build up inside plug-in cryostat due to leak in LN_2 and helium circuits, suitable rupture disks/pressure release valves will be installed at the current lead chamber ports to release the buildup of pressure and prevent the systems from any damage. With rupture discs/safety pressure release valves installed, maximum pressure inside the plug-in cylinder in the event of a leak will be limited to about 1.1 atmospheres.

Also the plug-in cryostat will be equipped with temperature sensors to monitor the temperature patterns, Venturi meters to measure the flow parameters, voltage traps, suitable grounding scheme etc.

IOP Conf. Series: Journal of Physics: Conf. Series **823** (2017) 012041 doi:10.1088/1742-6596/823/1/012041

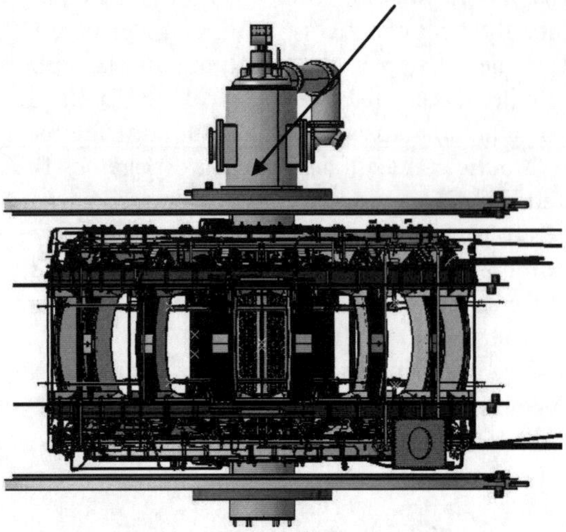

Figure 3. Elevation view of CCY Assembly with SST-1

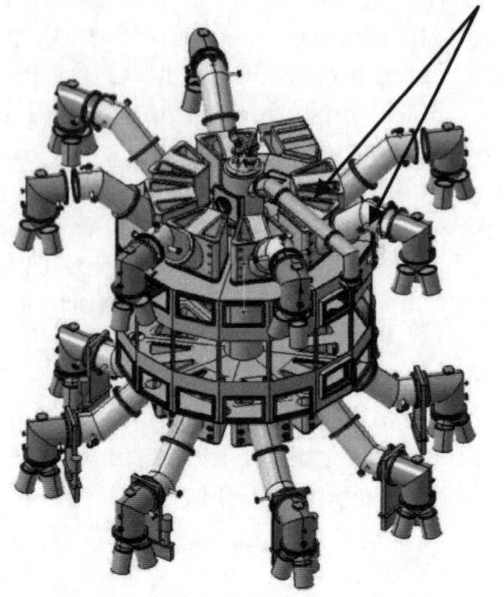

Figure 4. Isometric view of CCY with SST-1

5. Load acting on the plug-in Cryostat

The main loading acting on the plug-in cryostat are the (a) radiation/conduction heat loadings on the shield and coil support maintained at 80 K and 5 K with respect to the cylinder surface and base maintained at ambient temperature of 300 K, (b) the vacuum (external pressure) loading on cylinder and current lead chamber walls and (c) the dead load of coil on support structures, base supports etc.

(a) (i) Radiation heat loads on shields (300 K to 80 K): The total radiation heat loads on LN_2 shields Q_r from ambient cylinder and CLC surfaces are calculated analytically with Stefan Boltzman formula (equation 1) based on concentric cylinder assumption between shields and cylinder/CLC are found to be around 278 watt.

$$Q_r = \frac{\sigma A_{CY}\left(T_{CY}^4 - T_s^4\right)}{\frac{1}{\varepsilon_{CY}} + \left(\frac{A_{CY}}{A_s}\right)\left(\frac{1}{\varepsilon_s} - 1\right)} \quad \text{Watt ---------- (1) [3]}$$

where A_{CY}, A_S are surface area of cylinder/CLC and shield, T_{CY} and T_S ambient cylinder/CLC temperature of 300 K and 80 K, $\varepsilon_{CY} = 0.2$ emissivity of cylinder/CLC (buffed surfaces), $\varepsilon_S = 0.1$ emissivity of shields (electro-polished surfaces), $\sigma = 5.67 \times 10^{-8}$ W/m^2K^4 is Stefan Boltzman constant.

(ii) Conduction heat load through supports: The conduction heat load onto coil at 5 K from ambient temperature of 300 K through supports/insulators are estimated with FE package ANSYS, [4] using 2D, axis-symmetric thermal analysis with PLANE 55 elements considering temperature dependent thermal conductivity of the support/insulator materials. Figure 5 shows the CAD model and boundary conditions of the analysis. Figure 6 shows the temperature distribution on a support between 300 K to 5 K.

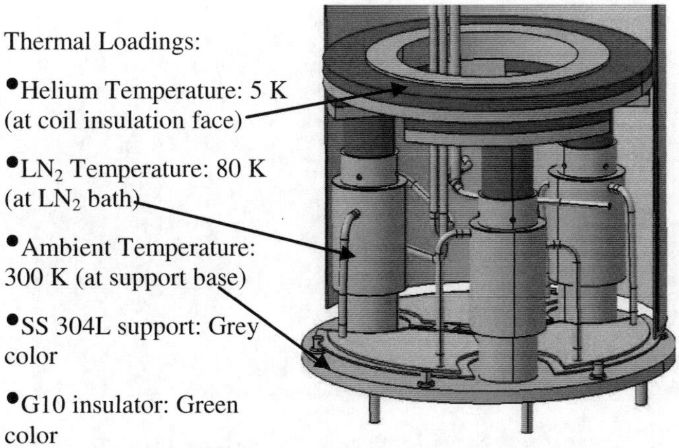

Figure 5. Thermal Loadings on Support

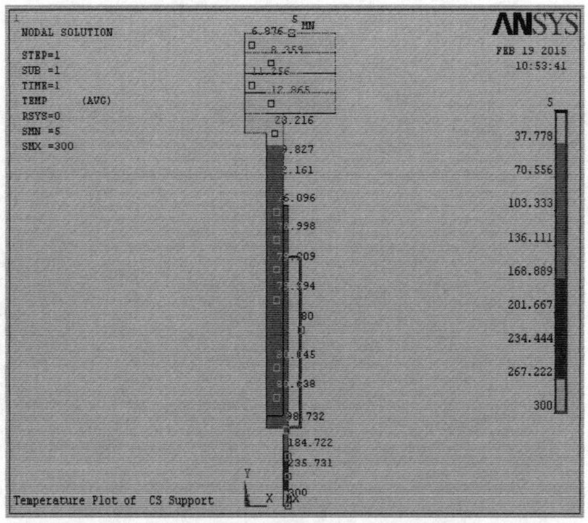

Figure 6. Temperature plot Support

Heat load on coil base support as estimated from ANSYS is Q = 0.5 Watt (per support)

For 3 Supports, $Q_{TOTAL} = 3 \times 0.5 = 1.5$ Watt

(b) Vacuum loading: The principal load acting on it is the vacuum loading i.e. external atmospheric pressure of 1 bar on the wall surfaces. The allowable external pressure is calculated using ASME code (ASME Section II part D, subpart 3) [5] for external loading. The cylinder/CLC being a thin one subjected to external pressure, the critical buckling pressures [6] are also calculated for the same. The allowable loading and stress parameters [7] are analytically estimated and tabulated in Table 1.

Table 1. Allowable Design Pressure & Buckling Pressure for CSCY & CLC for Vacuum Loading

Component Name	Material	Pressure Load Acting	Allowable design Pressure (N/mm^2)	Critical buckling Pressure (N/mm^2)	Remarks
Cryostat Cylinder (CS)	SS304L	Vacuum Load: External Pressure $p = 0.1\ N/mm^2$	0.3	0.35	Safe design
Current Lead Chamber (CLC)	SS304L	Vacuum Load: External Pressure $p = 0.1\ N/mm^2$	0.48	0.27	Safe design
Current Lead Chamber top elliptical Cover	SS304L	Vacuum Load: External Pressure $p = 0.1\ N/mm^2$	9 MPa (Maximum Stress developed)	----	Safe design

(c) Dead weight of coil: The load acting on the supports are the dead weight of the coil and accessories about 5 tons, we have kept a FOS of 2 to account for dynamic loading condition scenario, hence total acting on three supports is 10 tons. A 3D ANSYS structural analysis with SOLID 45 elements has been done for a single support assuming equal distribution of dynamic loading and support base pipe fixed at all directions on each support to estimate the deformation and stress developed in them. Maximum deflection obtained is 0.14 mm figure 7 while maximum Von Misses stress developed is 35 MPa figure 8 much less than allowable stress of 110 MPa for SS 304L. The corresponding maximum deflection and Von Misses stress values for base spacer with a total loading of 10 tons is 0.29 mm and 62.5 MPa respectively, as shown in figure 9 and figure 10.

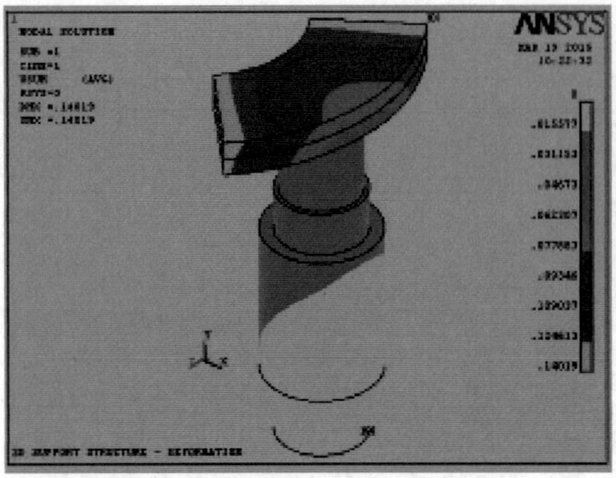

Figure 7. Deflection 0.14 mm (max.) (Support)

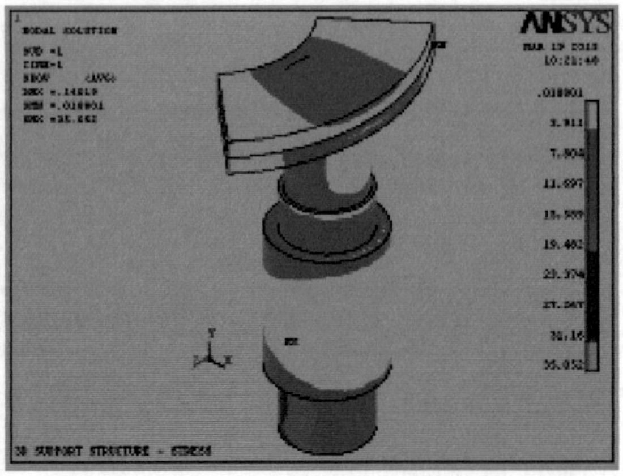

Figure 8. VM stress 35 MPa (max.) (Support)

Figure 9. Deflection 0.29 mm (max.) (Base spacer)

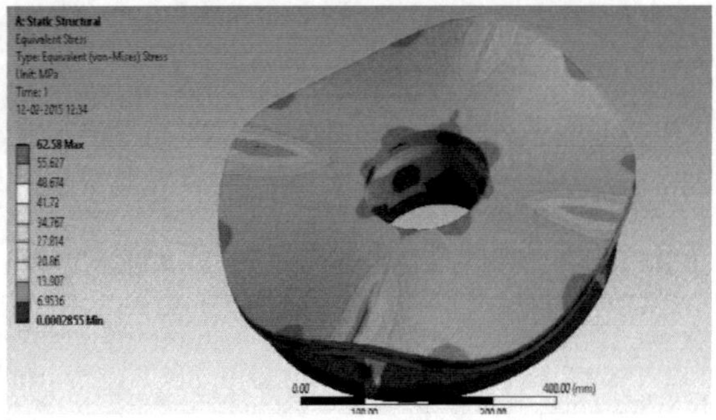

Figure 10. VM stress 62.5 MPa (max.) (Base spacer)

6. Present status

The plug-in cryostat cylinder has been fabricated, successfully inspected/tested at vendors site and has been delivered to IPR in January 2016. The functionality checks of the thermal shields at cryogenic LN_2 temperature and integrated vacuum leak testing of assembly system with a leak rate of 1.5×10^{-9} mbar l/s, has also been successfully completed. Figure 11(a) shows the thermal shock tests for the CS cylinder thermal shield figure 11(b) shows the thermal shock tests for the CLC thermal shield. Figure 12 shows the integrated leak testing of the plug-in cryostat assembly.

7. Conclusion and future works

The conceptual and engineering design of plug-in cryostat for superconducting CS is discussed in this paper with the following points

- Conceptual and engineering design for component of plug-in cryostat and both analytical, FEM structural analysis shows that the design is safe and acceptable.
- Thermal analysis shows minimum conductive heat transfer of 1.5 watt through three columnar supports from ambient temperatures, which is acceptable. Also radiation heat loads on shields are with in the design limits.
- The assembly sequence for the installation of plug-in cryostat into SST-1 cryostat bore considering the requirements and constraints involved has been discussed.

- The component has been fabricated, tested and delivered at IPR. Thermal shock tests, ground assembly test mock ups and integrated leak testing has also been successfully completed at IPR.
- The final assembly trials plug-in cryostat along with the superconducting coil and integration with SST-1 will be carried out in near future.

(a) (b)

Figure 11. Thermal Shock test of (a) CCY Shield, (b) CLC shield

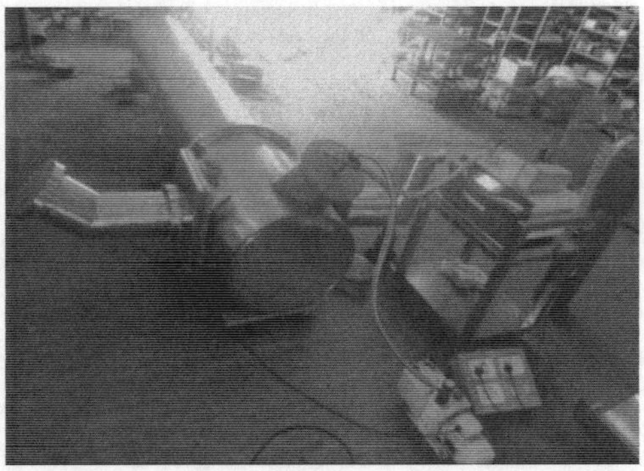

Figure 12. Integrated Vacuum leak testing CCY assembly

References

[1] Pradhan S, Sharma A N, Tanna V L, Khan Z, Prasad U, Doshi K, Raval D C, Pathan F, Gupta N C, Tank J, Gupta M K, Santra P, Biswas P, Parekh T, Masand H, Sharma D, Srivastava A, Patel H, 2012 *IEEE Transaction on Plasma Science* **40** Number 3 pp.614-621
[2] Santra P, Biswas P, Vasava K, Jayswal S, Parekh T, Chauhan P, Patel H ,Pradhan S, May 2015 *IPR Technical Report: IPR/TR-343/2015*
[3] R K Rajput, 2013 *Heat & Mass Transfer* pg. 715
[4] ANSYS users Manual, ANSYS Inc. PA, USA
[5] ASME Code Section II part D, subpart 3 Buckling of Thin Walled Circular Cylinder, NASA Report, August 1968
[6] Papadakis G, 2008 *International Journal of Solids and Structures* **45** Issue 20 pp 5308 – 5321
[7] Moss D R, 2005 *Pressure Vessel Design Manual*, 3rd Edition pg. 16, 25

IOP Conf. Series: Journal of Physics: Conf. Series **823** (2017) 012042 doi:10.1088/1742-6596/823/1/012042

Comparative analysis on flexibility requirements of typical Cryogenic Transfer lines

Mohit Jadon[1], Uday Kumar[1], Ketan Choukekar[1], Nitin Shah[1] and Biswanath Sarkar[1]

[1]ITER-India, Institute for Plasma Research, Sangath Skyz, Bhat – Koteshwar Road, Koteshwar - 380005, Gujarat, India.

E-mail: mohit.jadon@iter-india.org

Abstract. The cryogenic systems and their applications; primarily in large Fusion devices, utilize multiple cryogen transfer lines of various sizes and complexities to transfer cryogenic fluids from plant to the various user/ applications. These transfer lines are composed of various critical sections i.e. tee section, elbows, flexible components etc. The mechanical sustainability (under failure circumstances) of these transfer lines are primary requirement for safe operation of the system and applications. The transfer lines need to be designed for multiple design constraints conditions like line layout, support locations and space restrictions. The transfer lines are subjected to single load and multiple load combinations, such as operational loads, seismic loads, leak in insulation vacuum loads etc. [1]. The analytical calculations and flexibility analysis using professional software are performed for the typical transfer lines without any flexible component, the results were analysed for functional and mechanical load conditions. The failure modes were identified along the critical sections. The same transfer line was then refurbished with the flexible components and analysed for failure modes. The flexible components provide additional flexibility to the transfer line system and make it safe. The results obtained from the analytical calculations were compared with those obtained from the flexibility analysis software calculations. The optimization of the flexible component's size and selection was performed and components were selected to meet the design requirements as per code.

1. Introduction

The cryogenic systems and their applications, primarily in large Fusion devices, utilize multiple cryogen transfer lines of various sizes and complexities in terms of layout to transfer cryogenic fluids from plant to various user/ applications. The mechanical sustainability (under failure circumstances) of these transfer lines are primary requirement for safe operation of system and applications.

The Liquid Nitrogen carrying transfer line has been considered for the following study and analysed for 2 cases as follows:
case-1: without using flexible components
case-2: with flexible components

2. Design Criteria and constraints

The transfer lines need to be designed for multiple design constraint conditions, which are explained below:

i) Diameter constraint; the space restriction alongside the transfer line shall be complied.
ii) Flow rate constraint; the mass flow rates shall not be exceeded for line design.
iii) Routing constraint; the transfer line routing shall be complied in order to avoid clashes or interference with the very close adjacent equipment.
iv) Pressure drop constraints; the pressure drop values shall be within allowable limits.
v) Support locations; the external supports are available at limited locations and no scope is there for additional supports.

3. Analysis approach

i) Selection of segment of Liquid nitrogen cryogenic transfer line and modeling performed on CAESAR-II software.
ii) Finalization of applicable loads.
iii) Flexibility analysis performed based on loads without considering any thermal compensation device.
iv) Failure zones identified with respective stress values.
v) Flexible components such as metallic hoses and bellows placed at various locations.
vi) Flexibility analysis performed based on similar loads with flexible components.
vii) The number of flexible components were finalized based on the obtained stress values and allowable pressure drop values.

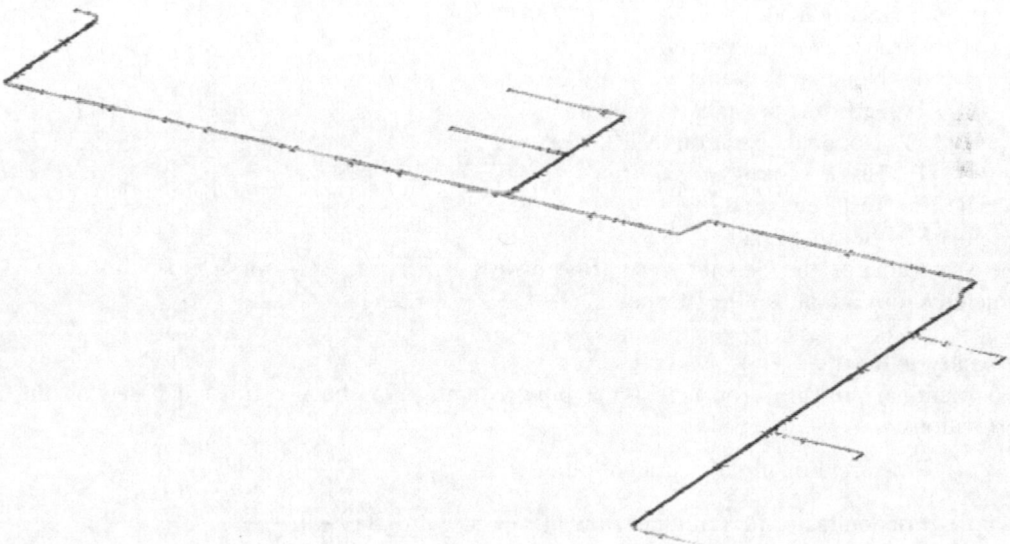

Figure 1. The layout and parameters of cryogenics transfer line.

4. Significance of load cases

i) LIV-2 is a failure case of ingress of atmospheric air due to rupture of outer vacuum jacket (OVJ).
ii) Seismic event non-nuclear (SL) may lead to operational failure of the system.
iii) The above cases along with normal operation and test conditions are considered for the safe design of transfer line.

5. Assumptions and specifications

i) Piping network is considered to be located in controlled environment, hence Snow and Wind loads are exempted.
ii) Bellows used in analysis are untied type bellows with internal sleeves, hence pressure drop is not considered in calculations.
iii) Factor of safety is taken as 1.5.
iv) OVJ Diameter: DN150, Process Pipe Main line Diameter: DN 80, Process Pipe Branch line Diameter: DN 50.
v) Material of pipe: ASTM A312.
vi) Design Code: EN 13480.

Table 1. Boundary conditions.

Load Case		T (K)	P (MPa)
G+NT+LIV-2	Process Pipe	80	1.1
	OVJ	252.1	0.01
G+NO	Process Pipe	80	0.39
	OVJ	323	0 (vacuum)
G+TP+TT	Process Pipe	323	1.573
	OVJ	323	0 (vacuum)
G+NT+LIV-2+SL	Process Pipe	80	1.1
	OVJ	252.1	0.05

where,

i) G : Gravity load
ii) NT : Nominal Temperature
iii) NP : Nominal Pressure
iv) LIV-2 : Loss of insulation vacuum
v) NO : Nominal operation (NT + NP)
vi) TP : Test Pressure
vii) TT : Test Temperature
viii) SL : Seismic Load

The spectrum for the Seismic case (SL) have been extracted from EN 1998-4 code (Design of Structures for Earthquake Resistance) .

6. Analysis Result

The frictional pressure drop across the pipe assembly has been estimated based on the following correlations:

i) Pressure drop through straight pipe [3], $\Delta P_{pipe} = \frac{\rho f l v^2}{2D}$ Pa (1)

ii) Friction factor for turbulent flow [4], $\frac{1}{\sqrt{f}} = -2log\left(\frac{Ra}{3.7d} + \frac{2.51}{Re\sqrt{f}}\right)$ (2)

iii) Pressure drop through pipe bends and fittings [3], $\Delta P_{bend} = \frac{K\rho v^2}{2}$ Pa (3)

iv) Pressure drop due to flexible elements [5], $\Delta P_f = \left(\lambda \frac{l_h}{d_h} + \xi_h\right)\rho v^2$ Pa (4)

v) Total pressure drop without flexible elements $\Delta P = \Delta P_{pipe} + \Delta P_{bend}$ Pa (5)

vi) Total pressure drop with flexible elements $\Delta P = \Delta P_{pipe} + \Delta P_{bend} + \Delta P_f$ Pa (6)

Table 2. Pressure Drop results.

Line description	Process pipe diameter	Mass flow rate (kg/s)	Operating Pressure (bar)	Operating Temperature (K)	Pressure drop without flexible elements (mbar)	Pressure drop with flexible elements (mbar)	Allowable pressure drop (mbar)
Main line	DN 80	2.5	3.6	81	252	302	450
Branch line	DN 50	2.5	3.6	81	197	250	300

Table 3. Analysis results.

Load Case	Allowable stress (MPa)	Maximum stress without flexible elements (MPa)	Maximum stress with flexible elements (MPa)	Maximum Deflection (mm)		
				X	Y	Z
G+NT+LIV-2	258	1201	198.6	-31.30	-11.75	20.66
G+NO	143.33	573	113	-37.20	-16.64	-16.90
G+TP+TT	204.25	29	142	1.44	-2.11	2.92
G+NT+LIV-2+SL	258	838.3	198.5	-0.15	-2.11	-0.11

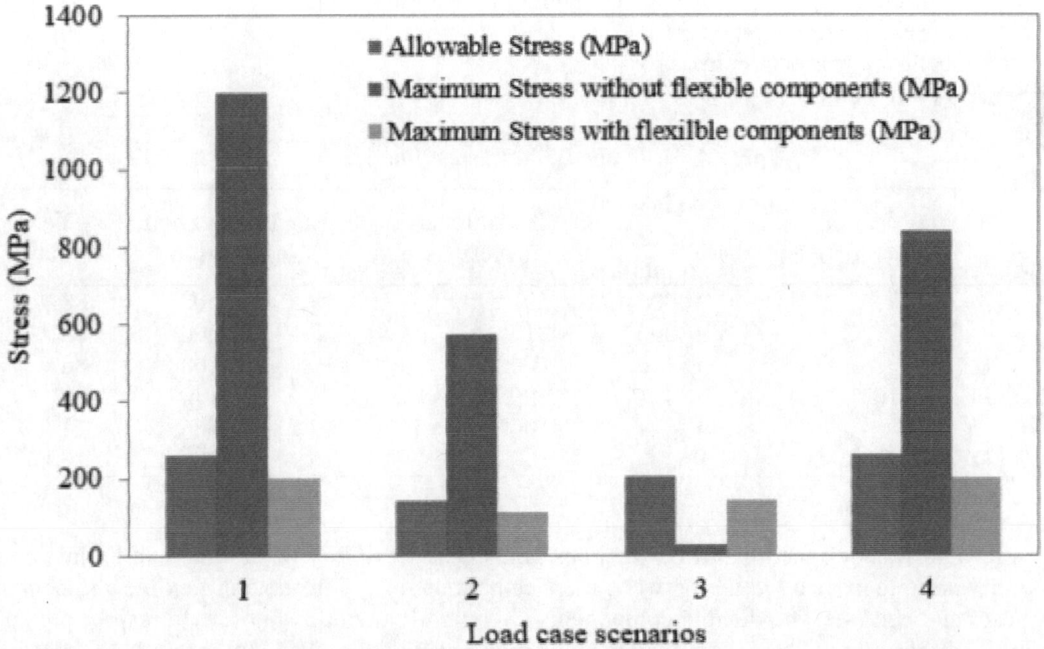

Figure 2. Graphical comparison of Maximum stress under conditions of usage of flexible components.

IOP Conf. Series: Journal of Physics: Conf. Series **823** (2017) 012042 doi:10.1088/1742-6596/823/1/012042

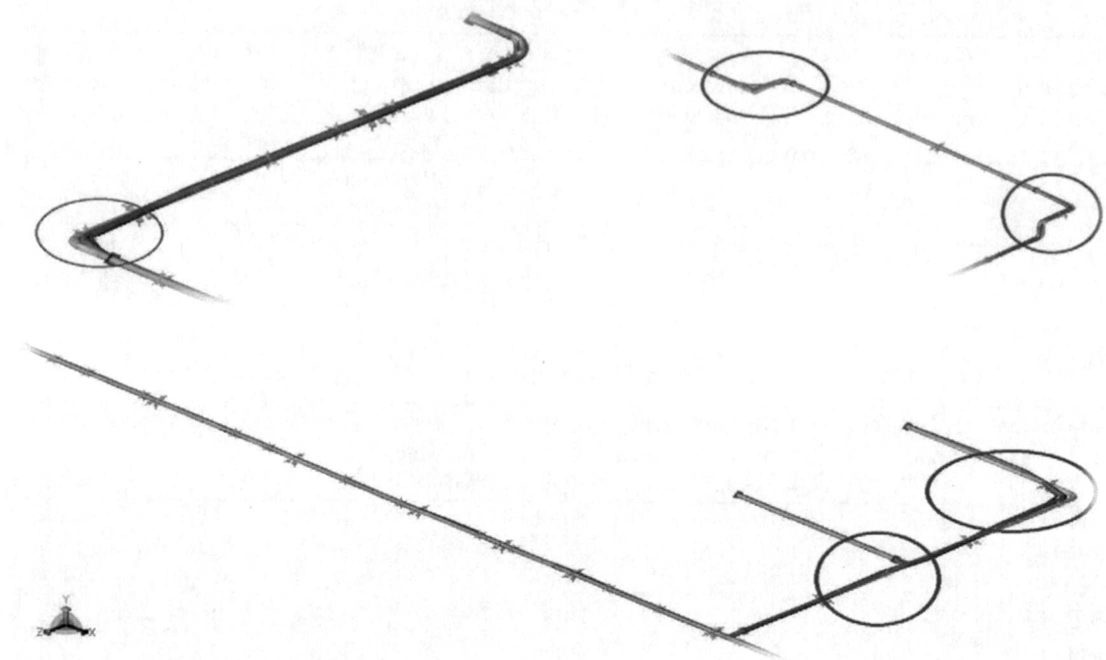

Figure 3. CAESAR II snapshots of the displacements in the critical sections of cryolines without use of flexible elements.

The following flexible components were selected after optimization of analysis:

i) No. of hoses on process pipe : 11
ii) No. of bellows on process pipe : 16
iii) No. of bellows on OVJ : 9

Table 4. Hose and bellows specifications.

		No. of component	Axial stiffness (n/mm)	Transitional stiffness (n/mm)	Eff. Id (mm)	Length (mm)	Dia. (mm)
Hose		2	10000	59	N/A	500	88.9
		9	10000	31.9	N/A	500	73
Bellow	PP	16	151.2	10000	105	168	88.9
			134	10000	75.5	140	60
			144	10000	93.2	165	73
	OVJ	9	96.5	10000	183	78	168.3

7. Conclusion

The Transfer line was designed based on the constraints. The flexibility of the line can be improved either by providing loop/ bend in the network if space permits or by introducing flexible components such as hoses and bellows. The flexible components are primarily avoided in order to respect pressure drop values and because of the higher maintenance required by them. The study concludes that it is important to achieve fine balance between the allowable pressure drop values and space constraints in order to select the flexible components.

Acknowledgement
We are sincerely thankful to Mr. Vikalp Sri Bachahan for his valuable contribution during the course of this work.

References

[1] Badgujar S, Benkheira L, Chalifour M, Forgeas A, Shah N, Vaghela H, and Sarkar B, "*Loads specification and embedded plate definition for the ITER cryoline system*", Cryogenic Engineering Conference and International Cryogenic Materials Conference, Tucson, Arizona; (2015)
[2] EN 10217-5 : 2004; *Seamless steel tubes for pressure purposes, technical delivery condition*
[3] *Flow of fluids through valves, fittings and pipe* – CRANE (Publication 410M (Metric edition))
[4] Colebrook C F "*Turbulent Flow in Pipes, with particular reference to the Transition Region between the smooth and Rough Pipe Laws*"
[5] WITZENMANN metal hoses manual (Section 7.2 Page 246-257)

IOP Conf. Series: Journal of Physics: Conf. Series **823** (2017) 012043 doi:10.1088/1742-6596/823/1/012043

Thermo-mechanical Design Methodology for ITER Cryo-distribution cold boxes

Vinit Shukla[1], Pratik Patel[1], Jotirmoy Das[1], Hitensinh Vaghela[1], Ritendra Bhattacharya[1], Nitin Shah[1], Ketan Choukekar[1], Hyun-Sik Chang[2], Biswanath Sarkar[1]

[1]ITER-India, Institute for Plasma Research, A-29 GIDC Electronics Estate, Sector-25, Gandhinagar - 382016, Gujarat, India
[2]ITER Organization, Route de Vinon-sur-Verdon, CS 90 046, 13067 St. Paul Lez Durance Cedex, France

E-mail: vinit.shukla@iter-india.org

Abstract. The ITER cryo-distribution (CD) system is in charge of proper distribution of the cryogen at required mass flow rate, pressure and temperature level to the users; namely the superconducting (SC) magnets and cryopumps (CPs). The CD system is also capable to use the magnet structures as a thermal buffer in order to operate the cryo-plant as much as possible at a steady state condition. A typical CD cold box is equipped with mainly liquid helium (LHe) bath, heat exchangers (HX's), cryogenic valves, filter, heaters, cold circulator, cold compressor and process piping. The various load combinations which are likely to occur during the life cycle of the CD cold boxes are imposed on the representative model and impacts on the system are analyzed. This study shows that break of insulation vacuum during nominal operation (NO) along with seismic event (Seismic Level-2) is the most stringent load combination having maximum stress of 224 MPa. However, NO+SMHV (Séismes Maximaux Historiquement Vraisemblables = Maximum Historically Probable Earthquakes) load combination is having the least safety margin and will lead the basis of the design of the CD system and its sub components. This paper presents and compares the results of different load combinations which are likely to occur on a typical CD cold box.

1. Introduction

During the intended life time of ITER , several loads are envisaged on the CD system [1], which are, gravity/assembly loads, interface loads, nominal pressure/temperature, test pressure/temperature, purge pressure, thermo-mechanical loads due to break of insulation vacuum (BIV), transport acceleration and seismic loads (SL). Single loads or combinations of them can act on the CD system and on its sub systems; therefore, it is very important to analyze the integrity of the system and components under their influence. A typical CD cold box, i.e. Auxiliary Cold Box (ACB) for the CPs which includes most of the components and largest in size is chosen for the study, and load combinations have been applied on it in order to understand their impacts. A complete 3-D model of ACB-CP [2] is developed in CATIA along with its internal piping, cryogenic control valves and internal components as shown in figure 1. This paper focuses on the understanding of the nature of the loads or their combinations for the ITER CD system as well as their impacts on the design.

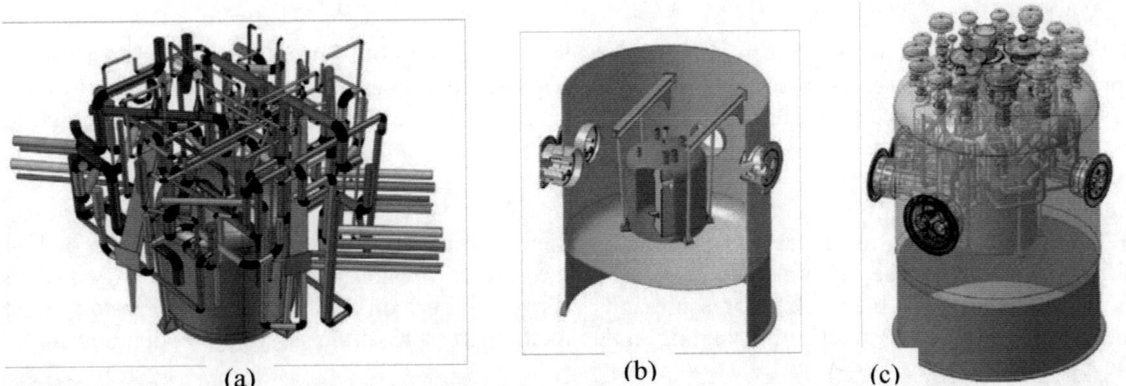

Figure 1. (a) 3D model of ACB-CP with internal piping routing, (b) liquid helium bath arrangement and outer vacuum jacket (OVJ) of ACB-CP.

2. Configuration of a typical CD cold box

ACB-CP is equipped with LHe bath with two submerged HX's, cryogenic control valves, cold circulator and process piping in order to supply the supercritical helium (SHe) at the required mass flow rate, pressure and temperature level. The cold circulator generates the required pressure head for the supply of the SHe to application. The thermal shield, surrounding the 4 K components in order to reduce the radiation heat load, is provided with 80 K gaseous helium (GHe).

3. Inputs and assumptions

3-D model of ACB-CP is developed in CATIA® [3] and taken as input for study of the load combinations. Thermal shields, internal components inside LHe bath i.e. heat exchangers, heaters have not been analysed individually but their gravity loads have been considered during the analysis. Forces and moments coming from the piping over the nozzles of LHe bath and on vacuum barriers are calculated from pipe stress software CAESAR® and are given as boundary condition to finite element software ANSYS® [4]. Process pipes at the interface position with the cryolines are anchored at vacuum barrier location which has to sustain the generated forces and moments. Detailed analysis of the vacuum barrier is not performed in the present study. Heat flux of 0.5 W/m^2 is given to 4.5 K components as a radiation heat load. Titanium alloy material, having high strength and low thermal conductivity, is taken for the tie rods used to support the LHe bath from OVJ. During the BIV-2 event the minimum temperature of the OVJ is calculated considering OVJ as lumped mass. The minimum temperature of OVJ is calculated as 136 K, considering breakage of LHe bath itself, and applied on OVJ during the analysis; however, there can be spatial distribution of temperature over OVJ. Transport acceleration, lifting and handling loads are not considered in this study. The relevant single loads [3] which are applicable to each expected modes of the ITER CD system are summarized below.

3.1. Nominal operating condition (NO)

Operating pressure (NP) and operating temperature (NT) is considered for the analysis during nominal operation. The OVJ of ACB-CP is kept under vacuum thereby exposed to a pressure difference of 0.1 MPa.

3.2. Test condition

All pressure equipment covered under the European Pressure Equipment Directive (PED) Essential Safety Requirements (ESR) are subjected to a pneumatic test pressure (TP) to be performed at room temperature (RT) in agreement with the safety French regulations. Test pressure, which is 1.43 times the design pressure, is applied.

IOP Conf. Series: Journal of Physics: Conf. Series **823** (2017) 012043 doi:10.1088/1742-6596/823/1/012043

3.3. *BIV-1*
BIV-1 is an accidental case due to a major air leak in the CD cold box cryostats during nominal operation. For this case, the pressure loads in the process components are considered equal to their maximum allowable working pressure and internal pressure of the CD cold box cryostats is equal to the atmospheric pressure (0.1 MPa).

3.4. *BIV-2*
BIV-2 is an accidental case due to a rupture of internal process piping/component during nominal operation. For this case, the pressure loads in the process components shall be considered equal to their maximum allowable working pressure and internal pressure of the CD cold box cryostats equal to 0.14 MPa (absolute value, corresponding to the opening pressure of the safety protection devices of the CD cold box cryostats).

3.5. *Incidental confinement pressure (ICP)*
As ACB-CP is located on level 3 of the Tokamak building, where maximum possible external pressure of 0.12 MPa on the OVJ is considered for the design.

3.6. *Seismic level 1/2/SMHV*
The basement of the ITER Tokamak building is constructed on seismic isolators which will have a particular response to a seismic event. For an SL-2 event, acceleration spectra [6] as shown in figure 2 are provided. SL-1 acceleration spectra are derived by dividing the SL-2 values by 4. To take into account the different location inside the Tokamak building, a conservative spectrum taking the envelope of the different nodes is considered for the analysis. The seismic response to SMHV event is obtained by multiplying the results from SL-2 by 0.73. 2% damping is considered for response spectrum analysis for each direction X, Y and Z. For simplicity full SL-1 analysis is not performed and results are obtained dividing the results from SL-2 by a factor of 3.

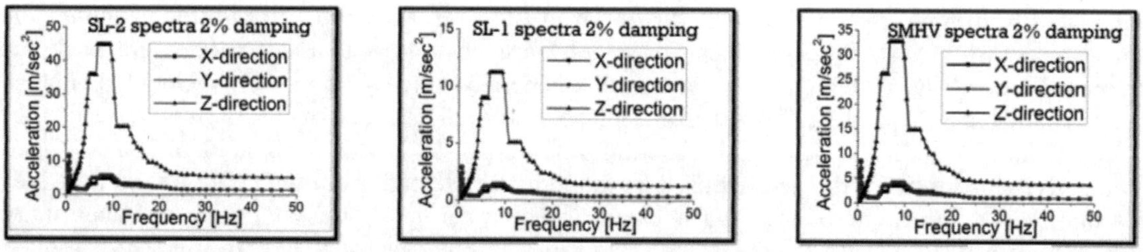

(a) (b) (c)

Figure 2. (a) Acceleration spectra for SL-2, (b) SL-1 and (c) for SMHV.

4. Methodology
ACB-CP is modelled in CATIA® [3] to carry out complete thermo-mechanical analysis using Finite Element Method. Internal piping with valves are modelled in CAESAR® software which gives extensive results based on European piping design code EN-13480. Cryogenic valves are modelled by CAESAR® as pipe element with anchoring points on the OVJ location and valve seat location connected with process piping. Internal piping layout inside the cold box is modelled in such a way that maintenance of HXs can be performed easily. LHe bath is supported from the OVJ with the help of two I-beams which cross section is 150 × 150 mm^2. Piping of the ACB-CP is modelled loop wise as shown in figure 3 and the all load cases are applied on it. Forces and moments at the anchor position are extracted from the pipe stress analysis and given as input for the analysis in finite element software ANSYS® [4]. Different load cases are categorized in different service limits [4] and are subjected to different allowable limits. The forces and moments on the valve seat location obtained from pipe stress

IOP Conf. Series: Journal of Physics: Conf. Series **823** (2017) 012043 doi:10.1088/1742-6596/823/1/012043

analysis software are applied to FE model of complete ACB-CP. For all the load combinations analysis is performed and allowable stresses are verified.

5. Results and discussion

5.1. CAESAR analysis results

Internal process pipes for ACB-CP are modelled in CAESAR®. Forces and moments at the anchor locations are extracted for all the load combinations. CAESAR® analysis for one loop is shown in figure 3(a). For some of the load combinations forces and moment at anchor locations are given in figure 3(b). The deformation at valve seat location are obtained in order to insure that valve deflection is within the allowable limit (2 mm) generally specified by the valve supplier and stresses in piping layout is within the allowable limits. CAESAR analysis also ensures that there is no clashes are foreseen between process pipes in any load combination.

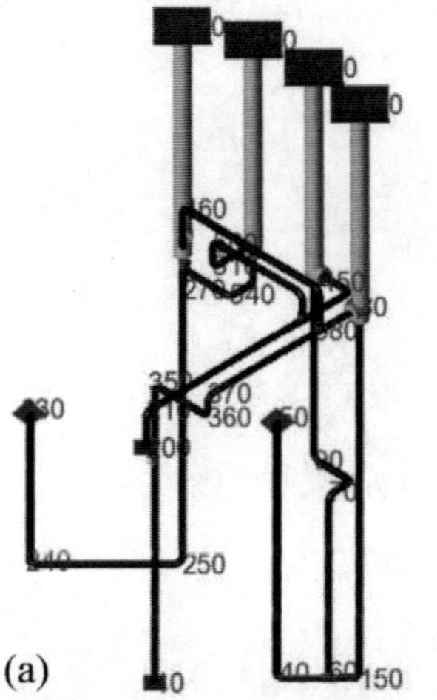

(a)

Node	130		190		330		580	
	NO	NO+ SL-2	NO	NO+ SL-2	NO	NO+ SL-2	NO	NO+ SL-2
Fx [N]	4	26	-108	513	131	622	-36	169
Fy [N]	-6	73	-90	428	-32	148	-11	52
Fz [N]	498	2362	13	62	-21	102	49	234
Mx [Nm]	-1379	1995	37	56	71	102	-131	189
My [Nm]	-270	390	-70	100	32	45	-23	33
Mz [Nm]	-21	38	-284	411	336	486	-123	183

(b)

Figure 3. (a) CAESAR analysis for loop F and (b) forces and moments at nodes for different load combinations.

5.2. ANSYS® analysis results

Forces and moments extracted from CAESAR are taken as input and suitable boundary conditions i.e. temperature, pressure, gravity, fixed support and heat flux etc. are applied while performing analysis in ANSYS® for each load combination. Steady state thermal, static structural, modal analysis and response spectrum analysis are performed for each load case. Analysis is performed considering all the parts as shell element except tie rods, supporting LHe bath, which are considered as solid element. For all the load combinations analysis is performed in ANSYS® [4] and safety margin is derived for each case. For all load combinations NO+SMHV comes out with least safety margin as 0.06, therefore results shown in this paper are for NO+SMHV load combination only. For the NO+SMHV case the equivalent stress obtained in ANSYS® [4] are shown in figure 4. Maximum stress in LHe bath and its support location is 178 MPa as shown in figure 4(a). For the same case maximum deformation in LHe bath support is 47.6 mm as shown in figure 4(b). Maximum deformation in case of NO+SMHV in the OVJ is 1mm as shown in figure 4(c) and equivalent stress in the OVJ is 141 MPa as shown in figure 4(c). Table 1 given below shows the summary of results, category of the load combinations, allowable limits and available safety margin for all the load cases.

IOP Conf. Series: Journal of Physics: Conf. Series **823** (2017) 012043 doi:10.1088/1742-6596/823/1/012043

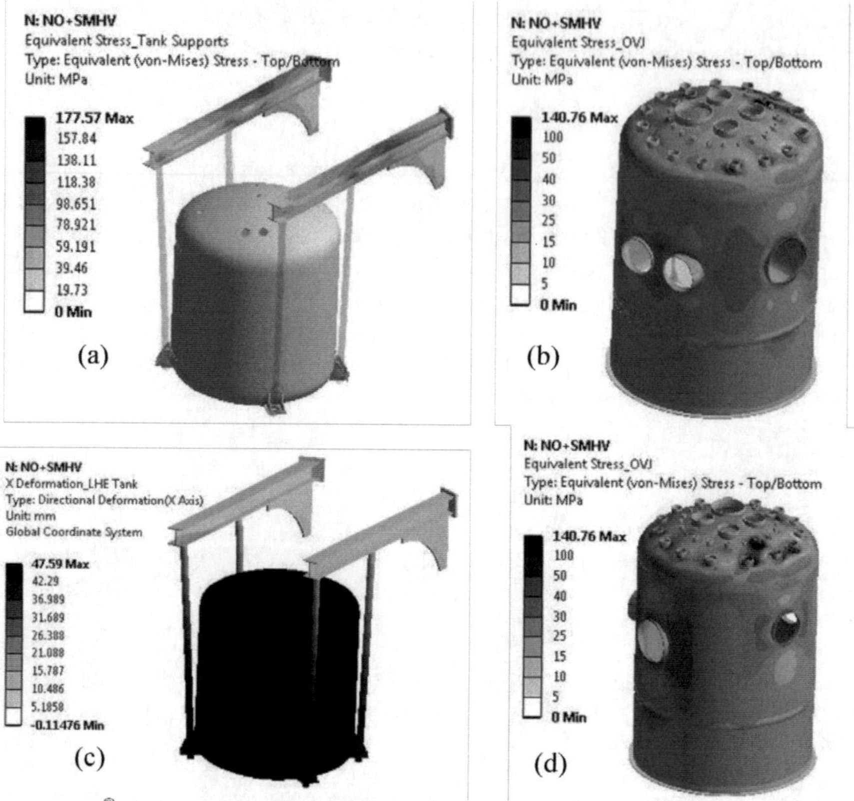

Figure 4. (a) ANSYS® simulation results equivalent stress on LHe support, (b) equivalent stress on OVJ, (c) deformation in LHe support and (d) and deformation in OVJ.

Table 1. Summary of results and available safety margin for all the load cases and their combinations.

Load combinations	Category	Allowable limit [MPa] [7]	Actual stress value [MPa]		Available Safety margin
			Pm	Pm+Pb	
Dead weight (G)	I	Pm<143	40	NA	2.58
TP + TT	I	Pl<172	85	NA	0.68
NP + NT (NO)	I	(Pm or Pl)+Pb <172	84	NA	0.70
NO + ICP	II	Pm<143	89	NA	0.61
TP+TT+SL-1	II	Pl<172	94	NA	0.52
NO+SL-1	II	(Pm or Pl)+Pb <172	92	NA	0.55
TP+TT+SMHV	III	Pm<172	161	175	0.07
BIV-1+SMHV	III	Pl<258	158	177	0.09
NO+SMHV	III	(Pm or Pl)+Pb <258	**163**	**194**	**0.06**
BIV-2+SMHV	III		163	191	0.06
NO + SL-2	IV	Pm<286	214	231	0.34
TP+TT+SL-2	IV	Pl<430	216	229	0.32
BIV-1+SL-2	IV	(Pm or Pl)+Pb <430	224	247	0.28
BIV-2+SL-2	IV		210	245	0.36

IOP Conf. Series: Journal of Physics: Conf. Series **823** (2017) 012043 doi:10.1088/1742-6596/823/1/012043

6. Conclusion

Present study encapsulates all the loads which are to be borne or likely to be borne by the ACB-CP, and thus establishes a well-defined methodology for the repeated analysis of ACB-CP or analysis for other CD boxes. The thermo-mechanical design and analysis for one representative ACB has revealed the load combinations which are really design drivers, e.g. NO+SMHV. During the design iterations of the ACBs concentration on the design drivers reduces the analysis efforts for the other CD boxes.

Acknowledgement

The authors are indebted to Mr. Gopal, Mr. Santosh and Mr. Vikalp for their contribution towards completion of this study. This study is conducted with the support of all cryogenic group members from ITER-India.

Disclaimer

The views and opinions expressed herein do not necessarily reflect those of the ITER organization.

References

[1] Serio L et al. 2011 *Proceedings of ICEC 23-ICMC* 833
[2] Internal technical report 2011 *Conceptual Design of ACBCP-II-ODA2EC7-v_1_1*
[3] Dassault Systems, 2012. CATIA® Version 5, Vélizy-Villacoublay, France
[4] ANSYS® Version 15, Canonsburg, PA, USA
[5] ITER Document, Loads Specification for ITER CD system-II-IWSX43Q-v_1_0
[6] Tender Design-Tokamak Complex 2012, Generation of the Floor Response Spectra ITER_D_3Z5N6P v1.3
[7] ITER Document, Allowable values and limits in service l_3G3SYJ_v3_1

IOP Conf. Series: Journal of Physics: Conf. Series **823** (2017) 012044 doi:10.1088/1742-6596/823/1/012044

Safety and Environment aspects of Tokamak- type Fusion Power Reactor- An Overview

Bharat Doshi[1] and D. Chenna Reddy

[1]Institute for Plasma Research, Near Indira Bridge, Bhat, Gandhinagar 382 428, India

E-mail: doshi@ipr.res.in

Abstract. Naturally occurring thermonuclear fusion reaction (of light atoms to form a heavier nucleus) in the sun and every star in the universe, releases incredible amounts of energy. Demonstrating the controlled and sustained reaction of deuterium-tritium plasma should enable the development of fusion as an energy source here on Earth. The promising fusion power reactors could be operated on the deuterium-tritium fuel cycle with fuel self-sufficiency. The potential impact of fusion power on the environment and the possible risks associated with operating large-scale fusion power plants is being studied by different countries. The results show that fusion can be a very safe and sustainable energy source. A fusion power plant possesses not only intrinsic advantages with respect to safety compared to other sources of energy, but also a negligible long term impact on the environment provided certain precautions are taken in its design. One of the important considerations is in the selection of low activation structural materials for reactor vessel. Selection of the materials for first wall and breeding blanket components is also important from safety issues. It is possible to fully benefit from the advantages of fusion energy if safety and environmental concerns are taken into account when considering the conceptual studies of a reactor design. The significant safety hazards are due to the tritium inventory and energetic neutron fluence induced activity in the reactor vessel, first wall components, blanket system etc. The potential of release of radioactivity under operational and accident conditions needs attention while designing the fusion reactor. Appropriate safety analysis for the quantification of the risk shall be done following different methods such as FFMEA (Functional Failure Modes and Effects Analysis) and HAZOP (Hazards and operability). Level of safety and safety classification such as nuclear safety and non-nuclear safety is very important for the FPR (Fusion Power Reactor). This paper describes an overview of safety and environmental merits of fusion power reactor, issues and design considerations and need for R&D on safety and environmental aspects of Tokamak type fusion reactor.

1. Introduction

Achieving controlled fusion could provide civilisation with a safe, sustainable, emission free and low cost energy source. Our ability to transition from fossil fuels to renewable sources of energy will likely determine the fate of the planet. Many countries are progressing towards the goal of using solar, wind, hydro (and possible future nuclear fusion) power. For example, China decided that solar and wind power would account for 20 % of China's total energy production by 2030. Denmark, which aims to completely eliminate its use of fossil fuels by 2050 and will rely on its cutting-edge wind power industry. Germany has focused on solar and wind power in its push to remake its electricity system, Brazil now derives more than 75 % of its electricity from hydro-power sources, Similarly India is also focusing on solar energy and energy from Thorium for securing India's sustainable energy future.

India is targeting to ensure that a significant fraction (say 50 %) of electricity supply coming from non-fossil sources like renewable (hydro included) and nuclear energy by the year 2035. To ensure long term energy resource security, our effort to develop fusion energy is also very crucial. An attractive fusion power plant will need to have major safety and environmental advantages, and prove to be economically viable compared with other sources of electrical energy, to meet the needs of the second half of this century and future. A swift transition from fission to fusion not only would allow us to escape the worst medium- to long-term environmental and social ramifications of climate change, it also would enable the creation of a more stable and credible equilibrium in a world with no nuclear weapons. A world without nuclear weapons but with fissile material will always be in fragile equilibrium; a world without both would be far more sustainable. First-generation tokamak type fusion reactors generating electricity for the grid will use the heavy hydrogen isotopes of deuterium and tritium as fuels; deuterium is abundant in nature and stable, while tritium is extremely rare and radioactive, with a half-life of 12.6 years.

2. Safety and Environment merits of Fusion Power Reactor

Fusion power will not create greenhouse gases, produce other harmful pollutants or result in long-lasting radioactive waste. A reactor using nuclear fusion to generate electricity is intrinsically safe: First, a runaway nuclear chain reaction cannot take place, under any circumstances. Second, no long-lived, highly radioactive products are created. Third, of those magnetic confinement fusion reactors that will require radioactive fuels such as tritium, both the radioactive fuel requirements and fuel half-life are orders of magnitude lower than their fission counterparts. Its fuel consumption will be extremely low. A 1000 megawatt electric fusion power station would consume ~100 kg of deuterium and three tonnes of lithium a year to generate 7 billion kilowatt-hours of power. The neutrons generated by DT fusion reaction will interact with the materials close to the reactor, but careful choice of these materials will ensure that no long-term legacy of radioactive waste is produced by fusion power. It is difficult to get the fusion reaction going in the first place, however it can be quickly stopped by eliminating the injection of fuel. After engineers learn how to control the first generation of fusion plasmas, from deuterium and tritium fuels, advanced second- or third-generation fuels could reduce radioactivity by orders of magnitude as shown in table 1. Radioactivity Levels in Fusion Power Plants are very low and decay rapidly after Shutdown. The fusion energy may have advantages of competitive cost of electricity (5 c/kWh); Steady-state operation; Low level waste; Public & worker safety and High availability [1].

The volume of gas in a fusion reactor will always be low. Any problem will always cool the plasma and stop reactions - so a runaway situation is impossible. Also the raw fuels for the reactor (deuterium and lithium) are not radioactive. Tritium is mildly radioactive having a short half-life of 12.6 years, but will be produced and used within the reactor. Consequently, no transport of radioactive fuels will be needed for a fusion power plant - and even the worst possible case accidents would not require evacuation of neighbouring populations. The inherent safety characteristics of a fusion reactor are due to the very low fuel inventory in the reactor during operation and to the rapid cooling that extinguishes the fusion reactions should a malfunction occur.

The main benefits of fusion are; there is an abundance of fuel in nature: deuterium is extracted from seawater and reserves are estimated at several millions of years; tritium is produced from lithium, which is highly abundant in the Earth's crust and in the oceans.

The safety of Tokamak type fusion reactor operations includes intrinsic termination of the fusion reaction (hence no runaway reactions) in case of power transient, excessive fuel supply ,loss of fuel supply, excessive heating, uncontrolled perturbation of the plasma (e.g. by ingress of impurities), low quantity of fuel in the plasma and hence the energy of the fusion reaction, low energy density and large surfaces allowing heat transfer and use of significant heat sink masses, vacuum-tight chambers providing sealed material confinement systems, absence of high-level waste or high residual energy; the deuterium-tritium reaction produces 14.1 MeV neutrons whose interaction with the materials

IOP Conf. Series: Journal of Physics: Conf. Series **823** (2017) 012044 doi:10.1088/1742-6596/823/1/012044

surrounding the plasma causes the progressive activation of these materials. However, after the plasma reaction is terminated, the residual power decreases rapidly.

Fusion is believed to have favourable safety and environmental characteristics:

- Reactors will be designed for complying with a maximise use of inherent fusion safety characteristics as intrinsic passive shutdown and fail safe termination of plasma.
- The worst accident initiated in a fusion plant could not result in the need for public evacuation from around the site.
- Waste from a fusion power plant would not become a burden on future generations.

3. Issues and Design requirements for S&E aspects of Fusion Power Reactor

Fusion's success as an energy source will depend on how the challenges to build and operate it safely and reliably can be met in a way that makes the cost of fusion electricity economically competitive. The deuterium-tritium reaction produces 14.1 MeV neutrons whose interaction with the materials surrounding the plasma causes the progressive activation of these materials. The large flux of energetic neutrons through the first wall and blanket is an issue. In the fusion reactor the damage per atom would be of the order of 80 dpa instead of 3 dpa (In case of ITER) and the Tritium consumption of about 55 kg for 1GW.year fusion power reactor. The need for developing radiation-resistant, low-activation material such as Ferritic steels, Vanadium alloys, SiC composites are essential. Tritium breeding and feasibility of the DT fusion fuel cycle are one of the main technical problems to be solved for the future fusion reactors. Dose rate in operating accessible zones of a fusion power reactor is a crucial point in calculating the Occupational Radiation Exposure for maintenance staff. It also plays a significant role in radiation shielding design.

Risk of Radioactive Tritium release:

The use of tritium induces a risk of atmospheric release of tritium in gaseous or in oxidised forms due to leaks (desorption or diffusion from the intact equipment under normal operation conditions or from the failed equipment under accidental conditions) into building rooms and then through leaks and through ventilation systems into the environment. Gaseous tritium can be oxidised and there is a risk of suspension of tritiated water aerosols in the building rooms or atmospheric release of tritiated water vapour in the rooms. Tritium contamination of water in the primary cooling systems of Tokamak in-vessel components is caused by the diffusion of tritium accumulating in the in-vessel components during plasma operation. The production of tritium by water activation is negligible in comparison with the diffusion phenomenon. Tritium is a radioelement emitting low-energy β radiation (less than 20 keV). Its specific activity is approximately 3.7×10^{14} Bq/g. The various form of tritium is gaseous tritium HT, DT or T_2. The oxidised form is tritiated water, HTO, DTO or T_2O.

The key safety requirements could be; the avoidance of sheltering or evacuation in the event of an accident; all waste generated in the fusion plant must be able to be disposed of as low level waste.

Specifically, concerning safety and environment, the following points needs to be addressed as a design requirement:

(1) It must be clearly shown that the worst possible fusion accident will constitute no major hazard to populations outside the plant perimeter that might result in evacuation.

(2) Radioactive wastes from the operation of a fusion plant should not require isolation from the environment for a geological time span and therefore should not constitute a burden for future generations.

(3) The worst possible accident would not be able to breach the confinement barriers. Even when a hypothesis is done that confinement barriers shall be breached, any accidental radioactive release from a fusion power station in this case cannot reach the level that would require the evacuation of the local community.

IOP Conf. Series: Journal of Physics: Conf. Series **823** (2017) 012044 doi:10.1088/1742-6596/823/1/012044

4. ITER an experimental Fusion Reactor-Example of S&E design

The main objective of ITER is to demonstrate the scientific and technical feasibility of a controlled fusion reaction capable of the production of approximately 500 MW of fusion power for durations of several hundred seconds. ITER is an experimental facility as shown in figure 1, whose purpose is to demonstrate the control of plasma and the fusion reaction using deuterium (D) and tritium (T) as fuel. It will allow the exploration of plasma scenarios and technological testing essential for the preparation of a future fusion reactor. Hence the goal of ITER during its nominal 20 years operating period is to:

- Achieve fusion plasmas with a ratio Q of fusion power produced to external power supplied (Q-factor ratio) ≥ 10 between 300 to 500 s duration.
- Test the feasibility of controlled ignition.
- Demonstrate the availability and integration of technologies required for a fusion power reactor (e.g. in-vessel components, superconducting magnets, remote maintenance systems, tritium production and fuel cycle).

ITER is an indispensable step for the pursuit of the program of development of fusion energy, with the final objective being to achieve a fusion power reactor. The main goal of ITER is to demonstrate the safety and environmental potential of fusion and thereby provide a good precedent for the safety of future fusion power reactors. Key aspects of the safety of ITER are effluents and emissions during normal operation, occupational safety of workers at the site, proper storage and treatment of radioactive materials generated during operation and decommissioning, and potential accidents and incidents. To ensure the safety of ITER, top-level safety objectives have been defined;

- General safety: to protect individuals, society and the environment; to ensure in normal operation that exposure to hazards within the premises and due to release of hazardous material from the premises is controlled, kept below prescribed limits and minimised; to prevent accidents with high confidence, to ensure that the consequences of more frequent events, if any, are minor; to ensure that the consequences of accidents are bounded and the likelihood is small.

Figure 1. 3D view of ITER

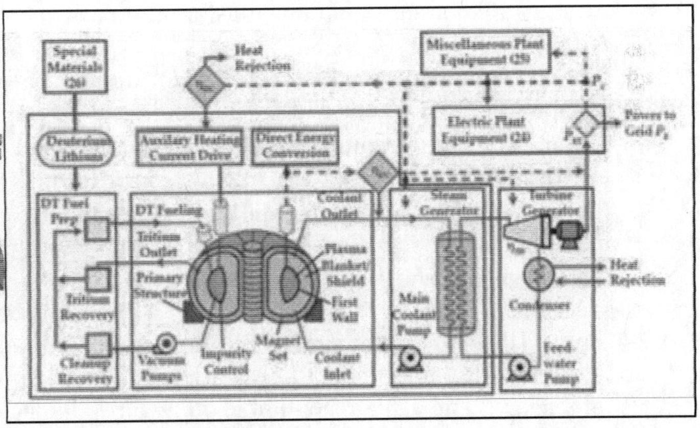

Figure 2. Schematic of a typical Fusion power Reactor

- No evacuation: to demonstrate that the favourable safety characteristics of fusion and appropriate safety approaches limit the hazards from internal accidents such that there is technical justification for not needing evacuation of the public.
- Waste reduction: to reduce radioactive waste hazards and volumes.

These objectives have been defined based on internationally recognized safety criteria and radiological limits following ICRP and IAEA recommendations, and in particular on the concept of defence in depth and the As Low As Reasonably Achievable (ALARA) principle.

IOP Conf. Series: Journal of Physics: Conf. Series **823** (2017) 012044 doi:10.1088/1742-6596/823/1/012044

Comprehensive safety and environmental assessments of the ITER design and operation have been performed under site specific assessments required for the licensing process in France. The result of the study has been reported in the Preliminary Safety Analysis Report.
ITER will be a precedent for future fusion licensing. Safety functions that have been defined for ITER differ slightly from fission approach since "limitation of exposure for workers and environment" is also considered as a safety function in the scope of Cadarache site licensing [2].
The potential nuclear risks induced by the DT fusion reaction in the ITER facility are following [3]:

- Risk of radioactive materials release (risk potentially impacting personnel, the public and the environment).
- Risk of external exposure (mainly concerning personnel).

The following radioactive materials are identified in the ITER facility:

- Tritium, inherent to deuterium-tritium fusion reactions and also produced by nuclear reactions in beryllium and in the lithium contained in materials used in TBM.
- Activated products that are mostly produced and located in the in-vessel components. Smaller quantities are also produced and located in the components outside the vacuum vessel or in the cooling systems of the in-vessel components.

With respect to the radiological risks identified, the following two safety functions have been adopted for the ITER installation:

- Confinement of radioactive material.
- Limitation of external exposure to ionizing radiation.

The risk of exposure to ionizing radiation potentially affects installation personnel. This risk is associated with neutrons, produced during fusion reaction and emitted from the plasma, the γ radiation emitted by activated products, X-rays emitted by some heating and current drive generators and the β radiation emitted by tritium.
Safety considerations for the ITER:

- Analysis of normal operation illustrating the measures to optimize the design, based on the ALARA principle.
- Preliminary safety studies relating to design basis accidents and beyond design basis situations.
- Preliminary safety studies relating to internal and external hazards and structural design justification with regard to the hazards considered.
- Preliminary assessment of radiological impact on workers and the environment in normal, incident and accident situations.
- Internal emergency plan design for ITER facility.

5. S&E approach in design of future Fusion Power Reactor

A fusion power plant possesses not only intrinsic advantages with respect to safety compared to other sources of energy, but also a negligible long term impact on the environment provided certain precautions are taken in its design and in the selection of structural materials [4]. Safety consists of taking measures to avert any hazards that could jeopardise a facility. In particular, this means preventing accidents and if they occur minimising their consequences. In the genesis of the idea of fusion reactors, the notion of safety appeared very early on in the literature. Thus, at the IAEA conference in 1971, the first TOKAMAK-type fusion power plant as well as a first safety assessment was presented [5]. The 1989 ESECOM Report, for example, drew up a first comparison of various possible reactor options with regard to safety [6]. A few years later, the Colombo Report stressed that fusion potentially possesses 'inherent environmental and safety advantages over all current alternatives for base load electricity generation' [7]. Through careful design, only a small fraction of neutrons are absorbed in structure and induce radioactivity. Low-activation material will generate low Rad-waste. Rad-waste generated in DT fusion is similar to advanced fuels (D-^{3}He). For liquid coolant/breeders (e.g. Li, LiPb), most of fusion energy (carried by neutrons) is directly deposited in

the coolant simplifying energy recovery. The safety relevance characteristics of D-T and advanced fusion fuels is shown in table 1.
The safety approach shall respect two major important points:

(1) The worst possible fusion accident will constitute no major hazard to populations outside the plant perimeter that might result in evacuation.
(2) Radioactive waste from the operation of a fusion plant should not constitute a burden for future generations.

The licensing process for Fusion Power Reactor (FPRs) may well be regulated with common administrative procedures and supportive licensing documentation as being done for Fission Reactors. From the owner's point of view, a FPR could be contracted as turnkey plant and the FPR supplier would be responsible for obtaining Design Approval once compliance with the country's legislation had been substantiated, and the owner would be responsible for the operating permit once the site had been selected.

Table 1. Comparison of characteristics of D-T and advanced fusion fuels with relevance to safety [8, 9].

Characteristic	D-T	D-D	D-^{3}He	D-^{6}Li	p-^{11}B
Main reaction products and energy yield (MeV)	n + ^{4}He + 17.6	p + T + 4.0 n + ^{3}He + 3.3	p + ^{4}He +18.3	2 ^{4}He + 22.4 p + ^{7}Li + 6.0 n + ^{7}Be + 3.4 p + T + ^{4}He + 2.6 n + ^{3}He + ^{4}He + 1.8	3 ^{4}He+8.7
Energy yield per reaction pair (MeV)	17.6	>3.65[a]	<18.3[b]	Not quantified	8.7
Fraction of energy carried by neutrons (%)	80	35	Small	Not quantified	None
Neutron energy (MeV)	14.0	2.45(14.1)[a]	2.45(14.1)[b]	2(14.1)	-
Ratio of activation in structure (%)	100	20 to 60	25	20 to 60	-
Radioactive fuel or ash	T	T	T	T, ^{7}Be	-
Total inventory of Tritium (kg)	10	0.1	0.01	0.001	-
Special fuel cycle requirements	Li blanket	None	Production fusion reactors for ^{3}He	None	Feasibility for ignition is doubtful

[a] The energy limit is due to contributions from D-T side reactions.
[b] contributions from D-D and D-T side reactions.

IOP Conf. Series: Journal of Physics: Conf. Series **823** (2017) 012044 doi:10.1088/1742-6596/823/1/012044

The radioactive material in a fusion reactor includes tritium gas, burned as a fuel, and activation products produced by high energy neutrons from the fusion reactions. Ensuring that these materials will not affect the environment and public, requires a strategy to minimize inventories, develop adequate containment and control, and eliminate potential release mechanisms. The accident with the greatest potential for a large radioactive release is a lithium fire. Less active forms of lithium, under consideration for use in fusion reactors, would eliminate this concern. Potential energy releases from large magnet systems, and the health effects of long-term exposure to magnetic fields are also concerns.

The Safety and Environment (S&E) objectives set by various study of conceptual designs of Fusion Power Reactors by EU, US and Japan are as follows[10]:

European Safety and Environmental Assessment of Fusion Power (SEAFP) looked at conceptual designs of fusion power stations and their safety and environmental assessments and set the S&E objectives:

- It must be clearly shown that the worst possible fusion accident will constitute no major hazard to populations outside the plant perimeter that might result in evacuation.
- Radioactive wastes from the operation of a fusion plant should not require isolation from the environment for a geological time span and therefore should not constitute a burden for future generations.

The US fusion S&E objectives:

- The avoidance of sheltering or evacuation in the event of an accident; all waste generated in the fusion plant must be able to be disposed of as low level waste (e.g. Class C by US rules).
- The facility must demonstrate that day to day activity in the local community is not disturbed and that workers must not be exposed to risks greater than from other power plants.

The Japanese fusion S&E objectives:

There is no official recommendation for the safety and environmental objectives for future fusion power plants. However, it is believed that for a fusion power plant to obtain social acceptability it is important to significantly improve environmental and safety aspects over those of fission power plants. IAEA public information sheets summarises how to promote safety in the nuclear installations [11], which perfectly applies to the future fusion reactors: safety analysis is conducted on the whole range of plant situations: normal operation, anticipated operational occurrences and possible accidents. By examining all these situations in detail, the robustness of the plant design and the effectiveness of the safety systems have to be demonstrated. A safely designed nuclear power plant is one that ensures basic functions at all times, even in an accident situation. All possible accident scenarios have to be taken into account at a very early stage in the design process. Appropriate safety analysis for the quantification of the risk shall be done following different methods such as FFMEA (Functional Failure Modes and Effects Analysis) and HAZOP (Hazards and operability). Safety analysis shall include systematic identification and ranking of the potential accident sequences within design basis and beyond design basis.

6. Safety classifications and its level

The safety classification for SSCs of FPR based on their functional boundary, like nuclear safety function and non-nuclear safety functions etc. is very important. Safety level shall be defined based on the radiation releases and shall be classified as class II, III or N2 and N3 as per ASN and ESPN respectively. Seismic categorization of SSCs of FPR is crucial for radiation releases. In case of FPR tritium release is more crucial. A set of top level system requirements and goals for system economics, safety and waste disposal, and reliability and availability needs to be established for demonstration and commercial fusion power plants. For all the possible version of future reactors it should be considered that a fusion plant is not only a fusion-machine reactor and the same importance should be paid to all

IOP Conf. Series: Journal of Physics: Conf. Series **823** (2017) 012044 doi:10.1088/1742-6596/823/1/012044

the nuclear buildings and to their integration and interfaces. The schematic of typical tokamak type fusion reactor is shown in figure 2.

7. Conclusion

Fusion power possesses numerous advantages with which to meet the challenge of safety and protection of the environment in the future. It appears possible to ensure negligible radioactive releases whatever type of accident is postulated, as well as a very small volume of wastes requiring permanent disposal. At the end of a fusion power station's working life the radio toxicity in the reactor chamber and other structural and waste materials will decay rapidly. In less than 100 years the residual activity of these materials would be less than the radio toxicity found in the waste from a conventional coal-fired power station. Fusion power will not burden society with a long-term toxic waste issue. However, to meet these objectives, this concern has to be taken into account in the early design stage of the reactor: the choice of material and appropriate safety systems. In addition, specific R&D needs to be performed in order to confirm the safety assessment and to develop efficient and reliable safety systems.

References

[1] ARIES Web Site: http:/aries.ucsd.edu/ARIES
[2] Rodríguez-Rodrigo L. et al.2006 *Design* **81** 1391
[3] ITER Preliminary Safety Report (RPrS)
[4] Marbach G, Cook I 1999 *Fusion Engineering and Design* **46** 243
[5] Steiner D, Emergency cooling and Radioactive-Waste-Disposal requirements for fusion reactors. *Report IAEA-CN-28: K-11*
[6] Holdren J P, Report of the Senior Committee on Environmental, Safety, and Economic Aspects of Magnetic Fusion Energy. *LLNL Report UCRL-53766, September1989*
[7] Colombo U, Evaluation of the Community Fusion Programme (1984–1990). *Research evaluation-Report No. 45, EUR 13104 EN, July 1990*
[8] Canon J B, Ed., Background Information and Tech. Basis for Assessment of Environmental Implications of Magnetic Fusion Energy, Illinois Univ., Urbana, *Rep. DOE/ER-0170 (1983)*
[9] Gordon J D, Samec T K, Freije S A, Hau B I, Viability of the B^{11} (p, α) 2α cycle, *Nuclear Technology /Fusion 42 pt.2 (1983)348*
[10] Gulden W et al. 2000 *Fusion Engineering and Design* **51–52** 419
[11] INTERNATIONAL ATOMIC ENERGY AGENCY Information Series, Division of Public Information 02-02467 / FS Series 2/06/E.

IOP Conf. Series: Journal of Physics: Conf. Series **823** (2017) 012045 doi:10.1088/1742-6596/823/1/012045

Indigenously developed bending strain setup for I-V characterization of superconducting tapes and wires

Arun Panchal[1], Anees Bano[1], Mahesh Ghate[1], Piyush Raj[1] and Subrata Pradhan[1]

[1]Institute for Plasma Research, Near Indira Bridge, Bhat, Gandhinagar 382 428, India

E-mail: arunpanchal@ipr.res.in

Abstract: An indigenously developed bending strain setup to examine the effect of pure bending on critical current of superconducting tapes and strands has been presented in this paper. This set up is capable of applying various bending radius in situ at cryogenic temperature with rack and pinion gear mechanism. The bending strain applied on samples can be controlled externally by rotational input which is transferred in the form of bending radius during experiments. The working principle, design and optimization of this set up have been discussed. The performance and validation of this setup has been done on various HTS tapes and copper strands at 77 K in actual experimental facility. Effect of bending radius (15.5 mm - 48 mm) i.e. strains and ramp rate (2 A/s – 8 A/s) is observed on current capability of various HTS Tapes. It is observed that in uniform bending condition, degradation in current carrying capacity BSCCO and Di-BSCCO (~ 30 %) is more as compare to YBCO (~ 2.75 %) at 77 K. The effect of pure mechanical strain has been experimentally observed and presented.

1. Introduction

Strain developed in the tape or wire due to bending when external load is applied on it axially is known as bending strain. This bending strain affects the current carrying capability of superconducting tapes and wires. Fabrication of superconducting magnets involves bending and winding of superconductors in desired configuration. It is critical to estimate applied bending strain during fabrication of superconducting magnet and keep it under critical limit. If bending strain exceeds its critical value, the current carrying capability of superconductor degrades [1-3]. Therefore, it is necessary to measure critical-current of individual wire at various bending to estimate minimum bending radius of wire during magnet fabrication. It is difficult to measure the current-voltage (I-V) characteristic of superconducting tapes and wires with different radius in-situ at cryogenic temperature during experiments. With this motivation, an indigenous setup is developed with the help of rack and pinion mechanism. The required bending radius to introduce desired strain was provided in-situ by the rotational motion of the pinion at different angles during experiments. A rack and pinion is a type of linear actuator that comprises a pair of gears which convert rotational motion into linear motion. A circular gear called “the pinion” engages teeth on a linear “gear” bar called “the rack”; rotational motion applied to the pinion causes the rack to move, thereby translating the rotational motion of the pinion into the linear motion of the rack. On completing one rotation (360°) of pinion, the rack travels by 42.4 mm in linear direction in this setup. The bending strain setup has been optimized and validated for its performance with number of trials at room temperature. The low temperature performance has been validated in LN_2 bath. Details of design, fabrication and integration of bending strain set up in experimental facility have been discussed in this paper. Effect of bending strain on critical current of

IOP Conf. Series: Journal of Physics: Conf. Series **823** (2017) 012045 doi:10.1088/1742-6596/823/1/012045

YBCO, BSCCO and Di-BSCCO has been measured in this setup. The sample length of tape for these measurements has been taken ~ 200 mm in which 10 mm length is used for gripping on each side. Voltage is measured between 40 mm at center of sample. During experiments, it is observed that with increasing bending radius of tape, its current carrying capability gradually decreased. The effects of bending strain on the current carrying capability of various HTS tapes have been presented in this paper.

2. Design of Bending Strain Setup

The bending setup consists of rack and pinion mechanism to translate rotational movement into linear movement which results in desired bending of superconductors. Following sections describes the components and design calculations for this set up.

2.1. Critical Parameters & Gear Terminology

Following parameters has been selected on the basis of availability of space inside the cryostat and structural requirements. Table 1 and figure 1 show important design parameters for rack & pinion setup.

Outside diameter (D_o) = 15 mm, Width of Rack and Pinion (b) = 10 mm, Number of teeth (N) = 18

Table 1. Important parameters of rack and pinion arrangement.

Parameter	Symbol	Formula	Value (mm)
Diametric Pitch	P_d	$N+2/D_o$	1.33
Module	m	P_d/π	0.42
Circular Pitch	P	π/P_d	2.36
Pitch Dia.	D	N/P_d	13.5
Addendum	a	$1/P_d$	0.75
Dedendum	b	$1.157/P_d$	0.87
Circular thickness	t	0.5 x P	1.18
Clearance	C	$0.157/P_d$	0.2
Working depth	hk	$2/P_d$	1.5

2.2. Diameter for Connecting Rod

Connecting rod has been used to give a rotational motion to pinion. One end of the connecting rod is attached with the pinion and the other end with the hand wheel which is provided at the top of the cryostat of self-field facility. Connecting rod transferred the motion to pinion on applying certain rotation to hand wheel. It has been designed to apply maximum 500 N forces on the basis of typical mechanical properties of superconducting strands and HTS tape.

$F = 500$ N, $r = 37.5$ mm

$$\Rightarrow \quad T = r \times F = 18750 \text{ N.mm [4,5]} \tag{1}$$

$\sigma = 515$ N/mm^2

$$\Rightarrow \quad \tau = 0.577 \times \text{UTS}(\sigma) = 297.155 \text{ N/mm}^2 \tag{2}$$

The selection of dimensions for connecting rod is done by following calculation,

$$T = \tau \times \pi D^3/16 \text{ [4,5]} \tag{3}$$

$$\Rightarrow \quad D^3 = 16 \times T / \tau \times \pi$$

$$\Rightarrow \quad D = 6.85 \text{ mm} \approx 7 \text{ mm}$$

where,

F = Max. Force apply on the rod in N, D = SS 304 Rod diameter in mm

r = Radius of wheel in mm, τ = Shear stress of SS304 in N/mm^2,

IOP Conf. Series: Journal of Physics: Conf. Series **823** (2017) 012045 doi:10.1088/1742-6596/823/1/012045

σ = Ultimate tensile stress of SS 304 in N/mm^2, T = Required torque to transmit force in N.mm,

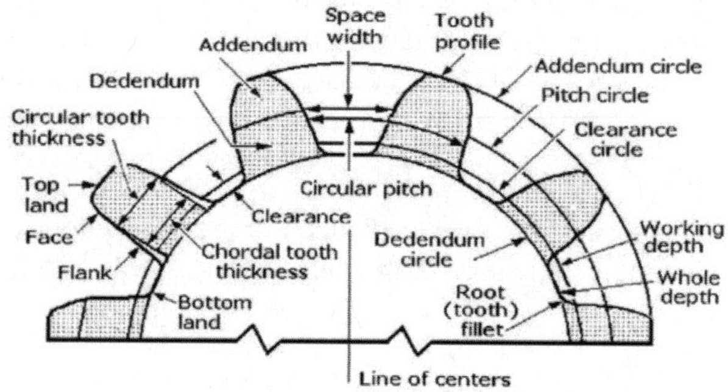

Figure1. Critical parameters and its terminology for gear.

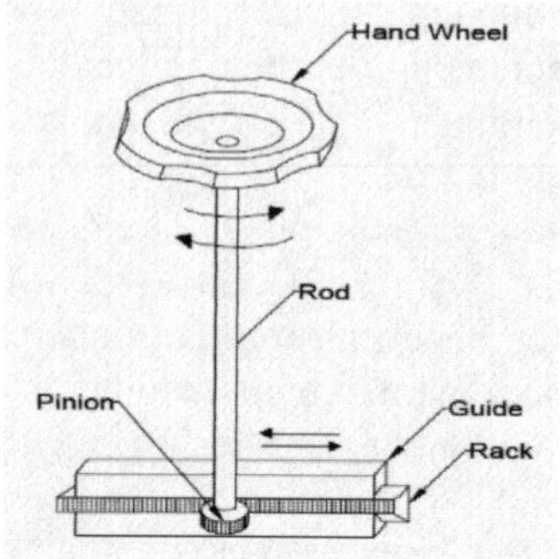

Figure 2. Schematic of rack and pinion.

2.3. Linear Displacement with respect to Angular Rotations
The linear movement of the rack is determined by the rotational motion of the pinion which is provided by hand wheels on the top of the setup as shown in figure 2. For one complete rotation (360°) of pinion, the rack will be displaced by 42.39 mm in linear direction. Similarly, for 10° of rotation, rack will be displaced by 1.178 mm in linear direction.

Pitch diameter of pinion D = 13.5 mm, displacement of rack at 360° rotation = π × D = 42.39 mm , similarly at 10° rotation, 42.4 × 10 / 360 = 1.178 mm

3. Test Setup for Demonstration
Prototype setup has been prepared for trials and validation of the rack and pinion arrangements before its installation inside self-field facility. It is prepared as per the dimensions of the actual experimental setup on the hylum base sheet to validate it's working as per requirement. It is validated for its required performance such as desired bending radius and estimated linear motion at room temperature. It is also observed there is no change at LN_2 temperature at smaller angles but at large rotation angle, there is an error of approximately 1% in linear displacement of rack.

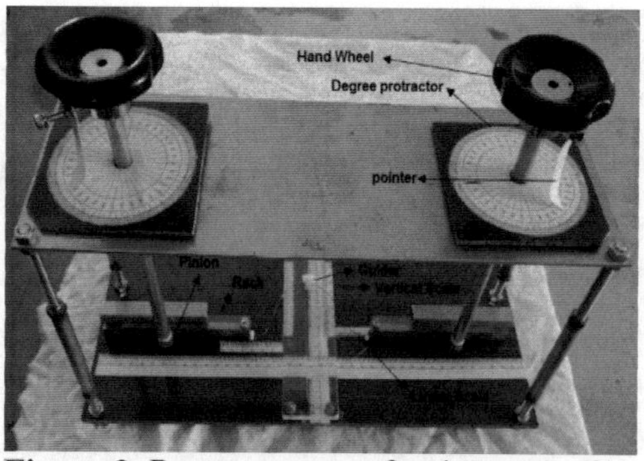

Figure 3. Prototype setup for demonstration

As shown in figure 3, linear scales have been used for calibration on the surface of the hylum sheet. The circular protractors have been used in the setup to estimate linear movement with respect to angle of rotation of pinion. Teflon bush and transparent acrylic sheet is used in the center of the two racks to provide smooth motion and direction to wires and tapes. The number trials have been done on copper wires and HTS tape for validation of this setup.

4. Experimental Setup

The experimental arrangement of strain setup inside cryostat is shown in figure 4. Rotation of hand wheels at required angle provides linear movement to rack. It results in bending of the tape or wire which is fixed at the end of the rack. The number of trials has been taken toverify if there is any error in linear movement of rack with respect to required rotation angle compared to calculated values. The critical current measurement for the HTS tapes such as YBCO, BSCCO and Di-BSCCO has been done starting from 60° at interval of 60° rotation up to 480° rotation of the pinion. Length of the HTS sample tapes is 180 mm. Voltage taps has been kept at 20 mm apart the center of the tape (distance between two voltages taps is 40 mm). Table 2 shows various bending radius of YBCO tape with respect to angle of rotation. Typically, it is observed that percentage bending strain in YBCO tape increases as bending radius decreases. As shown in table 2, at highest bending radius of 48.1 mm, typical strain is 0.33 %.While at the lowest bending radius of 15.7 mm, typical bending strain is 1.0 %.

Table 2. Typical bending radius and strain in tape

Rotation (degree)	Linear displacement (theoretical) (mm)	Linear displacement (avg. observed) (mm)	Bending Radius (mm)	Typical Strain (%)
60	7.0	7.25	48.1	0.33
120	14.1	14.5	36.3	0.44
180	21.2	21.5	30.5	0.48
240	28.2	28.25	27.2	0.56
300	35.3	35.0	25	0.64
360	42.4	42.0	19.5	0.82
420	49.4	49.5	17.4	0.92
480	56.5	56.25	15.7	1.0

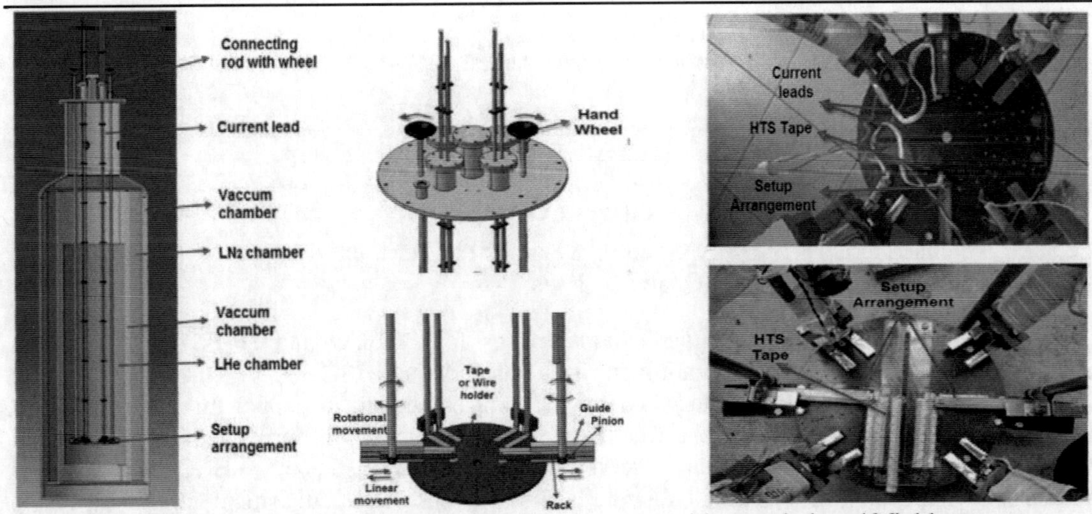

Figure 4. Assembly and detail view of bending strain setup with sample in self-field cryostat

5. Results and Discussion

Experimental measurements have been carried out on various HTS tapes in LN_2 Bath. The current-voltage characteristics of HTS tapes were studied at various bending radius for different ramp rate (2 A/s, 4 A/s, 6 A/s and 8 A/s). Figure 5(a), (b) and (c) shows I-V charectristic plots for BSCCO, Di-BSCCO and YBCO tape respectively at different strain for ramp rate of 8 A/s. Maximum current carried by the tape for various strain has been mentioned the bracket as shown in figure 5.

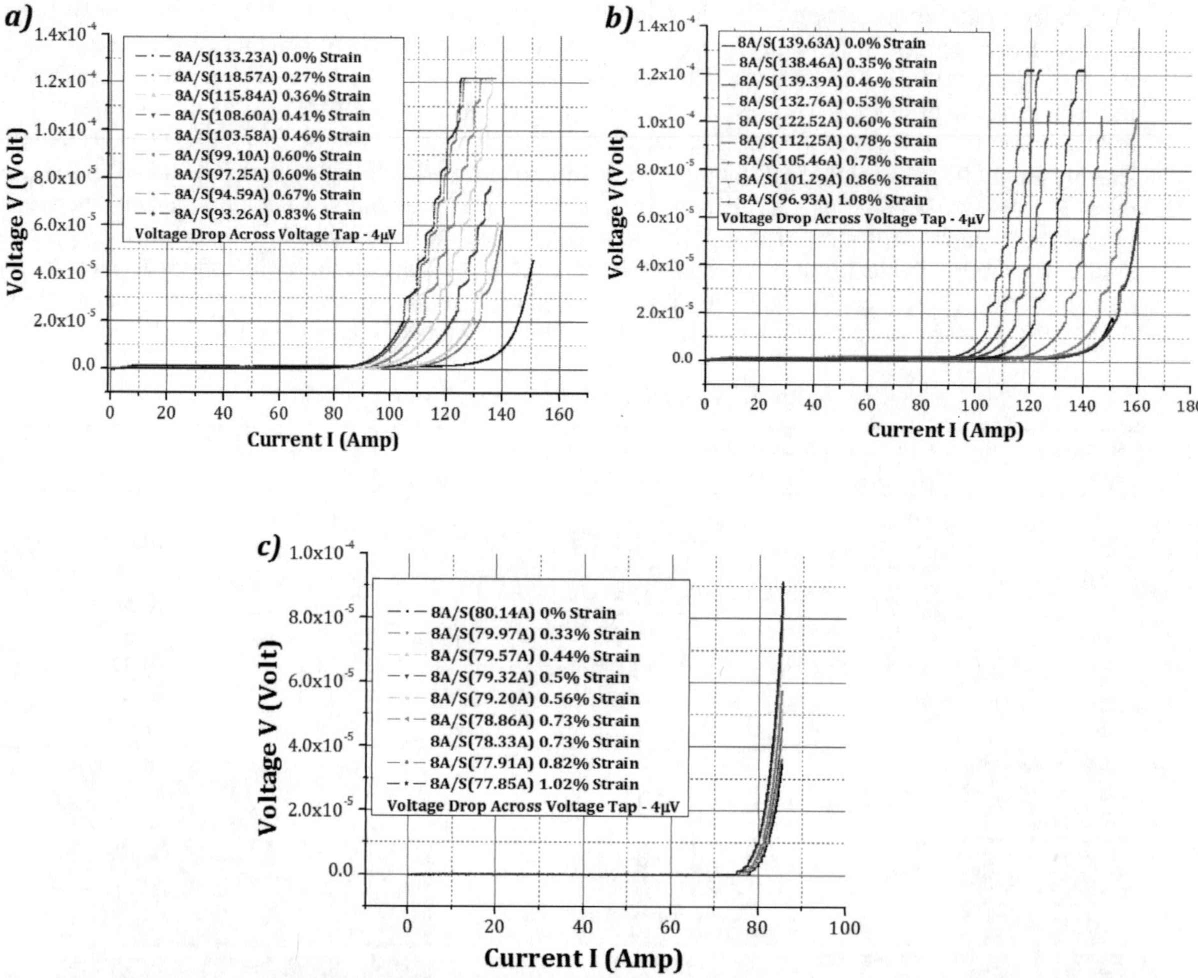

Figure 5. (a) I-V characteristics of BSCCO tape@ 8 A/s, (b) I-V characteristics of Di-BSCCO tape @ 8 A/s and (c) I-V characteristics of YBCO tape @ 8 A/s.

The steps observed during current-voltage characteristics for BSCCO and Di-BSCCO is due to material characteristics and method of manufacturing for HTS tapes. BSCCO and Di-BSCCO used in these experiments are of multi-filamentary types which are embedded in copper matrix. In BSCCO and Di-BSCCO interaction between those filaments results in fluctuation during current charging in experiments. Negligible steps has been observed in I-V characteristics of YBCO tape as it is manufactured by depositing thin film on substrate layers. It is observed that, critical current capability of BSCCO tape without any strain is 133 A which is reduced to 93 A with 0.83 % strain at ramp rate of 8 A/s. The critical current capability of Di-BSCCO tape without any strain is 139.6 A which reduced to 97 A with 1 % strain at ramp rate of 8 A/s. The critical current capability of YBCO tape without any strain is 80 A which is reduced to 77.8 A with 1 % strain at ramp rate of 8 A/s. In mentioned conditions, it is observed that in uniform bending condition, the degradation in current carrying capacity BSCOO and Di-BSCCO (~ 30 %) is more as compare to YBCO (~ 2.75 %) at 77 K.

6. Summary

An indigenous strain setup to provide in-situ bending radius at cryogenic temperature has been developed. The performance validation and its calibration have been established by number of trials at room temperature as well as in LN_2 bath. It has been successfully applied to measure the effect of bending strain on various HTS tapes. It is observed that current carrying capability of the various HTS tape has been degraded gradually as bending strain increased. The significant decrease (~ 30 %) has been observed in in current carrying capability of BSCCO and Di-BSCCO tape for typical bending strain of ~ 1%. YBCO tape has shown less sensitivity towards strain with respect to its current carrying capability.

References

[1] Minyi F, Jiangxing C, Zhengkuan J, Kumakura H, et al. 2004 *Physica C: Superconductivity* **53-57**
[2] Kuroda T, Itoh k, Kumakura H, Wada H and Katagiri K et al., 2005 *Physica C: Superconductivity* **425** 111-120
[3] Khan Z, Kundu A, Pradhan S 2013 *Pramana* **81** 667-676
[4] Khurmi R and Gupta J *Machine design*, (Ramnagar, New Delhi: Euresia publication house (PVT.) LTD) p 130-150
[5] Oberg E and Jones F et al., 2008 *Machinery's Handbook,* (Industrial press: NewYork Publication) p 2036-2040

IOP Conf. Series: Journal of Physics: Conf. Series **823** (2017) 012046 doi:10.1088/1742-6596/823/1/012046

Study of Transport and Micro-structural properties of Magnesium Di-Boride Strand under react and bend mode and bend and react mode

Ananya Kundu[1], Subrat Kumar Das[1], Anees Bano[1] and Subrata Pradhan[1]

[1]Institute for Plasma Research, Near Indira Bridge, Bhat, Gandhinagar 382 428, India

E-mail: ananya@ipr.res.in

Abstract. I-V characterization of commercial multi-filamentary Magnesium Di-Boride (MgB_2) wire of diameter 0.83 mm were studied in Cryocooler at self–field I-V characterization system under both react and bend mode and bend and react mode for a range of temperature 6 K - 25 K. This study is of practical technical relevance where the heat treatment of the superconducting wire makes the wire less flexible for winding in magnet and in other applications. In the present work the bending diameter was varied from 40 mm to 20 mm and for each case critical current (I_c) of the strand is measured for above range of temperature. A customized sample holder is fabricated and thermally anchored with the 2nd cold stage of Cryocooler. It is observed from the measurement that the strand is more susceptible to degradation for react and bend cases. The transport measurement of the strand was accompanied by SEM analyses of bend samples. Also the tensile strength of the raw strands and the heat treated strands were carried out at room temperature in Universal Testing Machine (UTM) to have an estimate about the limiting winding tension value during magnet fabrication.

1. Introduction

MgB_2 based superconducting strand is emerged as economically viable technical superconductor in several applications owing to its low raw material cost, moderately high Tc, lower anisotropy, promising critical current density and good mechanical strength.MgB_2 based magnet can be successfully implemented in cryogen free system at elevated temperature range of 10 K - 20 K. The suitability of conductor to be used in magnet highly depends on its performance during winding as the winding induced strain and tension may affect the current carrying capacity of the conductor [1-4]. Again the thermal stress and Hoop stress are added up to the total stress acting on the conductor in operational scenario. In the present work, the performance of commercially available multi-filamentary MgB_2 strand with Monel as sheath material is studied in both bent and react mode and react and bent mode for bending diameter 20 mm, 30 mm and 40 mm for temperature range 6 K – 25 K. Scanning electron microscope (SEM) analyses of the samples were carried out after transport measurement to examine the microstructure and filament breakage of the sample, if any. Also the tensile strength measurement of the heat treated and un-heat-treated strand is carried out at room temperature in universal testing machine. This measurement would help in deciding the limiting winding tension to be maintained during magnet fabrication.

2. Strand configuration

The MgB_2 strand used in the present work is M/s HyperTech make 36 filamentary strand manufactured through continuous filling and tube forming [CTFF] route [5]. The strand diameter is

0.83 mm with superconducting fill factor of 15 %. The SEM images of the strand and the typical heat treatment schedule of the strand are shown in figure 1 and figure 2 respectively. The more details of the strand are presented in table1.

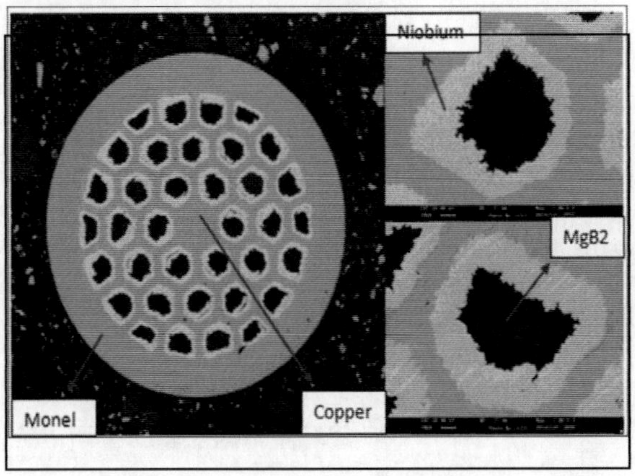

Figure 1. SEM image of MgB_2 strand.

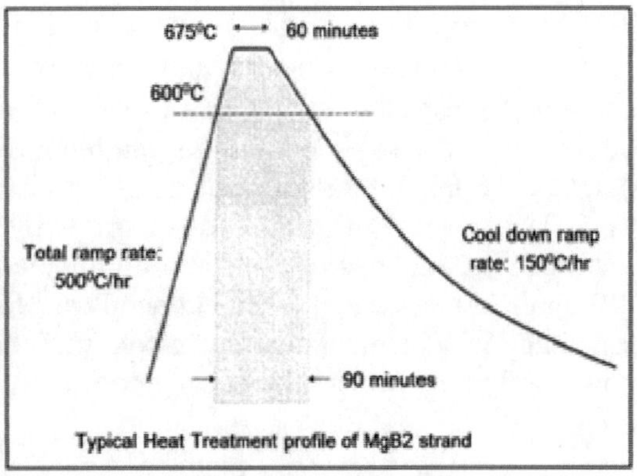

Figure 2. Typical heat treatment schedule of MgB_2 strand.

Table 1. MgB_2 Strand parameters.

Parameters	Values
Superconducting fill factor	15%
Copper Matrix Fraction	22 %
Niobium Barrier fraction	27 %
Monel Sheath Fraction	36 %

3. Experimental Configuration

I-V characterization of samples are carried out in react and bent mode and bent and react mode in pulse tube Cryocooler based facility employing four probe method maintaining 1 μV/cm criterion. A customized sample holder is fabricated to mount the bent sample. The sample holder is thermally anchored with 2nd cold stage of the cold head unit. The schematic of the sample holder and the image of the system are shown in figure 3 and figure 4 respectively. The sample temperature is increased from 6 K to 25 K using a calibrated cartridge heater mounted on the 2nd cold stage.

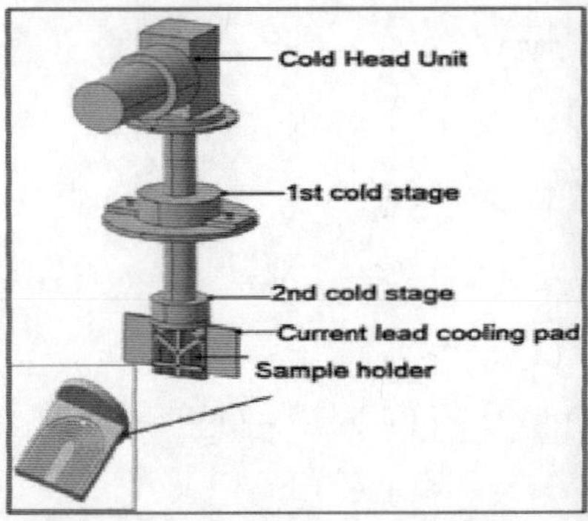

Figure 3. Schematic of sample holder with cold head.

Figure 4. Pulse Tube Cryocooler based self-field I-V characterization facility.

4. Results and Discussions

The bending induced strain is estimated to be 3.98 %, 2.69 % and 2.033 % for bending diameter of 20 mm, 30 mm and 40 mm respectively. The critical current density (J_c) of the strand decreases from 3.02 $\times 10^5$ A/cm^2 to 2.4 $\times$ 10^5 A/cm^2 at the minimum temperature of 6 K and from 1.47 $\times$ 10^5 A/cm^2 to 1.232 $\times$ 10^5 A/cm^2 at the maximum temperature of 25 K due to increment of bending strain from 2.033 % to 3.98 % respectively. In case of react and bend mode, J_c of the strand decreases from 1.947 $\times$ 10^5 A/cm^2 to 1.257 $\times$ 10^5 A/cm^2 at the minimum temperature of 6 K and 4.93 $\times$ 10^4 A/cm^2 to 3.32 $\times$ 10^4 A/cm^2 at the maximum temperature of 25 K on account of 2.033 % to 3.98 % increase in bending strain. The Jc value of the strand is expectedly measured to be ~ 1.55 times less in case of react and bent mode at 2.033 % strain and at minimum temperature of 6 K. Again the J_c values is measured to be ~1.9 times less in case of react and bend case for maximum bending strain of 3.98 % at 6 K. I-V plots as a function of temperature for different bending diameters in case of bend and react modes are shown in figure 5 (a), 5(b) and 5(c) respectively.

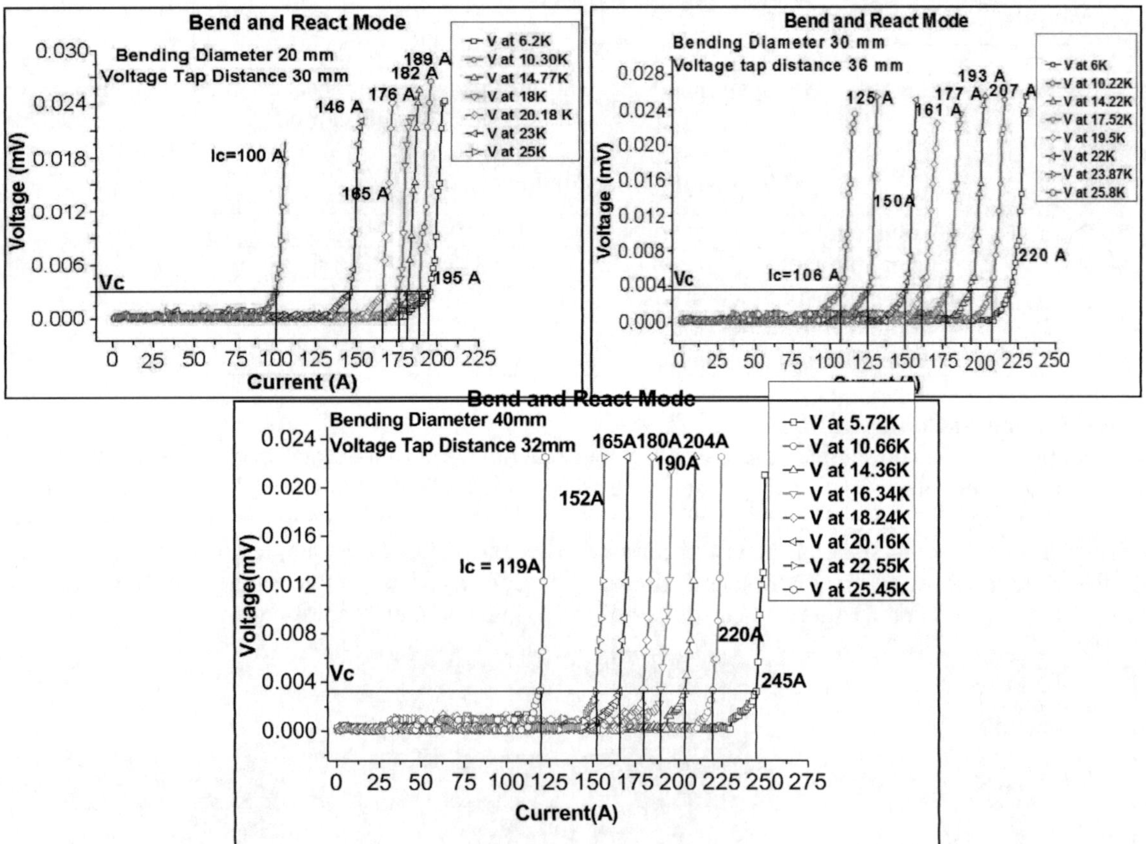

Figure 5: I-V characteristics of MgB_2 strands for bending diameter of (a) 20 mm, (b) 30 mm and (c) 40 mm in bend and react mode from 6 K to 25 K.

Similar analyses are carried out for react and bend mode for the same temperature range and are shown in figure 6 (a), 6 (b) and 6 (c) for varying bending diameter of 20 mm, 30 mm and 40 mm respectively. A comparison plot is presented in figure 7 summarizing the J_c values in both cases as a function of bending strain.

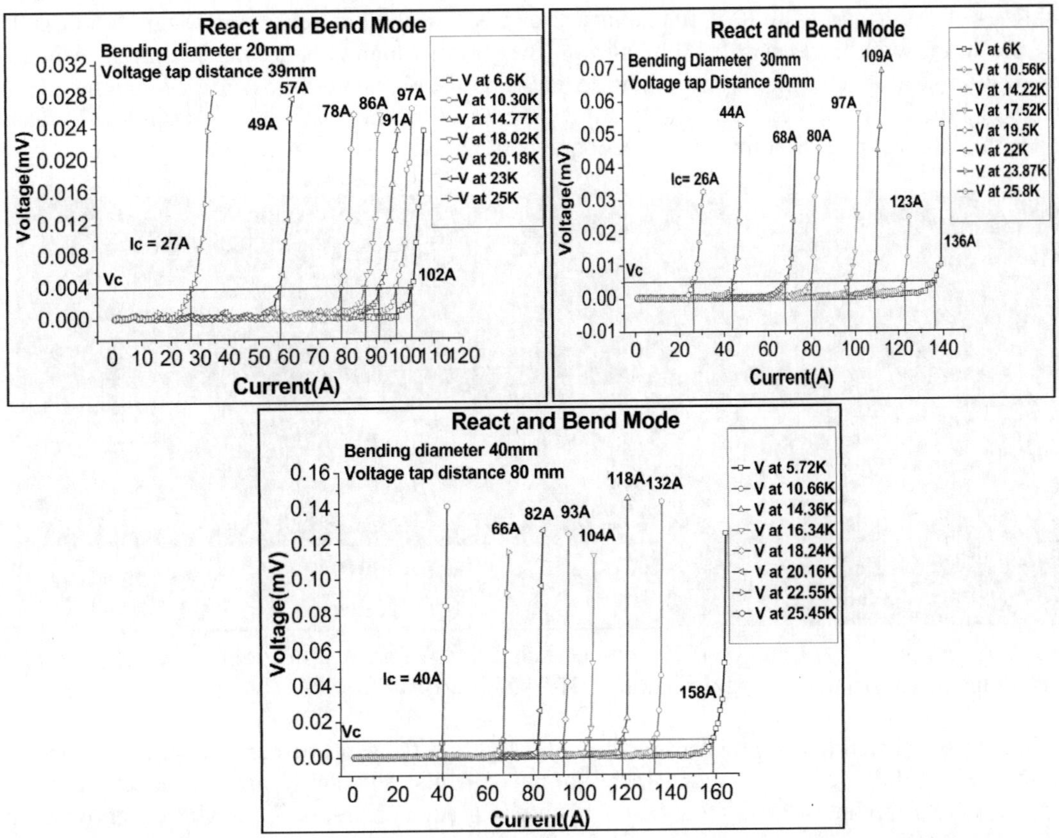

Figure 6. I-V characteristics of MgB_2 strands for bending diameter of (a) 20 mm, (b) 30 mm and (c) 40 mm in react and bend mode from 6 K to 25 K.

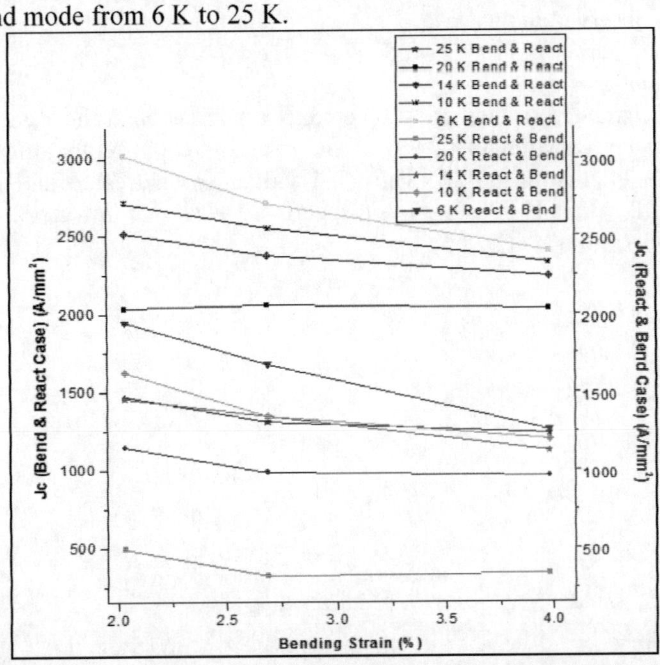

Figure 7. Comparison plot of J_c in bend and react mode and react and bend mode as a function of bending strain.

4.1. Microstructural analyses

SEM analyses of the strand as well as filaments are carried out to comprehend the degradation in Jc values of the strand with increasing bending strain. The strand which is bent with minimum bending diameter of 20 mm are etched with Nitric acid to examine the filaments in SEM. Few filaments are observed to be broken during bending as shown in figure 8 (a) and this may be the reason for decrease in Ic values in both bent and react mode and react and bent mode.

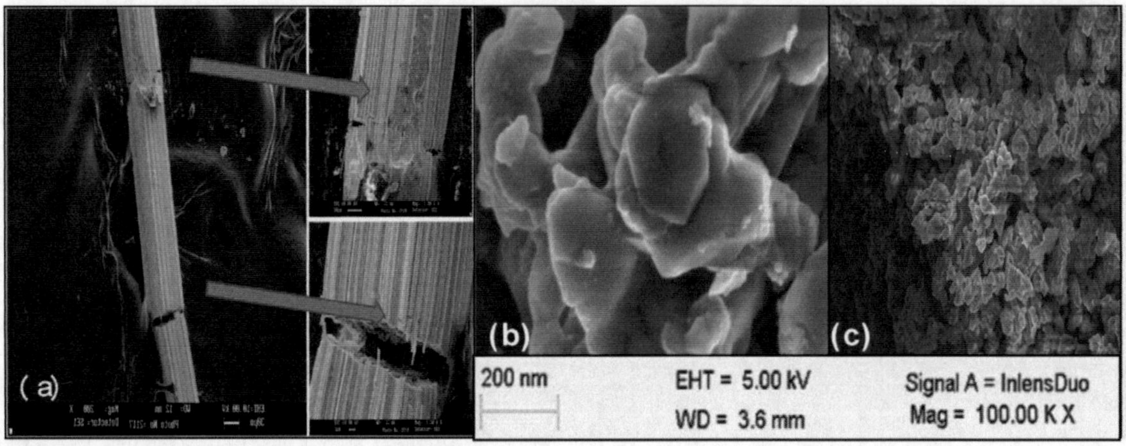

Figure 8. (a) Micro-structural analyses of broken filament, (b) Porosity in grain structure of MgB_2 volume in strand as shown under magnification of 100 kX and 10 kX in SEM.

SEM images of reacted MgB_2 powder are shown in figure 8 (b) and this confirms the presence of porosity in MgB_2 volume of the strand. Brittle nature of MgB_2 and lower grain connectivity may results in gradual de-linking of grains and hence results in gradual decrease in J_c values with bending radius. These causes for greater degradation of J_c in react and bend mode. At the same time in the bend and react mode links between in grains are formed after bending and during heat treatment. Hence lesser degradation is observed in this case.

4.2. Tensile strength measurement

The heat treated and untreated strands are subjected to axial tension at the strain rate of 2.5 mm/min to measure the ultimate tensile strength (UTS) of the strand at room temperature employing Universal testing machine (Model: Instron 5982). The UTS values for untreated and heat treated strand are measured to be 783.48 MPa and 308.31 MPa respectively as shown in figure 9 owing to the softening of Monel sheath material at 675 °C.

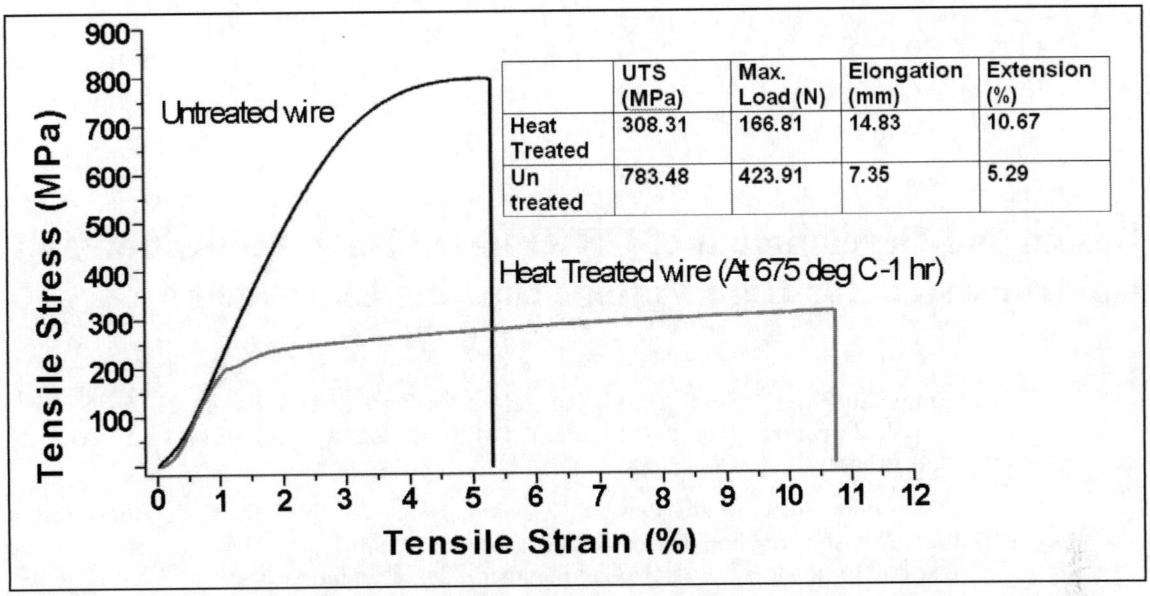

	UTS (MPa)	Max. Load (N)	Elongation (mm)	Extension (%)
Heat Treated	308.31	166.81	14.83	10.67
Un treated	783.48	423.91	7.35	5.29

Figure 9. Tensile strength measurement of heat treated and untreated MgB_2 strand.

5. Summary

In this work, commercially available multi-filamentary MgB_2 strands have been tested for their performance under the influence of finite bending strain in bend and react mode and react and bend mode. J_c value of the strand in self –field are found to be ~ 1.55 times less in case of react and bent mode at minimum applied strain of 2.033 % and at minimum temperature of 6 K. Again the J_c values is measured to be ~1.9 times less in case of react and bend case for maximum bending strain of 3.98% at 6 K. The degradation of J_c with bending strain is supported by the micro-structural analyses of strand and filament. The tensile strength measurement of heat treated as well as un-heat treated strands are carried out at room temperature in Universal testing machine to check the winding tension limit during coil fabrication employing the strand.

Acknowledgement

The authors would like to thank Mr. Narendra Chauhan and Mr. Jignesh P Chauhan for providing their valued support during micro-structural analyses and tensile strength measurement of the strands.

References

[1] Hanna M,Fang H, Zhou Y X, Alessandrini M, Putman P T and Salama K 2007*Journal of Materials Processing Technology* **181** 44

[2] Young E A, Falorio I, Beduz C, Bailey W O S and Yang Y 2014 *Journal of Physics: Conference Series* **507** 032061

[3] Nishijima G, Oguro H, Awaji S, Tsubouchi H, and Watanabe K 2009 *IEEE Trans. On Appl. Supercond.***19** 1112

[4] Badica P, Awaji S, Nishijima G, Oguro H, Sandim M J R, Cangani M P, Ghivelder L, Katagiri K and Watanabe K 2007 *Supercond. Sci. Techn.* **20** 273

[5] Suo H L et al.2007 *IEEE Trans. Appl. Supercond.* **17** 2822

IOP Conf. Series: Journal of Physics: Conf. Series **823** (2017) 012047 doi:10.1088/1742-6596/823/1/012047

Design and Development of CRIO based Data Acquisition and Control system for High Voltage Bushing Experiment

Himanshu Tyagi[1], Sejal Shah[1], Ratnakar Yadav[1], Kartik Patel[2], Hiren Mistri[2], Deepak Parmar[1], Dheeraj Sharma[2], Mainak Bandyopadhyay[1] and Arun K Chakraborty[2]

[1]ITER-India, Institute for Plasma Research, Block A, Sangath Skyz, Bhat-Motera Road, Koteshwar, Ahmedabad - 380005, India
[2]Institute for Plasma Research, Gandhinagar382 428, India

E-mail: htyagi@iter-india.org

Abstract. Experiment of Prototype High Voltage Bushing (PHVB) is an important step to validate the design of HV bushing (HVB) of DNB for its electrostatic performance. The experiment is performed considering scaled down configuration DNB HVB with similar operating parameters and maintaining comparable electric stresses. Applied voltage is restricted up to 60 kV considering it's scaled down configuration.To automate the entire experimental process and acquiring important experimental data for post analysis, a dedicated Data Acquisition and Control System (DACS) is designed. CRIO (compact reconfigurable input output) is a rugged, small sized hardware platform which combines the power of real time processor and FPGA bus. NI CRIO is now widely used as a controller for small scale experiments. For ensuring smooth control operation of the experiment; NI CRIO was selected as the controller for DACS of PHVB. In this paper the technical details of DACS design, implementation and test results are discussed.

1. Introduction

The Diagnostic Neutral Beam (DNB) [1] in ITER, provides 100 kV H° beam to support the charge exchange recombination spectroscopy for He ash measurement. DNB comprises of vacuum immersed RF source. All high voltage services to the source should be routed through vacuum feed through called High Voltage bushing (HVB) [2], which forms the interface between the dry air-insulated transmission line and the DNB injector which is connected directly to the torus primary vacuum.

To ensure the High voltage bushing design, prototype HVB (PHVB) [3] is designed and fabricated. PHVB will be subjected to similar test scenarios as in DNB such as HV power supply (HVPS) and vacuum environment. For automating the entire experimental set up, a dedicated Data acquisition and control system (DACS) is required. The DACS is responsible to control applied voltage on the PHVB and monitor and acquire the important parameters related to experiment.

In this experiment CRIO interfaces the HVPS and provides set points for current and voltage. The DACS keeps the set points for the voltage and current settings of the HVPS used in experiment Also it provides seamless control and acquisition for pulse durations of up to 3600 s. The voltage signal is generated as a ramp signal with upper voltage set point. The ramp rate applied is 1.5 kV/min and is user configurable.

The testing of PHVB has been successfully completed using the developed DACS.

IOP Conf. Series: Journal of Physics: Conf. Series **823** (2017) 012047 doi:10.1088/1742-6596/823/1/012047

2. Experimental System And DACS Requirement

The DACS interfaces with following systems in the experimental set up:

1) PHVB: The two insulators based bushing designed for 60 kV isolation
2) HV power supply: A power supply from Spellman with rating of 120 kV 50 mA
3) Vacuum system: Turbo molecular pumping station backed by rotary pump to achieve desire vacuum.

The interfacing system layout is shown in figure 1.

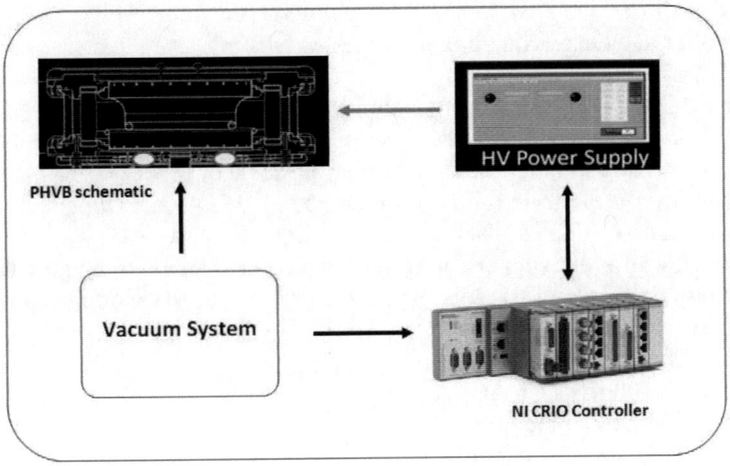

Figure 1. Overall Experimental system.

The DACS has an important responsibility to interface with all the components of the experimental system and to automate the entire experimental sequence.
Following are the main functions implemented by DACS:

1. To interface with HV and Vacuum systems
2. To provide a controlled variable voltage ramp from HV power supply. The voltage and time steps are both user configurable.
3. To acquire signals of applied voltage and leakage current through the system via PS for long durations of more than 3600 sec.
4. Categorize shots for the ease of user identification.

In view of these requirements, NI CRIO was selected as the top level controller for this experiment.

In terms of signals following acquisition and control signals are interfaced by DACS:

Table 1. Control and Acquisition list.

System	Acquisition signal	Control signal
HVPS	Voltage and Current (0 -10 V)	Digital: PS on and HV on (24V TTL) Analog: Voltage and current (0-10V)
Vacuum system	Pressure gauge: 0-10V (2)	

IOP Conf. Series: Journal of Physics: Conf. Series **823** (2017) 012047 doi:10.1088/1742-6596/823/1/012047

3. System architecture

The Control system is divided into two parts.

1. Real-Time controller and data acquisition modules.

2. Host Application Graphical User Interface.

The tasks done by control system includes initialization, start-up procedure, sequential operation of the power supply, data acquisition and data logging of various signals and implementation of machine safety interlocks for safe and reliable operation.

3.1. *Real Time Controller and Data acquisition modules*

Real Time Controller and Data acquisition modules consist of:

1. NI CRIO 9073, a RT controller
2. 2M gate FPGA, 8-slot chassis for custom I/O timing, control, and processing
3. 10/100BASE-T Ethernet port; RS232 serial port for connection to peripherals
4. Digital and Analog I/O modules to acquire the status of process parameters.

Analog I/O Modules are selected to accommodate 0-10V DC and digital I/O modules are selected to acquire TTL signals as well as 12-24V (For logic High) signals.

I/O Modules with different conversion times have been selected to acquire fast (time critical) and slow (normal or less time critical) signals. Some features of the I/O modules along with model no. are given below

1. NI 9201 8 Ch. 500 KS/s AI module.
2. NI 9263 4 Ch 100 KS/s/Ch AO module.
3. NI 9375 32 Ch DI/DO module

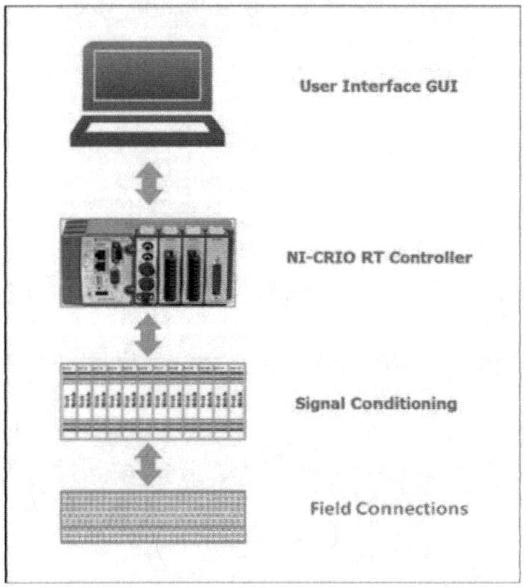

Figure 2. Schematic of Control system.

The overall schematic in figure 2 shows CRIO system with I/O modules mounted on both the chassis. Real Time Controller is programmed to handles the following tasks.

1. Data acquisition, processing and data logging.
2. Data sharing with Host PC for user interface and command execution.
3. Determinism to sample I/O channels at equal time slices to provide better controls. The actual current for the bushing has to be monitored at regular time interval for proper HV operation.
4. Managing FIFO to buffer and share data.

IOP Conf. Series: Journal of Physics: Conf. Series **823** (2017) 012047 doi:10.1088/1742-6596/823/1/012047

3.2. Graphical User Interface

The GUI (Graphical User interface) is developed to display the status of the power supplies as well as interlock signals. GUI is also providing interface to I/O modules through RT controller to execute commands. The GUI has separate windows to handle each task. To set the parameters during start up, to display status information of major subsystems during operation and to display flow, temperature and pressure interlock signals. The GUI has been developed in Labview and is running under Windows OS. The communication between CRIO system and windows PC is through Ethernet on TCP/IP. Figure 3 shows the GUI during operation. For communicating data between RT controller and host, NI network published shared variable [4] are used.

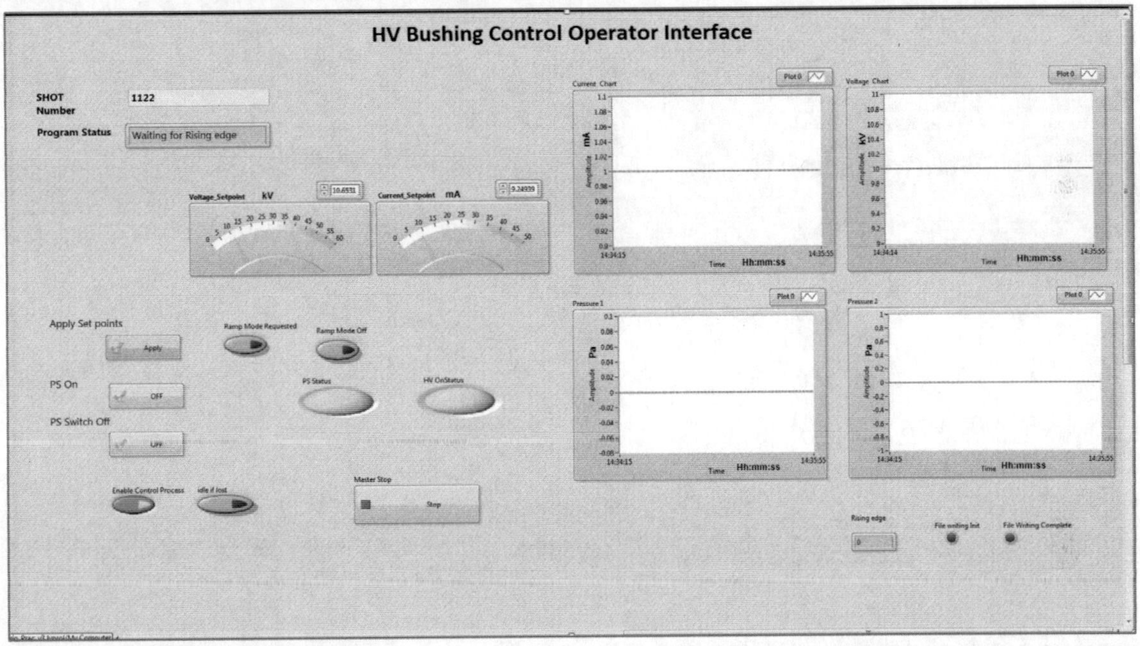

Figure 3. Graphical user interface.

3.3. Logic flow in software

Application software for data acquisition, sequential control and data logging has been developed in LabView Real-Time.NI CRIO can be programmed in two ways namely Scan mode and FPGA mode. Though the control program was initially developed in FPGA mode, currently scan mode version is being used.

In Scan mode CRIO works like a standard RT controller and the program flow occurs in between host and RT only. In FPGA mode, the FPGA in CRIO also comes into picture and program flow happens between, host, RT and FPGA. The FPGA mode is used only where the timing constraints are in order of less than 1us which is not the case in our experiment. The program has two parts Host VI and RT VI.

In Host VI the user events are logged like switching on the power supply, taking status of HVPS, online monitoring of current and voltage signals.

In RT VI the program is divided into 3 parts: Initialization, Critical loop and non-critical loop. The Critical loop handles the actual field signals for control and acquisition functions. In non-critical loop the data from host is exchanged and user events are identified. Both the loops are running at 200 ms and are timed while loop.

IOP Conf. Series: Journal of Physics: Conf. Series **823** (2017) 012047 doi:10.1088/1742-6596/823/1/012047

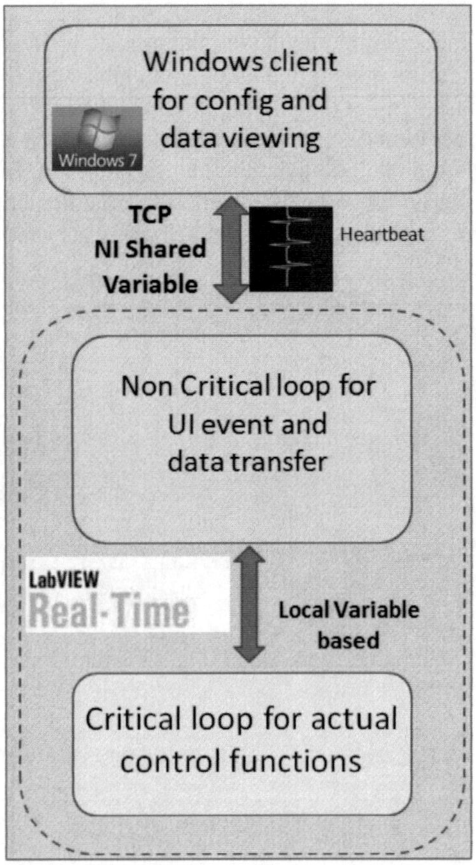

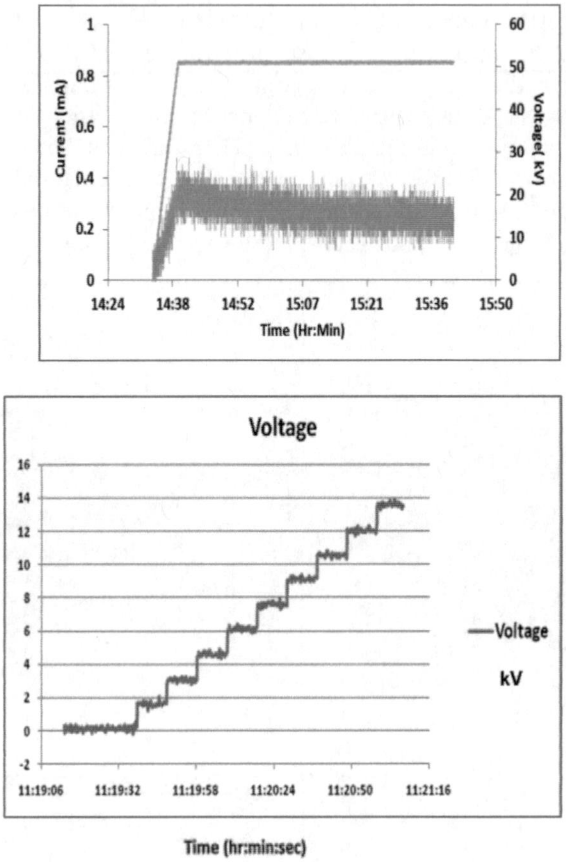

Figure 4. Software architecture of Real time software. **Figure 5.** Test results from the experiment.

NI Shared variables (SV) have been used to transfer data between RT and Host. During development of the control program the main challenge was posed in using the NI SV was to avoid any data and communication loss due to long duration of pulses. To realise this, a heartbeat signal was used between host and RT. A variable was constantly changed at host vi and monitored at RT vi. If the RT vi finds the variable unaltered for 400 ms duration, all the set points will be set at 0 and the system will be in safe state. The software architecture of the program is described in figure 4. The data logging function was realized at the host side. The signals of pressure, voltage and current were logged at sampling interval of 200 ms and saved in .csv files. For each experimental sequence a shot number is generated and file are saved as per shot numbers only for post analysis. Figure 5 shows the test results of signal acquired during the test.

4. Present status

The DACS is developed and has been installed successfully with PHVB experiments. DACS and power supply interface is established and operated for different ramp voltage from 1.5 kV/min to 8 kV/sec as per the requirement of PHVB operation. Further, the system is operated for long pulse durations (> 3600 s) and acquired data for individual shots. Being a FPGA and RT based system CRIO supports higher level of determinism for time critical loops.

IOP Conf. Series: Journal of Physics: Conf. Series **823** (2017) 012047 doi:10.1088/1742-6596/823/1/012047

References

[1] Chakraborty A et. al 2010 *IEEE Transactions on Plasma Science* **38** 248
[2] Shah S et al 2011 *Fusion Engineering and Design* **86** 892
[3] Shah S et al 2016 *Fusion Eng. Des.* **113** 6
[4] *NI LabVIEW User Manual*

IOP Conf. Series: Journal of Physics: Conf. Series **823** (2017) 012048 doi:10.1088/1742-6596/823/1/012048

Serial Interface through Stream Protocol on EPICS Platform for Distributed Control and Monitoring

Arnab Das Gupta[1], Amit K Srivastava[1], S Sunil[1] and Ziauddin Khan[1]
[1]Institute for Plasma Research, Near Indira Bridge, Bhat, Gandhinagar 382 428, India

E-mail: arnabdasg@ipr.res.in

Abstract. Remote operation of any equipment or device is implemented in distributed systems in order to control and proper monitoring of process values. For such remote operations, Experimental Physics and Industrial Control System (EPICS) is used as one of the important software tool for control and monitoring of a wide range of scientific parameters. A hardware interface is developed for implementation of EPICS software so that different equipment such as data converters, power supplies, pump controllers etc. could be remotely operated through stream protocol. EPICS base was setup on windows as well as Linux operating system for control and monitoring while EPICS modules such as asyn and stream device were used to interface the equipment with standard RS-232/RS-485 protocol. Stream Device protocol communicates with the serial line with an interface to asyn drivers. Graphical user interface and alarm handling were implemented with Motif Editor and Display Manager (MEDM) and Alarm Handler (ALH) command line channel access utility tools. This paper will describe the developed application which was tested with different equipment and devices serially interfaced to the PCs on a distributed network.

1. Introduction

EPICS is a widely used software for distributed soft real time control system application. It has been used in large physics experiments for better reproducibility and stable operation of various control systems like accelerators [1, 2]. This paper describes the development work that has been carried out for interfacing hardware such as microcontroller, power supplies and pump controllers on EPICS base built on a windows-7 based 64-bit operating system using Cygwin with GCC (GNU compiler collection) for lab based prototype experiments which demands remote operation, monitoring and control. The EPICS base version R3.14.12.4 and the EPICS windows tools are utilized to develop the interface and hardware operations. For serial communication asyn Driver was utilized as it is a general purpose facility for interfacing device specific code to low level communication drivers. Also StreamDevice has been used to communicate with asyn serial ports. StreamDevice is a generic EPICS device support for devices with a "byte stream" based communication interface. StreamDevice is used with an interface to asyn Driver, which implements the low-level support for those communication interfaces [3]. Here we have used the RS-232 communication [4,6] for the monitoring of different equipment such as atmega 2560 based microcontroller kit, an Agilent make programmable power supply and a turbo molecular vacuum pump controller. All the devices have been integrated to EPICS platform and remotely either monitored or controlled or both. The succeeding sections briefly describes the interfacing of these equipment. A block diagram is shown below to represent the schematic of the experimental set-up.

The block diagram shown in figure 1 is the layout of the devices connected in a distributed network. The Central computer has the EPICS IOCs (Input output controller) running on it and the data are transferred to and fro with the hardware through the Stream Device Communication. The other PCs in the network are monitoring and controlling the process variables of the applications on the main server through remote operation through caget, caput commands. The data obtained from the hardware can also be stored in a local as well as remote PC for offline analysis with python channel access for EPICS (pyepics). PyEpics is an interface for the Channel Access (CA) library of the Epics Control System to the Python Programming language. The pyepics package provides a base EPICS module to python, with methods for reading from and writing to EPICS Process Variables (PVs) via the CA protocol.

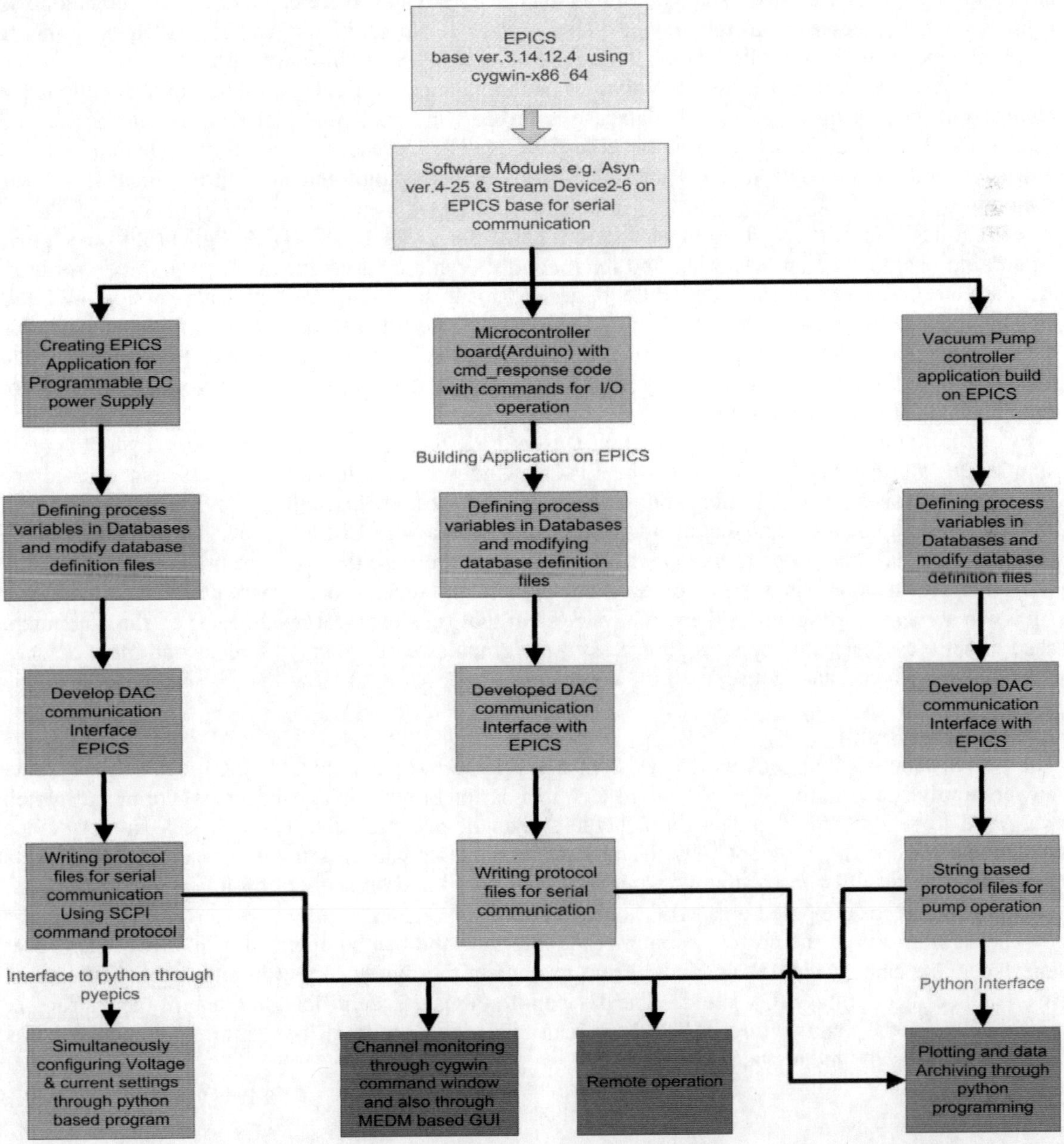

Figure 1. Block diagram of the system

2. Interfacing and Implementation Scheme

Interfacing of arduino with EPICS requires hierarchy of processes for Data Acquisition system (DAQ), all of first an EPICS base is required over which application utility for the DAQ process is developed. EPICS Base version 3.14.12.4 has been used here. For the communication between arduino and EPICS a database file, Stream Protocol, Asyn and IOC initializing commands also used to access the Input/output of the arduino.

StreamDevice is a kind of device support that communicates with the hardware in form of strings serially over the USB, RS-232, RS-485, GPIB, TCP/IP etc. interfaces, thus it provides the way of communication for controlling and monitoring various devices based upon above interfaces [6] and called as protocol files. Unlike developing a system with asyn drivers only, other device support are not required in this case for the hardware to be interfaced StreamDevice cannot be used standalone as it just sends and receives data from device over an interface, an additional package called as ASYN is required which actually establishes the interface between EPICS and the hardware.

A database file defines the Process Variables and the associated fields describing their functions for control and monitoring purpose. This database is linked with the IOC initializing commands and the Stream Protocol file, when IOC will start, the database file in accordance to StreamDevice protocol will send and receive data in the PV's which can then be monitored or control either locally or remotely.

EPICS has been interfaced to three different hardware's that are E3632A Agilent power supply, an arduino board based on ATmega 2560 microcontroller and a Agilent make Ion Pump control unit. All these are discussed below. The EPICS base used for this is 3.14.12.4 with asyn version 4-25 and stream device version 2-6. The records and databases were written to properly configure the process variables such as voltage and current. Motif Editor and Display Manager (MEDM) was used for the GUI (Graphical User Interface) display which is shown below to program the voltage and current levels. This is also verified through remote computer.

Motif Editor and Display Manager (MEDM) is an EPICS extension which provides a way to create simple OPI's [5,6]. MEDM is one of the EPICS windows tools for windows OS that is used for graphical analysis and for designing and implementing control screens, called displays that consist of a collection of graphical objects that display and change the values of EPICS process variables. MEDM is typical software that needs to be run with Xwin server otherwise there will be problem in saving the GUI that we made, xwin server comes with Cygwin package, (xorg-server under X11 package) An Xwin server is a program in the windows System that runs on local machines (i.e., the computers used directly by users) and handles all access to the graphics cards, display screens and input devices (typically a keyboard and mouse) on those computers.

3. Local Test Result

The power supply which was used is Agilent make E3632A programmable DC power supply. The power supply supports the SCPI (Standard Commands for Programmable Interface) command which was used to send or receive data through the stream device protocol. Through this stream device syntax written for the protocol files the process variables are accessed through EPICS. The plots shown are the result of the testing of equipment monitored and controlled through a local computer. The display screens are based on a MEDM based GUI.

The figure 2 shows the power supply module interface and can be operated using the MEDM user interface. One can set the voltage and current to operate the power supply in either constant current (CC) and constant voltage (CV) mode. The device info tab is the identification string of the instrument as provided by the manufacturer. This supply can either be operate in local and remote condition as per the options provided by the manufacturer.

IOP Conf. Series: Journal of Physics: Conf. Series **823** (2017) 012048 doi:10.1088/1742-6596/823/1/012048

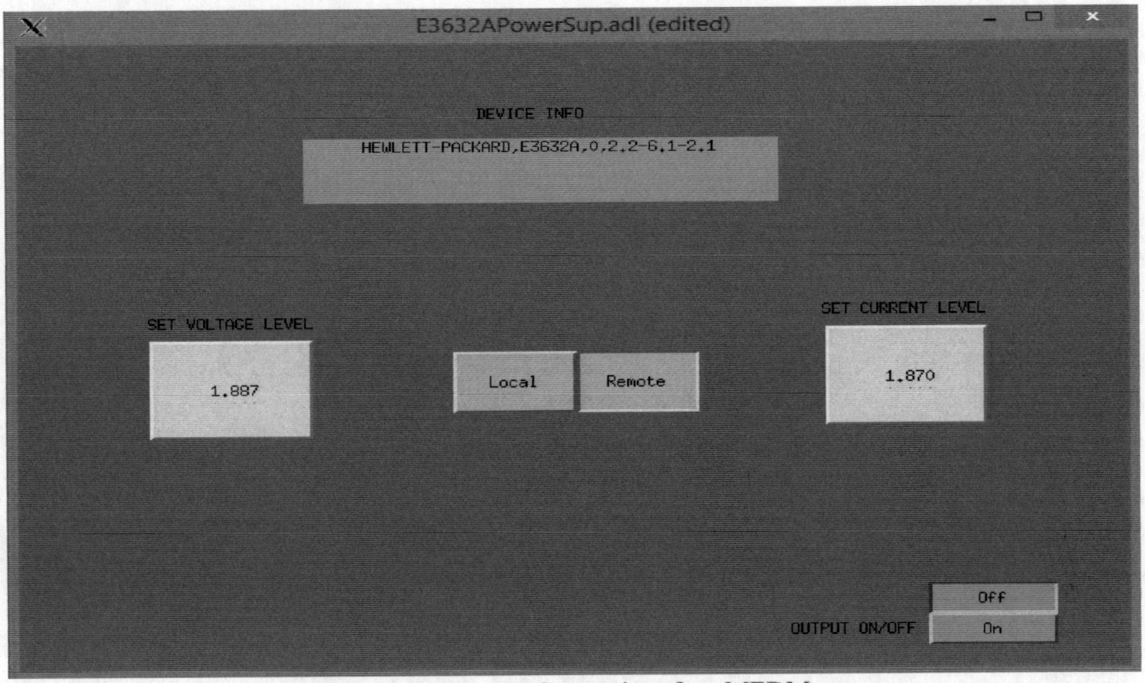

Figure 2. Power supply user interface MEDM screen

Next the arduino based microcontroller board was interfaced with EPICS. The controller was programmed with the support available from EPICS. The arduino sketch is a code or program for the arduino controller that defines the interface with EPICS and the operations performed from it. This sketch defines and maps PV from database such as analog read pin, analog write pin, digital write, etc. After building the application IOC on epics, defining the databases and process variables and writing the protocol files the connection was established. Since the code in the arduino has the defined set of variable from epics support to understand the stream protocol syntax, so the syntax were used in the same order to communicate with the I/Os. Temperature Sensor LM 35 and simulated signals from function generators were used to acquire the data using the I/O available on the board. The GUI developed is as shown below in figure 3. EPICS could not be used for arduino to acquire data at high sampling rate here as EPICS is used for supervisory control and monitoring for slow controls in the order of few milliseconds to seconds.

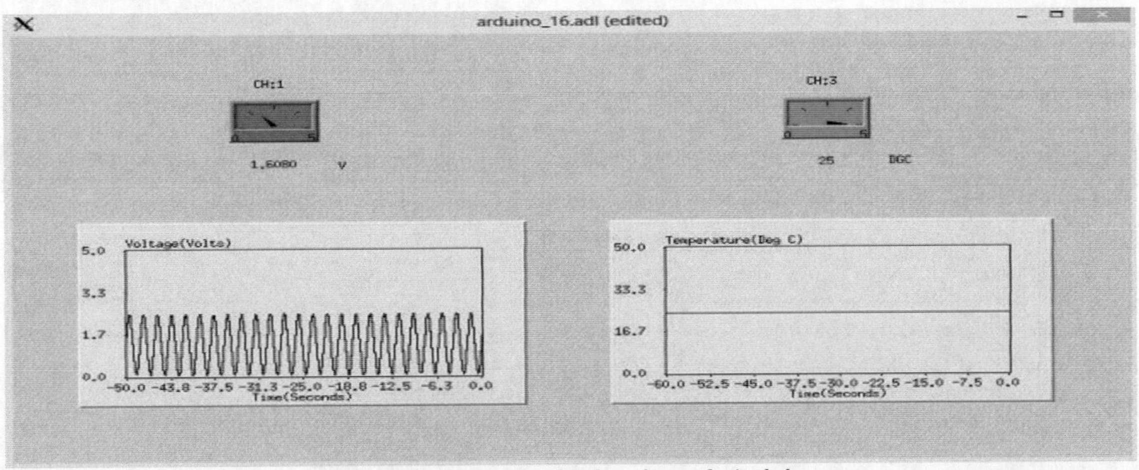

Figure 3. Data Monitoring through Arduino

Figure 4 below shows the GUI interface of an Agilent based Turbo molecular pump control unit was epics. The strings to access the device controller were converted to the corresponding hexadecimal values and were send to the pump controller for its operation using the protocol files. This allowed us a successful communication between the EPICS IOC and the controller.

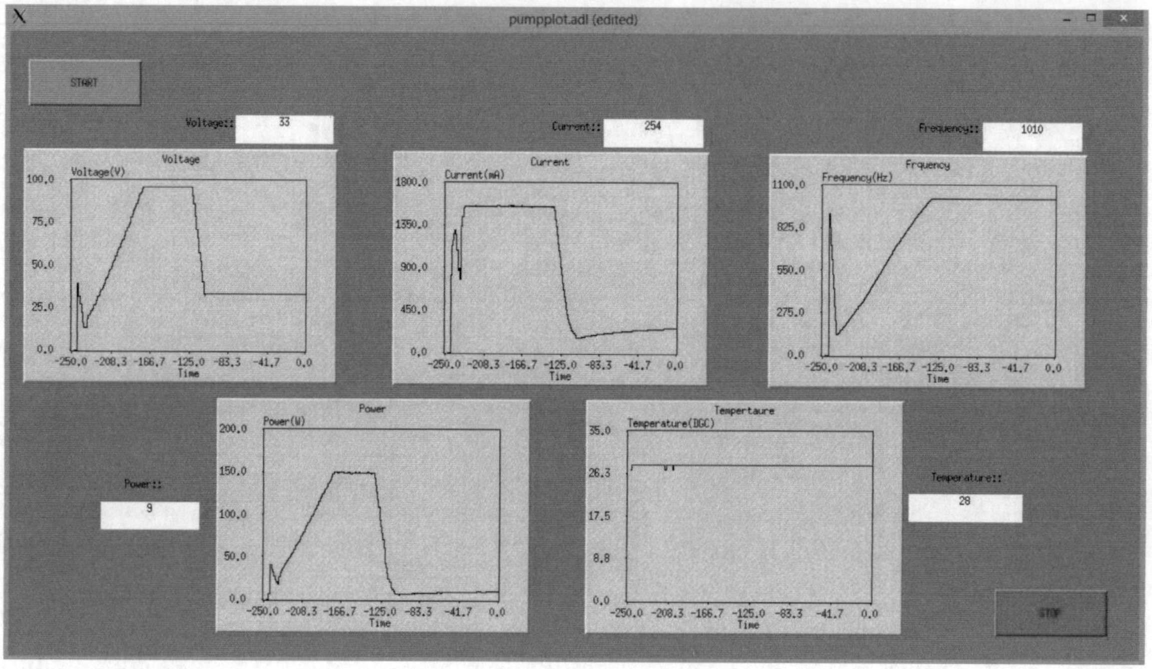

Figure 4. Pump controller MEDM OPI

4. Remote Testing Results

The system interfaced and developed with the EPICS platform was also tested on remote terminals connected in a distributed network. Here are the pictures shown for the vacuum pump controller parameters obtained remotely on different plotting platforms. EPICS has the advantage of broadcasting the process variables through UDP protocol and these protocol are captured for display and archiving on various client utilities. These are also implemented in the same fashion for the other two hardware setups as described earlier in section 3.

.

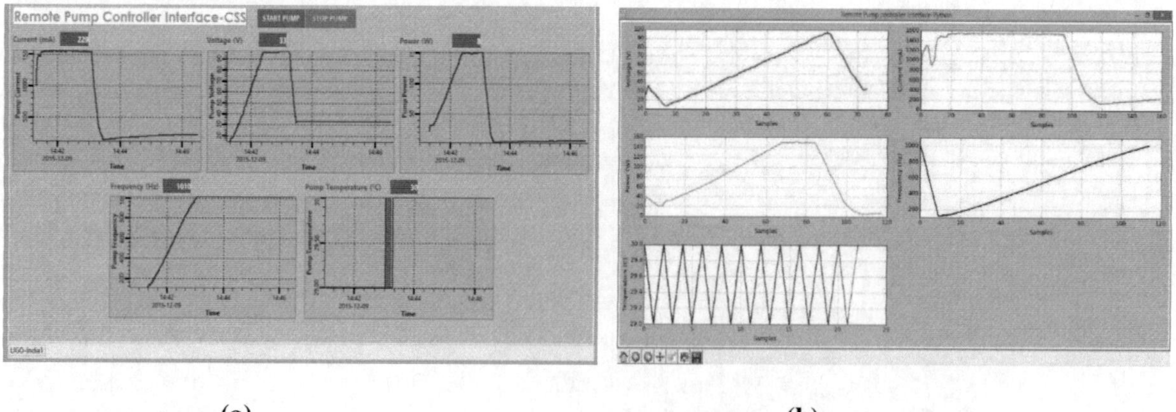

(a) **(b)**

Figure 5. Remote Control and monitoring utility using CSS and Python pyepics module

The plot shown here are the profile obtained on a remote connection through EPICS channel access. The screen shown here are displayed in control system studio and python software respectively and the plots for the MEDM are displayed earlier. The python software has been also utilized to store and archive the process variable data remotely.

5. Monitoring through ALARM-HANDLER

Alarm Handler is yet another tool from EPICS which provides an interactive graphical application used primarily by to display and monitor EPICS database alarm states [5, 6]. It serves as an interface between an operator and the EPICS database and it communicates with the database using channel access function calls.

The Alarm Handler monitors alarm conditions of EPICS database records. The primary responsibilities of the Alarm Handler (ALH) are to:

- Bring alarms to the operator's attention.
- Provide the operator guidance for handling of specific alarms
- Allow the operator to globally acknowledge alarms.
- Provide for logging alarms and display of the logged alarm history

Alarm has different types of severity which will inform about the status of the system and will signal the operator accordingly, alarms can be defined in terms of their severity NO_ALARM, MINOR, MAJOR, INVALID, etc. This configuration file required as an input must be saved with .alhConfig extension. It defines the Alarm Group structure and takes data in a flexible input format and can be prepared via any text editor. Here in the figure 6 below two PV's i.e. channel:ai0 and channel:ai1 under group 1 are being monitored. We can create further subgroups by writing in configuration file, whenever the values from data base record will increase its specified level, alarm handler will activate and gives continuous beep sound until the operator logs the error.

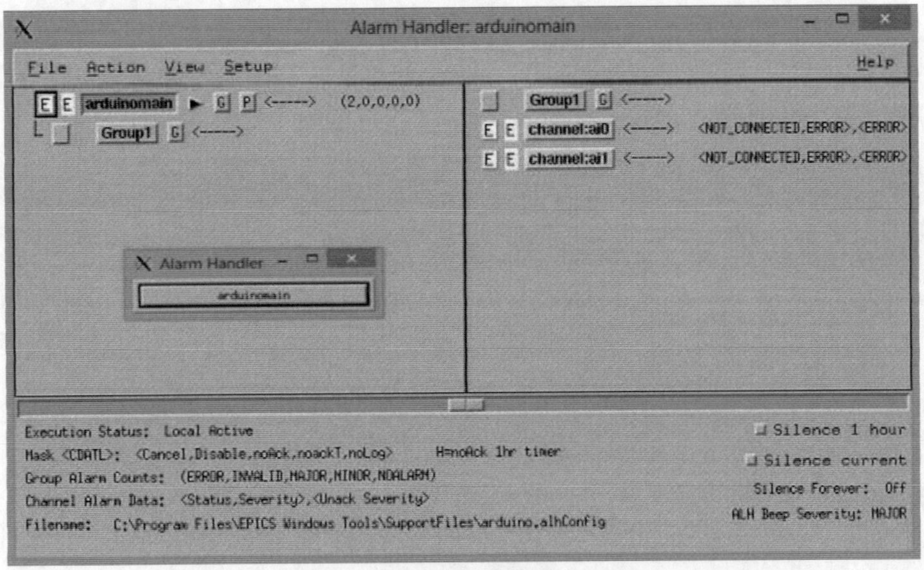

Figure 6. Alarm Handler window

6. Conclusion

The interface between different hardware and EPICS has been integrated and tested successfully. The physical variables have been monitored remotely with this interface. This work has been carried out as standalone development that can be extended to other hardware interfaces supporting such type of serial interface, and different protocol files can be developed accordingly as per application requirements in a distributed control for status monitoring and operation.

Acknowledgement

This work was supported by our organization, Institute for Plasma Research Gandhinagar as a developmental activity. We thank our colleagues who provided insight and expertise that greatly assisted in carrying out this work. We would also like to thank the EPICS community who have contributed in extending their support and answers to troubleshoot the errors, which we encountered during our work.

References

[1] Dalesio L R, Kraimer M R, Kozubal A J 1991 *Proceedings of International Conference on Accelerator and Large Experimental Physics Control Systems* p278
[2] Dalesio L R, Hill J O et. al 1994 *Nuclear Instruments and Methods in Physics Research A* **352** 179
[3] Andrews A et al. 2012 *Proceedings of IPAC2012* **C1205201** 4002
[4] Chevtsov P et al. 2001 *ICALEPCS* **C011127** 200
[5] Lange R, Anderson J B, Johnson A N, Kraimer M R, Norum W E, Dalesio L R, and Hill J O 2003 "EPICS: Recent developments and future perspectives", ICALEPCS, Gyeongju,Korea
[6] Carvalho P, Duarte A, Pereira T et al. 2011 *Fusion Engineering and Design* **86** 1085

IOP Conf. Series: Journal of Physics: Conf. Series **823** (2017) 012049 doi:10.1088/1742-6596/823/1/012049

Development of Data Acquisition Set-up for Steady-state Experiments

Amit K Srivastava[1], Arnab D Gupta[1], S Sunil[1] and Ziauddin Khan[1]

[1]Institute for Plasma Research, Near Indira Bridge, Bhat, Gandhinagar 382 428, India

E-mail: amit@ipr.res.in

Abstract. For short duration experiments, generally digitized data is transferred for processing and storage after the experiment whereas in case of steady-state experiment the data is acquired, processed, displayed and stored continuously in pipelined manner. This requires acquiring data through special techniques for storage and on-the-go viewing data to display the current data trends for various physical parameters. A small data acquisition set-up is developed for continuously acquiring signals from various physical parameters at different sampling rate for long duration experiment. This includes the hardware set-up for signal digitization, Field Programmable Gate Arrays (FPGA) based timing system for clock synchronization and event/trigger distribution, time slicing of data streams for storage of data chunks to enable viewing of data during acquisition and channel profile display through down sampling etc. In order to store a long data stream of indefinite/long time duration, the data stream is divided into data slices/chunks of user defined time duration. Data chunks avoid the problem of non-access of server data until the channel data file is closed at the end of the long duration experiment. A graphical user interface has been developed in LabVIEW application development environment for configuring the data acquisition hardware and storing data chunks on local machine as well as at remote data server through Python for further data access. The data plotting and analysis utilities have been developed with Python software, which provides tools for further data processing. This paper describes the development and implementation of data acquisition for steady-state experiment.

1. Introduction

With advancement in Physics research, Experiment duration is continuously moving towards the steady-state operation in pursue of occurrence of significant physical activity. The steady-state operation of an experiment imposes several technical challenges in the field of data acquisition [1,2]. The major problem is non-access of data at server until the channel data file at local machine is closed at the end of the long duration experiment. To overcome this problem we have adopted the concept of data time-slicing/chunking of user defined time duration for viewing data during acquisition [3]. This facilitates the on-line viewing of data at remote terminals and user may also retrieve the portion of acquired data for plotting and analysis with user-defined time window.

For such applications, steady-state data acquisition system [4] should have preferably Graphical User Interface (GUI) based data archival and retrieval system that should be configured as per experiment requirements. In our application we have developed a Python based data archival and retrieval utilities where the user may define diagnostics channels through GUI as per on-line viewing and analysis requirements. We have implemented the concept on a small-distributed data acquisition system interfaced to FPGA based central timing system and to DAQ server. The development work and test results are being discussed in following sections.

2. Hardware set-up

The objective of this experimental DAQ set-up is to facilitate continuous data acquisition for long pulse steady-state experiments with data viewing facility at any user/client machine as and when required.

2.1. Data Acquisition set-up:

The schematic of Data Acquisition Set-up is shown in figure 1. This has been developed to provide necessary facilities of distributed data acquisition system but at small scale. The important features of development work are as follows:

- Distributed set-up with three Data Acquisition System (DAS). One is having PXIe interface and others two have USB interface with Front-End Computer (FEC).
- Each DAS is interfaced through FEC to DAQ server and to other machines on Ethernet based network.
- FPGA based Timing System for clock synchronization and trigger distribution among all DAQ hardware.
- Storage of each channel data streams as single file as well as divided in data time-slices/chunks of user defined time duration for viewing data during acquisition.
- The concept has been tested with different acquisition software (LabVIEW, MATLAB, Keysight AMM) for hardware interface.
- Python software is being used for data time-slicing, for storage to remote DAQ server and as utility for plotting and analysis as per application requirements.

Figure 2 shows hardware set-up of DAS-3, which consists of Keysight make USB Digitizer interfaced to vacuum equipment.

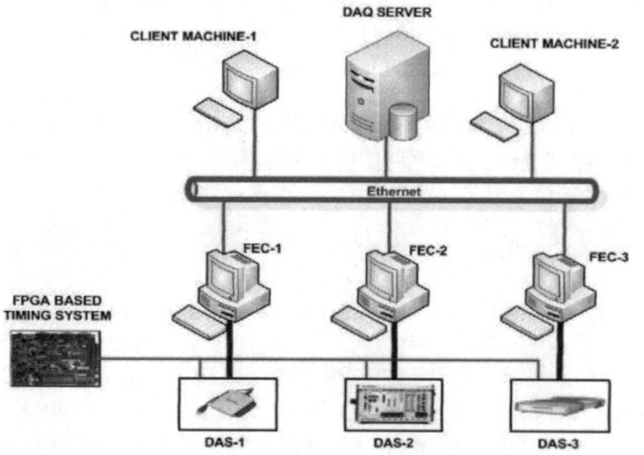

Figure 1. Schematic of Data Acquisition Set-up.

Figure 2. Experimental Set-up for DAS-3.

2.2. Timing Distribution System

A dedicated FPGA based Timing Distribution System has been developed using Xilinx Spartan-3E FPGA to distribute clock and trigger to DAQ hardware as per application requirements. The control and selection logic has been developed based on activities programmed for DAQ requirements.

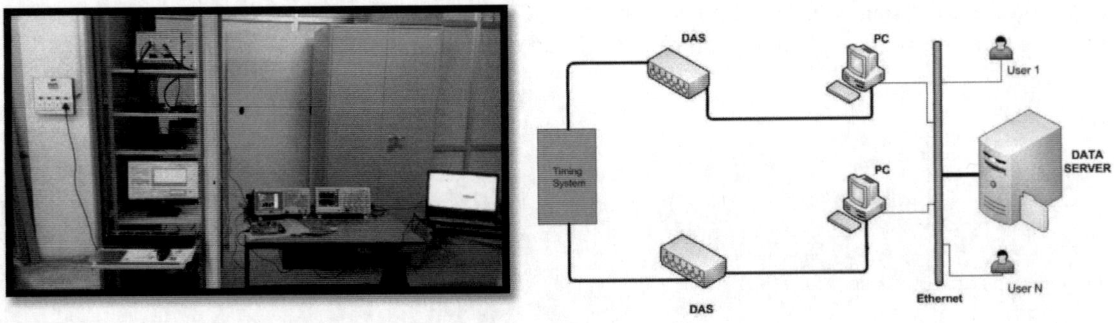

Figure 3. Timing System Interface.

2.3. *Distributed Data Acquisition*

All the three data acquisition hardware provides 16-bit analog to digital conversion interface supporting moderate sampling rate up to 250 KSPS per channel to diagnostics channels. The digitizers have been configured to different sampling rates as per analog signal requirements. All the three DAQ hardware have been arranged to form a distributed data acquisition system and are linked with Ethernet based network from FECs to DAQ server and to other remote machines.

3. Software development

3.1. *FEC Graphical User Interface*

Each FEC is interfaced to DAQ hardware through GUI that configures the hardware settings for acquisition, facilitates data file generation and data storage on local machine. GUIs have been developed in LabVIEW, MATLAB and custom-made application software (Keysight AMM software) to demonstrate the non-dependency of the concept on any particular acquisition software. DAS-1/DAS-2 has been interfaced with LabVIEW based GUI while Keysight AMM software and MATLAB have been used to configure and acquire data from DAS-3.

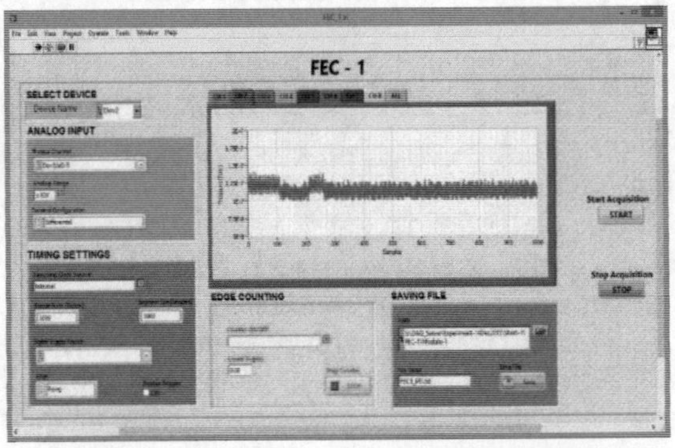

Figure 4. LabVIEW based GUI for DAS-1.

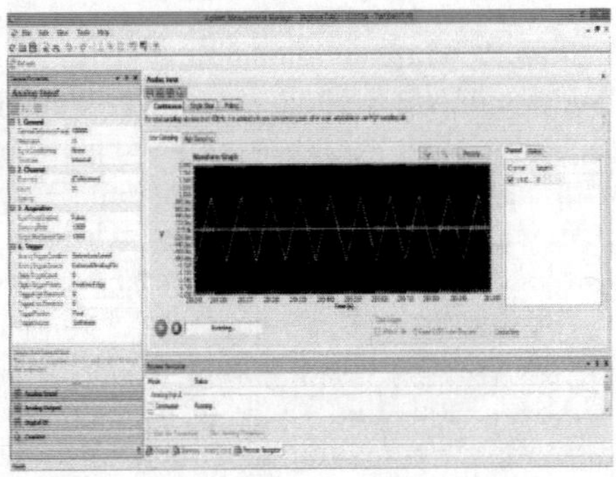

Figure 5. Agilent (AMM) GUI for DAS-3.

3.2. *Data Time-slicing*

To store data stream of long duration (in hours/days) experiment, the concept of time slicing of data stream has been adopted that divides the channel data in user defined chunks in terms of time or samples. Data chunks avoid the problem of non-access of server data until the channel data file is closed at local machine at the end of the experiment. A Python based GUI, as shown in Figure 6, has been developed for configuring an acquired channel data for data chunk size, data file location, server location etc. and it transfers the chunks (data chunk files) to DAQ server. This figure also shows uploading of recently formed data chunks to DAQ server in continuous data acquisition.

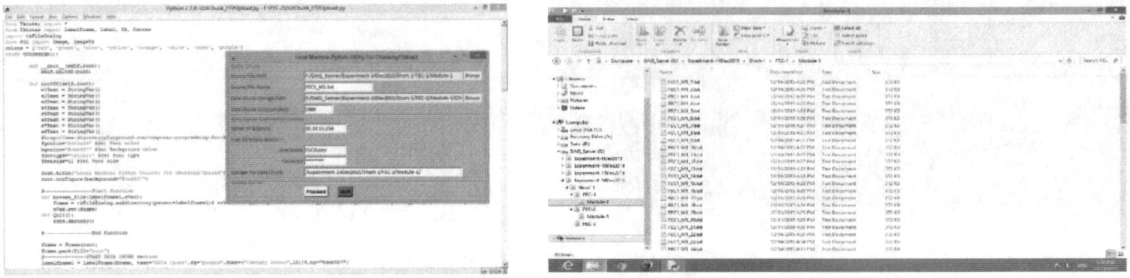

Figure 6. Configuring Data Time-slicing and uploading to DAQ Server.

3.3. Data chunks plotting at User/Client Machine

Data plotting utilities have been developed using Python to plot user-defined chunks for selected channels for on the go viewing of channel data. Down sampling methods have been applied to get the plot at same time scale for channel profile display.

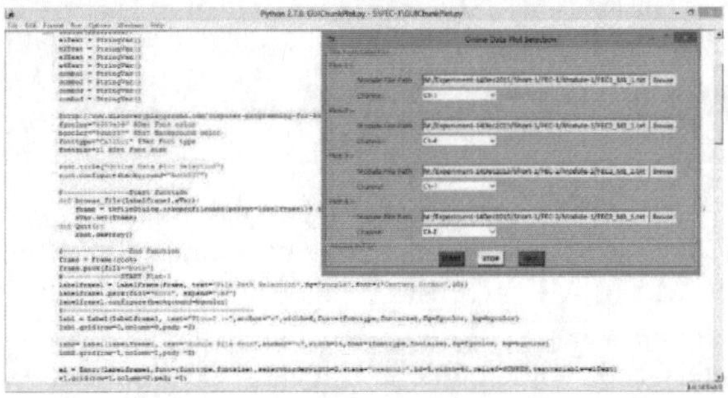

Figure 7. Data plotting utility developed using Python script for data chunks plotting.

4. Test Results

The developed hardware set-up has been successfully tested with supporting software for continuous acquisition with on-the-go viewing from remote machine in the network. We have tested our set-up with 32 channels distributed among three DAS supporting sampling rate from 5 Hz to 50 KHz per channel. Each DAS was tested with its FEC in stand-alone mode for data acquisition. Later on the whole data acquisition set-up was successfully tested for integrated operation for long duration.

All the three DAS generate channels data files continuously for active/configured channels. We have tested the set-up with LabVIEW (DAS-1 and DAS-2), MATLAB and Keysight AMM software (DAS-3) and the final result shows that adopted concept does not depend upon front-end application/acquisition software. The user-defined chunks files are generated from data channel files through Python based utilities and these data chunks files are uploaded to DAQ server from each DAS node. Figure 8 shows uploading of data chunks files to DAQ server in continuous data acquisition.

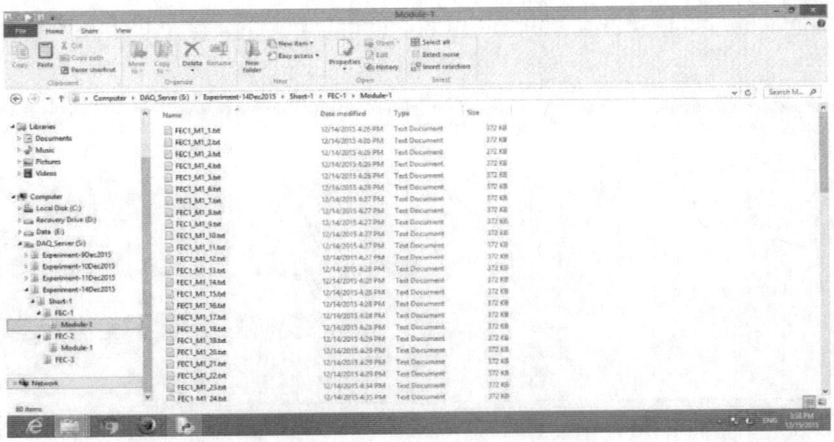

Figure 8. Uploading data chunks to DAQ Server in continuous data acquisition.

We have generated the clock signals for eight different frequency to cater the different sampling rates requirements for channel digitization. The input logic for clock selection has been tested through FPGA board switches as well as control word generated through PC. Clock and trigger signals are distributed to each DAS for time synchronization in data acquisition. All the clock signals are derived from base clock of FPGA board oscillator.

The developed software platform for the data retrieval and scientific data presentation has been successfully tested with mapping server data on client machine. The Python codes for data retrieval were tested for pulse mode (short duration) operation as well as continuous operation through data server. The developed Python based utilities were tested for raw data plotting as well as processed data plotting. The algorithm for data plotting has been operated at different user/client machines on the network. Figure 9 shows Graphical User Interface developed in Python language for data retrieval. This GUI facilitates interface for selection of specified channel(s) where user can select the Shot No., FEC, logical name of DAQ hardware module and the channel(s) for plotting. Also we have added down sampling feature, which may be enabled in GUI for signal processing. Similarly other features may be added in GUI as per application requirements.

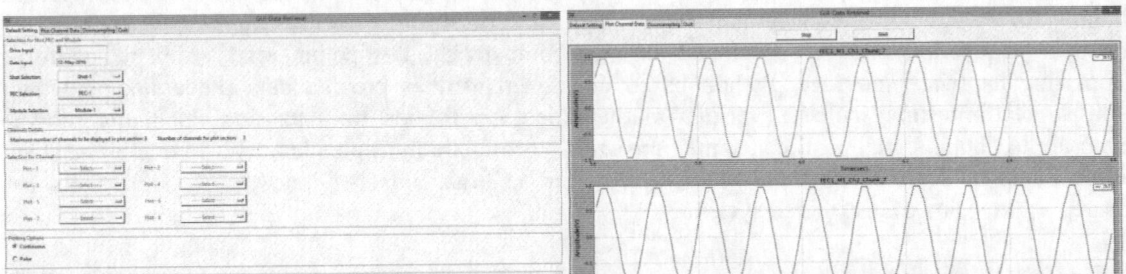

Figure 9. Graphical User Interface – Data Retrieval.

Figure 10 shows test results of data chunks formation and plot update for continuous on-line viewing of data. Here it has been observed that when plot for continuous acquisition is made initially only seven chunks are available in each channel and hence only seventh chunk (current chunk) is displayed in all four channels. The plot of the last chunk for all four channels in the directory is shown. The last chunk is initially seven. Then an eighth chunk file is added at the end while plot for

IOP Conf. Series: Journal of Physics: Conf. Series **823** (2017) 012049 doi:10.1088/1742-6596/823/1/012049

continuous chunks are being drawn, chunks' plot is updated to its eighth chunk file that was added in all channels.

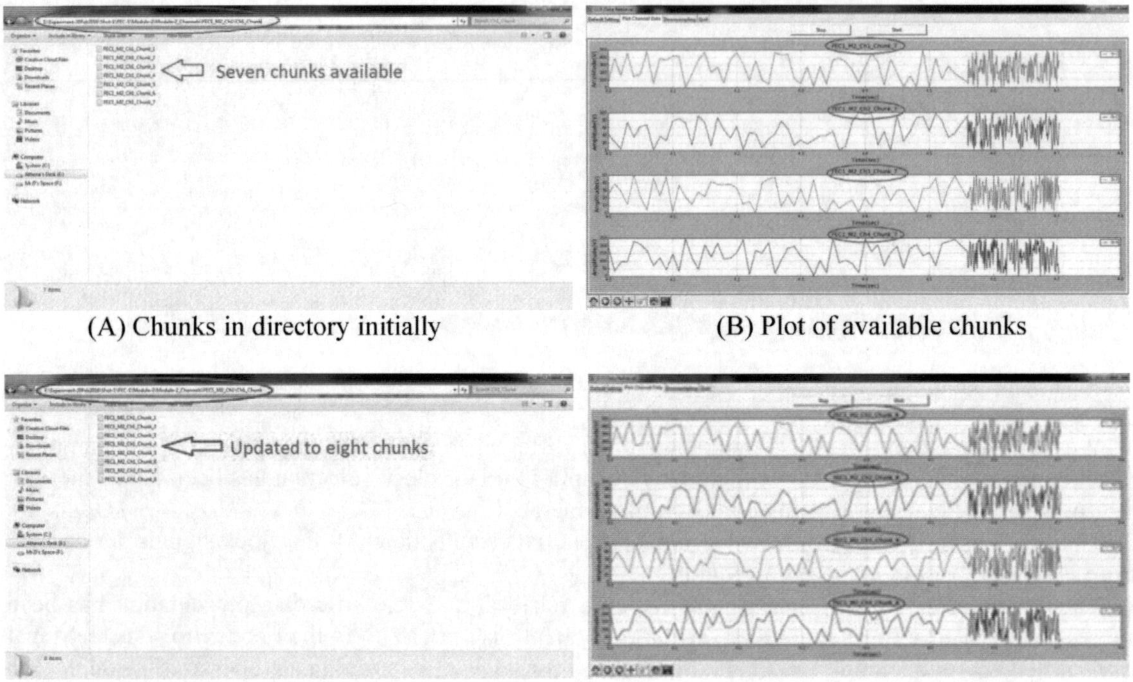

(A) Chunks in directory initially (B) Plot of available chunks

(C) Chunks updated in directory (D) Chunks' plot is updated

Figure 10. Continuous acquisition: Data Chunks formation and plot update

5. Conclusion

Developed set-up supports continuous data archival/retrieval through Time Slicing/Data chunking at multiple acquisition nodes. For on-the-go viewing of selected channels at different user/client machines, channels to be plotted can be configured at any user machines through GUI developed in Python. As the developed concept is not dependent on acquisition software, Graphical User Interface for data acquisition can be developed in any commercial/ non-commercial software platform supporting hardware interface. Python based developed utilities provide data processing and data plotting platform that includes various mathematical operations on data and decimation/down-sampling to obtain signal profile at same time scale to compare multiple channels. The present DAQ set-up could be scaled with more acquisition node and high end DAQ servers/Timing System for future requirements of distributed system.

Acknowledgement

The authors express their sincere thanks to all the group members for their kind co-operation.

References

[1] Jotaki E, Itoh S 1995 *Fusion Technology* **27** 171
[2] Van Haren P C, Oomens N A 1993 *Fusion Technology* **24** 391
[3] Nakanishi H, Emoto M, Nagayama Y et al. 2011 *Nuclear Fusion* **51** Issue 11 article id. 113014, 6 pp
[4] Nakanishi H N, Kojima M, Ohsuna M et al. 2003 *Fusion Engineering and Design* **66** 827

IOP Conf. Series: Journal of Physics: Conf. Series **823** (2017) 012050 doi:10.1088/1742-6596/823/1/012050

Manufacturing and assembly of IWS support rib and lower bracket for ITER vacuum vessel

R Laad[1], Y Sarvaiya[1], H A Pathak[1], J R Raval[2] and C H Choi[2]

[1] ITER-India, IPR, A-29, Electronic Estate, Sector-25, Gandhinagar: 382016, India
[2] ITER Organization, Route de Vinon-sur-Verdon – CS 90046, 13067 St Paul-lez-Durance Cedex, France.

E-mail: rahul.laad@iter-india.org

Abstract. ITER Vacuum Vessel (VV) is made of double walls connected by ribs structure and flexible housings. Space between these walls is filled up with In Wall Shielding (IWS) blocks to (1) shield neutrons streaming out of plasma and (2) reduce toroidal magnetic field ripple. These blocks will be connected to the VV through a supporting structure of Support Rib (SR) and Lower Bracket (LB) assembly. SR and LB are two independent components manufactured from SS 316L(N)-IG, Total 1584 support ribs and 3168 lower bracket of different sizes and shapes will be manufactured for the IWS. Two lower brackets will be welded with one support rib to make an assembly. The welding between SR and LB is a full penetration welding. Total 1584 assemblies of different sizes and shapes will be manufactured. Sufficient experience gained from manufacturing and testing of mock ups, final manufacturing of IWS support rib and lower bracket has been started at the site of IWS manufacturer M/s. Avasarala Technologies Limited (ATL). This paper will describe, optimization of water jet cutting speed on IWS material, selection criteria for K type weld joint, unique features of fixture of assembly, manufacturing of Mock ups, and welding processes with NDTs.

1. Introduction

Support Rib (SR) and Lower Bracket (LB) assembly is a supporting structure of ITER Vacuum vessel to support the IWS blocks. SR and LB are two independent components manufactured from SS316L (N)-IG material through waterjet cutting followed by CNC machining. Manufactured SR and LB components welded by combination of Tungsten Inert Gas (TIG) Welding and Shielded Metal Arc Welding (SMAW). 1584 assemblies of different size and shapes to be manufactured to support approximately 5000 blocks

Figure 1. SR+LB assembly with IWS blocks

IOP Conf. Series: Journal of Physics: Conf. Series **823** (2017) 012050 doi:10.1088/1742-6596/823/1/012050

2. Waterjet Cutting

Waterjet cutting is a cutting method used to cut wide variety of materials using a very high-pressure jet of water, or a mixture of water and an abrasive substance. When the cutting performed by only water jet is called pure waterjet cutting and when the cutting performed by a mixture of water and abrasive is called abrasive jet cutting or simply water jet cutting. The abrasive jet cutting is generally used to cut hard materials.

The main advantage of waterjet cutting is it doesn't superheat the area adjacent to the cut, hence it prevents Heat Affected Zone (HAZ) on the material. This is also a necessary requirement for IWS and its components.

The waterjet cutting is comparatively slow and costly process hence the optimization of waterjet cutting is required for IWS material to cost and schedule estimation for IWS.

3. Optimization of Waterjet Cutting Speed

During the optimization of waterjet cutting speed, considered the various parameters like maximum cutting speed, water jet pressure, nozzle diameter and type of abrasive powder etc. Figure 2 shows the waterjet cutting of sample piece with various cutting speeds and the result shows in table 1.

The waterjet machine used for waterjet cutting has the following specifications. The machines with these specifications are commonly used over the world to cut the various SS material.

Max. pressure: 3800 bar, Cutting pressure: 3600 bar, Material cutting thickness 40 mm.

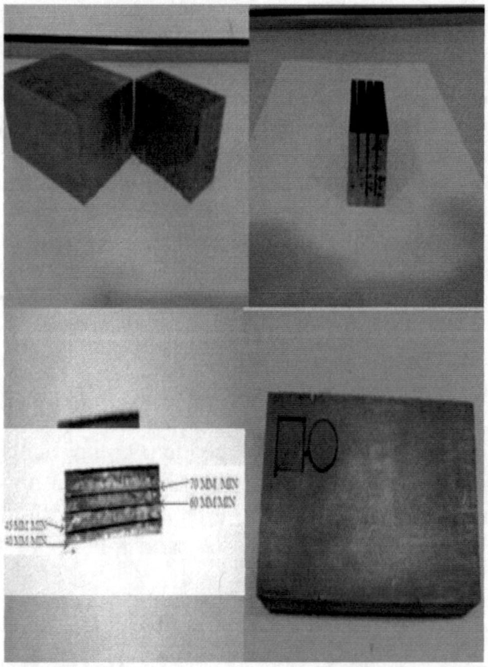

Figure 2. Waterjet cutting on a sample piece with various cutting speeds.

IOP Conf. Series: Journal of Physics: Conf. Series **823** (2017) 012050 doi:10.1088/1742-6596/823/1/012050

Table 1. Cutting results.

Sr No.	Max. cutting speed	Cutting result	Remark
1	70 mm/min	Material not cut properly	1.1 m
2	60 mm/min	Material not cut properly	0.2 m
3	45 mm/min	Material not cut properly	Hammering required to separate the material
4	40 mm/min	Material cut	Very less Hammering required to separate the material
5	30 mm/min	Material cut	No hammering required for straight profile but for circular profile hammering required.
6	25 mm/min (Optimized speed)	Material cut	No hammering required for straight and circular profile.

As a result of this study, Optimised waterjet cutting speed for IWS material is 25 mm/min.
The cutting area with optimised speed requires post CNC machining for desired surface roughness.

4. SR+LB design

The SR and LB assemblies will be subjected to following load conditions in vacuum vessel:

(1) Dead weight: weight of IWS block
(2) Electromagnetic load

Hence it is designed in such a way to take above loads. The selection of weld joint and welding process to bear the static and dynamic loads of IWS blocks are described in next section. The 3D models and 2D drawings of SR and LB assembly are shown in figures below.

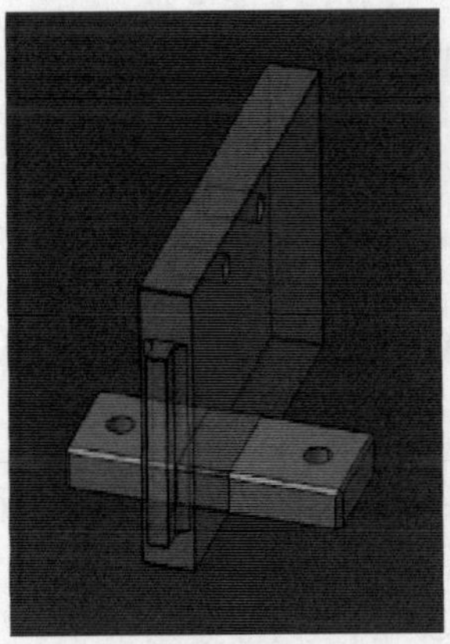

Figure 3. Old Design

Figure 4. New Design

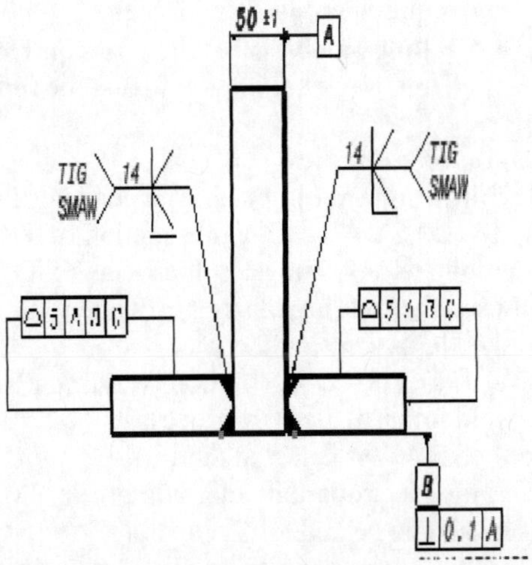

Figure. 4.1. New design drawing

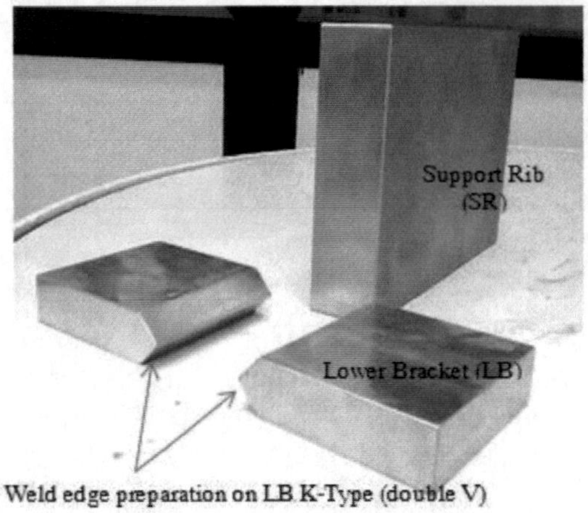

Figure 4.2. K type weld edge preparation on Lower bracket

5. Selection of Weld Joints and Welding Processes

The SR and LB are two independent components and welded together to support the IWS blocks in ITER-Vacuum vessel for life long. Hence it is necessary to select the weld joint and welding processes in such a way that it takes the static and dynamic load of IWS blocks.

5.1. Weld Joint

The following two weld options have been suggested by experts to support IWS blocks.

(1) K-type &
(2) Single side (Single Vee) type joints.

From the above options, K-Type weld joint was chosen for lower bracket with support rib due to following reasons;

- For strength wise, if it is a full penetration weld, both Single V groove and Double V groove joint have the same strength.
- As per Section IX of ASME, Single V groove is not acceptable for more than 16 mm thick, our requirement is 29 mm thick.
- Filling area will be 40 % less in Double V (K type) groove in comparison to single V groove. Therefore heat input will also be reduced, which further reduced the welding distortion.

5.2. Welding Process

The following two options have been suggested by experts to for welding of selected weld joint is
(1) TIG, (2) SMAW, (3) Combination of TIG and SMAW
A combination of Tungsten Inert Gas (TIG) welding at root and Shielded Metal Arc Welding (SMAW) is chosen for welding of assembly due to the following conditions
If only TIG is selected then:
(a) very slow process, (b) more welding distortion (c) Total 29 number of passes will be required to fill the weld joint (d) a costly process.
If only SMAW is selected then:
(a) As the electrode diameter is more in SMAW then the filling of root of selected joint is an issue and result can be lake of fusion at root.

5.3. Unique Fixture for SR +LB Assembly

The 1584 assemblies of different sizes and shapes will be manufactured within fastened tolerance to support IWS blocks in Vacuum vessel, Hence it is necessary to make an unique fixture to perform the selected welding process on different size and shapes assemblies.

Fixture is designed in such a way that whole assembly is held in single setup rigidly and welding is carried out on both sides of k-type weld joint. Lower bracket is arrested in fixture to avoid dis-location, and to minimize distortion during welding. Uniform purging of argon gas is planned by design of closed perforated envelope to avoid blow holes in the weld.

Figure 5. Welding fixtures for poloidal segment (PS) 1 and 2.

6. Welding Qualifications and NDT Criteria

Various mock ups have been manufactured to establish and validate the manufacturing processes, welding and inspection procedures. Process qualification documents WPS (welding procedure specification), PQR (procedure qualification record) and WPQR (welder performance qualification record) have been prepared and approved. Following figures show the manufactured mock-ups. Applicable NDTs are shown in table 2.

Figure 6. Weld edge preparation and welding.

Table 2. NDT of assembly

Sr. No.	NDT	Standards
1	UT (ultrasonic testing)	ASME section V article 4&5, and ASME section VIII div. 1
2	LPT (Liquid penetrant testing)	ASME section V article 6, and ASME section VIII div.1

With the gained experience from mock-ups, actual manufacturing of IWS has been started.

7. UT Challenges and Resolutions.

Ultrasonic Testing of austenitic stainless steel welds has been considered as difficult and very challenging due to high level of scatter and attenuation. It prevents to characterize defects due to low signal to noise ratio, beam splitting, skewing effects and spurious echo from weld. It is much more complicated to perform UT on typical T, K, Y type weld joints.

Hence UT is a very critical and challenging job for IWS SR and LB assembly. This is also impacting IWS and VV schedule. From number of discussions, it is concluded that a comprehensive UT examination and its comparison for welded samples with different probes and methods is required Based on number of discussions and study on coupons UT procedures has been finalized.

ATL successfully completed the UT of SR+LB assemblies of poloidal segment (PS)1 of vessel sector (VS) 6 & 5 and poloidal segment 2 & 4 of vessel sector 6.

8. Manufacturing Status of SR, LB and SR+LB

The continuous welding and inspection is going on to meet the agreed Factory acceptance (FA) date. Following figure shows the status of 164 assemblies out of 1584 assemblies. Status of FA of the completed assembly is shown table 3.

Table 3. Status of FA completed assemblies.

Sr. No.	Sector segment	Quantity
1	PS1VS6	48
2	PS1VS5	48
3	PS2VS6	54
4	PS4VS6	14

Acknowledgement

I acknowledge the support of members of IWS team of ITER-India, IO and ATL in various activities reported in this paper.

ITER Disclaimer

The views and opinions expressed herein do not necessarily reflect those of the ITER Organization.

References

[1] Pathak H A, Raval J R, Pull G S, Laad R, Wang X, Ioki K, Suresh PVS 2013 *IEEE SOFE* p113

IOP Conf. Series: Journal of Physics: Conf. Series **823** (2017) 012051 doi:10.1088/1742-6596/823/1/012051

Assembly & Metrology of First Wall Components of SST-1

Tejas Parekh[1], Prosenjit Santra[1], Prabal Biswas[1], Hiteshkumar Patel[1], Yuvakiran Paravastu[1], Snehal Jaiswal[1], Pradeep Chauhan[1], Gattu Ramesh Babu[1], Arun Prakash A[1], Dhaval Bhavsar[1], Dilip C Raval[1], Ziauddin Khan[1], Subrata Pradhan[1]

[1]Institute for Plasma Research, Near Indira Bridge, Bhat, Gandhinagar 382 428, India

E-mail: tejas@ipr.res.in

Abstract. First Wall Components (FWC) of SST-1 tokamak, which are in the immediate vicinity of plasma comprises of limiters, divertors, baffles, passive stabilizers are designed to operate long duration (1000 s) discharges of elongated plasma. All FWC consists of a copper alloy heat sink modules with SS cooling tubes brazed onto it, graphite tiles acting as armour material facing the plasma, and are mounted to the vacuum vessels with suitable Inconel support structures at ring & port locations. The FWC are very recently assembled and commissioned successfully inside the vacuum vessel of SST-1 undergoing a meticulous planning of assembly sequence, quality checks at every stage of the assembly process. This paper will present the metrology aspects & procedure of each FWC, both outside the vacuum vessel, and inside the vessel, assembly tolerances, tools, equipment and jig/fixtures, used at each stage of assembly, starting from location of support bases on vessel rings, fixing of copper modules on support structures, around 3800 graphite tile mounting on 136 copper modules with proper tightening torques, till final toroidal and poloidal geometry of the in-vessel components are obtained within acceptable limits, also ensuring electrical continuity of passive stabilizers to form a closed saddle loop, electrical isolation of passive stabilizers from vacuum vessel.

1. Introduction

After eleven (11) successful experimental campaigns since superconducting steady-state tokamak (SST-1) [1, 2] was commissioned, the first wall components (FWC) were installed in SST-1 vacuum vessel as one of the phase-1 up gradation objective during October 2014 to May 2015. During initial experimental campaigns, the machine was operated for circular plasma using two (02) sets of graphite based limiters as plasma facing material. Since the entire surface of SST-1 vacuum vessel was directly exposed and the impurities were realized due to plasma wall interaction, it was decided to integrate all the first wall components inside SST-1 vacuum vessel. First wall components (FWC) of SST-1 Tokamak consists of Inboard divertor plates (IDP), Outboard divertor plates (ODP), Inboard passive stabilizers (IPS), Outboard passive stabilizers (OPS) and main baffle (MBAF) [3,4]. All FWCs were required to be aligned along the magnetic field lines with positional accuracy in the range of 1-5 mm. Passive stabilizers were electrically isolated from in-vessel components and support structures in order to have a maximum electrical conductivity which was essential for slowing down the vertical instability growth rate to ensure the feasibility of active feedback control of the vertical instability. Similarly, these all FWCs were thermally connected with interment isolations for baking and cooling requirements.

IOP Conf. Series: Journal of Physics: Conf. Series **823** (2017) 012051 doi:10.1088/1742-6596/823/1/012051

2. Design of SST-1 first wall components

2.1. Design description

The SST-1 machine was designed for steady-state plasma operation of 1000 s. The main consideration in the design of FWC was to steady-state heat removal up to 0.6 MW/m^2 and to bake at 250 °C. The FWCs will also experience a large electromagnetic forces during disruptions due to halo currents and induced eddy currents for which they were designed to handle such forces. Also FWC support structures were designed to be flexible enough to limit the thermal stresses induced due to thermal expansion during baking of FWC. These FWC support structures were in modularity form for easy handling during assembly-disassembly of modules and to provide the flexibility during their alignment inside the vacuum vessel (VV) within the limited space. The supports were having two parts out of which one is welded with the vacuum vessel and other part is bolted with the welded part. This allows the flexibility for adjustments to be made for the proper alignment of FWC [3, 4]. 3-D model of FWCs inside SST-1 vacuum vessel is shown in figure 1.

Graphite tile was chosen as plasma facing material considering its good thermal properties, low atomic number. Copper-Zirconium (Cu-Zr) and Copper-Chromium-Zirconium (Cu-Cr-Zr) alloys plates embedded with SS 304L piping were used as back plate materials for proper heat transfer during baking and cooling operations (figure 2). Approximately 3800 numbers of tiles were mounted on 136 numbers of these copper alloys back-plates.

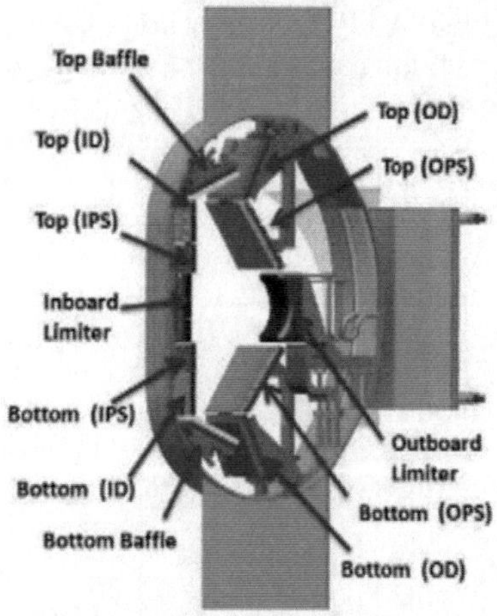

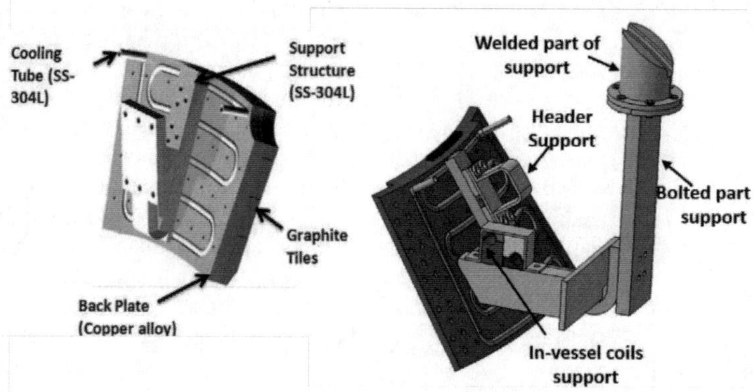

Figure 1. FWC arrangement in SST-1 vessel. **Figure 2.** Cad model of MABF & OPS modules.

2.2. Passive Stabilizers

Passive stabilizers comprised of conducting structures surrounding the plasma to reduce the growth rate of vertical displacement events (VDE). There were two pairs of stabilisers in SST-1 connected in saddle configuration with a current bridge, one on the inboard side and other on the outboard side. The cad model of a typical outboard passive stabilizer module is shown in figure 2. The inner top and bottom stabilizers consist of eight (08) modules each while that of outer top and bottom stabilizers made up of eighteen (16) modules each. The modules were joined with lap joints having lap area of 70 mm in toroidal direction and full width of the stabilizer in the poloidal direction. 99.99% alumina based ceramic bushings, washers and plates have been used to keep the stabilizers electrically isolated from the welded part of the support.

2.3. Divertor & Baffle

The inboard and outboard divertor plates were designed for the worst possible heat loads by assuming in-out asymmetry of 1:2 (Single null) for the inboard divertor and 1:4 (Double null) asymmetry for outboard divertor. A baffle has been incorporated in the design so as to form a closed divertor configuration, which helps in increasing the neutral pressure in the divertor region and thus improving the neutral particle recycling.

2.4. Poloidal Limiters

There were two outboard limiters placed 180° apart on radial port numbers R-3 and R-11. Similarly there were two inboard limiters are placed at the same toroidal locations. These limiters were placed to accommodate various plasma equilibrium configurations from circular to highly D-shaped plasmas. The inboard side limiter acts as a safety limiter placed 30 mm away from the separatrix. The front face of the limiter is flush with inner passive stabilizers and divertors.

2.5. Cooling and baking of FWC

All PFCs were cooled and baked through SS 304L tubes brazed in the copper-alloy back plates. Same hydraulic circuit was used for both for cooling and backing of FWC. De-ionized water at about 30 °C and 5 bar (a) pressure was used for cooling of FWC while hot nitrogen gas at about 350 °C and 4.5 bar (a) pressure was used for baking. The distribution headers were designed such that MBAF, IDP and IPS (top/bottom) were connected in series and fed through two supply headers each having eleven (11) parallel sub-connections while ODP and OPS (top/bottom) were connected in series and fed through another two supply headers, each having sixteen (16) parallel sub-connections. In order to avoid eddy current during plasma disruption and $j \times B$ forces acting on headers, all these modules and their headers were electrically isolated from each other [5, 6].

2.6. Tile design and mounting

One of the major constraints of FWC design was the tile attachment scheme. Even though graphite tiles could be brazed to copper back plate for high heat transfer, but the mechanically attached tile system was opted. For inner and outer divertors 25 mm thick graphite tiles were bolted to actively cooled copper back plate having 0.5 mm thick flexible graphite compliant layer. In the case of limiter, the tile thickness varies from 40 to 20 mm, as it was profiled toroidally. Passive plates were covered by 20 mm graphite tiles. This design has ensured the graphite surface temperature to be less than 1000 °C.

3. Metrology and Alignment procedures

In order to identify the possible issues and assembly sequences during commissioning inside SST-1 vacuum vessel, two stage metrology was planned and implemented. In 1st stage, the ground activities were performed while in 2nd stage the integration inside the vessel performed.

3.1. Stage 1: Ground activities

During 1st stage, the measurement of all back plates was carried out with the help of templates and laser scanning of surfaces by coordinate measuring machine (CMM). The point cloud data was generated for back plate surfaces using scanning technique and this data was compared with 3-D cad model of Catia V5 software. The maximum deviation in the profile of the modules was found to be 1-2 mm. All the graphite tiles were then assembled to their respective modules for identification of fabrication errors in tiles. Gap between the rows and columns of tiles and their final profiles were checked with the help of go-no-go gauges and required modifications were implemented. During this trial, the ground assembly of complete stabilizer circuit was performed in order to check the compatibility of lap joints as shown in figure 3.

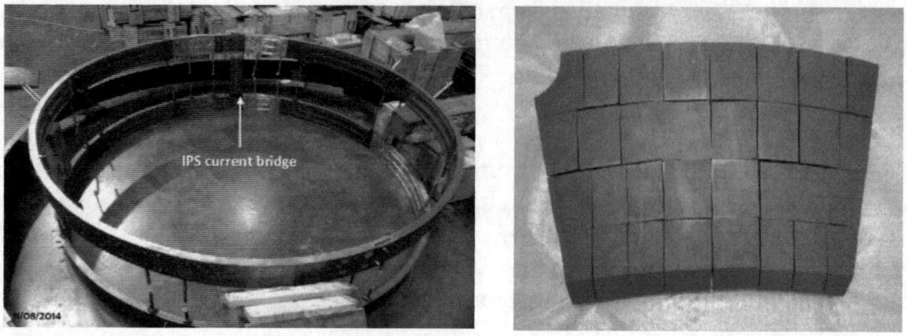

Figure 3. Ground assembly of Inner passive stabilizers & mock up tiles assembly.

3.2. Stage 2: In-vessel activities

In order to avoid the local heat load congregation and ablation of FWC materials, all FWCs have to be assembled with accuracy of 2-3 mm. Since in-vessel measurement of any object with such accuracy inside the tokamak device is difficult, the best suitable measurement process was identified by mock up assembly of 2 sets of FWC in the prototype octant of SST-1 vacuum vessel. The co-ordinates of points required for the positioning of welded supports were identified from engineering drawings. These points were marked on the vessel surface using theodolite based total station called Electronic Coordinate Determining System (ECDS). In house developed photogrammetry method has also been used for verifying these in-vessel marks. Thus by using both ECDS and photogrammetry, the reference imaginary axis was created and then the coordinate system was generated. The coordinates of centre points of all four sides of radial port were measured as shown in figure 4. Then the coordinates of in-vessel points with respect to centre points of radial port sides were measured using photogrammetry [7]. After many iterations, the theoretical coordinates of all the supports were marked on the vessel surface. Welded part of the each support was welded at their respective locations and further assembly of two complete poloidal rings was done in the prototype vessel. For further improvement, the precise templates were used for the final alignment of FWC surfaces.

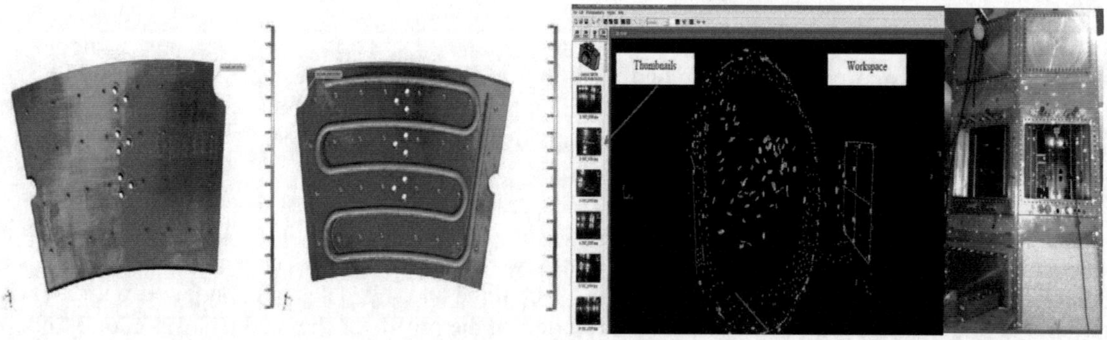

Figure 4. Comparison of cad model and point cloud data generated through scanning of surfaces through CMM (left) & user interface of photogrammetry software with prototype vessel (right)

Based on prototype assembly experience, the assembly and alignment of FWCs inside SST-1 VV was carried out considering inboard limiter as a reference. Back plate of the inboard limiter was aligned as per drawings with the help of ECDS and photogrammetry technique as discussed in section 2.2. Marking of IPS supports was carried out using precise templates having profile of IPS back plate with respect to limiter surface. For this, the supports were assembled with IPS back plate and then the surface of IPS back plate was aligned with respect to limiter surface and the location of support was marked on the vessel surface. IPS supports were welded on this marked location and IPS back plate was assembled. The marking of adjacent modules of IPS was comparatively easy as the complete IPS is connected through the lap joints. So adjacent IPS back plates were connected through laps and their supports were marked and welded on vessel surface. Complete assembly of IPS ring was completed by following this methodology.

IDP and baffles were also assembled by following the same technique. However these components were aligned with respect to more reference surfaces. Further, the reference of limiter back plate as well as the reference of IPS back plate were taken for the alignment of IDP back plate at limiter location. This procedure was followed for the marking of supports for all FWC back plates and complete assembly of back plates was completed. Suitable fixtures were developed for handling copper back plates and making adjustments for aligning the surfaces inside the vacuum vessel. Finally a precise check template were designed as per the theoretical cross-section of FWC to check the surfaces of assembled back plates. The complete assembly of FWCs along with their headers & sub-headers is shown in figure 5.

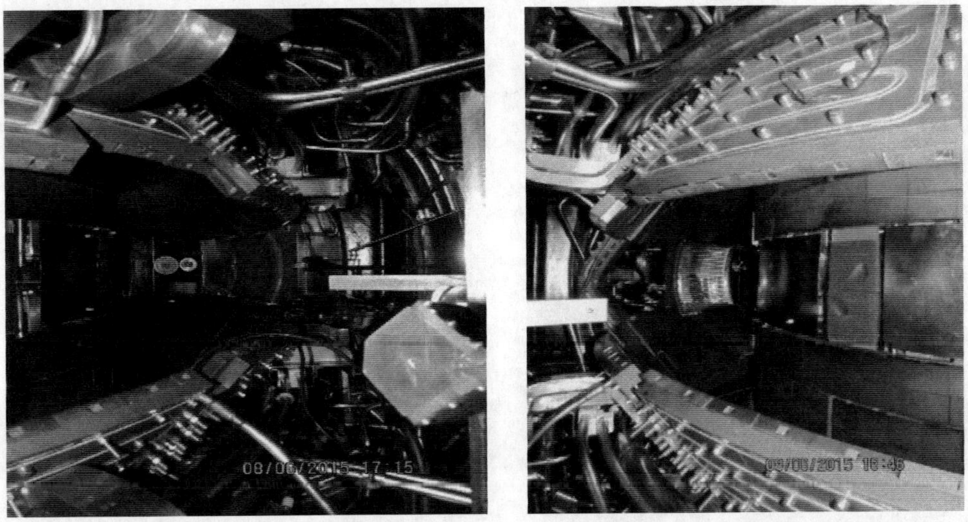

Figure 5. Inside view of vacuum vessel after complete assembly of FWCs.

4. Conclusion

Assembly of all FWC components was successfully completed within eight (08) months of time scale. The alignment of all the FWCs was done with the reference of inboard limiter surface. The inboard limiter was aligned as per the designed location with the help of in-housed developed photogrammetry and ECDS machine with an accuracy of 1-2 mm. Final deviations of FWC surfaces was found in the range of 3-5 mm. The first plasma of 33 KA with the duration of 160 ms was obtained after the installation of FWC in July, 2015. In the latest campaign of SST-1, 112 KA of plasma current was also obtained.

Acknowledgement

We would like to acknowledge, SST-1 vacuum group, Kirit Vasava (IPR Drafting section) for procurements of new components, leak testing activities, material identification & marking, baking of graphite tiles and making them available time to time for completing the assembly within schedule time of eight (08) months.

IOP Conf. Series: Journal of Physics: Conf. Series **823** (2017) 012051 doi:10.1088/1742-6596/823/1/012051

References

[1] Pradhan S and Team SST-1 mission 2010 *J. Fusion Res. Series* **9** 650
[2] Pradhan S, Sharma A N, Tanna V L, Khan Z, Prasad U et al. 2012 *IEEE Trans. Plasma Sci.* **40** 614
[3] Jacob S, Reddy D C, Choudhury P et al.1998 *Proceedings of the 1st IAEA Technical Committee Meeting on Steady State Operation of Tokamaks* (Hefei, China) p 63
[4] Prakash N R, Chaudhuri P et al. 2001 *Fusion Engineering and Design* **56-57** 355
[5] Khan Z, Paravastu Y and Pradhan S 2014 *Procedia Material Science* **6** 216
[6] Paravastu Y, Khan Z and Pradhan S 2015 *Fusion Engineering and Design* **98-99** 1375
[7] Babu G R, Khan Z and Pradhan S 2014 *IPR internal report* IPR/TR-297/2014

IOP Conf. Series: Journal of Physics: Conf. Series **823** (2017) 012052 doi:10.1088/1742-6596/823/1/012052

Effect of Geometrical Imperfection on Buckling Failure of ITER VVPSS Tank

Saroj Kumar Jha[1], Girish Kumar Gupta[1], Manish Kumar Pandey[1], Avik Bhattacharya[1], Gaurav Jogi[1], Anil Kumar Bhardwaj[1]

[1]ITER-India, Institute for Plasma Research, Block A, Sangath Skyz, Bhat-Motera Road, Koteshwar, Ahmedabad - 380005, Gujarat, India

E-mail: saroj.jha@iter-india.org

Abstract. The 'Vacuum Vessel Pressure Suppression System' (VVPSS) is part of ITER machine, which is designed to protect the ITER Vacuum Vessel and its connected systems, from an over-pressure situation. It is comprised of a partially evacuated tank of stainless steel approximately 46 m long and 6 m in diameter and thickness 30 mm. It is to hold approximately 675 tonnes of water at room temperature to condense the steam resulting from the adverse water leakage into the Vacuum Vessel chamber. For any vacuum vessel, geometrical imperfection has significant effect on buckling failure and structural integrity. Major geometrical imperfection in VVPSS tank depends on form tolerances. To study the effect of geometrical imperfection on buckling failure of VVPSS tank, finite element analysis (FEA) has been performed in line with ASME section VIII division 2 part 5 [1], 'design by analysis method'. Linear buckling analysis has been performed to get the buckled shape and displacement. Geometrical imperfection due to form tolerance is incorporated in FEA model of VVPSS tank by scaling the resulted buckled shape by a factor '60'. This buckled shape model is used as input geometry for plastic collapse and buckling failure assessment. Plastic collapse and buckling failure of VVPSS tank has been assessed by using the elastic–plastic analysis method. This analysis has been performed for different values of form tolerance. The results of analysis show that displacement and load proportionality factor (LPF) vary inversely with form tolerance. For higher values of form tolerance LPF reduces significantly with high values of displacement.

1. Introduction

The 'Vacuum Vessel Pressure Suppression System' is part of ITER machine. VVPSS is designed to protect the ITER Vacuum Vessel and its connected systems, from an over-pressure situation. It is comprised of a partially evacuated tank of stainless steel, with dimension approximately 46 m long and 6 m in diameter and thickness of 30 mm. Material of tank is SS304/304L dual mark. VVPSS Suppression Tank is connected to vacuum vessel by relief line which includes rupture disc and bleed valves. The VVPSS shall limit the Vacuum Vessel internal pressure, in the case of loss of coolant from the in-vessel components, to 0.15 MPa absolute by opening the rupture discs to let the steam from the vacuum vessel flow to the VVPSS Suppression Tank, where it can condense. In the case of small in-vessel water ingress or other loss of vacuum accidents, bleed valves open to connect the Vacuum Vessel to the VVPSS Suppression Tank.

IOP Conf. Series: Journal of Physics: Conf. Series **823** (2017) 012052 doi:10.1088/1742-6596/823/1/012052

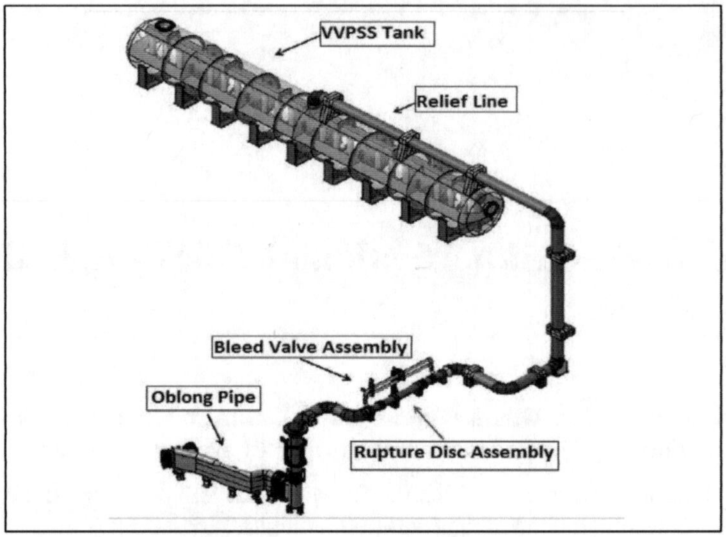

Figure 1. ITER VVPSS System

Geometrical imperfection has significant effect on buckling failure of any vacuum vessel. Major Geometrical imperfection comes from the form tolerance (Permissible out of roundness). Effect of geometrical imperfection on buckling failure of VVPSS suppression tank has been studied by FE analysis in line with ASME BPVC Section VIII Division 2. This Study has been performed for different values of form tolerance (60 mm, 40 mm, 30 mm and 15 mm) and results are shown in this paper.

2. FE Analysis

Linear static analysis, linear buckling analysis and nonlinear buckling analysis of ITER VVPSS Tank has been performed in ANSYS WORKBENCH 15.0. Analysis has been performed for Normal Operation Load case.

2.1. FE Model

VVPSS Tank and part of relief line has been considered for analysis. Geometric model has been prepared by creating a shell model by extracting mid surface from solid CATIA model of VVPSS Tank and relief line. Fillet joints have been modeled by MPC 184 rigid connection. To consider the inertia effect of tank internal, point masses has been applied at Centre of gravity of internal stiffeners.

Model has been meshed with SHELL281 elements and bonded contacts are defined.

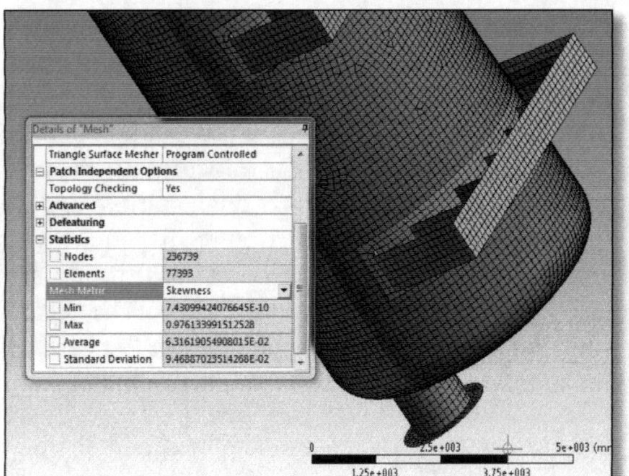

Figure 2.a. Mesh Details (Skewness)

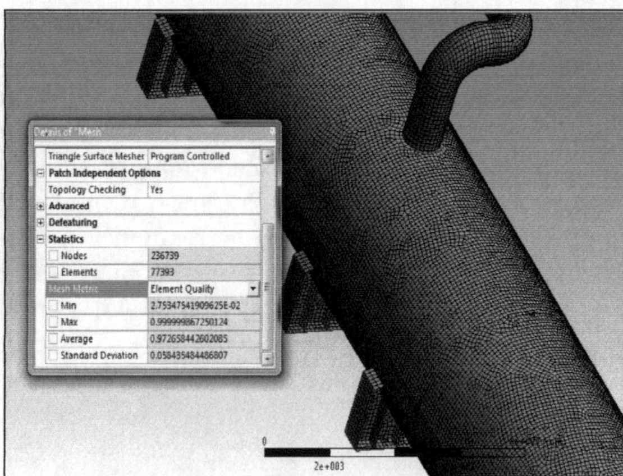

Figure 2.b. Mesh Details (Element Quality)

2.2. Loads
Analysis has been done for Normal operation load case. This load case includes dead weight of tank and relief line, external pressure of 0.1MPa. Tank has been considered half-filled and corresponding to that variable hydrostatic pressure of fluid has been applied to tank. The dead weight of the water has been modelled by modifying the density of the lower half of the tank.

2.3. Linear Analysis
Linear Static structural analysis and Linear Buckling analysis has been performed

First Linear static analysis has been performed for above said load case and plastic collapse and local failure of tank has been assessed as per [1]. Tank has been found safe against plastic collapse and local failure mode.

Linear Elastic buckling analysis has been performed to find the modes of buckling. First Buckling Mode has been amplified up to the form tolerance value to create deformed geometry for nonlinear analysis.

2.4. Non Linear Buckling Analysis
Nonlinear buckling analysis has been performed in line [1] by Elastic plastic method.

2.4.1. Geometrical Imperfection. Geometrical imperfection due to form tolerance of 60mm is incorporated in FE model. Deformation of first mode of linear buckling has been scaled up to 60 mm i.e. up to the form tolerance value. This deformed geometry has been used as input geometry for nonlinear buckling analysis.

2.4.2. Material Nonlinearity. Material Nonlinearity has been included by incorporating the true stress strain graph of material in analysis.

2.4.3. Load combination. Nonlinear buckling analysis has been done for Normal operation load case. The load combination has been followed from ASME section VII division 2. Buckling analysis has been performed up to 3 load proportionality factor (LPF).

Same analysis has been repeated for form tolerance value of 40 mm, 30 mm and 15 mm.

3. Results and Discussions

3.1. Results of Linear analysis
In linear static analysis membrane stress value is 39.9 MPa and sum of membrane and bending stresses are 67 Mpa. Allowable value for membrane stress is 138 MPa and that is for sum of membrane and bending stress is 207 MPa. Plastic Collapse and local failure has been assessed on the basis of [1] by linear elastic analysis.

In linear buckling analysis for the first mode of buckling, the buckling load factor is 15.02. Deformation of this mode has been scaled up to the form tolerance value to incorporate the geometrical imperfection in nonlinear buckling analysis as this is the most likely mode of buckling.

3.2. Results of nonlinear analysis
Nonlinear buckling analysis for different form tolerance values has been performed. The load proportionality factors, displacement and equivalent plastic strain has been noted for different values of form tolerance and presented in Table 1.

IOP Conf. Series: Journal of Physics: Conf. Series **823** (2017) 012052 doi:10.1088/1742-6596/823/1/012052

Table 1. Results for buckling analysis.

Form Tolerance	LPF	Displacement at LPF =1 (in mm)	Equivalent Plastic Strain at LPF =1	Displacement at Buckling Point (in mm)
15mm	2.13	11.3	0.00116	181.2
30mm	2.02	15.2	0.00126	242.3
40mm	1.83	22.71	0.00164	259.1
60mm	1.69	39.1	0.00244	272

Divergence of analysis has been observed for different form tolerance values. It has been observed that solution diverges with sudden rise in displacement value. This sudden rise in displacement signifies the buckling failure shown in Figure 3. The LPF has been noted at point of buckling. It has been observed that LPF varies inversely with form tolerance. It has been observed that LPF reduces significantly with increase in form tolerance shown in Figure 4.

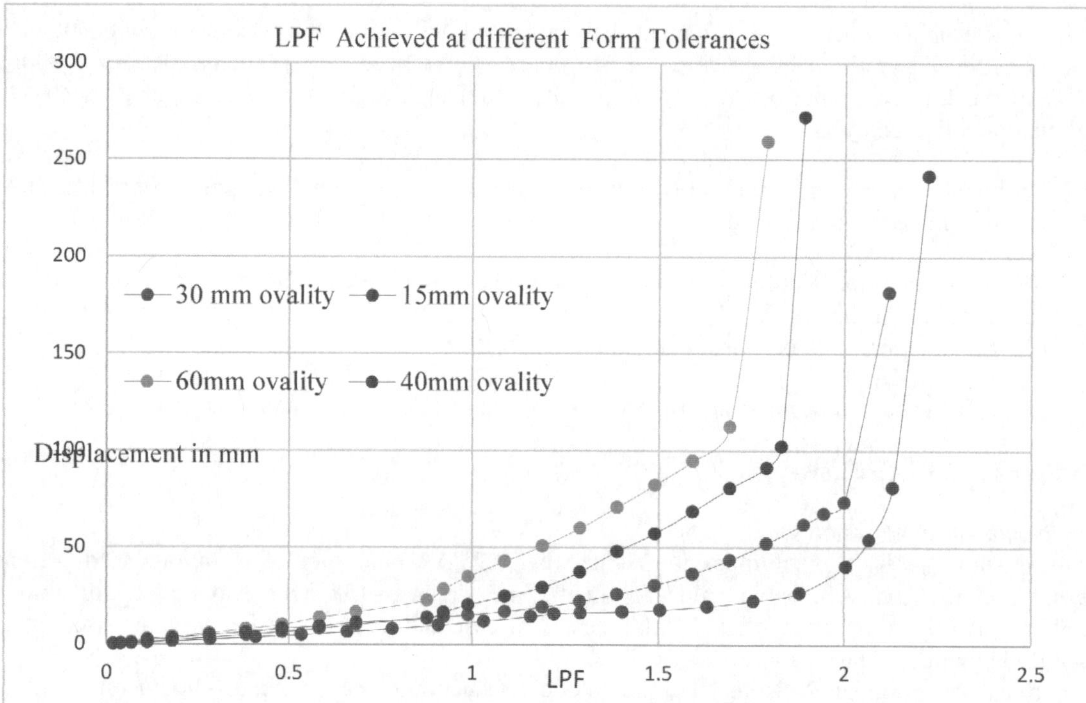

Figure 3. LPF achieved and buckling point at different form tolerances

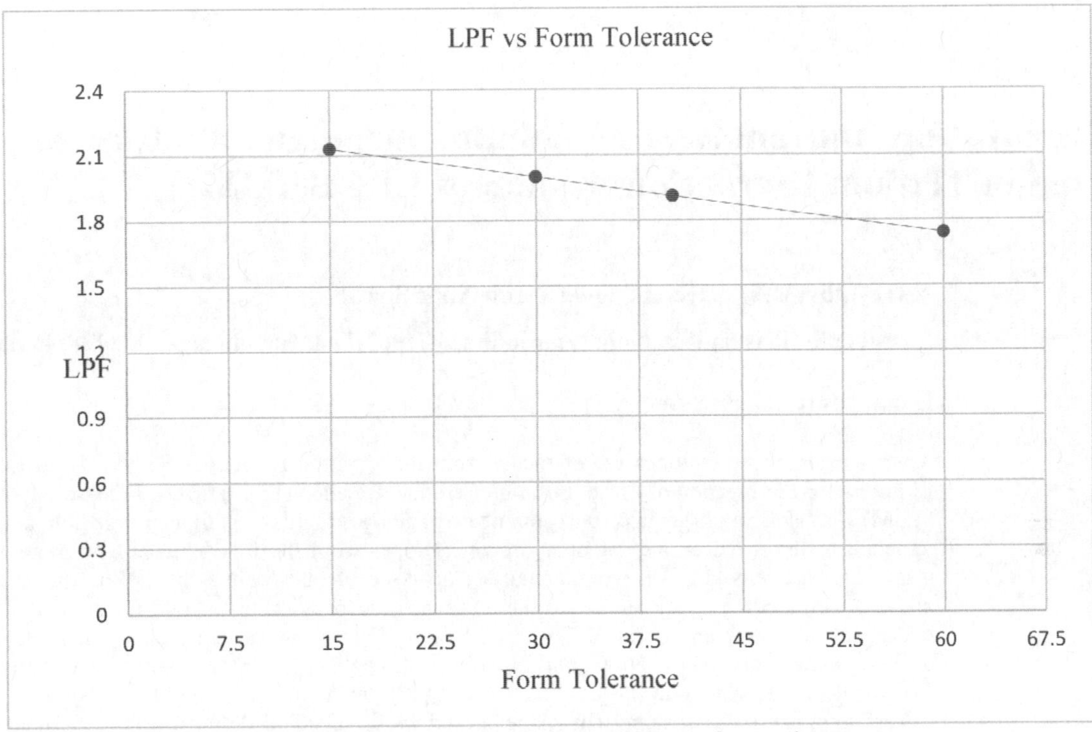

Figure 4. Variation of LPF with different form tolerances

4. Conclusion

The results of Study shows that load proportionality factor (LPF) vary inversely with form tolerance. Displacement and equivalent plastic strain Varies proportionally with form factor .For higher values of form tolerance LPF reduces significantly with high values of displacement.

Acknowledgement

This study was supported by Cryostat and VVPSS Group, ITER-India.

References

[1] ASME, "Rules for the Construction of Pressure Vessels," *ASME Boiler and Pressure Vessel Code, Section VIII, Division 2, Alternative Rules*, (2010).

IOP Conf. Series: Journal of Physics: Conf. Series **823** (2017) 012053 doi:10.1088/1742-6596/823/1/012053

Preparation and analysis of helium purge gas mixture to be used in Tritium Extraction System of LLCB TBM

V Gayathri Devi[1], Deepak Yadav[1] and Amit Sircar[1]

[1]Institute for Plasma Research, Near Indira Bridge, Bhat, Gandhinagar 382428, India

E-mail: asircar@ipr.res.in

Abstract. Hydrogen isotopes are extracted from the Ceramic Breeder (CB) and liquid Lead Lithium (Pb-Li) breeder of Lead Lithium Ceramic Breeder (LLCB) Test Blanket Module (TBM) with Helium purge gas. 1000 ppm of hydrogen gas is mixed with the purge helium gas to facilitate improved extraction of hydrogen isotopes from the breeder zones by hydrogen swamping reactions [1]. An experimental set up is developed for making up the purge gas mixture with a composition similar to the purge gas composition to be used for extraction of hydrogen isotopes from CB and Pb-Li of LLCB TBM. This is achieved by introducing different ppm levels (1000 - 5000 ppm) of hydrogen in helium gas by flow control mechanism. The analysis of the purge gas mixture is performed using a highly sensitive Gas Chromatography (GC) system. This paper describes the detailed design of the experimental set-up and results for the analysis of different concentrations of hydrogen in helium purge gas.

1. Introduction

Helium purge gas with different ppm levels of hydrogen gas (1000 – 5000 ppm) is prepared by mixing hydrogen gas (99.999 % pure) to helium gas (99.999 % pure) in the mixing chamber through partial pressure method. Schematic of the experimental setup for preparation of helium purge gas is shown in figure 1. In partial pressure method, hydrogen gas is filled in the mixing chamber at a known flow rate and pressure. Once the pressure of the gas in the chamber reaches the desired value, the flow of hydrogen is stopped and helium gas is passed at a known flow rate and pressure till the total pressure in the mixing chamber reaches the desired value. By varying the partial pressure of hydrogen in the mixing chamber, the concentration of hydrogen gas in helium purge gas can be varied.

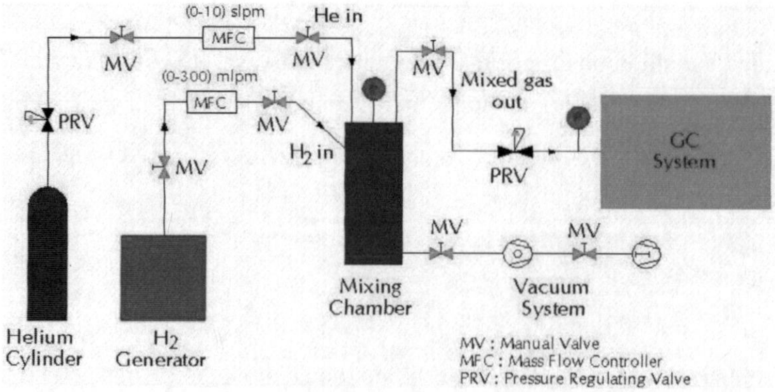

Figure 1. Schematic of experimental setup for helium purge gas preparation

2. Design of the mixing system

The first step in the design of the mixing system involves the process and mechanical design of the mixer chamber of dimensions 0.154 m (diameter) × 1.5 m (height). The process design is initially carried out based on the system requirement and design basis of the experiment. The mechanical design and analysis is carried out using ASME Sec. VIII, Div. 2, 2007 code. The mixer chamber is assembled with a hydrogen generator and helium supply with interconnecting SS 304L tubing. The mixing chamber is connected with a rotary vacuum pump for initial evacuation. The mixing system comprises of mixing chamber, pressure gauges, mass flow controllers for helium and hydrogen, valves and GC system

3. Experimental

The experiments have been started in continuous mode, where helium and hydrogen are continuously supplied with required pressures and flow rates. However, it has been observed that the concentration of hydrogen in the mixture steadily decreases and eventually becomes zero. This is due to the fact that partial pressure (hence, flow rate) of helium (~10^5 Pa) is very high compared to that of hydrogen (~10^2 Pa).Therefore, experiments are performed in semi continuous mode, where the mixing chamber is first evacuated and then filled out with hydrogen gas till the required pressure (to achieve required concentration) is obtained. This hydrogen is generated by a hydrogen generator using non-alkali electrolysis technology. The purity of hydrogen generated is 99.999 %. Mixing chamber is then filled out with helium till the total pressure in the chamber corresponds to desired concentration of hydrogen in helium gas. The helium purge gas mixture thus prepared is then sent to a GC system at 1×10^5 Pa for analysis [2]. Operating parameters and details for the helium purge gas preparation system are described in table 1.

Table 1. Operating parameters and details of helium purge gas preparation system.

Parameters	Values
Process gas	He gas + H_2 gas
Method	Semi-continuous, partial pressure method
Concentration range of hydrogen	1000 – 5000 ppm
Volumetric flow rate of hydrogen	10 – 500 ml/min
Pressure of mixing chamber	3×10^5 Pa
Volumetric flow rate of helium	10 – 50 l/min
Inlet pressure to GC	1×10^5 Pa
Analysis Technique	GC with ionization detector

4. Experimental results

A standard gas mixture containing hydrogen (9.4 ppm), oxygen (14.8 ppm) and nitrogen (11.9 ppm) in helium gas is used for calibrating the GC. A chromatogram is obtained for this standard gas mixture detecting hydrogenat retention time of 0.445 min followed by oxygen at 0.813 min and nitrogen at 1.592 min as shown in figure 2.

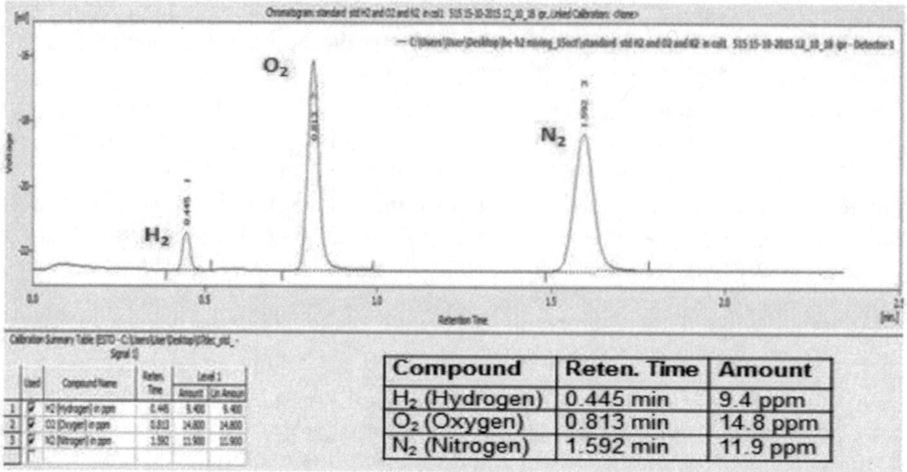

Figure 2. Chromatogram for standard gas mixture of H_2, O_2 and N_2 in helium gas

Helium purge gas with different concentration of hydrogen has been prepared. Chromatograms of helium purge gas with different concentration of hydrogen (1000, 2000, 3000, 4000 and 5000 ppm) are shown in figures 3(a) to 3(e).

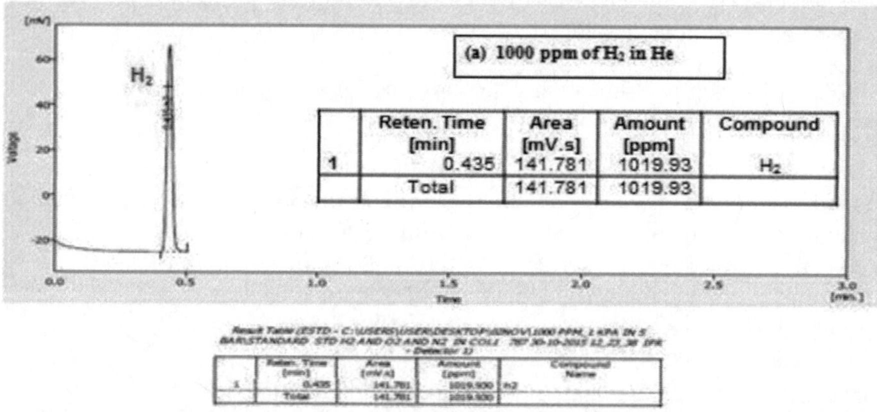

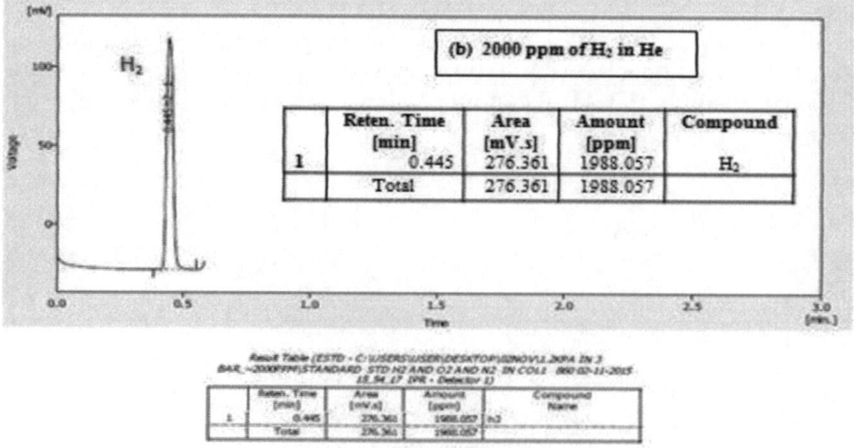

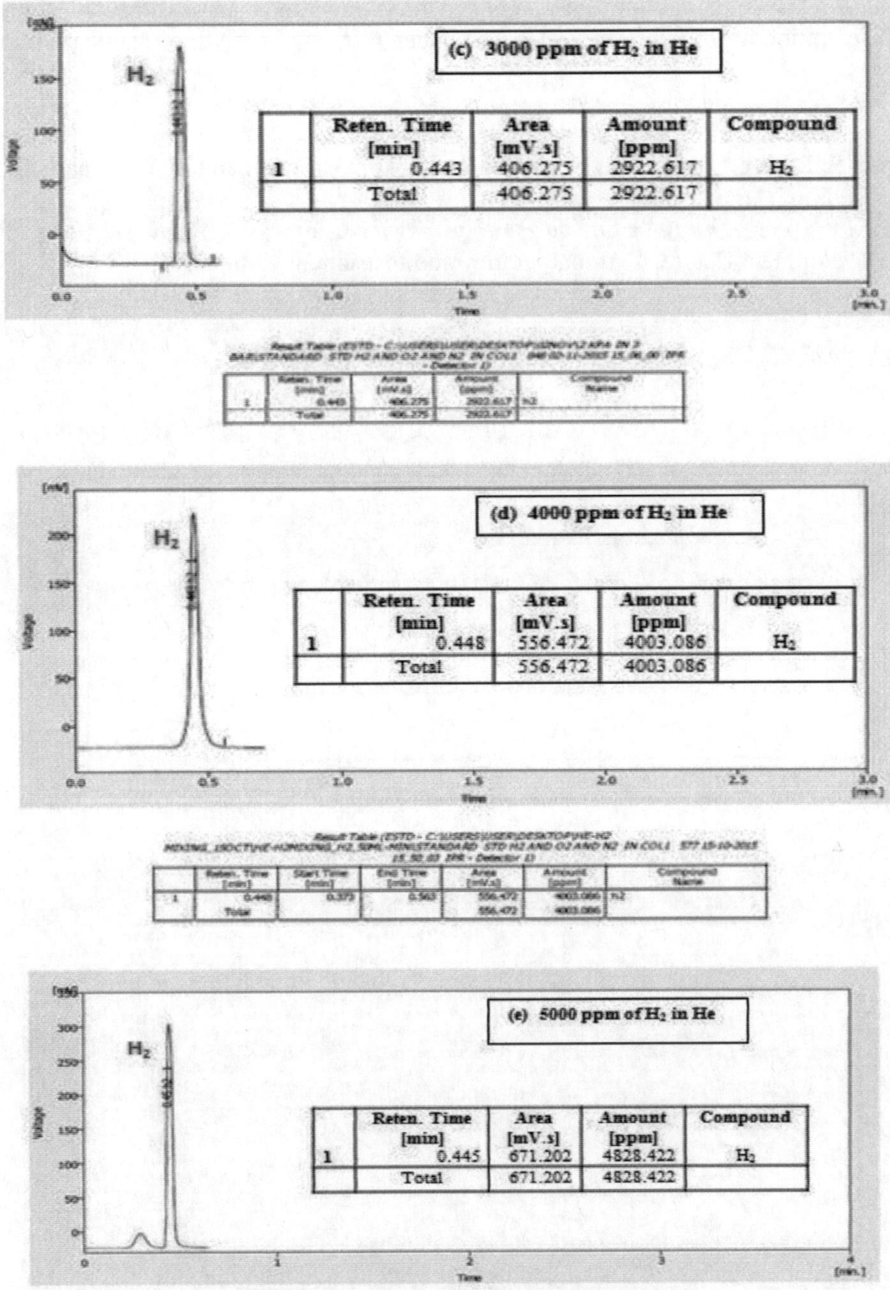

Figure 3. Chromatograms of helium gas with different concentrations (1000, 2000, 3000, 4000 and 5000 ppm) of hydrogen gas.

5. Conclusions

Different concentration of hydrogen (1000 – 5000 ppm) has been introduced in helium gas by varying partial pressure of hydrogen keeping total pressure ~ 3×10^5 Pa in a mixing chamber. The experiments are performed in semi continuous mode where hydrogen gas is first filled in the mixing chamber at desired pressure followed by helium gas. This gas mixture is then sent to the GC for analysis of hydrogen concentration in Helium. The hydrogen concentration is found to be stable (within ± 5 %) in this method.

References

[1] Chaudhuri P, Rajendra Kumar E, Sircar A, Ranjithkumar S, Chaudhari V, Danani C et al. 2012 *Fusion Eng. Des.* **87** 1009

[2] Shingari M K 1988 *Hand Book on Chromatography for Chemists & Engineers* (India: Research and Development Centre, Chromatography and Instrument Company)

IOP Conf. Series: Journal of Physics: Conf. Series **823** (2017) 012054 doi:10.1088/1742-6596/823/1/012054

Role of Outgassing of ITER Vacuum Vessel In-Wall Shielding Materials in Leak Detection of ITER Vacuum Vessel

A Maheshwari[1], H A Pathak[1], B K Mehta[1], G S Phull[1], R Laad[1], M S Shaikh[1], S George[2], K Joshi[2] and Z Khan[2]

[1]ITER-India, Institute for Plasma Research, Bhat, Gandhinagar 382 428, India
[2]Institute for Plasma Research, Bhat, Gandhinagar 382 428, India

E-mail: abham@iter-india.org

Abstract. ITER Vacuum Vessel is a torus-shaped, double wall structure. The space between the double walls of the VV is filled with In-Wall Shielding Blocks (IWS) and Water. The main purpose of IWS is to provide neutron shielding during ITER plasma operation and to reduce ripple of Toroidal Magnetic Field (TF). Although In-Wall Shield Blocks (IWS) will be submerged in water in between the walls of the ITER Vacuum Vessel (VV), Outgassing Rate (OGR) of IWS materials plays a significant role in leak detection of Vacuum Vessel of ITER. Thermal Outgassing Rate of a material critically depends on the Surface Roughness of material. During leak detection process using RGA equipped Leak detector and tracer gas Helium, there will be a spill over of mass 3 and mass 2 to mass 4 which creates a background reading. Helium background will have contribution of Hydrogen too. So it is necessary to ensure the low OGR of Hydrogen. To achieve an effective leak test it is required to obtain a background below 1×10^{-8} mbar l s^{-1} and hence the maximum Outgassing rate of IWS Materials should comply with the maximum Outgassing rate required for hydrogen i.e. 1×10^{-10} mbar l s^{-1} cm^{-2} at room temperature. As IWS Materials are special materials developed for ITER project, it is necessary to ensure the compliance of Outgassing rate with the requirement. There is a possibility of diffusing the gasses in material at the time of production. So, to validate the production process of materials as well as manufacturing of final product from this material, three coupons of each IWS material have been manufactured with the same technique which is being used in manufacturing of IWS blocks. Manufacturing records of these coupons have been approved by ITER-IO (International Organization). Outgassing rates of these coupons have been measured at room temperature and found in acceptable limit to obtain the required Helium Background. On the basis of these measurements, test reports have been generated and got approved by IO. This paper will describe the preparation, characteristics and cleaning procedure of samples, description of the system, Outgassing rate Measurement of these samples to ensure the accurate leak detection.

1. Introduction

ITER Vacuum Vessel ITER Vacuum Vessel is a torus shaped double wall structure. The space between double walls will be filled with In-Wall Shielding (IWS) Blocks and water. IWS Blocks will fill up about 60 % of volume between two shells. Schematic Diagram of In-Wall Shielding Block between the double walls of vessel is shown in figure-1. Main aims of these blocks are to provide neutron shielding during ITER Plasma Operation and to reduce the toroidal field ripple. There are certain requirements of high quality vacuum for the plasma which vacuum vessel along with IWS blocks must satisfy. If vacuum Vessel does not meet the requirement, vacuum quality will be investigated by Helium Leak Testing. IWS blocks are made of (a) Borated Steel SS 304 B4 (1-1.25 % Boron), SS 304 B7 (1.75-2.25 % Boron), Ferrite Steel SS 430 and Support Structures are made of SS 316 L (N)-IG. Selection of these materials is based on comprehensive assessment of the functional, operational, technological and radioprotection requirements.

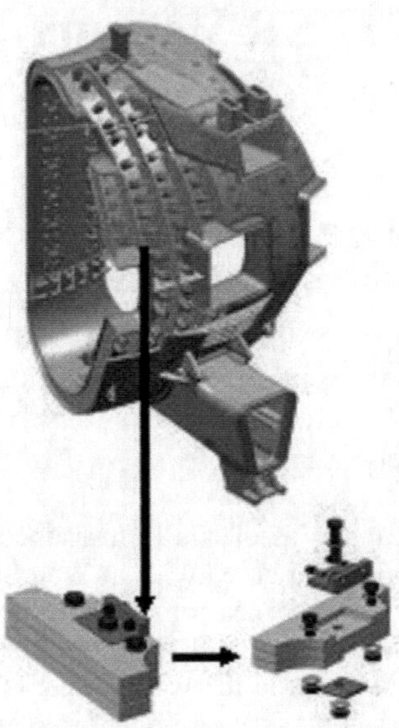

Figure 1. Schematic diagram of IWS Block.

2. Outgassing of IWS Blocks

Although IWS Blocks will remain in between the walls of vessel along with the water, yet play an import role in leak detection of ITER Vacuum Vessel. To provide high quality of vacuum and avoid any leak, Vacuum Vessel will be baked up to 200 °C after installing the In Wall Shielding Blocks in between the walls. It is not possible to eliminate or remove all the gas sources but it is very important to have a low Outgassing rate of IWS Materials and subsequently low Gas Load in Vessel [3]. The Outgassing of Hydrogen from Stainless Steel is the main gas load in Ultra High vacuum System [6] particularly in large systems. During manufacturing of IWS Materials it may possible that during rolling of material hydrogen gas can diffuse into the bulk of material which will result in the high outgassing of hydrogen in vacuum. Thus the hydrogen Outgassing is most significant limiting factor in reaching low Outgassing Rate in Stainless Steel vacuum Systems. By reducing the hydrogen content in bulk, hydrogen outgassing can be reduced. Water Vapor can be removed by baking the system. Outgassing results mainly from release gas molecules/atoms adsorbed on the surface and diffusive release of impurities present in the sub surface. IWS Materials undergo many operations like water jet cutting and CNC machining which may change the surface property of material and consequently the Outgassing rate. Thus the measurement of Outgassing Rate of IWS samples validates the manufacturing process of material as well as the component. Acceptable limit for Outgassing Rate of IWS Materials is $\leq 1 \times 10^{-10}$ mbar l s^{-1} cm^{-2} [1,2]

3. Samples for Outgassing rate Measurement

Measurement of Outgassing rate of IWS plates is not feasible so this measurement has been carried out on samples of IWS materials. These samples have been manufactured using the same techniques which are going to be used for production of actual IWS components. Three (03) samples of IWS Materials SS 304 B4, SS 304 B7, SS 316 L (N)-IG and SS 430 have been manufactured for the

IOP Conf. Series: Journal of Physics: Conf. Series **823** (2017) 012054 doi:10.1088/1742-6596/823/1/012054

measurement. IWS blocks are not accessible for maintenance and repair after fabrication of vacuum vessel. Hence, manufacturing and assembly procedures must be validated for IWS operational conditions. Surface area of each sample was ~ 200 cm^2. Surface Roughness plays a crucial role in determination of Outgassing rate of material. IWS components will have machined and un-machined surface both. Ratio (f) of surface area of as received surface (roughness about 12.5 μ Ra maximum) and area of machined surface (roughness about 8.3 μ Ra maximum) is on an average 1.25 and same has been applied to Outgassing Samples. For plain stainless steel plates, Outgassing rate after baking for 48 hours at 200 °C will not significantly vary for surface having roughness 12.5 μ Ra and 8.3 μ Ra. Hence, IN-DA and IO agreed to carry out Out-gassing measurements for IWS material coupons having different values of (f). [4]

4. System Design

A system with known conductance has been designed to measure the Outgassing rate of IWS Materials which is shown in figure 2. This system can measure total gas load due to thermal out-gassing and composition of various gases evolving from the material sample when exposed to different baking temperatures as a function of exposure time in vacuum. Material of system is SS 304L. Complete system is baked up to 200 °C. The acceptance limit Outgassing rate of IWS materials as per In Wall Shielding Procurement Arrangement is $< 1 \times 10^{-10}$ mbar l s^{-1} cm^{-2} [1]. Before starting the measurement, system has been calibrated. This includes Gauge Calibration and Conductance Calibration. A schematic diagram of the system to measure the Thermal Outgassing rate of ITER VV-IWS Materials is shown in figure 2.

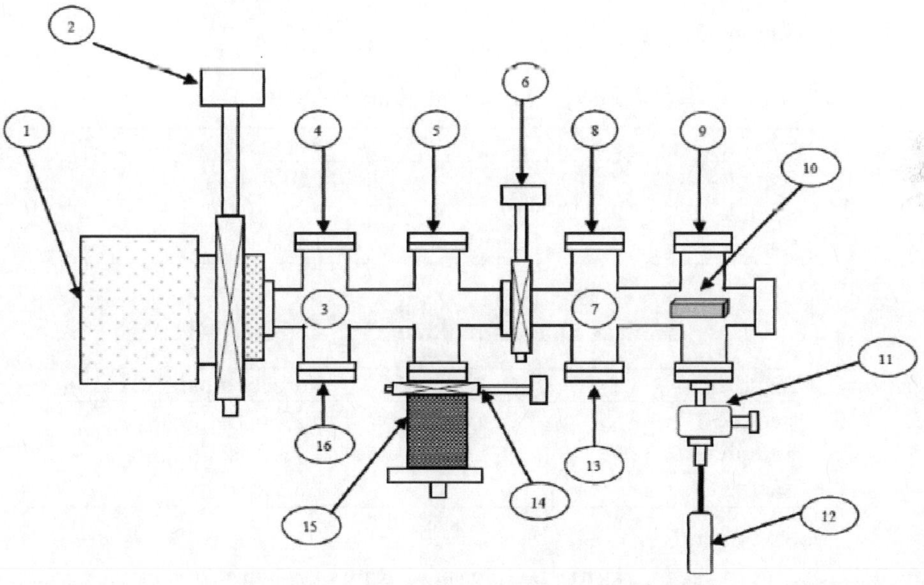

Figure 2. Design of Outgassing System.

1. Ion pump, 2. UHV Gate Valve, 3. Pumping chamber, 4. Pressure Gauge (Range: 1000 mbar to 1×10^{-9} mbar), 5. Residual Gas Analyzer (Range: 1 – 200 amu), 6. UHV Gate Valve, 7. Sample chamber, 8. Pressure Gauge (Range: 1000 mbar to 1×10^{-9} mbar), 9. Feed through for Temperature Measurement, 10.Sample coupon, 11. Variable leak valve, 12. Pressure gauge (Range: 2 bar to 1×10^{-2} mbar), 13. Extra port for venting valve, 14. UHV Gate Valve, 15.Turbo molecular pump, 16. Extra port.

IOP Conf. Series: Journal of Physics: Conf. Series **823** (2017) 012054 doi:10.1088/1742-6596/823/1/012054

4. Method of Measurement of Outgassing rate

The basic parameter measured in the system is gas load. The gas load is measured by measuring the pressure difference across a known conductance. The system measures the flow of the gas evolving from the sample material due to Thermal Out-gassing and the specific Thermal Out-gassing rate Q_{th} is given by

$$Q_{th} = C\ \Delta P/A \tag{1}$$

where, C = Known conductance (l/s)

ΔP = Pressure difference across the conductance (mbar)

A= Area of the desorbing surface- measurable physical area of the sample (cm^2).

The measured Outgassing rate is the net rate of the difference between the intrinsic Outgassing Rate of the surface and the re-adsorption rate [5].

Outgassing rates of different samples of IWS Materials measured at Room Temperature is listed in table 1, table 2, table 3 and table 4 which satisfies the acceptable criterion.

Table 1. Outgassing Rate of SS 304 B7

Samples	Values (mbar l s^{-1} cm^{-2})
Sample-1	2.30×10^{-11}
Sample-2	3.48×10^{-11}
Sample-3	1.65×10^{-11}

Table 2. Outgassing Rate of SS 304 B4

Samples	Values (mbar l s^{-1} cm^{-2})
Sample-1	1.01×10^{-11}
Sample-2	3.48×10^{-11}
Sample-3	4.27×10^{-11}

Table 3. Outgassing Rate of SS 316 L (N)-IG

Samples	Values (mbar l s^{-1} cm^{-2})
Sample-1	6.22×10^{-11}
Sample-2	3.78×10^{-11}
Sample-3	2.88×10^{-11}

Table 4. Outgassing Rate of SS 430

Samples	Values (mbar l s^{-1} cm^{-2})
Sample-1	3.04×10^{-11}
Sample-2	3.75×10^{-11}
Sample-3	3.90×10^{-11}

Gas Load in Vacuum vessel comes from all the sources which can be found in system. Those sources need to be considered separately. Hence, total Surface area of Shield Blocks and Interspace of vessel for one sector is calculated which comes out approximately 2000 m^2. Total Gas load can be

IOP Conf. Series: Journal of Physics: Conf. Series **823** (2017) 012054 doi:10.1088/1742-6596/823/1/012054

determined by multiplying the Outgassing rate of Materials with surface area covered. As determined from experiments, order of Outgassing Rate of IWS materials is same so to estimate the Gas load, Outgassing rate is averaged out and multiplied by surface area and hence the Gas Load for One Vessel Sector comes out as 9.3×10^{-6} mbar l s^{-1}.

5. Conclusion

Outgassing rate of IWS Materials in vacuum under ITER operating conditions gives an insight how IWS blocks will perform in leak detection of ITER Vacuum Vessel. Diffusion of hydrogen in bulk of material and the surface roughness during the production of actual component from these materials impact the Outgassing rate. Measurement of Outgassing rate and its compliance with the acceptable limit validates the both the production processes and materials as well as components from vacuum vessel point of view.

References

[1] Ioki K, Choi C H, Daly E, Dani S, Davis J, Giraud B, Hamlyn-Harris C, Jun C, R Le Barbier, J-M Martinez , Preble J, Terasawa A, Utin Yu, Wang X ITER Vacuum vessel and Vacuum requirements in *IVS 2012*.
[2] *ITER Vacuum handbook* Appendix-12
[3] Sefer A, Bojan E 2012 *Materials and Technology* **46**
[4] Mathewson A G *Vacuum* **24** 505
[5] IN DA Procedure for outgassing system for IWS (IDM no 4LQJP5)
[6] Redhead P A, *Extreme high vacuum, Proceedings* of the CERN Acceleration School, Snekersten, Denmark, CERN

IOP Conf. Series: Journal of Physics: Conf. Series **823** (2017) 012055 doi:10.1088/1742-6596/823/1/012055

Preliminary Design Development of ITER X-ray Survey Spectrometer

Sanjeev Varshney[1], Siddharth Kumar[1], Sapna Mishra[1], Namita Yadav[1], P V Subhash[1], T S Chaitanya[1], Shivakant Jha[1], Vinay Kumar[1], Robin Barnsley[2], Philippe Bernascolle[2], Natalia Casal[2], Gunter Bertschinger[3], Stefan Simrock[2], Jean-Marc Drevon[2] and Michael Walsh[2]

[1]ITER-India, Institute for Plasma Research, Bhat, Gandhinagar – 382 428, India
[2]ITER-Organization, Route de Vinon sur Verdon - CS 90046, 13067 St. Paul-Lez-Durance Cedex, France
[3]Forschungszentrum Jülich GmbH, Institute of Energy and Climate Research, Plasma Physics, 52425 Jülich, Germany

E-mail: sanjeev@iter-india.org

Abstract. The preliminary design of XRCS Survey spectrometer for ITER has been developed addressing many challenges such as designing a ~ 8.0 m long, vacuum extending sight-tube that interfaces crystal spectrometer, placed in the port-cell, with equatorial port-plug (EPP-11) while allowing ~ 50 mm machine movements, and optimizing neutron shield design so that systems can fit into the available space and still the shutdown dose rates (SDDR) remains within the safe limits. The design detailing has been done for the sight-tube and its components addressing the ITER specific requirements. Engineering and neutronic analysis are performed for estimating the thermal displacement, stresses in the front-end components, neutron flux on the sight-tube components, SDDRs in the interspace region etc.

1. Introduction

In order to reliably operate the ITER fusion reactor and make physics measurements, a large number of plasma and first wall diagnostics have been envisaged [1] which will not only provide signals for control of machine operations but also supply prime information on large scale fusion plasmas. Similar to JET X-ray crystal spectrometers (XRCS) [2], which operated successfully in D-T phase, advanced design of crystal spectrometers are in development to work in harsh radiation environment and to satisfy the ITER measurement requirements [3, 4]. ITER-India, the domestic agency of ITER in India, is developing X-ray Crystal Spectrometers for ITER. These are based on X-ray spectroscopy of Hydrogen or Helium like ions of low to high Z impurities present or introduced into the plasma.

The XRCS-Survey [4], a broad-band (0.1 – 10 nm) X-ray spectrometer providing single line of sight through the core plasma, is one of the first plasma diagnostics to help the start-up of the plasma operations. The spectrometer design has been made employing seven crystals placed in modified Bragg configuration. All crystals are bent to tight radius of curvature to disperse X-rays over less number of detectors. The seven channels, based on sensitivity of the detectors to detect X-rays, have been placed inside two adjoining vacuum chambers quite far from the ITER equatorial port isolated by a vacuum gate valve. The primary function of Survey spectrometer is to measure plasma impurities, in real time, due to various in-vessel components exposed to the plasma or from plasma dopants. The main operational role of this diagnostics is in ITER machine protection and basic plasma control. The preliminary developments of plant I&C system to control the functions of the diagnostics in conjunction with ITER CODAC system and handling of the acquired data are reported elsewhere [5].

The first preliminary design review of XRCS-Survey system has been successfully held in December 2015. The review was focused on the front part of the XRCS Survey spectrometer called sight-tube, which is an ITER's safety important component (SIC) due to its confinement function to the radioactive tritium and dust generated by plasma operation. This paper mainly focuses on the design and analysis of sight-tube. Discussion includes simulated performance against the ITER requirements. The full spectrometer design and analysis is currently under development. In this contribution, few of the design issues related to the spectrometer part are also included.

2. Sight-tube design and load assessment

The XRCS Survey spectrometer comprises, as shown in figure 1, a sight-tube from the equatorial port closure plate to the spectrometer chamber having crystals and detectors in the port-cell. The sight-tube is a nearly 160 mm diameter, 8 m long tube made of SS 316L (N). Its primary function is to provide access to X-rays upto the spectrometer. It is operated under vacuum to minimize attenuation of low energy X-rays. One end of the sight-tube is rigidly attached to the port-plug closure plate and other with the spectrometer chamber. Front of the tube is allowed to move with the vacuum vessel during baking and EM-events by using a double walled bellow. A double containment gate valve is mounted near the port to allow isolation of the torus vacuum. The sight-tube is supported near this point from the interspace support structure (ISS). On the other end, a second bellow decouples the spectrometer chamber from the sight-tube and a second gate valve enables removal of complete spectrometer along with the port-cell support structure. Remotely operated quick connectors are considered allowing fast connection and disconnection in the situation of port-plug installation and removal respectively, thereby, minimizing doses to the personnel.

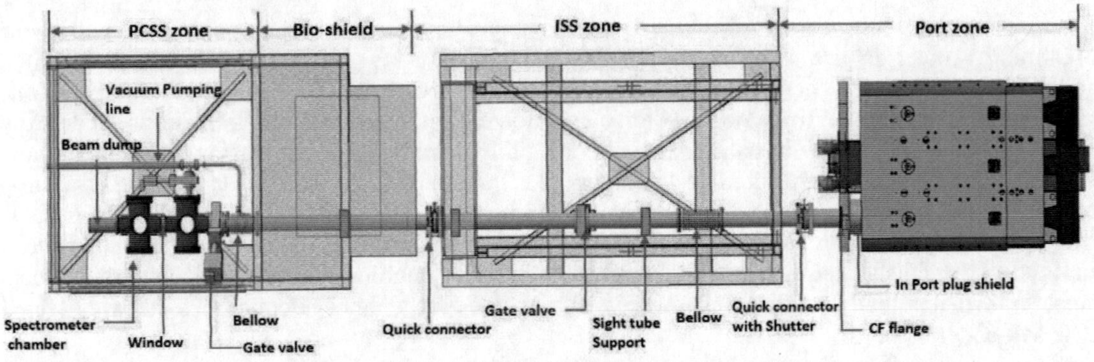

Figure 1. XRCS Survey sight-tube in the interspace and port-cell of equatorial port, EP-11. The spectrometer chamber is in port-cell support structure (PCSS) zone.

The full sight-tube is surrounded by neutron shielding, supported independently from ISS, which is required to minimize activation of the nearby components, and hence bring down the shut-down dose rate

IOP Conf. Series: Journal of Physics: Conf. Series **823** (2017) 012055 doi:10.1088/1742-6596/823/1/012055

to the safe level. Tungsten collimators are placed inside the sight-tube to lower the neutron flux to the diffractors. A shielding is also placed inside of port-plug around the aperture in closure plate. The material of all the elements in the sight-tube is non-magnetic, low carbon stainless steel. To achieve ultra-high vacuum, the tube will be baked to 240 °C using external heaters and will be operated at 100 °C during plasma operations. As per ITER perspective, the sight-tube is a Protection Important Component (PIC) classified as Safety Class-1 (SIC-1) for its confinement function, Quality Class-1 (QC-1) and Vacuum Class-1 (VQC-1A) component. To meet the French regulatory requirements, nuclear code RCC-MR is followed for design, manufacturing and qualification of SIC-1 sight-tube components.

The impact of various loads (direct and/or interface) acting on the sight-tube are also analyzed. The sight-tube experiences thermal stress due to baking of the vacuum vessel, port plug and tube itself, which are heated up to 240 °C but at relatively slower rate of 5 °C/h. In the equilibrium phase, thermal analysis showed significant displacement (~ 24 mm) and higher stresses (~ 2000 MPa) near both connecting ends of the sight-tube. Notably, port-plug thermal movements are also of the same order (~ 36 mm in radial and vertical directions). Figure 2 shows temperature pattern on the sight-tube covered with shield in the normal operation scenario when vessel and sight-tube is maintained at 100°C. Moreover, analysis recommends ~ 50 mm gap between external shield and port-plug flange to compensate thermal movements and to avoid undesired temperature rise of the shield.

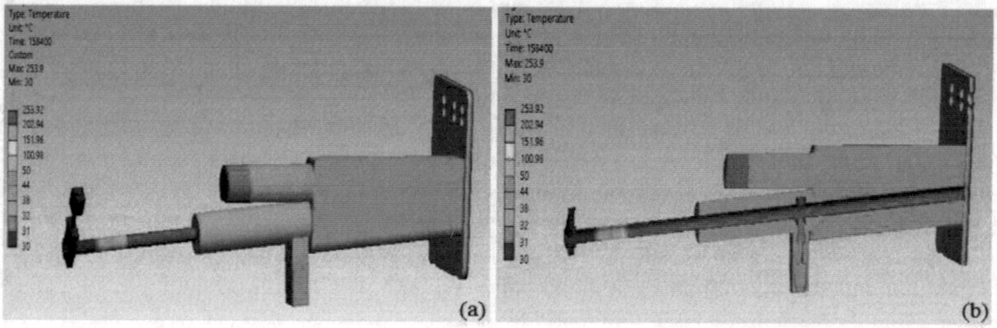

Figure 2. (a, b) Temperature pattern on the sight-tube and neutron shielding around the sight-tube.

Impact on sight-tube components due to electromagnetic events such as major plasma disruptions (MDs) and vertical displacement events (VDEs) is analyzed using ANSYS-MAG. A global ITER EM model [7] together with the sight-tube model is used to calculate force and moments acting on the components. Figure 3(a) & (b) shows EM force and moments at the connection with closure plate. Among all the major disruption events analyzed, the exponentially disrupting (16 ms) major disruption in upward direction exerts highest 45 N force on the sight-tube. In-port plug neutron shield experiences max EM force ~ 150 N. Seismic Level -2 which is classified as a category IV event (RCC-MR class D) is the most stringent loading condition among the various Seismic conditions. The effects of these loads have been analyzed and published elsewhere [8]. Detailed analysis is ongoing for structural integrity under the impact of different load combinations to access P-type and S-type damage conditions that are defined by RCC-MR code.

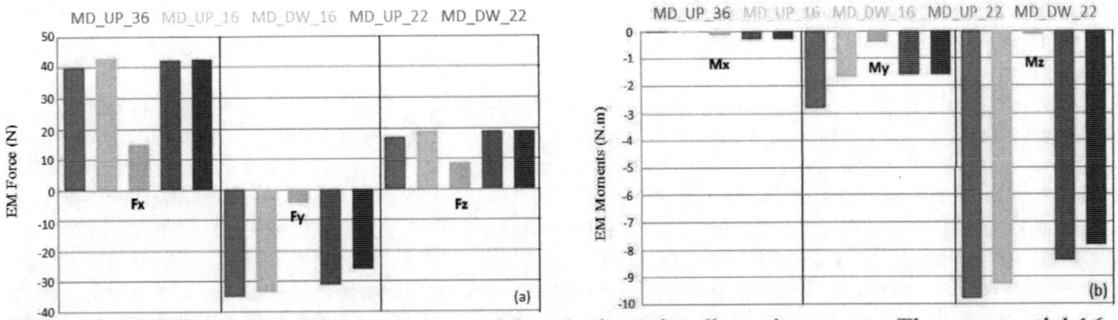

Figure 3. (a) EM force and (b) moment on sight-tube in major disruption events. The exponential 16 ms major disruption in upward direction exerts highest force on the sight-tube.

The effect on line-of-sight (LOS) due to variation in environmental temperature (25 to 50 °C) in the port cell area is also investigated. The support structure of the chambers is fixed to the port-cell support structure and is fixed to the chambers on simple mount. Analysis showed the steady state is achieved after 10,000 s. Displacement is observed mostly in upward direction (table 1). The results indicate that LOS displacements due to uncontrolled temperature variations are of the order of 100 μ (in vertical). In other directions the displacements remains much less. In case of controlled environment (30 to 32 °C) in which if the temperature is allowed to fluctuate only a degree, the LOS displacements in all directions would be ~ 20 μ.

Table 1. Displacements of the line of sight due to varying environment in the port cell.

Observation point	Environmental condition	Displacement (mm)		
		X	Y	Z
Crystal in camber-1	Uncontrolled	0.1	7.85e-003	0.1
	Controlled	3.56e-002	2.86e-003	3.55e-002
Crystal in chamber-2	Uncontrolled	-6.19e-002	6.52e-003	0.11
	Controlled	-2.19e-002	2.31e-003	3.92e-002

3. Sight-tube neutronics analysis

MCNP neutronics analysis has been carried out for the sight-tube in order to optimize the design of the neutron shield. The earlier work carried out with ATILLA code is reported elsewhere [9]. Fusion neutrons (14 MeV) along with X-rays will stream through sight-tube and gaps in the port-plug: from the plasma source to interspace and even beyond the bio-shield reaching upto the spectrometer in the port-cell. These neutrons will interact with the sight tube and surrounding port plug components in the interspace and activate them. As a result of activation, radioactive nuclides are produced which decay by emitting gamma, consequently creating a radiation environment until the nuclides decay completely. Personnel access to the equatorial interspace is envisaged for the purpose of diagnostic or port-plug maintenance. The dose rate to the personnel entering a port plug interspace area must be below 100 μSv/h at 10^6 s after the end of the Safety Assessment-2 scenario. Post DT operation, the interspace region will see heavy gamma flux generated from the decaying radionuclides. To limit the gamma flux, the formation of radionuclides has to be controlled at the first instance. Therefore, system construction requires appropriate material for the sight-tube components as well as radiation shields to shield the neutrons streaming through the diagnostic openings. To accomplish this, iterative neutronics analysis and shield optimization has been done.

The ITER C-lite neutronics model [10] is used for the analysis of sight-tube. It is grossly modified to include all details of the port plug, diagnostic apertures and the XRCS Survey diagnostic system as shown in figure 4. The neutron transport calculations are performed using Monte-Carlo N-Particle code (MCNP) version 5. For the cross sections FENDL-2.1 [11] and EAF-2007 [12] nuclear data libraries are used. Shutdown dose rate calculations have been performed using the D1S (Direct one step) method.

The neutronic analysis is done for two separate sight tube compositions, SS 316L (N) and Aluminum alloy-2024. The composition chosen for the external shields, as shown in figure 5, is a homogeneous mixture of pure aluminum and boron carbide: 50 % - 50 % volumetric fraction of both the materials. Aluminum is preferred because it is light and has relatively lower activation. The composition taken for in-port plug shield, which is inside ultra-high vacuum, is SS and boron carbide.

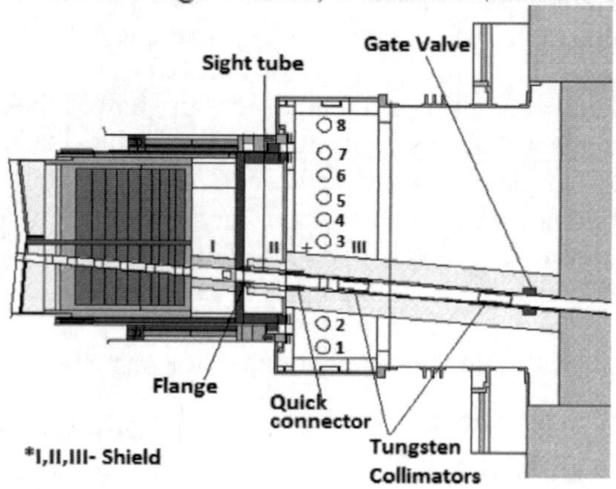

Figure 4. XRCS Survey sight-tube model upto bio-shield showing internal and external components of the sight-tube.

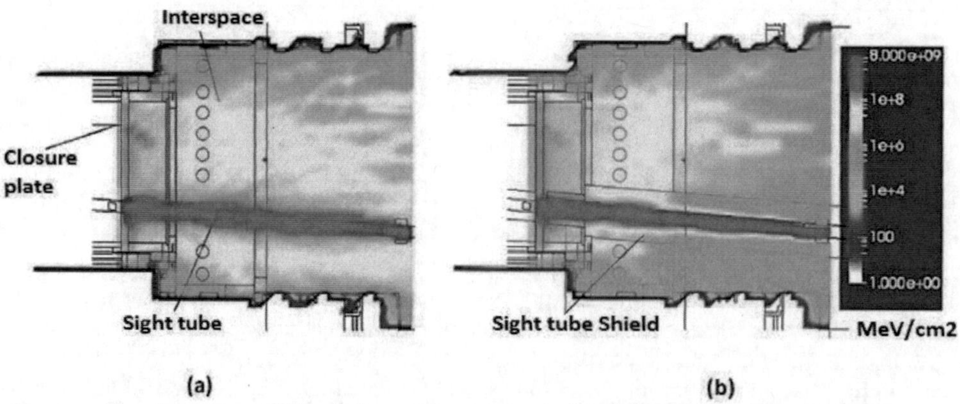

Figure 5. Neutron flux (energy weighted) in port interspace using MCNP (a) without shield, (b) with shields.

IOP Conf. Series: Journal of Physics: Conf. Series **823** (2017) 012055 doi:10.1088/1742-6596/823/1/012055

The analysis showed that neutron flux streaming out of the sight-tube is nearly same for the two materials evaluated for the sight-tube construction (table 1). This is due to the thickness of the sight-tube which is quite small (5 mm). The effectiveness of the sight-tube shielding for lowering of streaming neutron flux in the interspace can be seen from the figure 5, which shows contour plot of the neutron flux binned in different energies.

Neutron flux reduces to half by use of shields as provided in table 2. Without any shield the flux reaching the bio-shield is ~ 1×10^8 n/cm^2/s, which is reduced by a factor of 100 with neutron shield. This has positively reflected on SDDR values in the interspace that are given in table 2. The dose rate values are found upto four times higher for SS sight-tube as compared to Al sight-tube. Effectively, with use of the designed shields, the interspace dose rate reduces upto four times (as low as ~10 μSv/h) providing safe environment for personnel to access the interspace region for system maintenance.

Table 2. Neutron flux in the tallies placed in interspace (figure 4) and SDDR at 12 days after shutdown for SS and Al sight-tubes.

Tally Cell	Neutron flux (n/cm^2.s)				SDDR (μSv/hr)			
	Without Shield		With Shield		SS316L(N)		Al-2024	
	Al-	SS316-	Al-	SS316-	Without	With	Without	With
1.	5.91E+06	5.78E+06	3.16E+06	3.10E+06	14.32	2.71	4.14	1.16
2.	9.88E+06	9.96E+06	4.77E+06	4.76E+06	26.31	3.17	6.11	2.85
3.	1.78E+07	1.76E+07	1.36E+07	1.38E+07	23.58	11.25	12.6	9.96
4.	1.65E+07	1.66E+07	1.44E+07	1.43E+07	28.65	13.11	11.25	8.11
5.	1.50E+07	1.50E+07	1.34E+07	1.36E+07	25.68	13.67	10.7	10.8
6.	1.28E+07	1.28E+07	1.17E+07	1.18E+07	21.66	10.40	9.40	9.04
7.	8.83E+06	8.97E+06	7.97E+06	7.99E+06	10.78	5.61	8.50	4.33
8.	4.74E+06	4.57E+06	3.90E+06	3.78E+06	4.82	2.72	2.96	1.75
9.	1.13E+07	1.13E+07	7.49E+06	7.63E+06	19.0	6.95	11.12	3.87
10.	1.87E+07	1.91E+07	1.66E+07	1.68E+07	27.03	17.56	18.32	12.92
11.	1.68E+07	1.65E+07	1.53E+07	1.52E+07	17.88	12.63	13.08	9.96
12.	1.04E+07	1.05E+07	9.69E+06	9.36E+06	10.79	7.77	6.64	6.34

4. Conclusions

The identification of components and the design layout of the sight-tube for XRCS Survey spectrometer is completed. The analysis of impact of various loads on the sight-tube components has been carried out too. It was shown that thermal stresses on the front end components and total displacements could be significant and these have been considered in detailing the motion compensator. EM loads are not significant on the sight-tube. However, forces on in-port plug shield are significant. Neutronics analysis done for Al and SS sight-tubes has shown superiority of the Al-2024 for constructing sight-tube as SDDR is found upto 4 times lower in the interspace for this material.

The views and opinions expressed herein do not necessarily reflect those of the ITER organization.

Acknowledgement
The C-lite MCNP model used in the analysis is developed under a collaborative effort between the FDS team of ASIPP China, University of Wisconsin-Madison, ENEA Frascati, CCFE UK, JAEA Naka, and the ITER Organization.

References
[1] Donné A J H et al. 2007 *Nucl. Fusion* **47** S337
[2] Barnsley R et al. 1992 *Rev. Sci. Instrum* **63** 5023
[3] Beiersdorfer P et al. 2010 *J. Phys. B: Atom. Mol. and Opt. Phys.* **43** 144008
[4] Drevon J M et al, 2013 *Fusion Engineering and Design,* **88** 1306
[5] Varshney S K et al. 2012 *Rev. Sci. Instrum.* **83** 10E126
[6] Varshney S K et al. 2016 *Fusion Engineering and Design,* **112** 887
[7] Pak S et al. 2013 *Fusion Engineering and Design,* **88** 3224
[8] Kumar S et al. 2015 *Proc. of the IEEE 26th Symposium on Fusion Engineering (SOFE)*
[9] Subhash P V et al. 2013 *Transactions of American Nucl. Soc. Tran.,* ***108**, 1149*
[10] Zhang S et al. 2016 *Fusion Engineering and Design,* **113** 126 and *references therein*
[11] Fusion Evaluated Nuclear Data Library-2.1, *https://www-nds.iaea.org/fendl21/fendl21mc.html*
[12] Forrest R A et al. 2007 *UKAEA report FUS 535*

IOP Conf. Series: Journal of Physics: Conf. Series **823** (2017) 012056 doi:10.1088/1742-6596/823/1/012056

Observation of plasma shift in SST-1 using optical imaging diagnostics

Manoj Kumar[1], Chesta Parmar[1], Vishnu Chaudhary[1], Ajai Kumar[1] and SST-1 team[1]

[1]Institute for Plasma Research, Near Indira Bridge, Bhat, Gandhinagar 382 428, India

E-mail: mkg@ipr.res.in

Abstract. A tangential viewing optical imaging system at SST-1 is used to observe the plasma shift both vertical and horizontal during experimental campaigns. The images from the plasma are transferred through optical imaging fiber and coupled to a CCD camera which operates at 31 frames/s. The data from the CCD camera is transferred through gigabit Ethernet cable to acquisition PC placed in diagnostics lab. The whole system is fully automated for operation and data acquisition of the imaging data. The complete imaging system will be explained in this presentation. With this optical imaging system, the shift in plasma position both in vertical and horizontal direction is observed. The plasma shape and diameter can also be estimated with this system. The estimated diameter during some of the plasma shots is ~ 50 cm and shape is circular. The data from this diagnostics is very useful from the operation point of view of the machine.

1. Introduction

The tokamak plasma emits radiation in wide spectrum originating from X-rays to infrared. Useful information can be extracted from these radiations like plasma temperature measurements, runaway measurement, impurity measurements, plasma wall interaction etc. Visible light imaging in tokamak is very useful tool for studying the interaction between the edge plasma and plasma facing components which are well known as plasma wall interactions. The temperature in the edge of the plasma is in the range of 10-100 eV. Visible light emission is dominant radiation in the edge plasma region, and it can provide some relative information to study different plasma behaviours, including instabilities [1,2], turbulence [3-5] and interactions [5-8] in the boundary region. So the optical imaging diagnostics are widely used in toroidal plasma devices, such as tokamaks, spherical tori, helical systems and so on [9]. In addition, the visible emissions are also related to the plasma position [10]. Thus the availability of an optical imaging diagnostics would provide an opportunity to evaluate the plasma position in SST-1. We have used optical imaging diagnostics to measure the plasma position both in horizontal and vertical directions .These measurements are reported in this paper.

2. Optical System setup and Installation

SST-1 is a mid-size tokamak with major radius 1.1 m and minor radius 0.2 m. Currently it is being operated in limiter configuration. The plasma current of ~ 74 kA and duration of around 450 ms has been achieved. The visible imaging diagnostics has been installed to measure plasma position, shape and size (figure 1). In this paper results regarding position of plasma will be presented. The visible imaging diagnostics consist of camera system (consisting of CCD camera, fiber optic connectors,

IOP Conf. Series: Journal of Physics: Conf. Series **823** (2017) 012056 doi:10.1088/1742-6596/823/1/012056

wound imaging optical fiber, video transmitter and receiver), trigger distribution box, data acquisition system (consist of PC, frame grabber card) and the database server. The optical system is triggered by the trigger circuit controlled by central control server. The computer system has the temporary storage of the data acquired during the discharge which is then transferred to the data server where it is stored permanently as well as it can be accessed by the other users.

This section explains about the various components of the optical imaging diagnostics and the data acquisition system. The diagnostic consists of two CCD cameras. One of the cameras is positioned in radial direction for wide angle view while the other camera is positioned in tangential direction to view poloidal cross-section. A micro head CCD camera (Toshiba: IK-UM51H) is placed at the radial direction. The camera has two components (a) Component head (2) control unit. The miniature camera head of 40 mm length and 12 mm diameter is connected to the camera control unit using 5 m long cable. The camera control unit (CCU) supports interlace scanning PAL (25 fps) and NTSC (30 fps) camera interfacing with external trigger for synchronization purpose. The CCU gives facility to program the exposure time (100 μs to 20 ms) using computer via RS232 interfacing option. The cameras are operated in harsh environment; the CCU uses shielded cables for data connectivity to camera for signal integrity and higher signal to noise ratio (46 dB). The light collected through the camera is sent to the video input convertor SI TECH BIT DRIVER 2809-2810 VIDEO INPUT EDP EQUIPMENT and the output is taken as an optical signal which is given to the computer via IMAQ A6822.

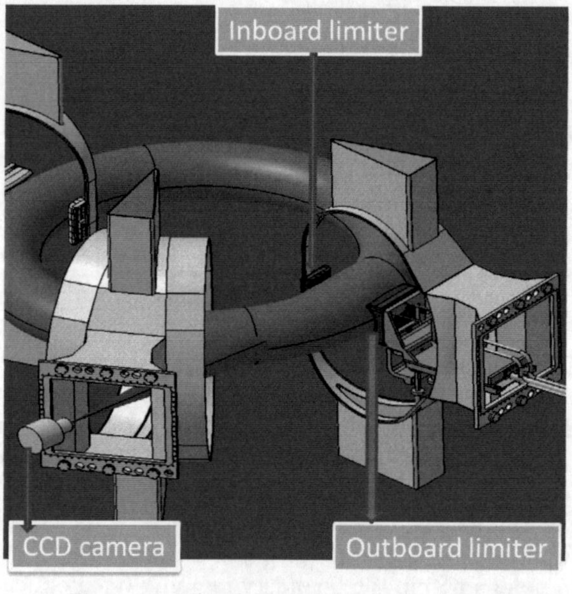

Figure 1. Viewing geometry of camera.

The other camera viewing tangentially is Digital monochrome progressive scan camera (CM 140-GE). It supports 31 frames/second in full resolution at continuous scan with external trigger. The camera comes with 1.4 megapixel resolution and GigE vision interface. All the subsystems are integrated and operated in a sequential manner.

The camera works on the burst mode wherein the trigger from the SST central control starts the burst mode operation acquiring trigger at -200 ms before the plasma shot. Acquisition is done using LabVIEW software. In order to minimize the effect of strong magnetic field, the camera is placed at a distance from the vessel and the data is transferred through (610 mm long) wound optical imaging fiber bundle (Schott wound Fiber bundle with NA 0.3 and a resolution of 45 lp/mm). The fiber connects the camera through an image coupler consisting of relay lens (Schott IG 1660 1:1). The other end of the fiber is attached with 8.5 mm CCTV lens. Data acquired is stored in the PC kept at the control room.

3. Results

Figure 2 shows a typical SST-1 plasma shot which contains some of the important diagnostics signal. In the first row current profile I_p (kA), second row internal loop (V), third row hard X-ray (a.u.) and in the fourth row hydrogen alpha (a.u.) measurements are shown. In this paper we present the data from optical imaging diagnostics for this particular plasma shot. Figure 4 shows the view of vessel during GDC from the tangentially viewing camera. SST-1 has a movable limiter and currently the separation between inboard and outboard limiter is 50 cm. The green horizontal line in figure 4 is at mid plane of the machine which is being taken as a reference to measure the plasma position.

The poloidal cross section of the plasma images is shown in figure 5. The sequence of images at different times shows the evolution of plasma with respect to time. Each frame is 31 ms. The poloidal cross section of the images show bright visible emission ring at the edges and a dark zone in the center. Our camera is sensitive only in the visible region of the spectrum so no visible radiation is obtained from the central zone of the poloidal cross section of the plasma. Since the temperature at the edges is low ~10-100 eV, this zone is dominated by the visible emission. In the center (core) of the plasma temperature is high, radiation in the UV, VUV, X-rays is emitted which is not in the detection range of our camera. In this work this visible emission ring at the edges is used to estimate the plasma movement. The line averaged emission profile of these images is generated. The line profile shows two peaks and the center of these two peaks is taken as plasma center. As mentioned earlier the geometric center figure 3, figure 4 (taken during GDC) is compared with the plasma center and thus the plasma movement is estimated.

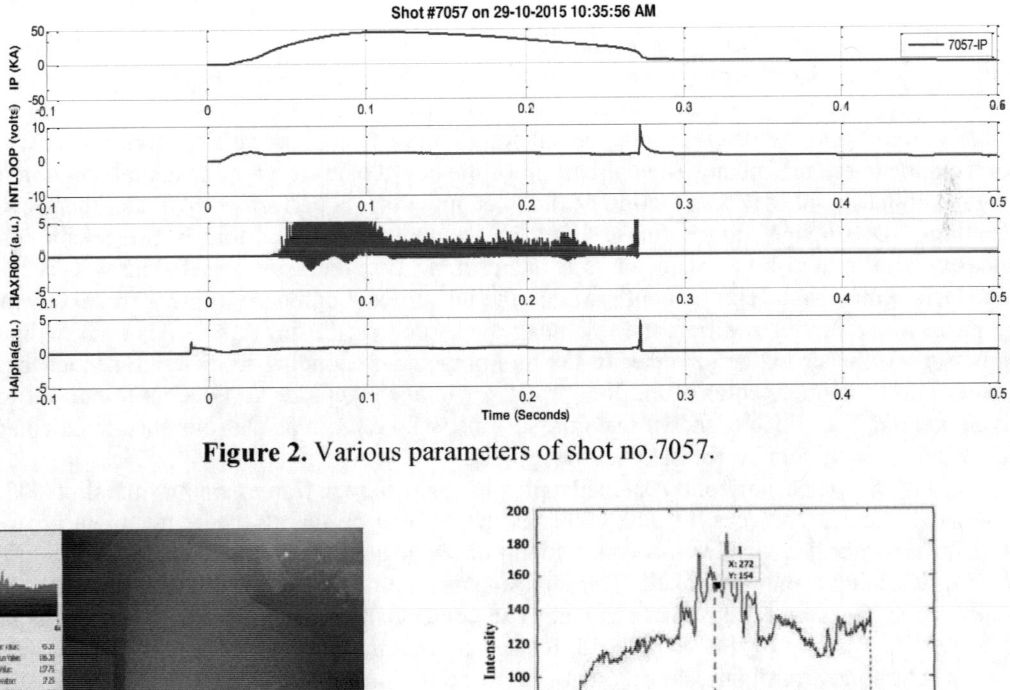

Figure 2. Various parameters of shot no.7057.

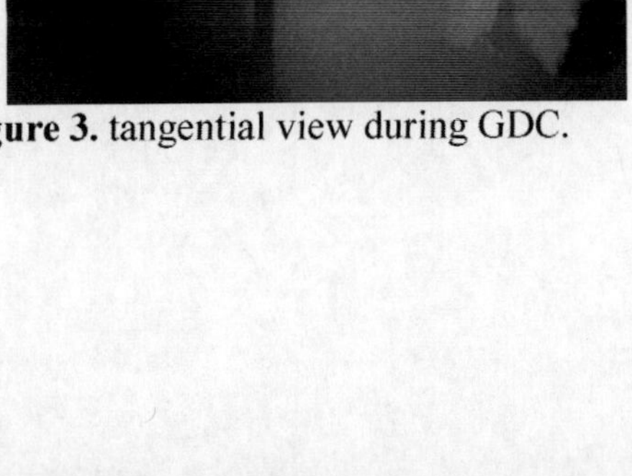

Figure 3. tangential view during GDC.

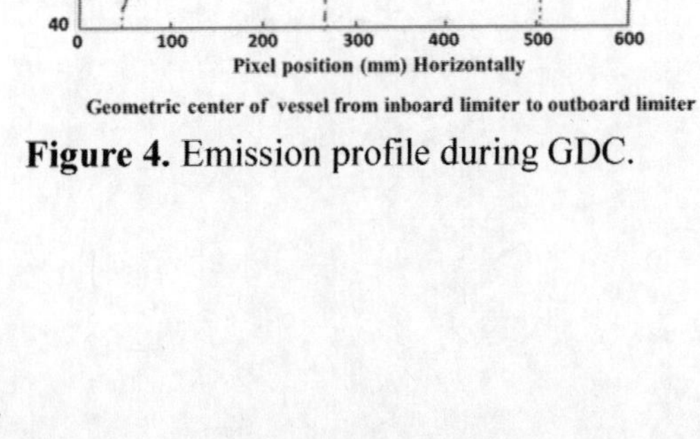

Figure 4. Emission profile during GDC.

IOP Conf. Series: Journal of Physics: Conf. Series **823** (2017) 012056 doi:10.1088/1742-6596/823/1/012056

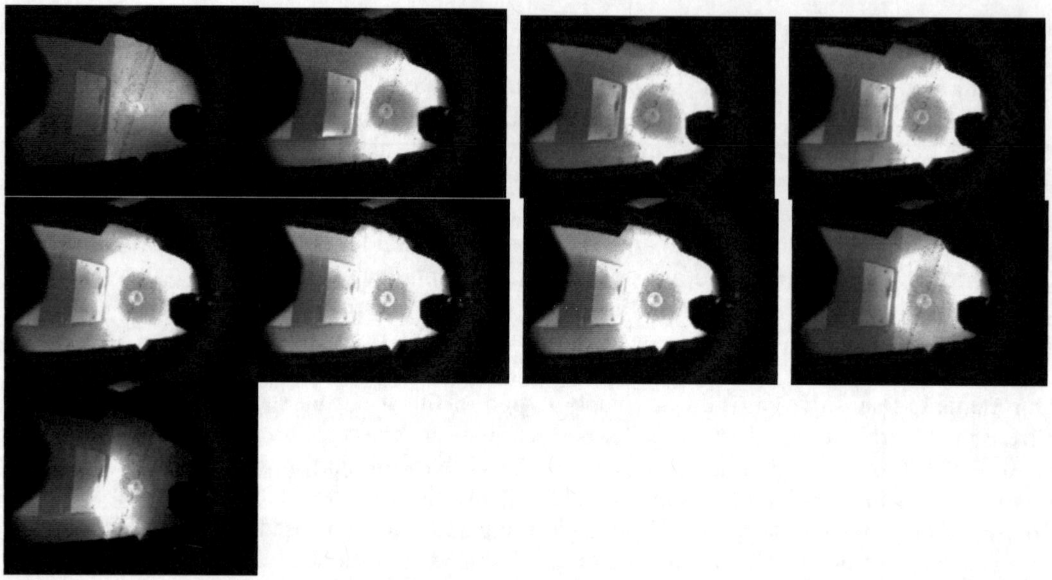

Figure 5. Poloidal cross section of sequence of images for shot no.7057.

3.1. Estimation of horizontal plasma shift

The plasma horizontal displacement, estimated from the distribution of the visible emission, can be obtained by processing the discharge images. The line-averaged intensity of each line along the vertical direction of the images can be calculated as a function of the vertical coordinate Z, where $Z=0$ corresponds to the mid-plane. The calibration of the coordinates to give the actual location of the line-averaged intensity at the cross section of the vacuum vessel is performed by using the glow discharge cleaning (GDC) images. Since during GDC the vacuum vessel is uniformly filled with emission, the intensity profile across the diameter is a flat profile. The centre of this flat profile is taken as the geometric centre. In the current configuration the inboard and outboard limiters are 50 cm apart.
To measure the plasma position, the raw images are processed using MATLAB to draw line averaged emission profile across the diameter in the mid-plane corresponding to different sequences of plasma image. The geometric center of the line profile of image taken during GDC is found out to be at the pixel value of 272 which is shown in figure 4. This will be taken as the reference to calculate the shift in plasma position both horizontally and vertically.

The line averaged horizontal intensity profile for different frames of plasma shot 7057 is being calculated from raw images. Plasma center is determined by taking the center of the two peaks. As mentioned earlier the, visible emission is found in the edge of the poloidal cross section of the plasma and thus forming a emission shell. The emission profile of plasma, in the mid plane, is having two peaks at the edges and a dip in the center. The center of these two peaks is the plasma center. The difference in position of this center with respect to geometric center gives the change in position of plasma. The following figure shows one of the shot of complete duration from the time of the plasma evolution till its disruption. In figure 5, the radial shift of plasma is estimated.

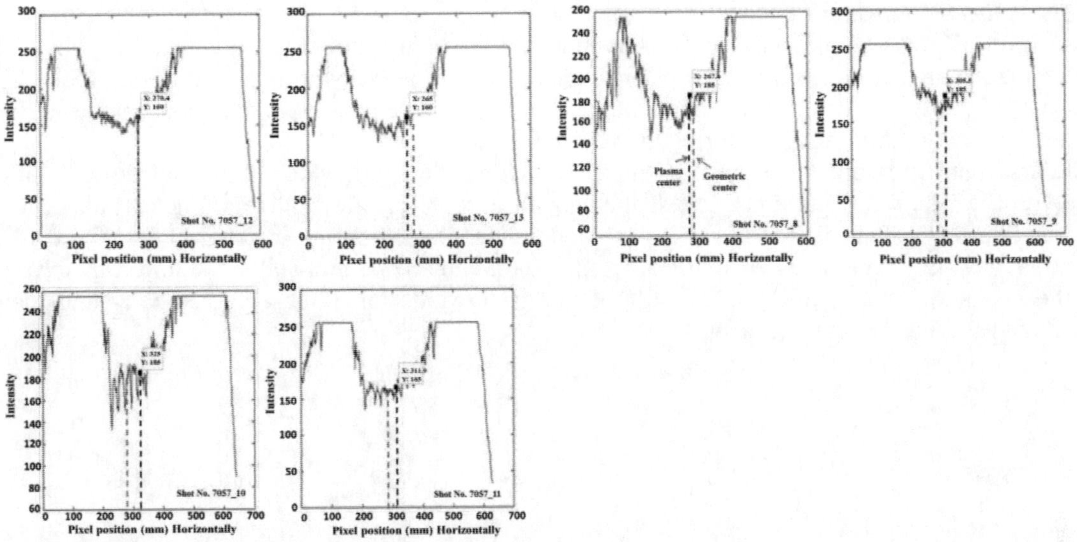

Figure 6. Observing horizontal Plasma shift during shot 7057.

The figure 6 shows line profiles of plasma shots and its center during shot 7057. From the line profile measurement for plasma center obtained is 276.4 mm, 308.5 mm, 325 mm, 311.9 mm, 270.4 mm, 265.5 mm sequentially. The red dotted line represents the geometric center of the plasma and the blue line represents the plasma center. From the movement of blue line with respect to red line the shift in plasma position is estimated shown in figure 6.

In order to verify the reliability of the results obtained from the imaging system, a detailed comparison between the results estimated from the imaging system and those given by the magnetic diagnostics is shown in figure 7. The red circles gives the results estimated by VLIS, while the green line joining these circles denotes the spline function The blue line shows the results measured with magnetic loop. The discrepancy between two curves at 100 ms (ramp-up phase) and 250 ms (decline phase) is around ~ 2 cm. The differences at other times are within 1.0 cm. In conclusion, the horizontal displacement estimated by the optical imaging diagnostics is essentially in accordance with those obtained using the magnetic loop. The reliability of optical imaging diagnostics in estimating the plasma displacement has been proved in the analysis of hundreds of discharges. Additionally we had tried to fit spline on the emission profile and it matches quite well. This fitting was tried in number of shots and most of the time it shows a good match.

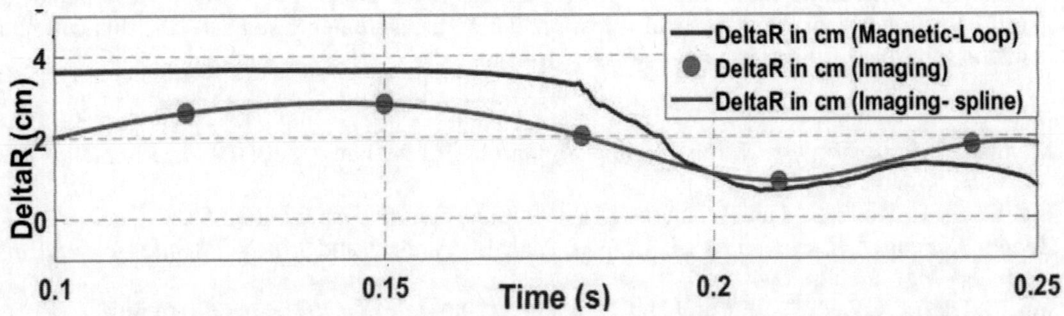

Figure 7. Comparison of imaging and magnetic diagnostics measurement.

3.2. Estimation of vertical plasma shift

The vertical shift was also measured in the shot number 7057 in the same way explained for horizontal shift. To measure the vertical shift of the plasma, the horizontal intensity profile for different sequence of plasma images is calculated. The centers of each frame are compared with the geometric center in order to measure the shift observed during the plasma discharge.

The first frame in figure 8 shows the geometric center as 386.7 mm which is calculated by plotting the line profile. The following figure shows the geometric center as the red dotted line and the center of profile in each frame is depicted by the black dotted line. On comparing 9th, 10th, 11th, 12th and 13th frames that have the center as 356.5 mm, 352.5 mm, 363.5 mm, 379.5 mm and 375.5 mm respectively with the geometric center, the gradual shift of plasma towards top is observed initially and then concentrated at center during later stage.

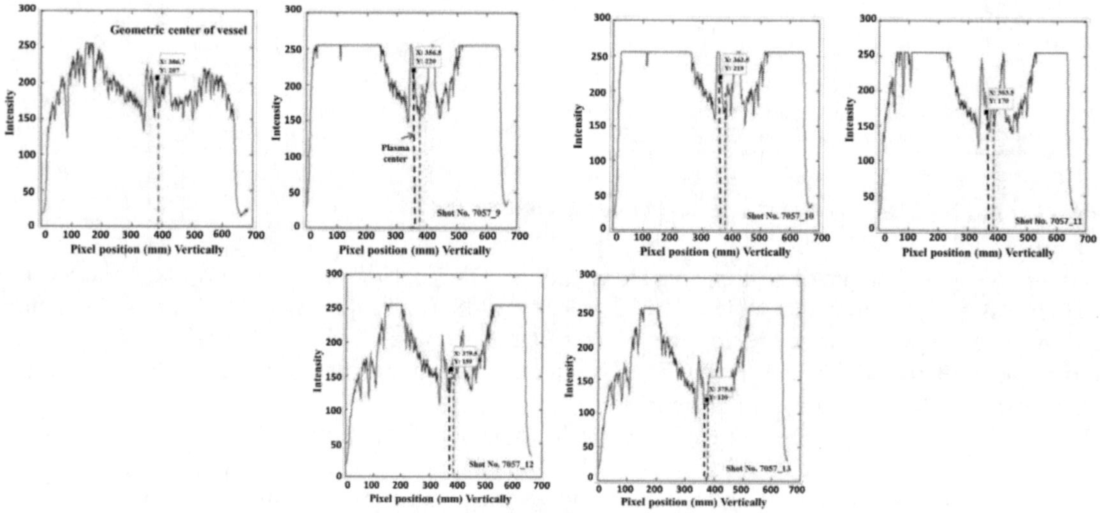

Figure 8. Observing Vertical Plasma shift during shot 7057.

The similar shift is also observed in Shot No. 7032 and few others. 9th frame (with center 353.5 mm), 10th frame (center 361.5 mm), 11th frame (center 368.5 mm), and frame 12 (with center 371.4 mm) is compared with the geometric center that confirms the similar vertical movement.

4. Conclusions

The visible imaging system was successfully developed and installed on SST-1. The complete description of the system and its operation was presented in the paper along with measurement of plasma shift both in horizontal and vertical direction. These measurements suggest that this can be a possible method for measuring plasma position in a tokamak.

References

[1] Mengzhou Z, Zhuang Ge, Wang Z, Ding Y, Gao L, Hu X, Pan Y, 2010 *Plasma Science and Technology* **12** 641.
[2] Ingesson L C, Koning J J and Donne A J H, 1992 *Rev. Sci. Instrument.* **63** 560.
[3] Zheng Y, Feng Z, Luo C, Liu Li, Li Wei, Yan L, Yang Q and Liu Y *Plasma Science and Technology* **6** Number 4 1254
[4] Iraji D, Diallo A, Fasoli A, Furno I and Shibaev S, 2008 *Rev. Sci. Instrum.* **79** 10F508
[5] Jia M, Yang Q, Zhong F, Zeng A, Lu H, Shu S ,2015 *Plasma Science and Technology* **17** No.12 1276
[6] Maqueda R J, Wurden G A, 2008 *Nuclear Fusion* **39** No. 5 2354.
[7] Feng Y, Goree J and Bin L, 2007 *Rev. Sci. Instrum.* **78** 537.

IOP Conf. Series: Journal of Physics: Conf. Series **823** (2017) 012056 doi:10.1088/1742-6596/823/1/012056

[8] Chung J, Lee D K, Seo D, et al. , 2008 *Rev. Sci. Instrum.*, **79**: 10F510.
[9] Zheng Y, Feng Z, Luo C, et al., 2004 *Plasma Sci. Tech.* **6** 2353.
[10] Maqueda R J, Wurden G A, 1999 *Nucl. Fusion* **39** 629.

IOP Conf. Series: Journal of Physics: Conf. Series **823** (2017) 012057 doi:10.1088/1742-6596/823/1/012057

Preliminary Results from Electron Cyclotron Measurements at SST-1

Varsha Siju[1], Praveena Shukla[1], Jayesh Raval[1], Shankara Joisa.Y[1], S. K. Pathak[1]

[1]Institute for Plasma Research, Bhat, Gandhinagar 382 428, India.

Email: varsha@ipr.res.in

Abstract: An 8-channel heterodyne radiometer system is developed and installed for the measurements of second harmonic electron cyclotron emission (ECE) at magnetic field of 1.5 T at SST-1. This system covers a spectral range of 75.4 to 84.5 GHz at a spatial resolution of less than 1 cm, sensitivity of 9.51×10^6 V/W. The calculated noise temperature of the system is 1.66 eV. The system is calibrated using Hot/cold technique. This paper presents the preliminary observations of the heterodyne radiometer system at SST-1 and the measured radiation temperature.

1. Introduction

Electron Cyclotron emission (ECE) is a standard diagnostic in present day tokamak for plasma electron temperature measurements [1]. The use of multichannel super heterodyne receivers has made it possible to measure emission at high spatial and temporal resolutions. Since the ECE from optically thick plasma can be treated as blackbody radiation, the amplitude of ECE is proportional to the electron temperature. Furthermore, the frequency of the ECE is proportional to the magnetic field strength since the ECE is emitted due to gyro motion of electrons. In general, distribution of magnetic field strength in a torus experimental device is a function of radial position, and the radial temperature distribution can thus be measured by resolving the ECE signals with electron cyclotron frequency or its harmonics [2].

The present paper describes the design of the radiometer system for second harmonic temperature measurements for optically thick plasma at SST-1 at toroidal magnetic field of 1.5 T. The preliminary observation of the raw signal and comparison with SXR diagnostics is also described.

2. Experimental Setup

SST-1 is a medium sized fully super conducting tokamak with major and minor radius of 1.1 m and 0.20m respectively. The ECE frequency spectrum and corresponding optical thickness for SST-1 parameters having toroidal magnetic field B_T = 1.5 T, electron temperature T_e= 300 eV, and electron density $n_e = 2 \times 10^{19}\ m^{-3}$ have been computed and shown in figure 1a and 1b.The optical thickness for X-mode emission ($n \geq 2$) can be computed as below

$$\tau_n^x = \frac{\pi^2\, n^{2(n-1)}\, R}{2^{(n-1)}(n-1)!\lambda} A_n \left(\frac{\omega_p}{\omega_c}\right)^2 \left(\frac{K_B T_e}{m_e c^2}\right)^{n-1} \quad (1)$$

IOP Conf. Series: Journal of Physics: Conf. Series **823** (2017) 012057 doi:10.1088/1742-6596/823/1/012057

where R is the major radius and the factor A_n is a constant calculated in Born [3]. For a tenuous plasma the constant A_n is approximately 1.From the graph it can be seen that 2nd harmonic ECE frequency (fce2) spectrum ranges from 72 GHz to 102 GHz, for which, the plasma is optically thick and favorable for electron temperature measurements. In our initial design we are covering only half of the entire plasma cross section with 75.4 GHz - 84.5 GHz, frequency range that lies on the low field side. However, we are working on the process of designing a radiometer that would also be covering the other side i.e. high field side of the SST-1 plasma cross section. Henceforth we would be able to resolve the plasma electron temperature on high field side as well as low field side.

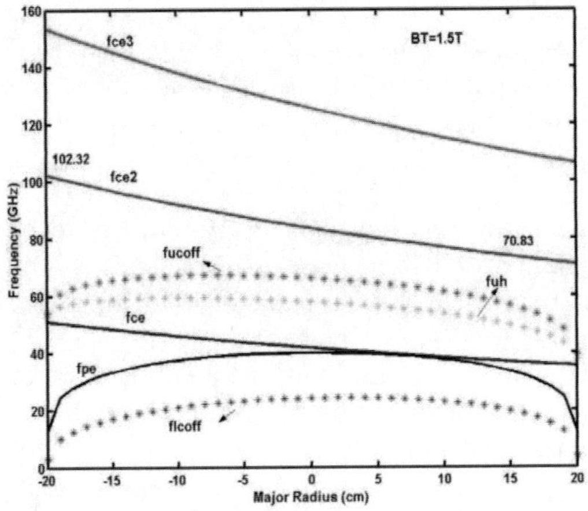

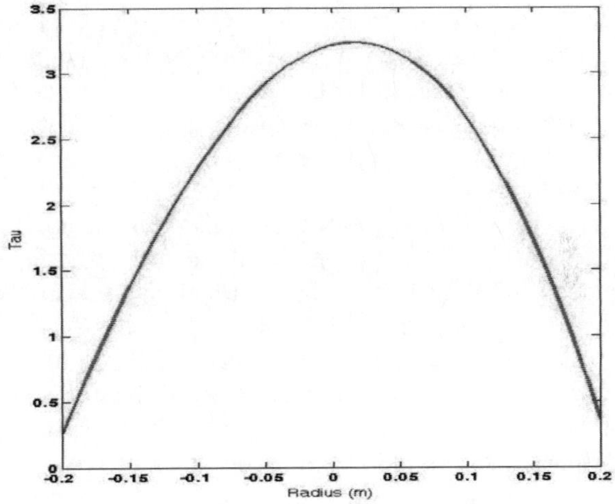

Figure 1a. Frequency plot for SST-1 at B_T = 1.5 T **Figure 1b**.Optical Thickness at B_T= 1.5 T

2.1. Radiometer system - Design

The designed E-Band Radiometer system, as shown in figure 2, is a heterodyne system with an input RF of 74-86 GHz down-converted to an IF of 1-12 GHz. Further the electronics amplifiers and data acquisition system are used for data sampling and filtering.

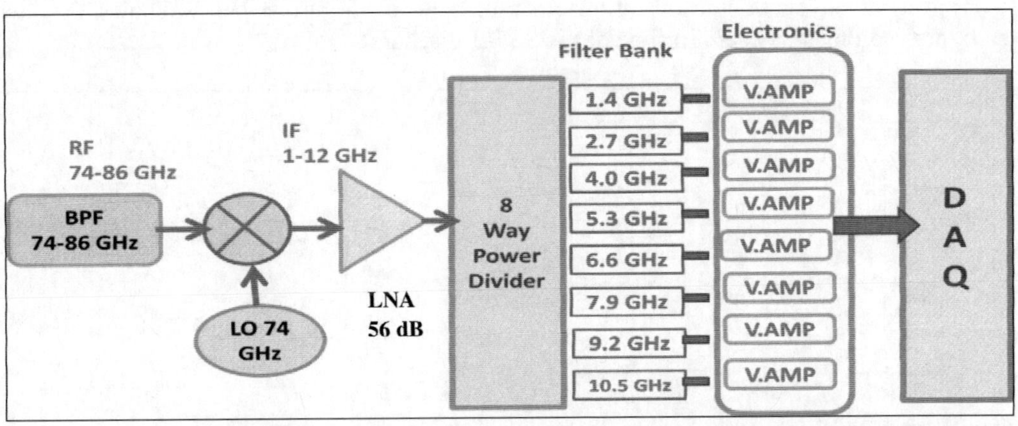

Figure 2. Block diagram of the designed radiometer system.

The EC emission is collected by a horn Antenna with gain of 23 dBi and transmitted by an oversized waveguide (X-Band) to the diagnostics room that is situated at a distance of 9.5 m from the tokamak machine. The emission that reaches the front end of the radiometer system is passed through a band pass filter allowing a spectrum of 74-86 GHz only. The filtered radiation is passed on to the down converting unit which consists of a balanced mixer and a fixed frequency local oscillator at 74 GHz, translating the RF spectrum to an IF spectrum of 1- 12 GHz.

The IF of 1-12 GHz is further amplified using Low Noise amplifiers with a total gain of 56 dB. This amplified signal is split into 8 channels using an 8-way power divider. Frequency selective cavity filters at frequencies 1.4 GHz to 10.5 GHz, with a step size of 1.3 GHz and a bandwidth of 400 MHz, follows each channel of the power divider. The signal is detected using a zero bias Schottky detector.

Electronic modules follow the voltage output of the detectors for signal conditioning and filtering. The electronic modules are divided into 3 parts: (1) A signal conditioning system (2) Two stage amplification (3) Filtering and optical isolation. The signal from the electronic modules is differentially carried to the DAQ system using a Multicore cable.

The calculated spatial resolution of the system is 0.65 cm while its temporal resolution is 1ms. This will be enhanced further. The calculated system noise temperature is of 1.66 eV that depends on the overall noise figure of the cascaded components of the system.

3. Calibration

In order to obtain the electron temperature from ECE measurements the receiver system has to be calibrated. The ideal way to calibrate the radiometer system would be to use a black body at the temperature close to the plasma temperature. Unfortunately, such black bodies with such high and known temperatures do not exist.

We have developed a silicon carbide based black body source that can withstand a temperature of 800 °C and is used for the calibration of the radiometer system. This black body at 600 °C acts as the hot body while the room temperature acts as the cold body source. An in situ arrangement for calibration is required to avoid any kind of errors during plasma electron temperature measurement. However, such an in situ kind of calibration is not possible here now, so the system is calibrated in lab including the transmission line losses; window losses and free space propagation losses that are calculated experimentally. The Y-factor so obtained by the calibration process is used to determine the plasma electron temperature.

4. Experimental Results

The radiometer system is horizontally aligned to a quartz window that receives the ECE radiation in the E-band. The system measures a part of the second harmonic i.e. 75.4 to 84.5 GHz on the low field side covering a radial location from - 1.02 cm to 12.13 cm. Figure 3 below shows the raw data received by the radiometer system for one of the SST-1 discharges.

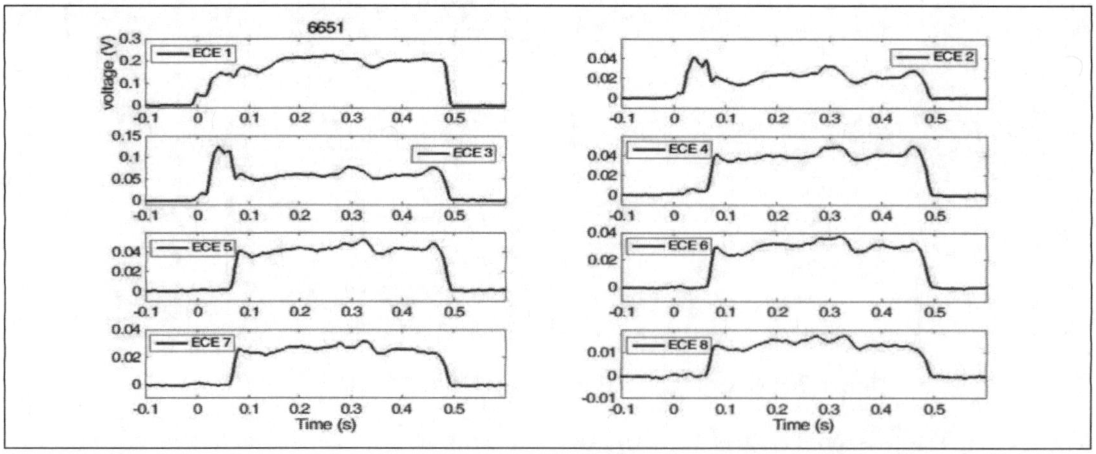

Figure.3 Raw data from ECE radiometer wherein ECE 1 to ECE 8 corresponds to radial locations 12.13cm, 10.06 cm, 8.06 cm, 6.13cm, 4.24 cm, 2.44 cm, 0.68 cm, - 1.02 cm respectively.

The depicted raw data is converted to plasma temperature using the calibration factor as described in section 2.2.The temporal and radial evolution of the measured radiation temperature, through ECE radiometer, for shot 7030 are shown in the figures below.

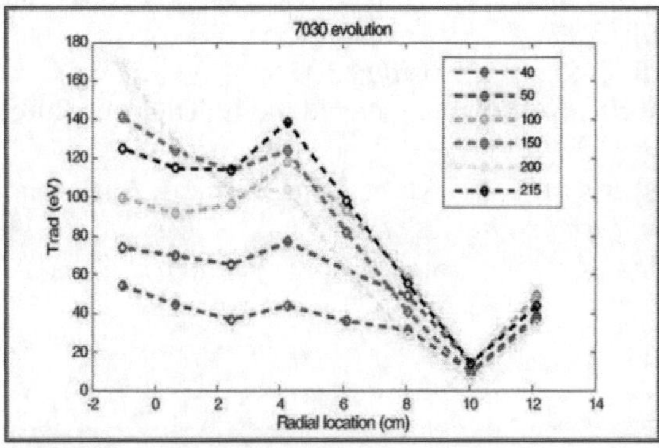

Figure 4a. Temporal evolution of T_{rad}.

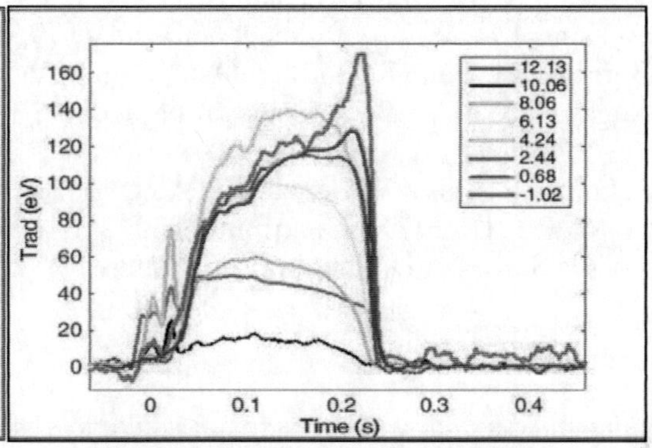

Figure 4b. Radial evolution of T_{rad}.

Since the measured density and temperature are of the order of $1 \times 10^{13}\ cm^{-3}$ and 200 eV, this plasma does not satisfy the optical thickness conditions for electron temperature measurements using ECE Radiometer system. Hence the measured temperature, as given by the ECE measurements, is the radiation temperature.

The comparison of the radiation temperature so measured is done with another temperature measuring diagnostics i.e. SXR that is available at SST-1. SXR diagnostic is designed to measure the chord averaged electron temperature of the central plasma region using the absorption foil thickness principle. Figure 5b shows the trace of the temperature data obtained by SXR diagnostic within an error of 30 %. The results measured by both the diagnostics are shown in figure 5a and 5b for one of the discharges at SST-1. However in most of the discharges the ECE measured temperature is found higher than that measured by SXR. Analysis is being carried out in this regard.

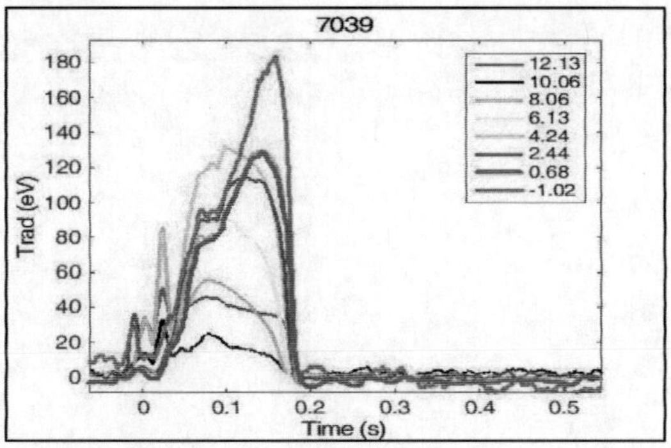

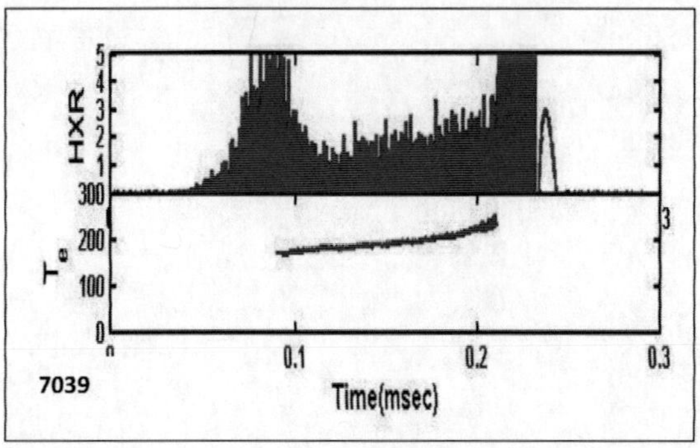

Figure 5a. T_{rad} measured by ECE radiometer. **Figure 5b**.T_e measured by SXR diagnostics.

5. Conclusion

An 8-channel heterodyne ECE radiometer is installed at SST-1 for temperature measurements. The design and calibration process for the diagnostic is explained. The observations and the measured radiation temperature are also shown with relevant comparison with SXR diagnostics. The measured radiation temperature using ECE diagnostics is around 180-200 eV.

IOP Conf. Series: Journal of Physics: Conf. Series **823** (2017) 012057 doi:10.1088/1742-6596/823/1/012057

References

[1] Van Gelder J F M 1996 Thesis on *"Electron cyclotron Wave absorption Experiment in tokamak Plasmas"* University of Utrecht

[2] Yuichiro K, Takuya S, Atsushi M, Naoki I, Yuya Y, Soichiro Y, Yoshio N, Seung H J, Myeun K and Kazuo K 2008 *Rev.of Sci.Instrum* **79** 10F115

[3] Bornatici M, Cano R, Barbieri O de, Engelmann R *1983 Nucl. Fusion* **23** 1153

[4] Nagayama Y, Kawahata K, Inagaki S, Kubo S, Narihara K, Ohyabu N and the LHD Group 2003 *J. Plasma Fusion Res.* **79** 601

[5] Schmuck S, Fessey J, Gerbaud T, Alper B, Beurskens M N A, E de la Luna, Sirinelli A, Zerbini M, and JET-EFDA Contributors 2012 *Rev. Sci. Instrum.* **83** 125101

[6] Varsha S, Kumar D, Shukla P and Pathak S K 2014 *Rev. Sci. Instrum.* **85**(5) 053503.

IOP Conf. Series: Journal of Physics: Conf. Series **823** (2017) 012058 doi:10.1088/1742-6596/823/1/012058

Design and Development of Amplitude and phase measurement of RF signal with Digital I-Q Demodulator

Dipal Soni[1], Kumar Rajnish[1], Sriprakash Verma[1], Hriday Patel[1], Rajesh Trivedi[1] and Aparajita Mukherjee[1]

[1]ITER-India, Institute for Plasma Research, Near Indira Bridge, Bhat, Gandhinagar 382 428, Gujarat, India

E-mail: dipal.soni@iter-india.org

Abstract. ITER-India, working as a nodal agency from India for ITER project [1], is responsible to deliver one of the packages, called Ion Cyclotron Heating & Current Drive (ICH & CD) - Radio Frequency Power Sources (RFPS). RFPS is having two cascaded amplifier chains (10 kW, 130 kW & 1.5 MW) combined to get 2.5 MW RF power output. Directional couplers are inserted at the output of each stage to extract forward power and reflected power as samples for measurement of amplitude and phase. Using passive mixer, forward power and reflected power are down converted to 1MHz Intermediate frequency (IF). This IF signal is used as an input to the Digital IQ Demodulator (DIQDM). DIQDM is realized using National Instruments make PXI hardware & LabVIEW software tool. In this paper, Amplitude and Phase measurement of RF signal with DIQDM technique is described. Also test results with dummy signals and signal generated from low power RF systems is discussed here.

1. Introduction

The Ion Cyclotron Heating and Current Drive (ICH & CD) system provides two essential functions: plasma heating and on-axis current drive and several auxiliary functions: plasma and burn control, wall conditioning at full field, plasma breakdown and assisted current rise. This system is to be designed to couple 20 MW of power in quasi-CW operation (pulses up to 3600 s), for a variety of ITER plasma scenarios. Power will be launched to the two antennas (placed inside the machine) from 8 numbers of sources via long coaxial transmission lines.

Each RF source (RFPS) consists of 2 RF chains, combined to generate output power of 2.5 MW @ 2.0 VSWR. A single chain of 1.5 MW generator consist low power components like RF switch, voltage controlled electronic attenuator, voltage control electronic phase shifter, monitoring and control components, Solid State Power Amplifier (SSPA/HPA-1) and high power amplifiers HPA-2 and HPA-3 [2]. The output of Master Synthesizer is split into two; each of them will drive individual chain of the amplifier. To control, monitor and acquire the different parameters of RF Source, Local Control Unit (LCU) is developed. The local control and monitoring unit is the heart of the whole system which will operate the system as per experimental requirement and ensure safe operation of amplifier with dynamic plasma load condition. As each chain is cascaded amplifier, Amplitude and Phase at the output of each stage of amplifier would be measured and monitored in the local control unit and same information would be passed to ITER central control room. The feedback control loops would be used for maintaining the Amplitude and Phase at requested value. For Amplitude and Phase measurement I-Q based RF monitoring and control system will be used. Figure 1 shows the measurement schema of amplitude and phase at each stage of RF source.

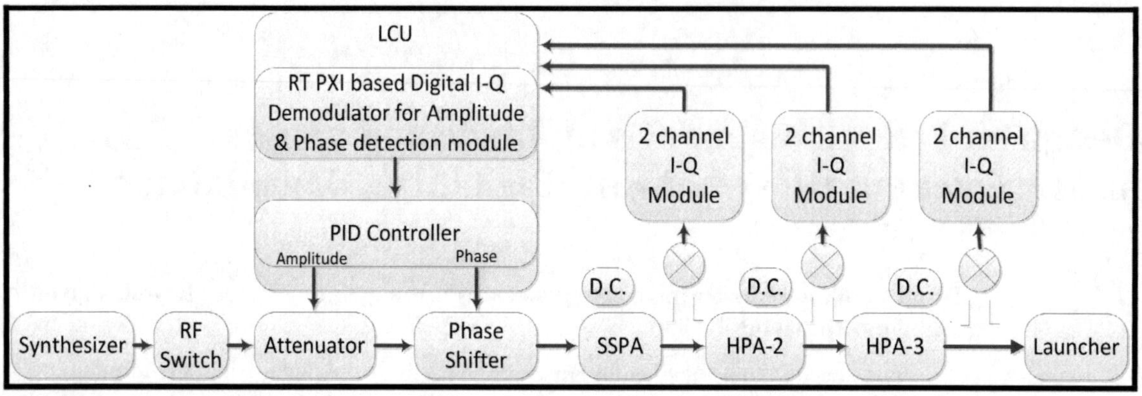

Figure 1. Amplitude and Phase Measurement schema.

2. Theory of I-Q Demodulator

The implementation methodology of digital I-Q technique to measure Amplitude and phase of RF signal is shown in figure 2. At output of each stage of amplifier directional coupler is connected for measurement of forward power and reflected power. The output of directional coupler is connected to mixer and mixer having 1 MHz band pass filter, which gives Intermediate frequency (IF) of 1 MHz down converted signal. After that Analog to Digital Converter (ADC) will sample the IF signal with 4 times higher clock frequency and get the I1, Q1, I2 and Q2 samples from IF signal. Averaging of I1 & I2 gives I component and averaging of Q1 & Q2 gives Q component [3]. Figure 3 shows the sampling of IF signal and figure 4 shows the Cartesian representation for I-Q sample. Amplitude and phase can be calculated through mathematical formulas after getting I and Q components [4].

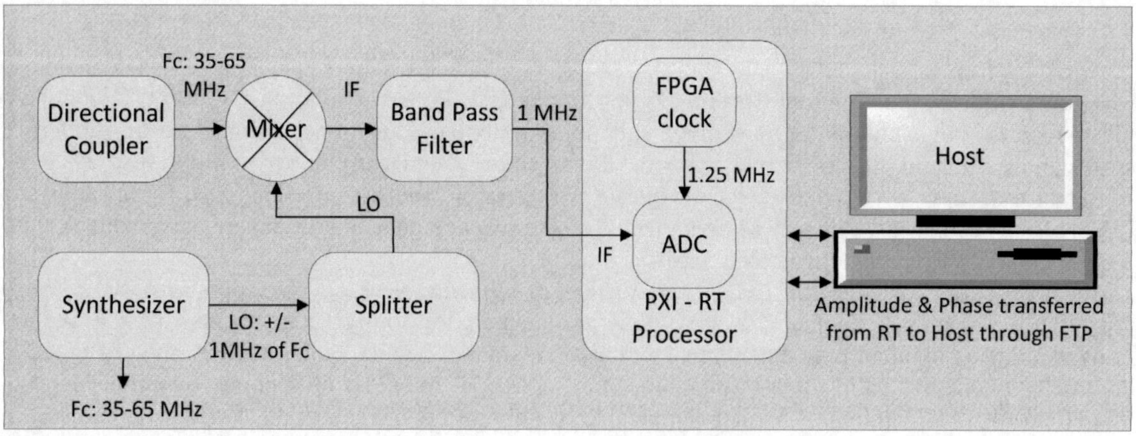

Figure 2. Architecture of Amplitude and Phase measurement.

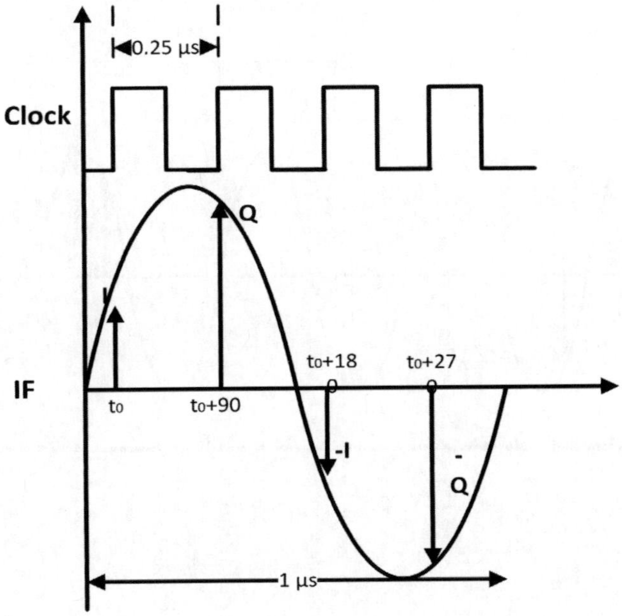

Figure 3. Actual sampling of RF Signal

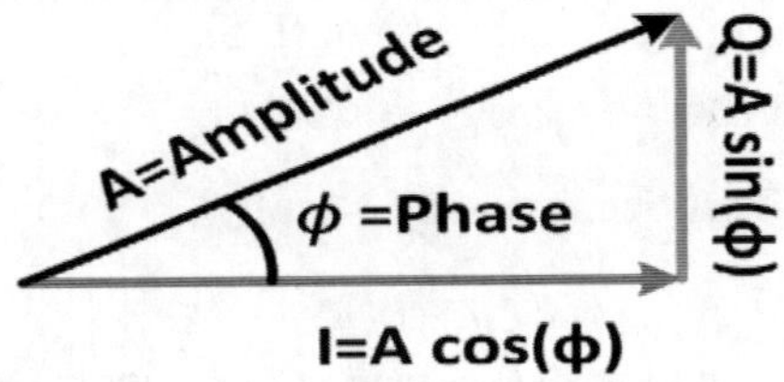

Figure 4. Cartesian representation for I-Q

$$A = \sqrt{I^2 + Q^2} \qquad (2.1)$$

$$\phi = \tan^{-1}(\tfrac{Q}{I}) \qquad (2.2)$$

FPGA-7841R is used for clock generation. Analog I/O module, PXI- 6133 S is used for acquisition of the IF and RT 8108 controller is used to make averaging and calculation for amplitude and phase measurement from I and Q component of IF input. Clock frequency of 800 kHz (1.25 µs) is generated through FPGA for sample the down converted RF signal (IF) of 1 MHz. The time period between consecutive ADC samples is kept 1.25 µs which corresponds to exactly 90° of next cycle of IF of 1 MHz, The ADC output provides a data stream consisting of the repeating pattern of measurements of I, Q, -I, and -Q. Thus, at each 5 µs I & Q sample for 1 cycle of IF is available. Sampling of the RF signal with clock is shown in figure 5. As described in figure 3, sampling frequency of IF of 1 MHz should be 0.25 µs exactly but same can be achieved by sampling it with 1.25 µs considering IF frequency is fixed throughout the experiment and its amplitude and phase is fixed for at least 5cycle of IF i.e. for 5 µs. For completing this initial experiment, available hardware PXI-6133 S is used that can support sampling rate at 1.25 µs. I and Q values within this data stream are separated through software. The sign inversion and averaging on these samples are carried out at RT controller to get I and Q of the input IF Signal for each 5 cycles. With each set of data of I and Q, amplitude and phase of individual channel is calculated using RT controller. After calculation, data will be stored in file and that file is transferred to host system via FTP protocol for offline analysis. Also each data is shared between RT and host through shared variable using shared variable engine protocol of LabVIEW software to display it at Host screen.

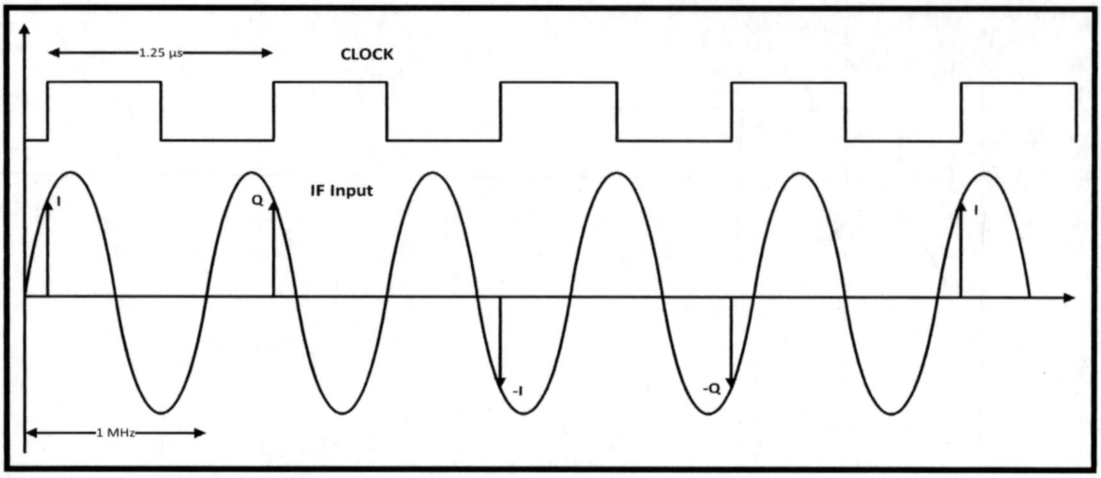

Figure 5. Sampling of IF Signal

3. Test Scenario and Result

3.1. Asynchronous Mode

Two independent Function Generators (FG) are used for this test set up. One of them is used as source for clock frequency (800 kHz) for ADC and other FG is used as IF input of 1MHz. Amplitude of IF input is kept 3Vp-p. As amplitude and frequency of IF input is fixed, I and Q value is expected to be constant and not variable. However, experimental results show the variation of I & Q as shown in figure 6.

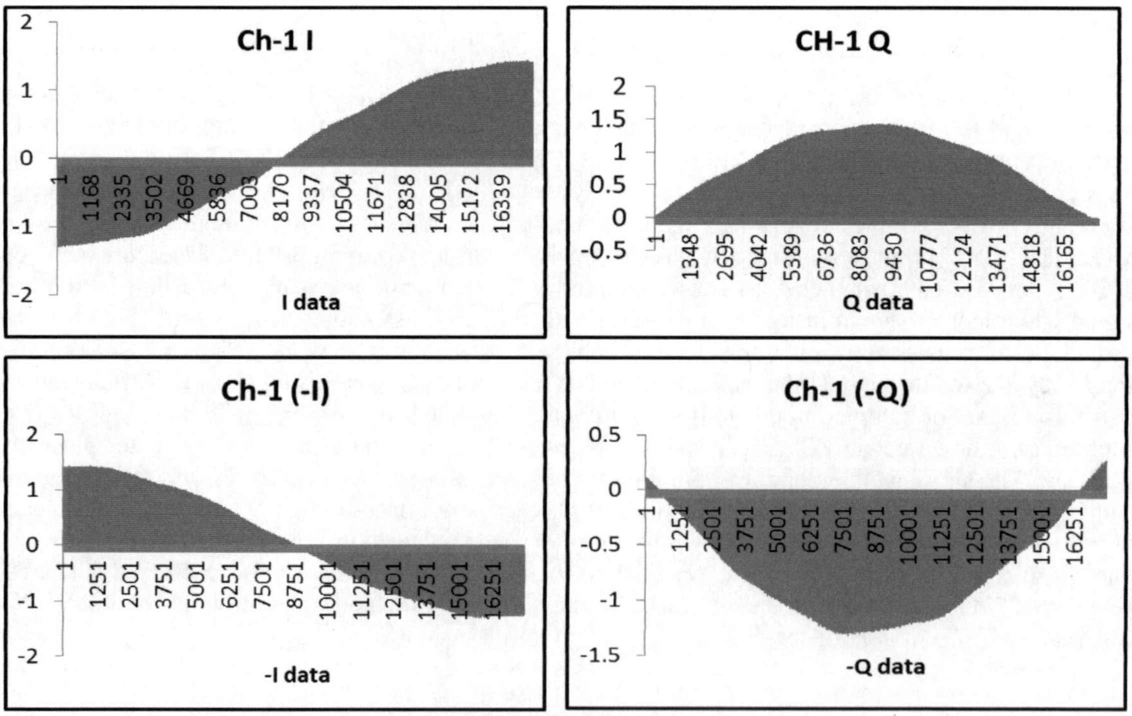

Figure 6. I-Q samples with Asynchronous mode of operation

Observation: As I/p signal & sampling clock are derived from two independent FG, they are not operated in synchronised mode. The desynchronization in terms of pico-second (~ 7.5 ps) order will causes the variation of I and Q samples with respect to time.

3.2. Synchronous Mode

Two independent function generators are used for this test set up also, but with this experiment both the function generators are synchronized with standard procedure defined by Tektronix. One of function generator is used for sampling clock generation of 800 kHz and other FG is used as test IF signal. Input frequency of IF signal is 1 MHz and amplitude is 4V p-p. Figure 7 shows the sampled IQ data which are fixed with respect to time.

Observation: Samples of I and Q are not varying with respect to time as input signal and acquisition clock is synchronized. This implies that synchronization of sampling frequency and IF signal are required otherwise, instantaneous sample of I-Q should be processed to get correct information of amplitude and phase.

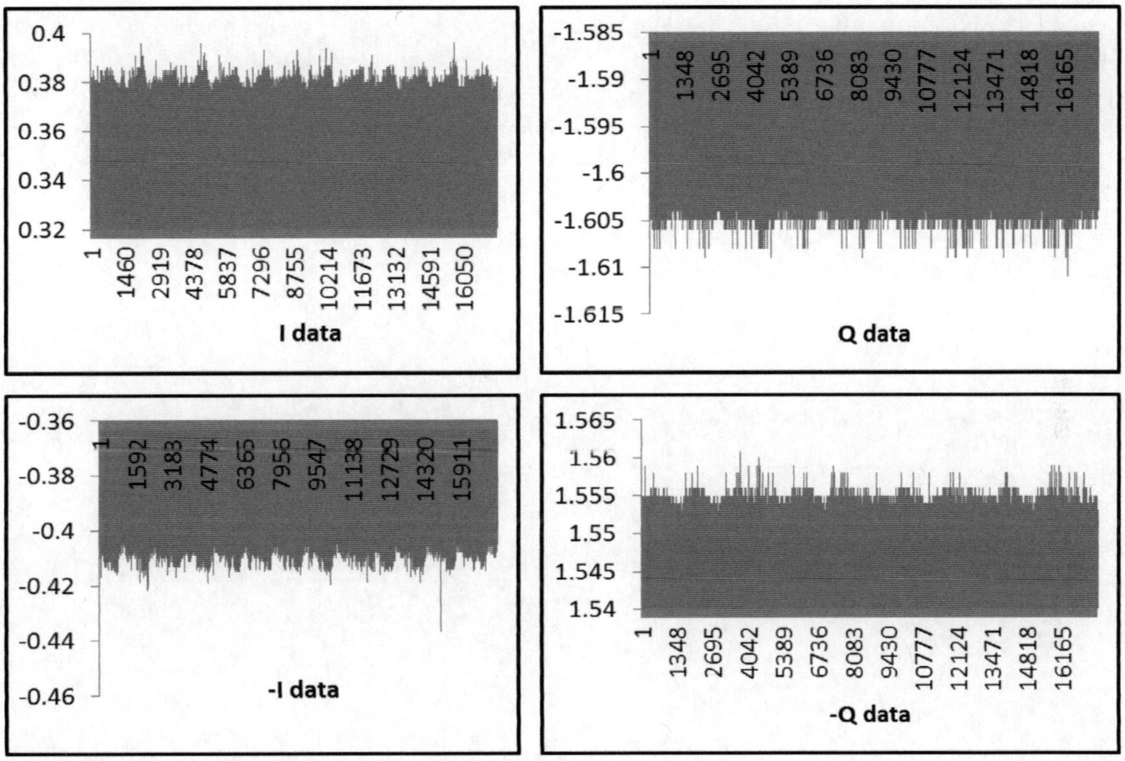

Figure 7. I-Q sample with synchronous mode

3.3. Actual Test Result

Here, FPGA is used to generate clock for acquisition and it is not synchronized with input signal. Amplitude of simulated IF signal is given to two channels of Digitizer board and amplitude for channel-1 & channel- 2 is set to 2 V @ 1 MHz using Function Generator. Phase setting between Ch1 & Ch2 from Function generator is 170°. Figure 8 shows the measured value of amplitude and relative phase between two i/p channels. This shows that both measurements are stable and capable to measure the phase with the accuracy of 1degree.

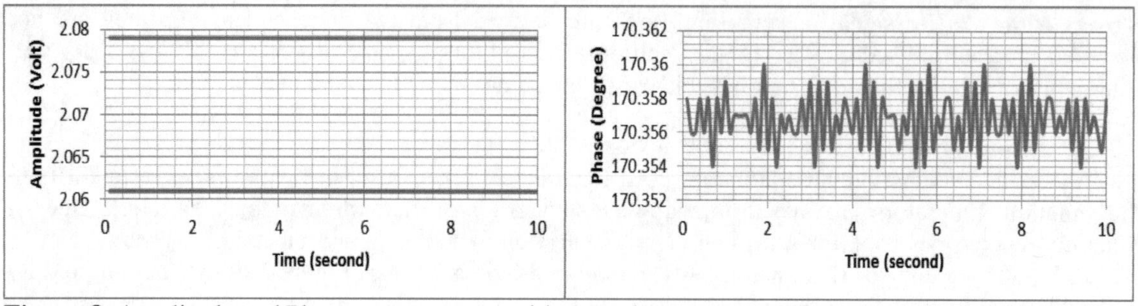

Figure 8. Amplitude and Phase measurement with Actual setup

4. Conclusion

Design concept is validated with this initial experimentation for measurement of amplitude and phase using digital I-Q demodulator technique. It is planned to incorporate this measurement technique with LCU where all required hardware are available and used for real time acquisition and control of RF source.

References

[1] Ikeda K et.al. 2007 *Nuclear Fusion* **47** S285
[2] Mukherjee A et al. 2013 *Fusion Science & Technology* **Vol 65** 120
[3] Ziomek C and Corrdoura P 1995 *Proceedings of Particle Accelerator Conference* **4** 2663

IOP Conf. Series: Journal of Physics: Conf. Series **823** (2017) 012059 doi:10.1088/1742-6596/823/1/012059

Design of new central solenoid for SST-1

Upendra Prasad[1], Subrata Pradhan[1], Mahesh Ghate[1], Piyush Raj[1], V L Tanna[1], Ziauddin Khan[1], Swati Roy[1], Prosenjit Santra[1], Prabal Biswas[1], A N Sharma[1], Yohan Khristi[1], Deven Kanaber[1] and Pankaj Varmora[1]

[1]HBNI, Institute for Plasma Research, Near Indira Bridge, Bhat, Gandhinagar 382 428, India

E-mail: upendra@ipr.res.in

Abstract. The key role of central solenoid (CS) magnet of a Tokamak is for gas breakdown, ramp up and maintaining of plasma current. The magnetic flux change in CS along with other PF coils generates magnetic null and induces electric field in toroidal direction. The induced toroidal electric field accelerates the residual electrons which collide with the neutrals and an avalanche takes place which led to the net plasma in the vacuum vessel of a Tokamak. In order to maximize the CS volt-sec capability, the higher magnetic field with a greater magnetic flux linkage is necessary. In order to facilitate all these requirements of SST-1 a new superconducting CS has been designed for SST-1. The design of new central solenoid has two bases; first one is physics and second is smart engineering in limited bore diameter of ~ 655 mm. The physics basis of the design includes volt-sec storage capacity of ~ 0.8 volt-sec, magnetic field null around 0.2 m over major radius of 1.1 m and toroidal electric field of ~ 0.3 volt/m. The engineering design of new CS consists of Nb3Sn cable in conduit conductor (CICC) of operating current of 14 kA @ 4.5 K at 6 T, consolidated winding pack, smart quench detection system, protection system, housing cryostat and conductor terminations and joint design. The winding pack consists of 576 numbers of turns distributed in four layers with 0.75 mm FRP tape soaked with cyanide Easter based epoxy resin turn insulation and 3 mm of ground insulation. The interlayer low resistance (~1 nΩ) terminal praying hand joints at 14 kA at 4.5 K has been designed for making winding pack continuous. The total height of winding pack is 2500 mm. The stored energy of this winding pack is ~ 3 MJ at 14 kA of operating current. The expected heat load at cryogenic temperature is ~10 W per layer, which requires helium mass flow rate of 1.4 g/s at 1.4 bars @ 4.5 K. The typical diameter and height of housing cryostat are 650 mm and 2563 mm with 80 K shield respectively. The protection system consists of SS310 made array of dump resistor of 20 mΩ. The detail physics and engineering design of new superconducting CS of SST-1 will be discussed in this presentation.

1. Introduction

The cross-sectional view of SST-1 is shown in figure 1. The central solenoid (CS) plays a key role for plasma breakdown and ramp up of plasma current. The CS is required to generate the higher magnetic field, which can produce sufficient volt-sec with a greater magnetic flux linkage at plasma location. In order to achieve this requirement of a Tokamak, the CS coil is charged up to the full current and then discharged to induce a toroidal loop voltage required to initiate and ramp up of the plasma. The volt-sec storage capability of SST-1 CS coil has been estimated using analytical formula and the magnetic null evolution with other transformer coils using IPR code [1]. The magnetic null evolution, volt-sec estimation are discussed in section 2 and 3. The engineering design of winding pack, conductor, insulation, joints, cryogenic heat loads, current lead design, housing cryostat vacuum system design are discussed in section 4. The quench detection and protection system along with conclusion are discussed in section 5 and 6 of this paper respectively.

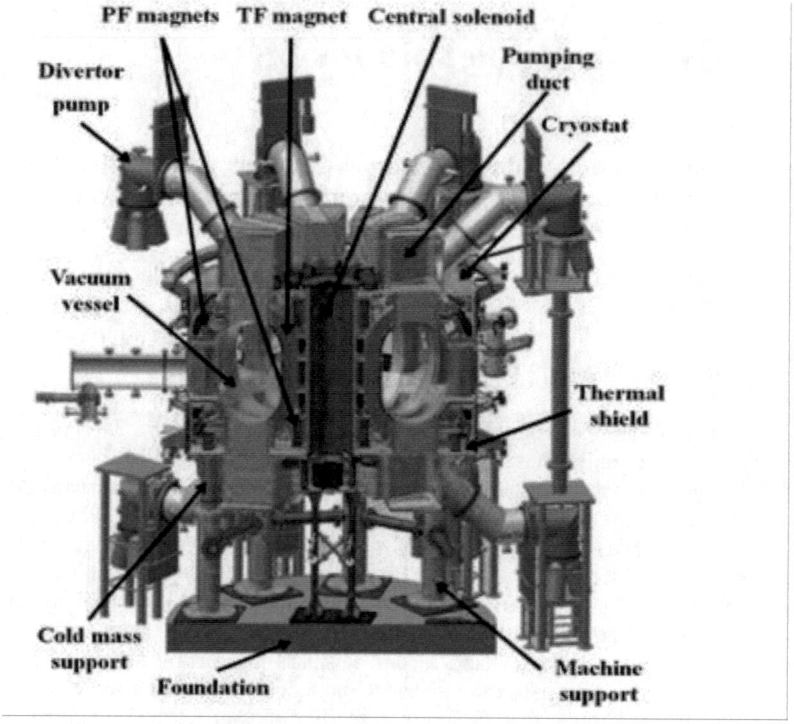

Figure 1. Cross-sectional view of SST-1 machine.

2. Magnetic null and volt-sec

The evolution of magnetic null is the primary requirement for plasma initiation inside the vacuum vessel of the machine. The vacuum magnetic null evolution for a set of coil currents (NI) and geometrical parameters mentioned in Table 1 is shown in figure 2. The vacuum magnetic null has been calculated using Green function technique. This technique uses poloidal magnetic flux at location (R', Z') due to a magnetic coil placed at (R, Z) is given by equation 1

$$\psi = \mu_0 \sum_{i=0}^{n_{coils}} I_i \, G(R, Z : R', Z') \tag{1}$$

In this equation μ_0 is vacuum permeability, I_i is coil current and n is the number of coils. Green function is given by equation 2

$$G(R, Z : R', Z') = \frac{1}{\pi} * \frac{\sqrt{RR'}}{k} \left[\left(1 - 0.5 * k^2\right) * K(k) - E(k) \right] \tag{2}$$

In this equation, k is the modulus of elliptic integrals is given by equation 3

$$k^2 = \frac{4RR'}{(R-R')^2 + (Z-Z')^2} \tag{3}$$

$K(k)$, and $E(k)$ are elliptic integral of first and second kind.
The radial (B_R) and axial (B_Z) magnetic fields are calculated using equation 4 and 5

$$B_R = \mu_0 I \frac{(Z-Z')k}{4\,R(RR')^{\frac{1}{2}}}\left(-K(k) + \frac{R^2+R'^2+(Z-Z')^2}{((R-R')^2+(Z-Z')^2)^{\frac{1}{2}}}E(k)\right) \quad (4)$$

$$B_Z = \mu_0 I \frac{k}{4\,(RR')^{\frac{1}{2}}}\left(K(k) + \frac{R'^2-R^2+(Z-Z')^2}{((R-R')^2+(Z-Z')^2)^{\frac{1}{2}}}E(k)\right) \quad (5)$$

It can be seen from this figure that the error field at null location is ~2 G.

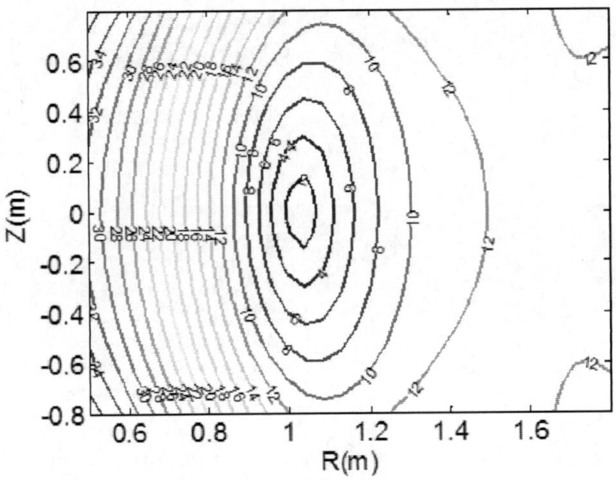

Figure 2. Vacuum magnetic null of SST-1 machine.

The volt-sec storage capability [2] of this CS is ~ 0.9 at 14 kA @ 4 T.

3. Engineering design of central solenoid

The winding pack (WP) was designed with four layers of insulated cable-in-conduit conductor (CICC) due to bore diameter limitation of 600 mm as shown in figure 3.

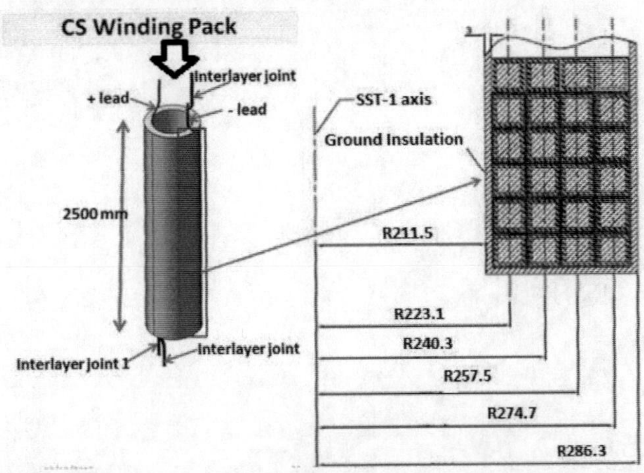

Figure 3. Cross-sectional view of central solenoid of SST-1.

The inner and outer radii of CC WP are 211 and 285 mm respectively having height of 2482 mm. Nb_3Sn superconducting strands based CICC was chosen as base conductor for this WP. The conductor parameters are provided in table 2. The insulation system consists of cyanide Easter resin soaked FRP tape wounded over Kapton as a background for turn and layer insulation. Few parameters of CS winding pack are shown in figure 3.

Table 1. Coil design parameters.

Parameters	CS	TR#2	TR#3	TR#4
NI (MA)	8.064	0.56	0.042	0.014
Mean Radius(R) (mm)	248.9	587	2449	2469
Radial thickness (ΔR)(mm)	68.8	195	58	19
Height(Z) form center(mm)	0	± 1480	± 1305	± 730

Table 2. Conductor design parameters.

Parameters	Values
Operating current (I_{OP})	15 kA
Strand diameter	0.8 mm
No. of SC strands	120
No. of Cu strands	60
Twist pitch length	20 mm,70 mm, 120 mm, 240 mm
Cabling configuration	(((2+1)×3)×4)×5
Void fraction	~ 34.5%
Jacket material	316 LN
Jacket thickness	1.5 mm
Conductor size	15 mm × 15 mm

3.1. Interlayer joint and current leads

The inter-layer praying hand joint [3] was designed for central solenoid winding pack for powering with two leads and single power supply (PS). The joint length equal to the last twist pitch length of ~ 300 mm CICC was chosen to ensure the SC strands electrical contacts with the septum material. Oxygen-free high thermal conductivity (OFHC) copper with residual-resistivity ratio (RRR) greater than 100 as septum material, Ag5Sn95 and Pb40Sn60 solder as pre-soldering and final soldering for joint making were chosen for this joint. The joint design schematic and few parameters are shown in figure 4 and in Table 3 respectively. Brass and MgB_2 superconductor joint [4] based current leads with optimum cryogenic consumption was also designed for new CS winding pack. The schematic of such current lead and its basic parameters are shown in figure 5 and Table 4 respectively.

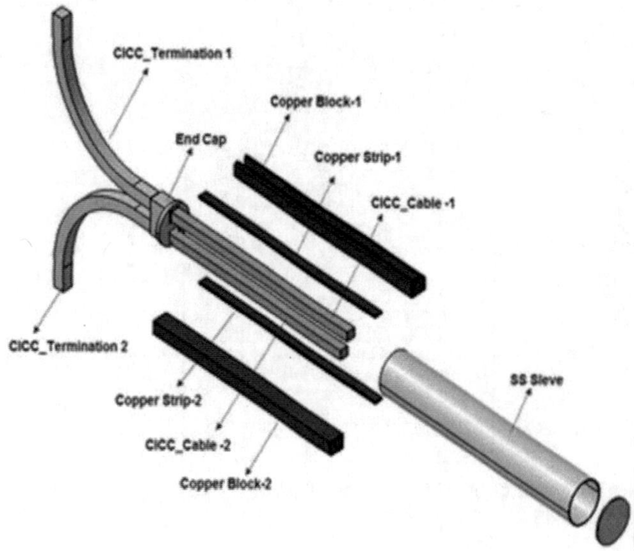

Figure 4. Schematic of Interlayer joint.

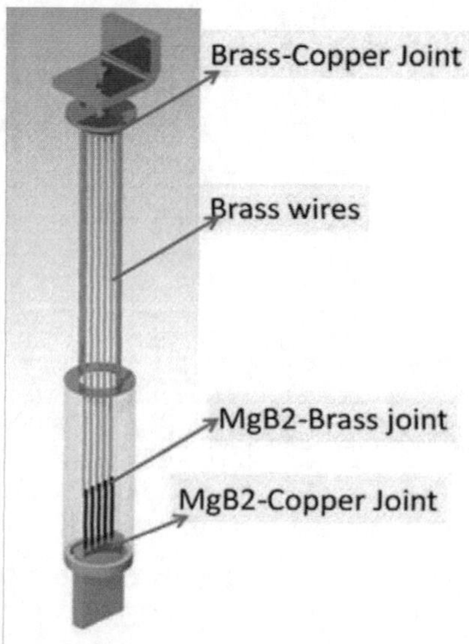

Figure 5. Schematic of current lead.

Table 3. Joint design parameters.

Parameters	Values
Operating temperature	4.5 K
Operating pressure	1.4 bar
Helium Mass flow rate	0.7 g/s
Joint resistance	0.5 nΩ
AC loss	2 W
Cryo-Stability	1.22 kW
Electromagnetic load	735 N

Table 4. Current lead design parameters.

Parameters	Values
Operating current (I_{OP})	15 kA
Brass cross-section (A_C)	1584 mm^2
Length of brass HX	750 mm
Shape parameter ($I\ell/A_C$)	1.4 bar
Heat load at I_{OP}	~30 W/lead
Heat load at standby	~3.50 W/lead
Helium mass flow rate at I_{OP} =0	~20 g-s^{-1}
Helium mass flow rate at I_{OP} = 15 kA	~0.58 g-s^{-1}
Burn-out time	~5 min
MgB_2 length	400 mm
MgB_2 area	301.62 mm^2
Voltage drop across the lead	~148 mV

3.2. Quench detection and protection

The quench detection system [5] with inductive voltage cancellation and resistive threshold voltage was designed for CS WP. The quench detection system was designed for a threshold voltage of 100 mV and 100 ms time for quench declaration. The dump resistor of 20 mΩ was used for safe extraction of stored electrical energy of 3 MJ. The schematic and designed parameters of dump circuit and dump resister is shown in figure 6 and figure 7.

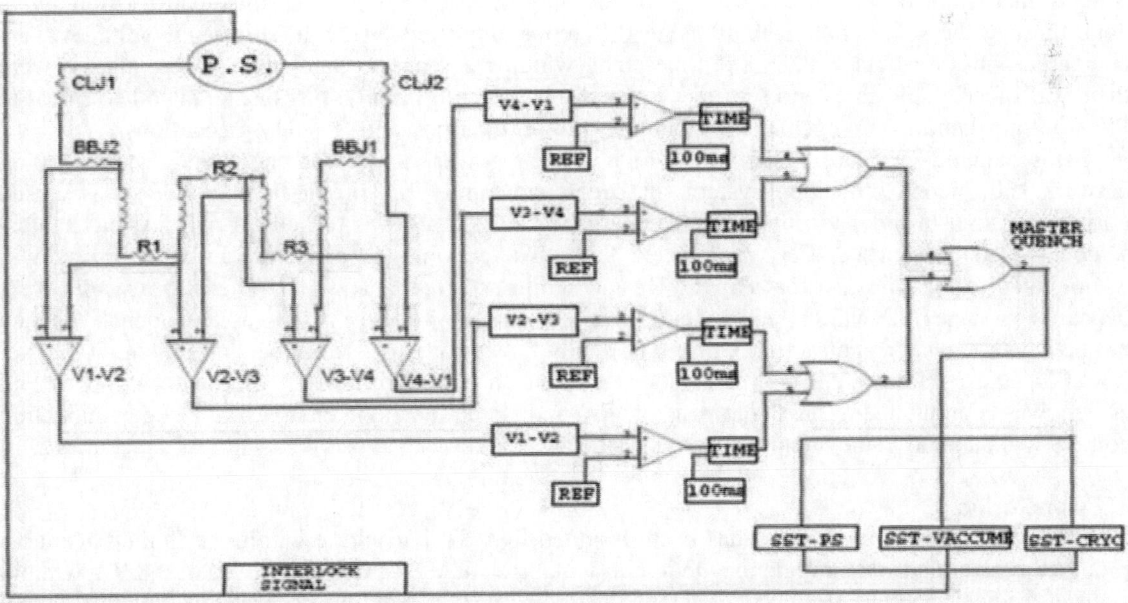

Figure 6. Quench detection schematic for SST-1 CS.

IOP Conf. Series: Journal of Physics: Conf. Series **823** (2017) 012059 doi:10.1088/1742-6596/823/1/012059

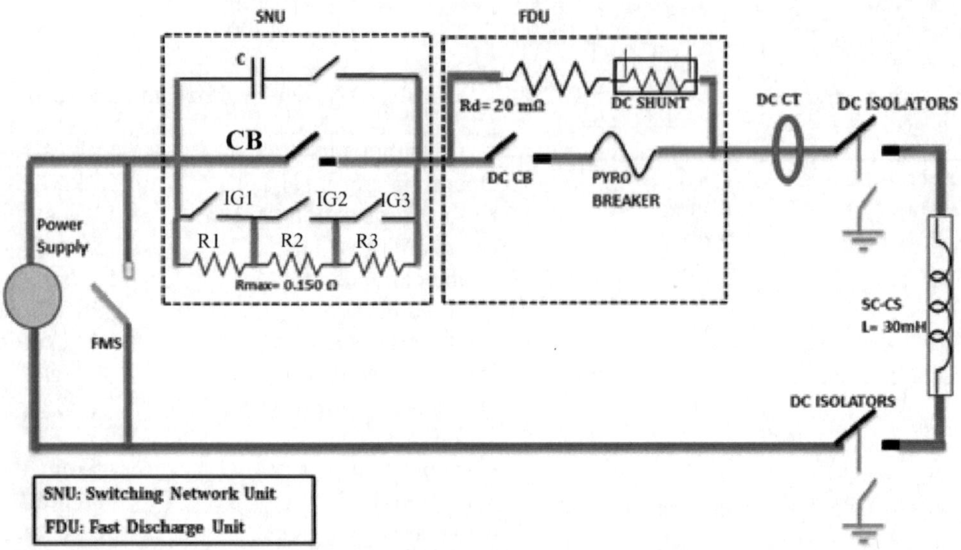

Figure 7. Schematic of quench protection circuit.

The quench detection system consists of four layers of CS described by four inductors with four joints, R_1, R_2 ,R_3 and bus-bar joint (BBJ) connected to a power supply (PS) via current lead joints (CLJ). V_1, V_2, V_3 and V_4 represent the resistive voltages across the 1st, 2nd, 3rd and 4th layer of CS. In order to protect these voltage taps form induced inductive voltage, a multi-level protection circuit consists of RC filter, metal oxide varistors and back-to-back Zener diodes were used. These voltages were further isolated and filtered using isolation amplifier and filters. The noise cancellations were done through the subtraction scheme using difference amplifiers. All four difference voltages were compared with threshold voltage and time using window comparator and timer. These signals were then evaluated by OR gates and sent to latch and relay circuit. Finally, this master signal was sent to PS to initiate dumping, cryogenic and vacuum group for their preventive safety precautions.

The schematic in figure 7 shows the high current power supply for energizing, wave shaping network and quench protection system for superconducting CS. Circuit breaker (CB) was closed during charging. In order to initiate the discharge, power supply was turned off and fast mechanical switch (FMS) was closed. CB was opened through a pre-charged capacitor and the current was commutated to the blip resistors R_1, R_2, R_3 and ignitrons IG_1 IG_2 and IG_3. The blip resistors were bypassed in stages to achieve required slopes in CS current waveforms. In case of quench, quench protection system was made active. Current normally flew through closed direct current circuit breaker (DC CB) and Pyro breakers for redundancy. As quench occur, either of these switches opens and current was commutated to the dump resistor (R_D) and stored magnetic energy of the superconducting coil CS was dumped in the dump resistor as heat.

4. Conclusion

Nb_3Sn based new central solenoid has been designed for SST-1 machine within the limited available room temperature bore to provide magnetic null with error field of 2 G and volt-sec of 0.8 Wb without altering original machine design parameters. The consolidated winding pack consists of four layers with 576 number of turns connected in series with four inter-layer sub-nano ohm resistance joints, one pair of MgB_2 based current leads and can produce magnetic field ~ 4 T at 4.5 K @ 14 kA operating current. Smart quench detection and protection system with shunt resistance of 20 mΩ has also been designed for this central solenoid.

Acknowledgement
The author is thankful to Dr. Srinivasan for his kind help during calculation of magnetic vacuum null for Tokamak like condition. The author is also thankful to all the technical staffs of SST-1 magnets, magnet technology division and SST-1 cryogenics, who have directly or indirectly contributed during evolution of the design of this superconducting central solenoid.

References
[1] Johnson J L, Dalhed H E, Greene J M, Grimm R C, Hsieh Y , Jardin S C, Manickam J, Okabayashi M, Storer R G, Todd ,A M , Voss D E, AND Weimer K E, 1979 *Journal of Computational Physics* **32** 212
[2] Prasad U, Srinivasan R, Pradhan S et al. 2016 *IEEE Transaction on Applied superconductivity* **26** 4200904 (pp 4)
[3] Prasad U, Panchal A, Raj P and Pradhan S 2015 *Internal report on Development of Prototype Praying hand Nb_3Sn CICC joint for new CS of SST-1,* IPR/TR-381
[4] Prasad U, Tanna V L and Pradhan S 2016 *Plasma and fusion science from fundamental to technological application, Apple academic press,* **part III** 22
[5] Sharma A N, Hansalia C J, Yeole Y, Bansal G, Pradhan S, Moitra A, Saxena Y C 2005 *Fusion Engineering and Design* **74** 819

IOP Conf. Series: Journal of Physics: Conf. Series **823** (2017) 012060 doi:10.1088/1742-6596/823/1/012060

Overall behaviour of PFC integrated SST-1 vacuum system

Ziauddin Khan[1], Dilip C Raval[1], Yuvakiran Paravasu[1], Pratibha Semwal[1], Kalpeshkumar R Dhanani[1], Siju George[1], Mohammad Shoaib[1], Arun Prakash[1], Gattu R Babu[1], Prashant Thankey[1], Firozkhan S Pathan[1] and Subrata Pradhan[1]

[1]Institute for Plasma Research, Near Indira Bridge, Bhat, Gandhinagar 382 428, India

E-mail: ziauddin@ipr.res.in

Abstract. As a part of phase-I up-gradation of Steady-state Superconducting Tokamak (SST-1), Graphite Plasma Facing Components (PFCs) have been integrated inside SST-1 vacuum vessel as a first wall (FW) during Nov 14 and May 2015. The SST-1 FW has a total surface area of the installed PFCs exposed to plasma is ~ 40 m^2 which is nearly 50% of the total surface area of stainless steel vacuum chamber (~75 m^2). The volume of the vessel within the PFCs is ~ 16 m^3. After the integration of PFCs, the entire vessel as well as the PFC cooling/baking circuits has been qualified with an integrated helium leak tightness of < 1.0×10^{-8} mbar l/s. The pumping system of the SST-1 vacuum vessel comprises of one number of Roots' pump, four numbers of turbomolecular pumps and a cryopump. After the initial pump down, the PFCs were baked at 250 °C for nearly 20 hours employing hot nitrogen gas to remove the absorbed water vapours. Thereafter, Helium glow discharges cleaning were carried out towards the removal of surface impurities. The pump down characteristics of SST-1 vacuum chamber and the changes in the residual gaseous impurities after the installation of the PFCs will be discussed in this paper.

1. Introduction

SST-1 Tokamak (figure 1) was successfully commissioned in 2012 [1-3] and the first plasma was achieved [4] in June 2013 with graphite poloidal limiters and SS 304L as the vessel wall material. Due to plasma wall interactions, high-Z impurities may be released from the vessel wall which in turn cools the plasma by radiation loss. In order to reduce this effect, in 2nd phase of SST-1 refurbishment, PFCs were installed in the system.

PFCs of SST-1 Tokamak [5,6] consists of Inboard divertor plates (IDP), Outboard divertor plates (ODP), Inboard passive stabilizers (IPS), Outboard passive stabilizers (OPS), main baffle (MBAF) along with their headers & sub-headers. Each IDP, ODP, IPS, MBAF and OPS has top and bottom modules mounted in the main vacuum vessel (VV) as shown in figure 2. All PFC's are structurally continuous in toroidal direction. Graphite was chosen as Plasma facing material considering its good thermal properties and low atomic mass. Copper Zirconium (Cu-Zr) & Copper Chromium Zirconium (Cu-Cr-Zr) alloys plates embedded with SS 304L piping were used as back plate materials for proper heat transfer during baking and cooling operations. Approximately 3800 tiles were mounted on 132 numbers of these copper alloys back-plates. Gas-to-gas heat exchange method was adapted to heat nitrogen gas which is then pressurized using a dedicated gas blower system [7] to bake the vacuum vessel as well PFCs. Some of the other major parameters of the SST-1 tokamak are shown in table 1.

Figure 1. Snap shot of SST-1 machine.

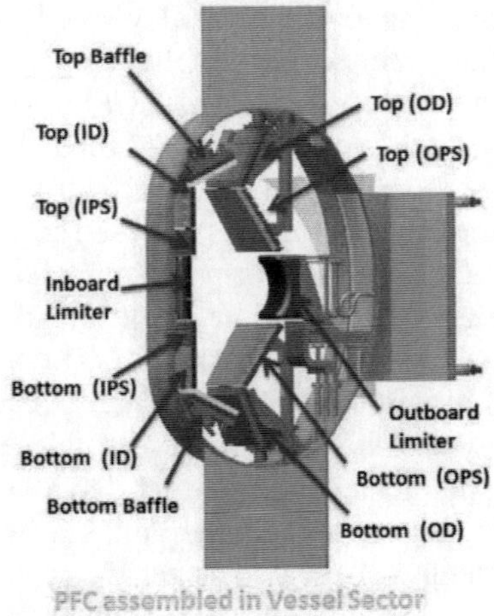

Figure 2. View of PFC inside the SST-1 main vessel.

Table 1. Major SST-1 machine parameters.

Parameters	Values
Major radius	1.1 m
Minor radius	0.2 m
SS surface area of VV	75 m^2
Exposed surface area of PFC	40 m^2
Plasma species	Hydrogen
Volume enclosed by PFC	16 m^3
Ultimate vacuum in VV	~1.0×10^{-8} mbar
Operating pressure range	5.0×10^{-5} mbar (max)

All individual PFCs were baked at a temperature of 250 °C for 8 hours flattop using hot nitrogen gas at the working pressure of 4.5 bar(a) under ultra-high vacuum (UHV) condition for validation. Other PFC auxiliary components like headers, sub-headers and isolators were also tested at their functional conditions to verify their functionality and to ensure operation conformity. Strict metrology and QA/QC plans were structured and executed to integrate the PFC components inside the vacuum vessel. During the installation of PFCs along with their supports, Electronic Co-ordinate Determining System (ECDS) and Photogrammetric measurements were carried out to accuracy of 0.8 mm over a length of 500 mm length whereas an accuracy of 1.9 mm over a length of 1600 mm. Initial reference points with XYZ co-ordinates were established inside the vacuum vessel using ECDS instrument. Then many coded and un-coded targets were mounted on the VV surface at these reference points for Photogrammetric measurements. These points were then scanned using DSLR camera and AUSTRALIS 7.0 software through which 3-D models were generated and the measurement was completed.

The complete operation and the performance of the PFC integrated SST-1 vacuum system is categorised as the pumping and baking of VV & PFCs and the gas puffing for glow discharge cleaning (GDC) and plasma break down. These two aspects are discussed in subsequent sections.

2. Pumping and baking of VV & PFCs

After the complete assembly of PFC components along with their headers and sub-headers inside the vacuum vessel, the integrated PFC system was helium leak tested at room temperature (RT) for the leak tightness of $< 1.0 \times 10^{-8}$ mbar l/s. The rectangular baking channels embedded inside the vessel surface for baking the VV were also leak tested at this background. Then the entire VV surface, PFCs including all subsystems exposed to inner side were cleaned properly as per UHV protocol and then the VV was closed for pumping activity. SST-1 VV has 16 numbers of large size rectangular flanges and 32 numbers of large size triangular flanges. All these large sized noncircular flanges were closed using demountable aluminium wire-seals of 2.6 mm diameter. All other CF openings were closed using aluminium / copper gaskets. A leak tightness of $\sim 1.0 \times 10^{-9}$ mbar l/s was achieved in the wire-seals with the applied force per unit length of 160–200 kN/m. Due to this force, a compression up to 25% was achieved [8], which have led to such leak tightness. After baking of VV at 130 °C, it was observed that few aluminium wire-seals showed the leak of $\sim 1.0 \times 10^{-6}$ mbar l/s which after retightening got sealed as per the requirements. The VV was initially pumped down from atmosphere to 1.0×10^{-3} mbar using a single roots pumping station of 1800 m^3/h pumping speed. Electro-pneumatic gate valves with proper isolation were used between the pumping station and the VV in order to provide protection against eddy current during plasma disturbance. Also these valves provide protection for transition from low vacuum to high vacuum operation. Two turbomolecular pumps (TMPs) with a net pumping speed of 3250 l/s (N_2 gas) were mounted at the diagonally opposite radial ports (R-3 & R-11) of VV [9], each having an electro-pneumatic gate valve. Eight numbers of TMPs were also installed at the top diverter while another eight numbers of TMPs were at the bottom divertors for diverter pumping due to long pulse plasma operations. Each divertor TMP set has a net pumping speed of 1750 l/s (N_2 gas). All these eighteen sets of TMPs were situated at a place where the transverse magnetic field is less than 30 gauss. A residual gas analyzer (RGA) and some vacuum gauges were mounted in one of the pumping lines. RGA was operated time to time for measuring the partial pressures of different gases as well as the leak tightness of the vacuum vessel. Also a cryopump with a net pumping speed of 1400 l/s (water vapour) was installed at the radial port R-7. Only four TMPs and the cryopump were under operation for pumping purpose.

After achieving a base pressure of 1.5×10^{-5} mbar inside the VV, the initial baking of SST-1 VV alone was carried out by flowing hot nitrogen gas at an inlet pressure of 2.5 bar(a) and a supply temperature of 230 °C through the rectangular channels embedded into the inner surfaces of VV. The temperature of the supply gas was controlled using AC drive and PLC based PID controller. Nearly three hundred fifty numbers of K-type Kapton insulated thermocouples were mounted at the different locations of VV and PFCs through copper block brazed to the inner surfaces. Temperature signals were acquired through D-type 50-pins feed-through using MASIBUS make Data loggers. Temperature data from Data loggers was stored in PC using CITECT SCADA based software via RS-485 Interface. The vessel was maintained at 130 ± 10 °C for few hours during which the temperature of PFCs was raised to 120 ± 5 °C due to radiation and remains stable. Since the baking temperature of PFC was insufficient for the removal of impurities adsorbed in it, the baking of PFC at higher temperature was carried out. Baking of VV was stopped and then the hot nitrogen gas at an inlet pressure of 2.5 bar(a) and supply temperature of 380 °C was passed through PFCs headers [10]. PFCs temperature was reached to 240 ± 10 °C within 8 hours of baking and maintained at flattop for 20 hours. During this period, the temperature of VV was rising continuously due to radiation. Since SST-1 VV is a continuous torus and has a thermal stress limitation, its temperature has to be maintained below 150 °C. In order to maintain the vessel temperature below 150 °C, the room temperature nitrogen gas was flown through the baking channels of VV continuously till the baking of PFC was under operation. During baking, the pumping of VV was carried out with four TMPs sets and a cryopump. An ultimate pressure of 6.0×10^{-8} mbar was achieved inside VV after few days of pumping which was followed with baking of both VV & PFCs. Figure 3 shows the pump down curve of VV along with the temperature curves of VV & PFCs maintained at required values. The partial pressures of the gases present inside VV during baking is shown in figure 4. It shows that all the adsorbed gases were releasing from the inner surfaces.

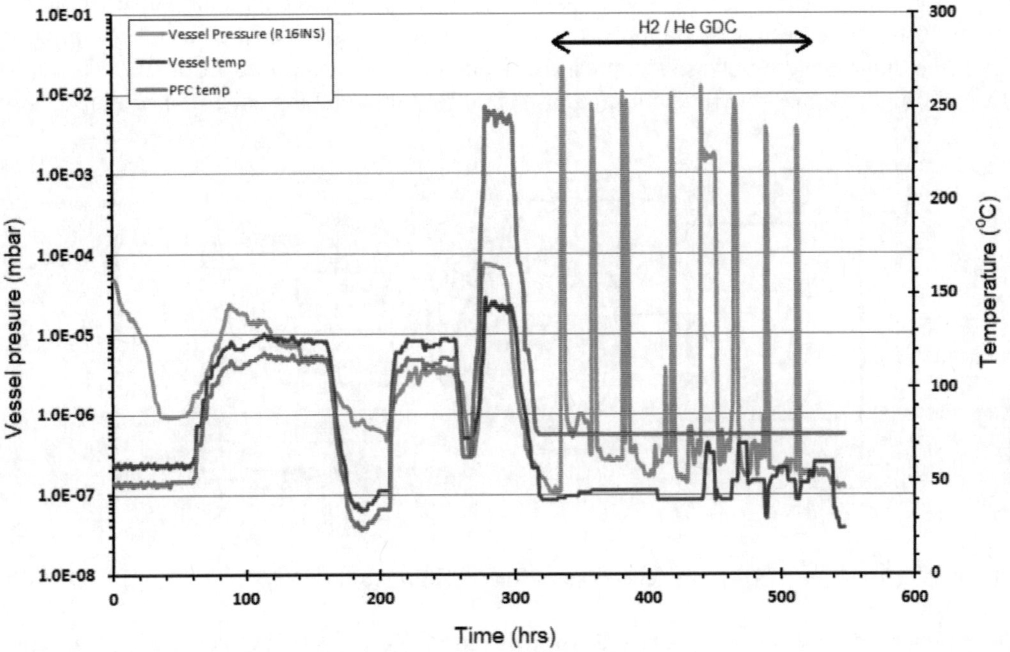

Figure 3. SST-1 VV pressure during baking and other operations.

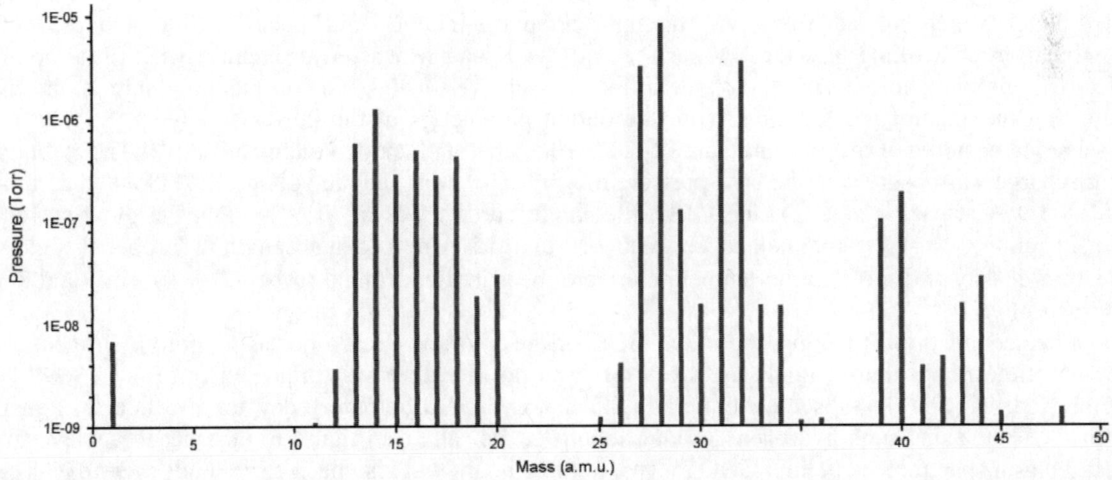

Figure 4. Partial pressure of different gases released from the inner surfaces of VV & PFCs during baking.

3. Gas puffing for GDC and plasma break down

SST-1 gas feed system was designed and installed in order to cater the initial base pressure requirements during GDC and prefilling for plasma breakdown operation. The schematic of SST-1 gas feed system for prefilling and GDC is shown in figure 5. Two piezoelectric valves were mounted at diagonally opposite locations of SST-1 machine RPs (R-6 & R-14) to maintain the overall balance of gas distribution inside the VV. These piezoelectric valves start functioning above 25 VDC and work up to 100 VDC when triggered with different shape of pulses. Gas feed pulse shape can also be adjusted as per requirement using PCI eXtensions for Instrumentation-based control system. The pressure of 1.0×10^{-5} mbar was achieved inside the VV when a rectangular pulse of 100 VDC amplitude of 10 ms time period was applied to the piezoelectric valves at a constant gas pressure of 2.0 bar(g) at the inlet of these valves. The pressure inside the VV was varied from 1.0×10^{-5} mbar to 5.0×10^{-5} mbar by increasing the time duration of gas puffing up to 50 ms under the similar condition.

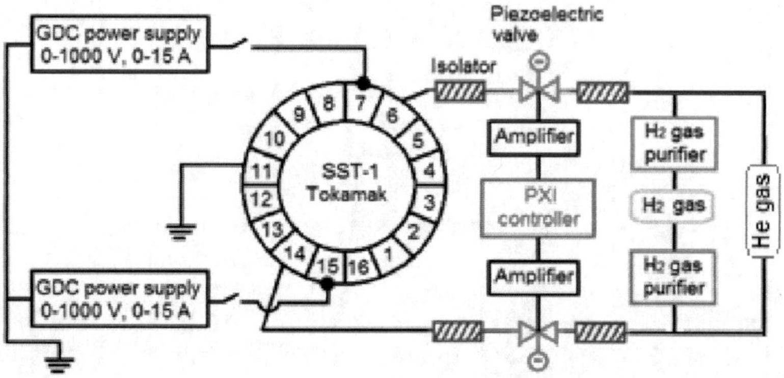

Figure 5. The schematic of SST-1 gas feed system for prefilling and GDC.

Glow discharge cleaning was selected for the conditioning of VV surfaces before plasma breakdown operation so that the impurities like oxygen (O_2) and carbon (C) gets reduced. Two numbers of electro-polished stainless steel rectangular plates (350 mm × 150 mm × 25 mm) were installed at the diagonally opposite radial ports (R-7 & R-15) outward radial locations of VV. These two plates were isolated from VV through ceramic isolators and behaved like anodes. The combination of working gas, the VV surface and the electrode via power supply system sets up an electrical circuit. The uniform and stable glow was achieved using a single pumping station and by adjusting the amount gas fed along with the output parameters of the power supply. In SST-1, the discharged occurred at the base pressure of 1.0×10^{-2} mbar and anode voltage of 880 V. The stability in discharge was observed at the base pressure of 1.0×10^{-3} mbar, anode voltage of 280 V and current of 2.8-3.0 A at each anode. During GDC, the anode current was adjusted so that the glow current density at the wall was maintained between 0.1 to 0.15 A/m^2. Using Langmuir probe, the glow electron density and the electron temperature were measured and found to be 4.0×10^{13} m^{-3} and 3.5 eV respectively.

After completion of baking, the initial GDC of the VV was carried out using pure hydrogen gas for few times. High purity grade hydrogen gas was purified to a very high purity of ~ 99.9999 % before fed into SST-1. Subsequently helium GDC was carried out every day for two hours. A snap shot for GDC using both hydrogen gas and helium gas is shown in figure 6. In order to achieve the better plasma parameters, helium GDC for two to three hours was also been carried out every day. The partial pressures of the different gases present inside the VV after GDC is shown in figure 7. It shows that the water vapour as well as the oxygen impurity was removed to very low partial pressure. Only trace of nitrogen gas was observed. Since a large numbers of isolators and welding of PFC headers and sub-headers have exposed to high temperature for long duration, there might be a probability of development of minor leak which was contributing to the existence of nitrogen inside the VV.

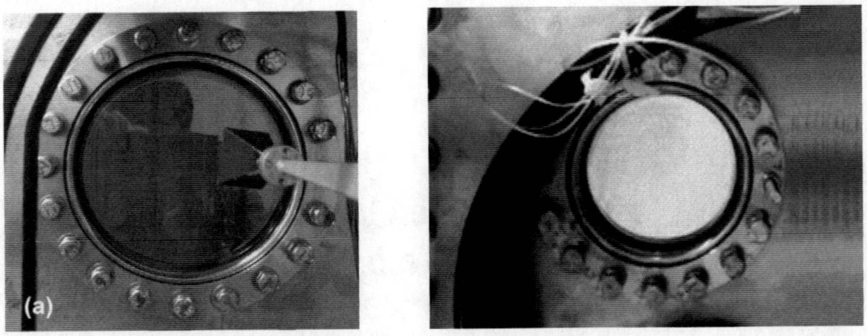

Figure 6. The snap shot of glow during GDC using (a) pure hydrogen gas (b) pure helium gas.

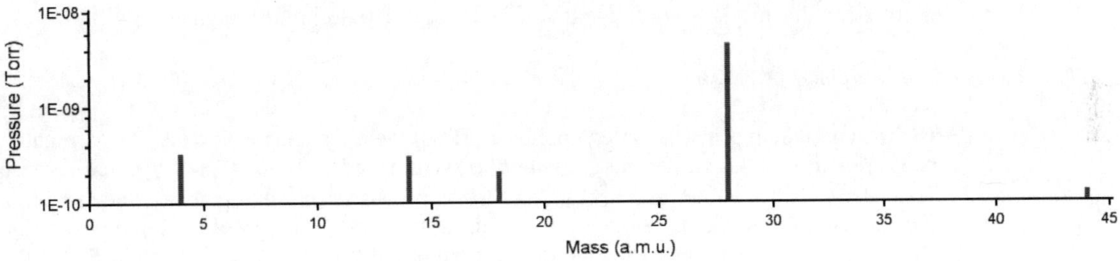

Figure 7. Partial pressure of different gases inside the VV after GDC.

4. Conclusion

During this experimental campaign of October 2015, the circular plasma breakdown of 65 kA current for 250 ms was demonstrated. The pumping system, baking system, gas feed system and GDC have been successfully and continuously operated for months without any fault. No major problems or serious deviation were observed which could affect the plasma operation. We have also successfully installed and commissioned the PFCs components as per their design requirements.

Acknowledgement

We acknowledge other SST-1 divisions like Central control, Magnet division and Cryo division, Power supply for extending their valuable time and support for experimental preparation.

References

[1] Pradhan S and Team SST-1 mission 2010 *J. Fusion Res. Series* **9** 650
[2] Pradhan S, Sharma A N, Tanna V L, Khan Z, Prasad U et al. 2012 *IEEE Trans. Plasma Sci.* **40** 614
[3] Khan Z, Pathan F, Gorge S, Dhanani K R, Yuvakiran P et al. 2014 *IEEE Transaction on Plasma Science* **42** 1006
[4] Pradhan S, Khan Z, Tanna V L et al. 2015 *Nucl. Fusion* **55** 104009
[5] Jacob S et al. 1996 *Journal of Nuclear Materials* **233-237** 655
[6] Khan Z, Paravastu Y and Pradhan S 2014 *Procedia Materials Science* **6** 216
[7] Khan Z, Pathan F S, Pravastu Y, George S et al. 2013 *Plasma Science and Technology* **15** 157
[8] Khan Z and Pathak H A 2013 *Advanced Materials Manufacturing & Characterization* **3** 511
[9] Khan Z, Pathan F S, George S, Dhanani K R, Paravastu Y et al. 2014 *Fusion Enginnering and Design* **89** 49
[10] Paravastu Y, Khan Z and Pradhan S 2015 *Fusion Engineering and Design* **98-99** 1375

IOP Conf. Series: Journal of Physics: Conf. Series **823** (2017) 012061 doi:10.1088/1742-6596/823/1/012061

Baking and helium glow discharge cleaning of SST-1 Tokamak with graphite plasma facing components

P Semwal[1], Z Khan[1], D C Raval[1], K R Dhanani[1], S George[1], Y Paravastu[1], A Prakash[1], P Thankey[1], G Ramesh[1], M S Khan[1], P Saikia[1] and S Pradhan[1]

[1]Institute for Plasma Research, Near Indira Bridge, Bhat, Gandhinagar 382 428, India

E-mail: prats@ipr.res.in

Abstract. Graphite plasma facing components (PFCs) were installed inside the SST-1 vacuum vessel. Prior to installation, all the graphite tiles were baked at 1000 °C in a vacuum furnace operated below 1.0×10^{-5} mbar. However due to the porous structure of graphite, they absorb a significant amount of water vapour from air during the installation process. Rapid desorption of this water vapour requires high temperature bake-out of the PFCs at ≥ 250 °C. In SST-1 the PFCs were baked at 250 °C using hot nitrogen gas facility to remove the absorbed water vapour. Also device with large graphite surface area has the disadvantage that a large quantity of hydrogen gets trapped inside it during plasma discharges which makes density control difficult. Helium glow discharge cleaning (He-GDC) effectively removes this stored hydrogen as well as other impurities like oxygen and hydrocarbon within few nano-meters from the surface by particle induced desorption. Before plasma operation in SST-1 tokamak, both baking of PFCs and He-GDC were carried out so that these impurities were removed effectively. The mean desorption yield of hydrogen was found to be 0.24. In this paper the results of baking and He-GDC experiments of SST-1 will be presented in detail.

1. Introduction

The interaction of energetic plasma particles with the walls of plasma chamber is inevitable in fusion devices resulting in the release of several low-Z and high-Z impurities into the plasma through processes such as sputtering, evaporation etc. The presence of impurities has two major harmful effects on plasma evolution. The first one is the power radiation which results in cooling of the main plasma thereby limiting the achievement of high central temperatures needed for fusion and the second one is the dilution of hydrogenic fuel which results in a reduction of the fusion power. Various wall conditioning methods such as baking, DC or RF glow discharges cleaning etc. are employed in tokamaks to remove low-Z impurities such as water vapour, carbon and oxygen from the surfaces of PFCs before initiation of plasma experiments. On the other hand, high-Z impurities are controlled by coating the walls with low-Z materials like carbon, boron or lithium by various wall coating methods.

One of the most undesired low-Z impurities in fusion devices with graphite covered first wall is oxygen. The main source of oxygen is the water vapour outgassing from the inner surfaces of the vacuum vessel [1]. As per studies [2], when energetic oxygen ions impact onto a pure carbon surface, it is retained up to a concentration of 0.25 oxygen per carbon atom within the implantation zone and then it is re-emitted in the form of CO and CO_2 with a yield close to unity even at room temperature (RT). Some of these re-emitted ions have high thermal energy (~ 0.25 eV) and therefore have a high probability of penetration into the main plasma causing power losses. Thus the efficient removal of water vapour is essential to achieve the desired plasma conditions.

Further hydrogen gets retained in graphite surfaces of PFCs during plasma discharges. Hydrogen isotopes from the plasma interact with graphite walls and get retained in it in the form of saturated layer by four processes. These processes are (a) saturation of the implant zone, (b) co-deposition with sputtered carbon on surfaces, (c) absorption on internal porosity and (d) transgranular diffusion with trapping. This saturated layer will dominate the recycling and has no effect on the inventory due to its limited thickness and hence it is possible to deplete such saturated layer by use of helium glow discharge cleaning (He-GDC). On the other hand, the co-deposited layer has limitless ability to store hydrogen isotopes and is very difficult to remove. Since hydrogen recycling makes the density control difficult, it is necessary to clean the near-surface saturated layer by He-GDC [3].

2. SST-1 Tokamak and its plasma facing components

The main vacuum vessel (VV) of SST-1 tokamak [4, 5] is fabricated from SS 304L material in D-shaped torus form. It has 16 rectangular radial ports, 16 triangular top vertical ports, 16 triangular bottom vertical ports, 16 interconnecting rings and 16 vessel sector rings. SS U-channels are welded inside the vacuum vessel through which pressurised hot nitrogen gas is passed for baking of the vessel. Some of the other major parameters of SST-1 Tokamak are shown in table 1.

Table 1. Major SST-1 machine parameters.

Parameters	Values
Major radius	1.1 m
Minor radius	0.2 m
SS surface area of VV	75 m^2
Exposed surface area of PFC	40 m^2
Plasma species	Hydrogen
Volume enclosed by PFC	16 m^3
Ultimate pressure in VV	~1.0×10^{-8} mbar
Operating pressure range	5.0×10^{-5} mbar (max)

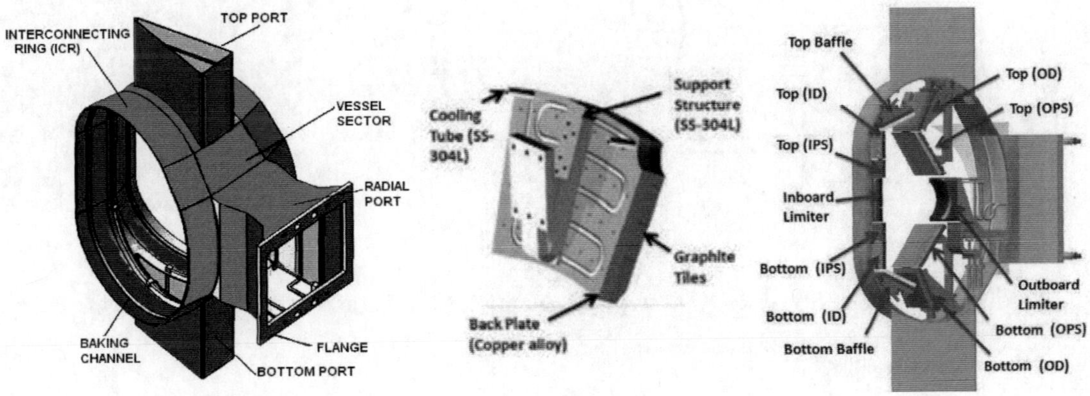

(a) (b) (c)

Figure 1. (a) SST-1 Tokamak schematic with baking channels, (b) Baking/cooling piping on copper back plate of PFC module (c) Schematic of SST-1 vessel module with PFCs.

Plasma facing components (PFCs) of SST-1 Tokamak [6] consist of Inboard divertor plates (IDP), Outboard divertor plates (ODP), Inboard passive stabilizers (IPS), Outboard passive stabilizers (OPS) and main baffle (MBAF). Each IDP, ODP, IPS, MBAF and OPS has top and bottom modules mounted inside the VV as shown in figure 1. All PFC's are structurally continuous in toroidal direction and were fabricated from Cu-Zr & Cu-Cr-Zr alloys back plates embedded with SS 304L piping for proper heat transfer during baking and cooling operations. Approximately 3800 graphite tiles are mounted on 132 numbers of these copper alloys back-plates.

3. Baking and glow discharge cleaning of SST-1 vacuum vessel

Baking of VV and PFCs was carried out in a controlled manner by flowing hot nitrogen gas at required temperature through the respective baking channels in a closed loop. A mass flow rate of 1.074 kg/s and 0.8 kg/s at 4.5 bar(a) of hot nitrogen gas was required to maintain VV at 150 °C and PFCs at 250 °C respectively [7]. During baking of VV and PFCs, the pumping lines are maintained at 100 °C using heating pads. K-type thermocouples were mounted on VV, PFCs tiles and the pumping lines to monitor and maintain their temperatures continuously.

A schematic of DC glow discharge cleaning system of SST-1 is shown in figure 2(a). Two anodes fabricated from electro-polished SS 304L (175 mm × 125 mm × 20 mm) are mounted in diagonally opposite radial ports R-7 & R-16. Each anode is insulated from VV with two Al_2O_3 ceramic rods and maintained at a minimum distance of 100 mm from VV wall as shown in figure 2(b). Each GDC electrode is powered by an independent power supply having variable voltage (0-1000 VDC) and an adjustable current range (0-15 A). Piezo-electric gas feed valves are used to feed pure hydrogen/helium gas or their mixture into VV from two diagonally opposite radial ports R-3 and R-11 for GDC as well as for prefilling. A double Langmuir probe (LP) with two cylindrical tungsten tips of diameter 1.0 mm and length 3.0 mm is used to measure the temperature and electron density during GDC. The cleaning effect due to GDC is studied using an online Residual Gas Analyser (RGA).

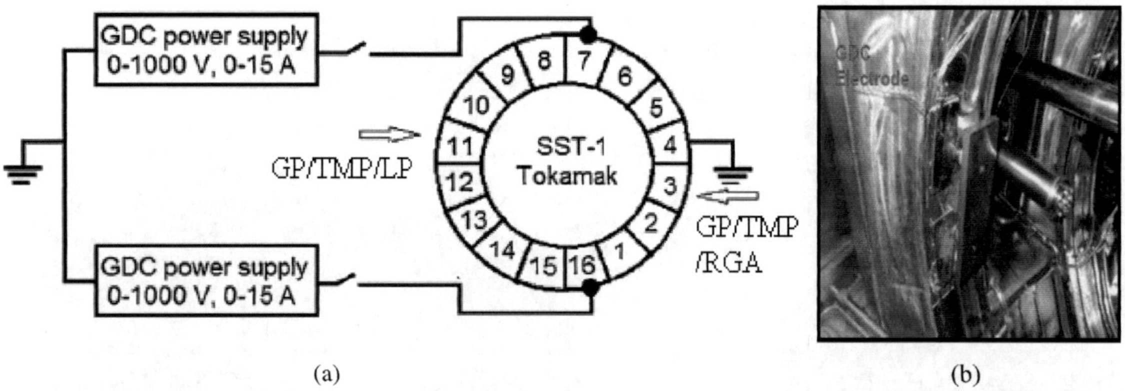

Figure 2. (a) Schematic of SST-1 GDC system (b) Snapshot of GDC anode mounted inside the vessel.

4. Results and Discussions

4.1. *Effect of baking*

RGA scan of VV before bake out at a total pressure of 6.17×10^{-7} mbar is shown in figure 3(a). Most dominant peak is observed at mass no. 18 which indicates the presence of large amount of water vapour inside the vessel. In order to remove it, the VV and PFCs were baked at an average temperature of 120 °C and 230 °C respectively for about 20 hrs. A snapshot of temperatures achieved in VV and PFCs at a given time is shown in figure 4. After baking, the ultimate vacuum achieved at room temperature (RT) inside VV was 1.07×10^{-7} with six (06) numbers of Turbomolecular pumps

(TMP) having a total effective pumping speed of 10,140 l/s (for N2). RGA spectra of the vessel acquired after baking is shown in figure 3(b). It can be seen that there is a decrease in the partial pressure of mass no. 18 by a factor of ~ 10 after bake out.

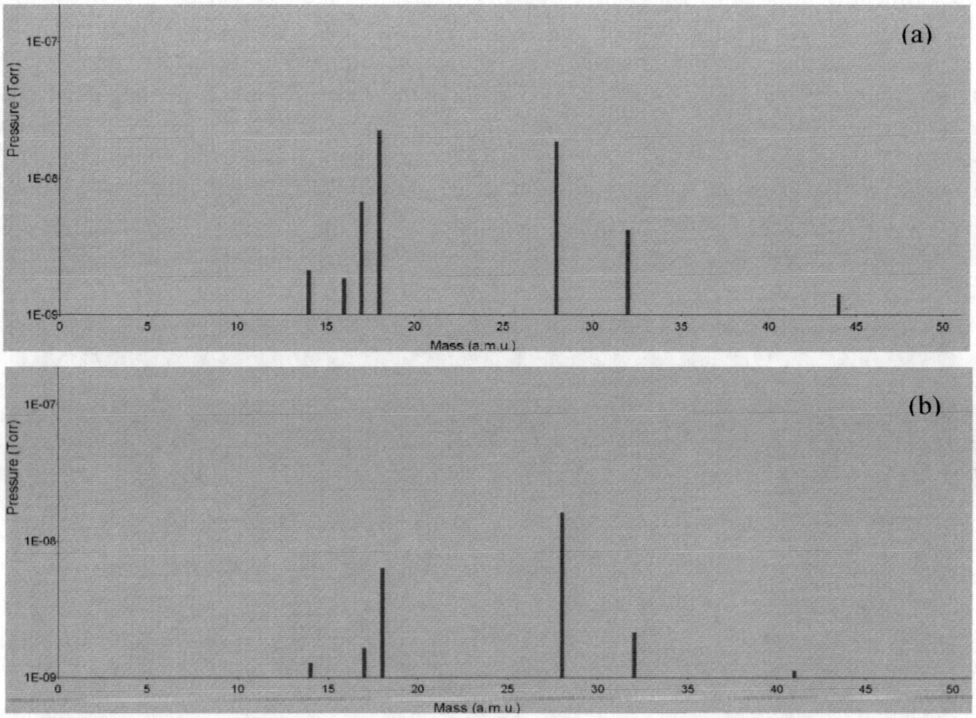

Figure 3. (a) RGA scan of unbaked VV, (b) RGA scan of VV at RT after bake out.

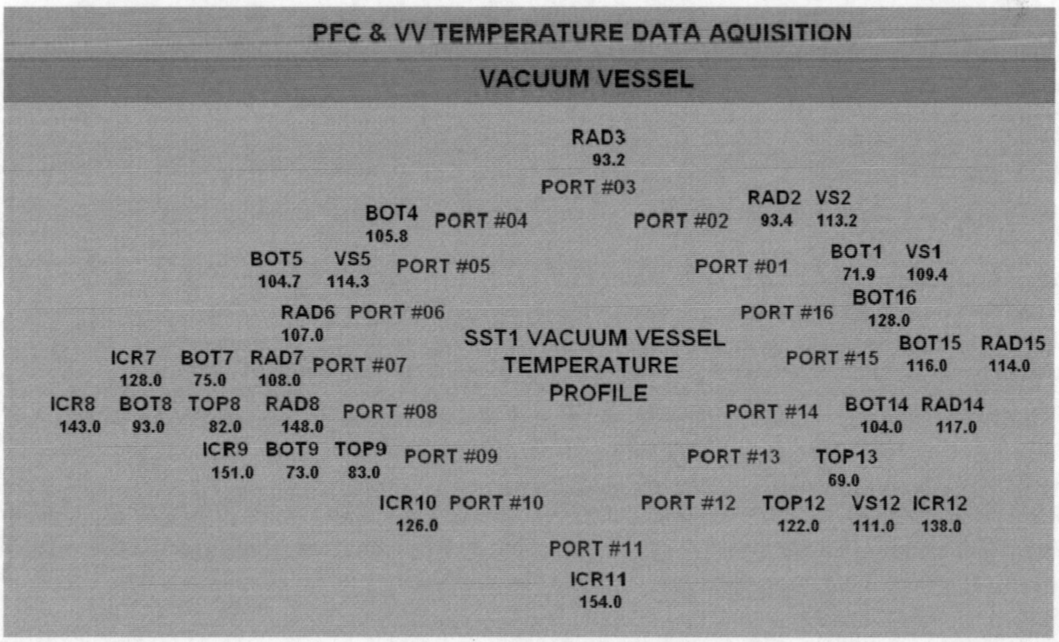

Figure 4. Screen shot of temperatures achieved in VV.

4.2. Effect of He-GDC

DC glow discharge cleaning was carried out for three (03) hours every day before, after and during the plasma experiment to remove hydrogen gas retained in the carbon walls. A total pressure of 2.0-3.0 × 10^{-3} mbar inside the VV and a current density of 0.1 A/m^2 was maintained during He-GDC. A snapshot captured during He-GDC in SST-1 is shown in figure 5. The partial pressures (PP) of different gases is measured during GDC with a differentially pumped RGA connected at one of the VV pumping lines with the correction factor of 100 times with respect to VV PP. A trend of hydrogen and helium partial pressures measured during GDC is shown in figure 6. It shows that during He-GDC, hydrogen partial pressure is increased by a factor of ~ 10 indicating the desorption of hydrogen from the walls. For such condition, hydrogen removal rate (R_{H2}) is estimated to be 0.698 mbar l/s based on an average change of 2.17 × 10^{-4} mbar in the hydrogen partial pressure.

Figure 5. Photograph of He-GDC inside SST-1 VV.

The number of hydrogen molecules released (ΔN) during GDC can be calculated using the formula

$$\Delta N = \left(\frac{R_{H2} \times \Delta t}{R \times T}\right) N_A \tag{1}$$

where Δt = 12600 s is the cleaning duration, R = 8.314 Pa.m^3/k/mole is the universal gas constant and T = 300 K is the gas temperature.

The total quantity of He ions injected during GDC can be estimated using the formula

$$Q_{He+} = \left(\frac{I_a}{e}\right)(1-\alpha) f \Delta t \tag{2}$$

where I_a is the total anode current, e is the electron charge, α is the He+ ion reflectance of the wall and f is the atom amount in a He atom (f = 1) [8]. For SST-1, Q_{He+} value is estimated to be 7.1 × 10^{23} ions which in turn gives the hydrogen desorption yield ($\delta_{H2} = \Delta N/Q_{He+}$) to be 0.24.

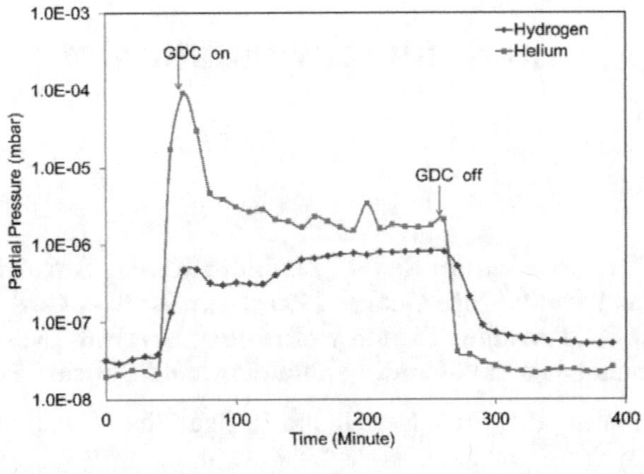

Figure 6. Hydrogen/Helium partial pressure trend during GDC.

RGA scan of VV after the GDC at a base pressure 3.24 × 10^{-7} mbar is shown in figure 7. The reduction in partial pressure of mass no. 18 could be observed. The peak at mass no. 4 indicates the presence of residual helium gas inside the VV after GDC. The electron temperature and density measured during GDC with Langmuir probe are found to be ~ 2-3 eV and (2-3) × 10^{8} cm^{-3}.

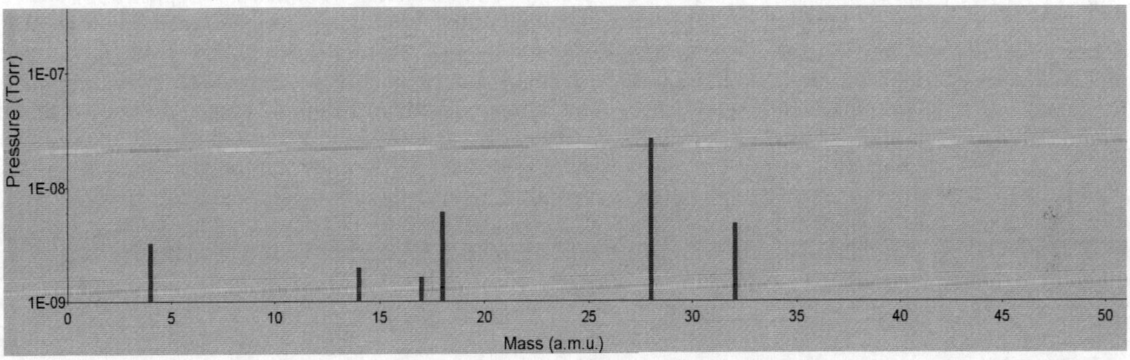

Figure 7. RGA scan of SST-1 VV after He-GDC

5. Conclusion

SST-1 vacuum vessel and graphite based PFCs were baked successfully at temperatures of 150 °C and 250 °C respectively for few days. This baking has been found effective in removing the water vapour absorbed in the bulk of graphite of PFCs. Also a routine helium glow discharge cleaning before, after and between few plasma shots was found to be very effective in reducing the hydrogen retention in the graphite. A hydrogen desorption yield of ~0.24 was estimated for SST-1 tokamak with He-GDC.

References

[1] Waelbroeck F G, Banno T, Esser H G et al. 1989 *J. Nucl. Mater.* **162–164** 496
[2] Winter J 1996 *Plasma Phys. Control. Fusion* **38** 1503
[3] A Causey R 2002 *J. of Nuc. Mater.* **300** 91
[4] Pradhan S and SST-1 Mission Team, 2010 *J. Fusion Res.Series* **9** 650
[5] Pradhan S, Khan Z, Tanna V L et al. 2015 *Nucl. Fusion* **55** 104009
[6] Jacob S, Khirwadkar S S, Sinha P, Pathak H A et al. 1996 *J. Nuc. Mater.* **233-237** 655
[7] Khan Z, Paravastu Y and Pradhan S 2014 *Proc. Mat. Sci.* **6** 216
[8] Zhi-wen W, Dong-hai Y and En-yao W 2002 *Plas. Science Tech.* **4** 1565

IOP Conf. Series: Journal of Physics: Conf. Series **823** (2017) 012062 doi:10.1088/1742-6596/823/1/012062

Commissioning and experimental validation of SST-1 plasma facing components

Yuvakiran Paravastu[1], Dilip Raval[1], Ziauddin Khan[1], Hitesh Patel[1], Prabal Biswas[1], Tejas Parekh[1], Siju George[1], Prosenjit Santra[1], Gattu Ramesh[1], ArunPrakash A[1], Prashant Thankey[1], Pratibha Semwal[1], Kalpeshkumar R Dhanani[1], Snehal Jaiswal[1], Pradeep Chauhan[1] and Subrata Pradhan[1]

[1]Institute for Plasma Research, Near Indira Bridge, Bhat, Gandhinagar 382 428, India

E-mail: pyuva@ipr.res.in

Abstract. Plasma facing components of SST-1 are designed to withstand an input heat load of 1.0 MW/m^2. They protect vacuum vessel, auxiliary heating source i.e. RF antennas, NBI and other in-vessel diagnostic from the plasma particles and high radiative heat loads. PFC's are positioned symmetric to mid-plane to accommodate with circular, single and double null configuration. Graphite is used as plasma facing material, back made of copper alloy and SS cooling/baking tubes are brazed on copper alloy back plates for efficient heat removal of incident heat flux. Benchmarking of PFC assembly was first carried out in prototype vacuum vessel of SST-1 to develop understanding and methodology of co-ordinate measurements. Based on such hands-on-experience, the final assembly of PFC's in vacuum vessel of SST-1 was carried out. Initially, PFC's are to be baked at 250 °C for wall conditioning followed with cooling for heat removal of incident heat flux during long pulse plasma operation. For this purpose, the supply and return headers are designed and installed inside the vacuum vessel in such a way that it will cater water as well as hot nitrogen gas depending up on the cycle. This paper will discuss the successful installation of PFC's and its plasma operation respecting all design criteria.

1. Introduction

SST-1 Tokamak was successfully commissioned in 2012 [1-3] and the first plasma was achieved in June 2013 with poloidal limiters having SS 304Las a vessel wall material. Due to plasma wall interactions, high-Z impurities released from the vessel walls which in turn cools the plasma by radiation loss. In order to reduce this effect, PFC components were installed insideSST-1 main vacuum vessel (VV) in 2nd phase of SST-1 up-gradation.

Plasma facing components (PFCs) of SST-1 Tokamak [4] (figure 1 and figure 2) consist of Inboard divertor plates (IDP), Outboard divertor plates (ODP), Inboard passive stabilizers (IPS), Outboard passive Stabilizers (OPS), Main baffle (MBAF), Inboard limiters (I-Lim) and Outboard limiters (O-Lim).Each IDP, ODP, IPS, OPS and MBAF has top and bottom modules which enable SST-1 to produce elongated D-shaped double null divertor plasma. Divertors are specially designed for steady-state removal of heat energy escaping out of the plasma core region and placed to intercept the magnetic flux surfaces immediately outside the last closed flux surface of the plasma core region. Thus most of the plasma particles escaping out of the last closed flux surface are diverted so as to dump their energy on to the divertors. Pair of inboard and outboard divertor are installed in SST-1 to extract dumped energy of particle escaping out of the plasma. Main baffle is positioned in between inboard and outboard divertor structure to provide closed divertor geometry which ensures high neutral pressure in the divertor region for effective pumping and reduction in electron temperature due

IOP Conf. Series: Journal of Physics: Conf. Series **823** (2017) 012062 doi:10.1088/1742-6596/823/1/012062

to high recycling. Set of inboard and outboard poloidal limiters are installed diagonally opposite in SST-1 which assists plasma breakdown, current ramp-up and ramp-down during steady state operation. Outboard limiter is designed to move which provides effective protection to RF antenna and other in-vessel components during steady state operation and events like VDEs and disruption. The elongated plasma column in tokamak is intrinsically unstable to vertical displacements which leads to plasma disruption and also causes large halo currents to flow in nearby structures. It is very necessary to stabilize this instability in plasma by active or passive feedback control. Having a conducting structure in inboard side and outboard side surrounding the plasma like Inboard and outboard passive stabilizers suffice the requirement of passive feedback control to stabilize vertical displacements in plasma. Some of the major parameters of SST-1 tokamak are listed in table 1.

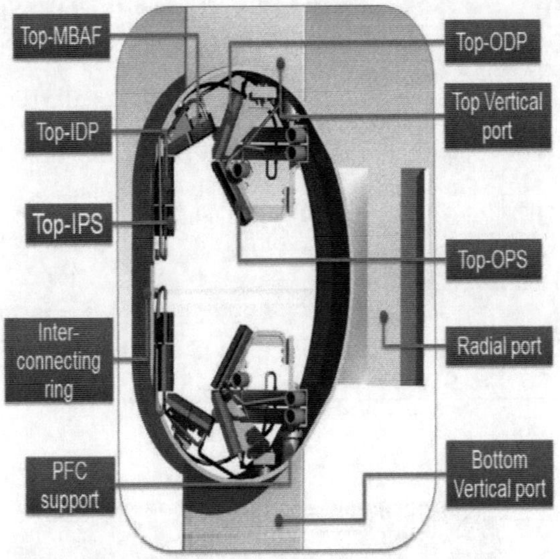

Figure 1. Elevation view of PFC inside vacuum vessel of SST-1

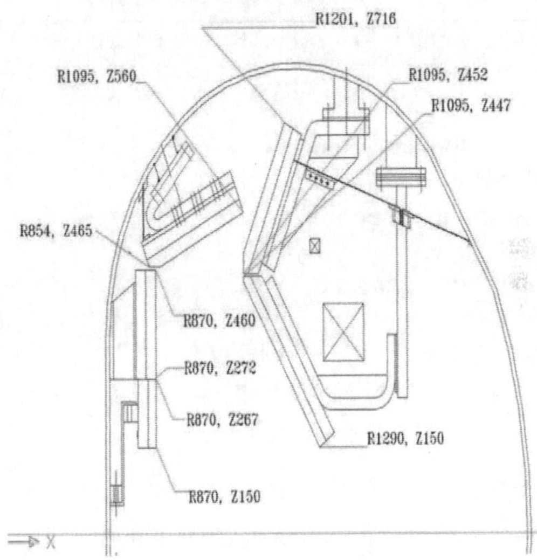

Figure 2. Elevation view showing coordinate of PFC with respect to machine axis.

Table 1. Major SST-1 machine parameters.

Parameters	**Values**
Major radius	1.1 m
Minor radius	0.2 m
SS surface area of VV	75 m^2
Exposed surface area of PFC	40 m^2
Plasma species	Hydrogen
Volume enclosed by PFC	16 m^3
Steady State Heat Flux	
Main Baffle	0.25 MW/m^2
In / Outboard Passive Stabilizer	0.25 MW/m^2
In / Outboard Divertor Plate	0.6 MW/m^2
In / Outboard Poloidal Limiter	1.0 MW/m^2

Each PFC module was fabricated from copper alloys (CuCrZr and CuZr) back plates on which graphite tiles were mechanically attached using graphoil in between so that the incident heat flux could be removed very fast. Passive stabilizers require high electrical conductivity in order to reduce the growth of vertical instability which has taken care by using CuZr back plate. This alloy gives good electrical conductivity without suffering degradation in mechanical properties at higher temperature ~350 °C. The normal incident peak heat flux on each PFCs is different due to provided angle/inclination to each of them. The components like ODP, OPS, IDP and IPS were mounted parallel to VV due to space constraints which still have average incident heat flux in tolerable limits.

IOP Conf. Series: Journal of Physics: Conf. Series **823** (2017) 012062 doi:10.1088/1742-6596/823/1/012062

SS pipe of 10.0 mm internal diameter (ID) and 1.0 mm thickness was brazed at the backside of these modules through which water will be flown to remove such incident heat flux. The layout of one of the module is shown in figure 3. Since SST-1 Tokamak is designed to operate for 1000 s in double-null configuration, the temperature of PFC modules may rise from room temperature (RT) to beyond the copper alloy melting point. CuCrZr and CuZr copper alloys degrade at 400 °C while the electrical property of CuZr degrades at 150 °C. Therefore, it is necessary to maintain the temperature of these components well below these limits.PFCs of SST-1 machine have 139 numbers of copper alloy back plates.The geometrical parameters for each PFCs is listed in table 2.

Table 2. Geometrical parameters of Plasma facing components of SST-1.

Module	IDP	ODP	IPS	OPS	MBAF	ILIM	OLIM
Poloidal Length (m)	0.187	0.267	0.117	0.330	0.264	0.290	0.290
Toroidal Length (m)	0.644	0.460	0.648	0.415	0.375	0.180	0.310
No. of Modules (Top + Bottom)	16	32	20	32	32	02	02
Area of each module (m^2)	0.126	0.128	0.073	0.143	0.092	0.052	0.090
Length of brazed Tube (m)	2.62	3.09	1.70	2.90	1.85	1.45	1.16
Copper Alloy	CuCrZr	CuCrZr	CuZr	CuZr	CuCrZr	CuCrZr	CuCrZr
Nos. of Graphite tile	480	1024	256	960	1024	48	72

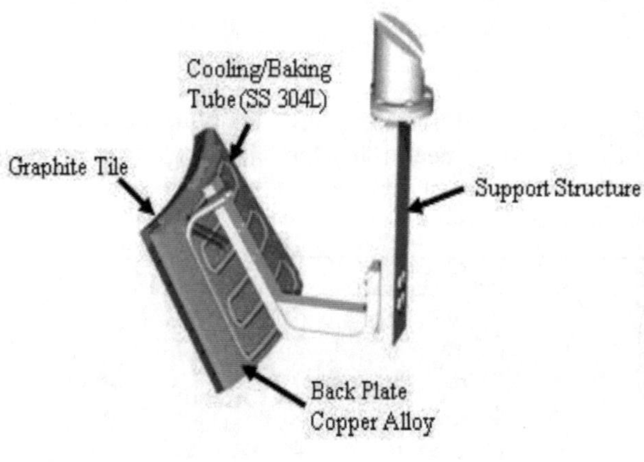

Figure 3. Typical PFC module

Figure 4. Typical PFC module

2. Metrology and assembly of PFC on SST-1 prototype vacuum vessel

For accurate assembly of PFCs into SST-1 machine, the methodology of assembling was developed and implemented in SST-1 Prototype for feasibility study. Electronic co-ordinate determining system (ECDS) and Photogrammetric measurement were carried out during the prototype assembly. The initial measurement of XYZ co-ordinate was carried out by ECDS instrument to establish the reference point for in-vessel measurements. Many coded and un-coded targets were mounted on the surfaces of VV as shown in figure 4.The interested area is scanned using DSLR camera and then 3-D models were generated using AUSTRALIS 7.0 software. For comparison of measurements, some common points were marked and measured by both ECDS and Photogrammetry. An accuracy of 0.8 mm over 500mm and 1.9 mm over 1600 mm was achieved in measurements using Photogrammetry relative to ECDS. A plot depicting the error in measurements between ECDS and Photogrammetry is shown in figure 5. Mock up assembly of back plate and support structure of PFC in prototype vacuum vessel is shown in figure 6.

IOP Conf. Series: Journal of Physics: Conf. Series **823** (2017) 012062 doi:10.1088/1742-6596/823/1/012062

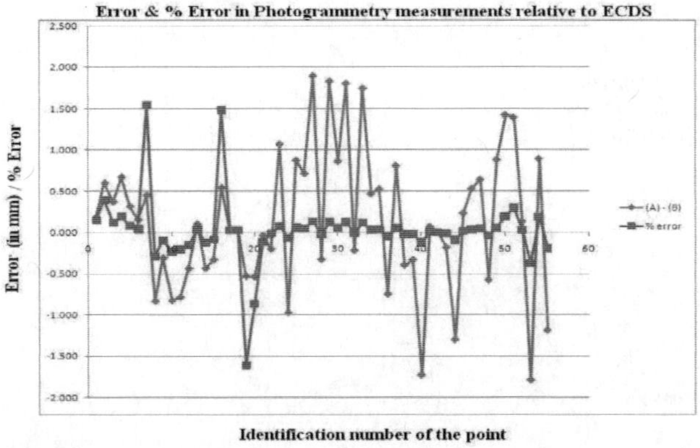

Figure 5. Error and % error in Photogrammetry measurements relative to ECDS

Figure 6. Back plate and support structures are installed in prototype Vessel.

3. Engineering validation of hydraulic scheme and PFC modules.

Hydraulic scheme for PFC baking / cooling was designed [5] considering the constrains like (a) the space available inside the VV, (b) the reduction of number of weld joints & their accessibility for repair and (c) the uniform flow distribution. The entire hydraulic scheme was distributed through two supply headers at the top and other two supply headers at the bottom. Each top and bottom configuration has a return header. Top / bottom MBAF, IDP and IPS were in series and fed through top two supply headers each having 11 parallel sub-connections while top / bottom ODP and OPS were in parallel and fed through the bottom two supply headers each having 16 parallel sub-connections as shown in figure 7. Prior to assembly of PFCs inside SST-1 VV, all copper alloy back plates along with their header piping, bellows and isolators were qualified for leak tightness at the operating temperature of 250 °C. A dedicated test facility was developed in which a number PFCs, their headers along with associated components as per design scheme were baked under high vacuum condition and qualified for final assembly into SST-1 VV as shown in figure 8.For baking the high temperature nitrogen gas at 250 °C 2.5 bar(a) pressure was flown through the circuit. A number of K-type thermocouples were mounted at different places of the entire scheme so that the temperature distributions though out the system could be monitored and maintained accordingly.

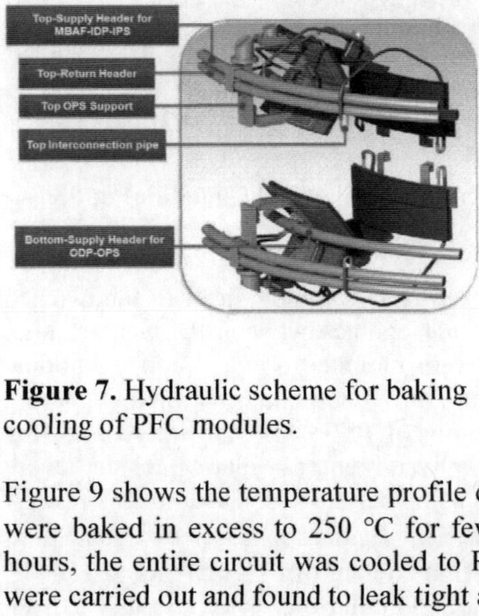

Figure 7. Hydraulic scheme for baking cooling of PFC modules.

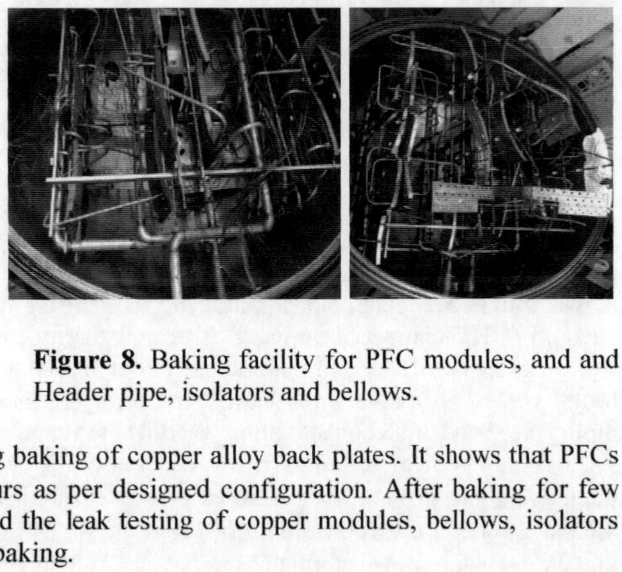

Figure 8. Baking facility for PFC modules, and and Header pipe, isolators and bellows.

Figure 9 shows the temperature profile during baking of copper alloy back plates. It shows that PFCs were baked in excess to 250 °C for few hours as per designed configuration. After baking for few hours, the entire circuit was cooled to RT and the leak testing of copper modules, bellows, isolators were carried out and found to leak tight after baking.

IOP Conf. Series: Journal of Physics: Conf. Series **823** (2017) 012062 doi:10.1088/1742-6596/823/1/012062

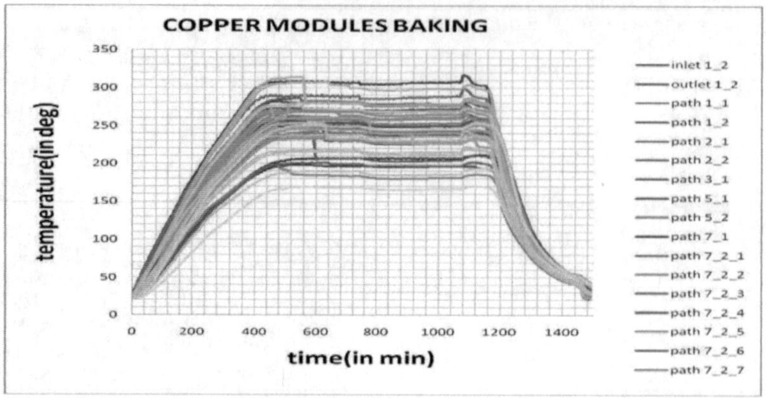

Figure 9. Temperature plot of PFC copper alloy back plates.

4. Assembly and integration of PFC in SST-1 vacuum vessel

Support structures of all the modules as shown in figure 1 were marked in SST-1 main vacuum vessel and installed sequentially. Mechanical templates and Go and Not-go gauges were used to ensure the alignment and the space between each module during assembly. During assembly, combined ECDS and Photogrammetric measurements were carried out for confirming the coordinates. Appropriate mechanical spacers and shims were used wherever required to correct alignment of the components with respect to other modules and machine axis. Figure 10 and figure 11 shows assembled PFCs in SST-1 machine.

Figure 10. Full assembly of copper module in SST-1 main vessel.

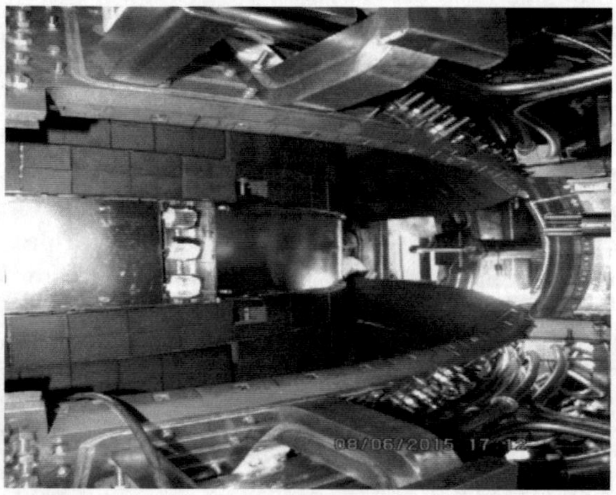

Figure 11. Final assembly of PFC modules in SST-1 main vessel.

Graphite tiles were mechanically bolted on copper alloy back plates. Graphoil sheet is sandwiched between the back plate and graphite tile to increase the thermal contact between them. Ultra-high vacuum (UHV)compatible mineral insulated thermocouples were mounted on each copper module and on graphite tiles to maintain and monitor the temperature. During assembly, the baking/cooling headers were also installed back-to-back prior to final assembly of PFC. Also the interconnecting piping were laid and connected from modules to respective supply and return headers. Each and every weld joints were leak tested in both vacuum and sniffer modes using helium gas. Vacuum leak test for the leak tightness of ~ 1.0×10^{-8} mbar l/s was ensured while the leak tightness ~ 1.0×10^{-6} mbar l/s at 7.6 bar(g) pressure was ensured for Sniffer mode. During the first attempt of PFCs baking at 250 °C, the VV temperature [6] started rising beyond 160 °C due to radiation. In order to avoid the VV from additional thermal stress, the VV temperature was restricted below 150 °C. Nitrogen gas baking system is than modified, in which a bypass line from condenser is separately laid and connected to vessel supply line. During baking, room temperature nitrogen gas will be supplied to vacuum vessel simultaneously high temperature gas for PFC. In order to bake PFC to 250 °C hot nitrogen gas is passed at 2.5 bar(a) having supply temperature of 385 °C and mass flow rate of 0.95 kg/s. Figure 12 shows assembled Plasma Facing Components in SST-1 tokamak were successfully baked to 250 °C for a duration of 170 hours.

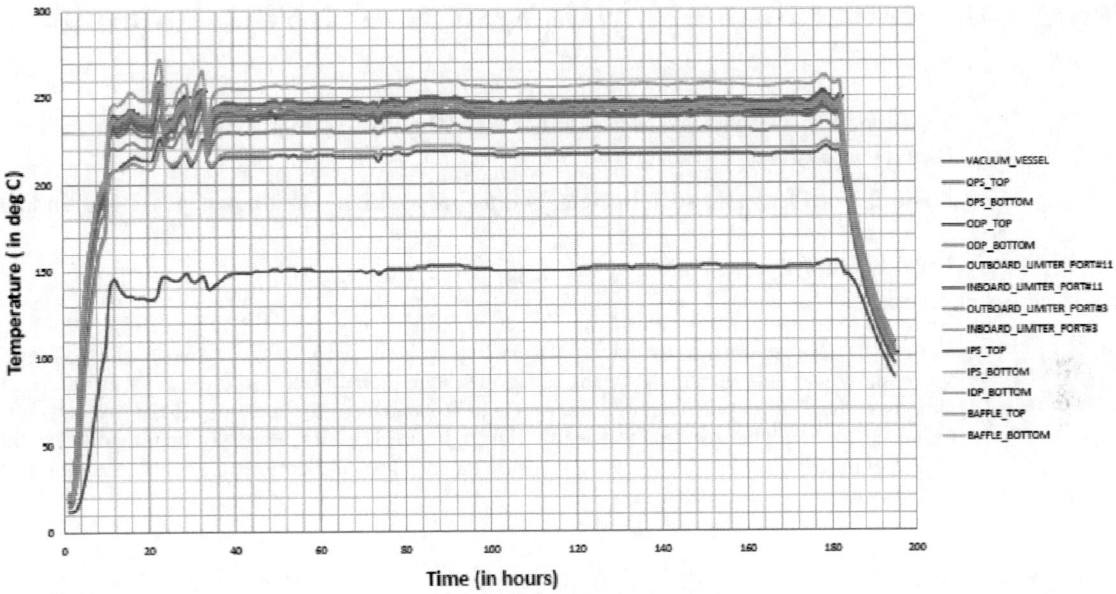

Figure 12. Temperature profile of various PFC modules during baking.

5. Conclusion

Assembly of complete PFCs system was successfully carried out in SST-1 tokamak with an accuracy of ± 3.6 mm. Also the performance of PFCs was satisfactorily demonstrated under Ultra High Vacuum conditions. The temperature distribution among PFCs was found to be within the limit of ± 12 °C. During and after baking, all the weld joints, isolators and flexible bellows were found to the leak tight as per design requirements. With integrated PFCs, we have demonstrated a plasma break down of 65 kA plasma current for 250 ms duration.

Acknowledgement

The work presented here is the efforts of the SST-1 PFC team in IPR. Author expresses sincere appreciation to all the members of the team.

References

[1] Pradhan S and Team SST-1 mission 2010 *J. Fusion Res. Series* **9** 650.
[2] Pradhan S, Sharma A N, Tanna V L, Khan Z, Prasad U et al. 2012 *IEEE Trans. Plasma Sci.* **40** 614
[3] Khan Z, Pathan F, Gorge S, Dhanani K R, Yuvakiran P et al. 2014 *IEEE Transaction on Plasma Science* **42** 1006
[4] Jacob S et al. 1996 *Journal of Nuclear Materials* **233-237** 655
[5] Paravastu Y, Khan Z and Pradhan S 2015 *Fusion Engineering and Design* **98-99** 1375
[6] Khan Z, Paravastu Y and Pradhan S 2014 *Procedia Materials Science* **6** 216

IOP Conf. Series: Journal of Physics: Conf. Series **823** (2017) 012063 doi:10.1088/1742-6596/823/1/012063

Design and Integration of SMBI System for SST-1 Tokamak

S George[1], Y Paravastu[1], M S Khan[1], K R Dhanani[1], D C Raval[1], Z Khan[1], S Banerjee[1] and S Pradhan[1]

[1]Institute for Plasma Research, Near Indira Bridge, Bhat, Gandhinagar 382 428, India

E-mail: siju@ipr.res.in

Abstract. Supersonic molecular beam injection (SMBI) is one of the most effective fuelling methods for injecting neutral particles at very high velocity into the plasma core. Due to higher speed and lower divergence, the beam penetrates several centimetres into the plasma and hence increases the fuelling efficiency. For SST-1 Tokamak, two types of SMBI systems are proposed. One will be installed in the low field side (LFS) while other will be integrated in the high field side (HFS). This paper describes the design, fabrication and implementation of SMBI system in SST-1 Tokamak.

1. Introduction

The currently adopted fueling techniques in Tokamak devices includes conventional gas puffing (GP), supersonic molecular beam injection (SMBI) and pellet injection (PI) etc. Among these methods SMBI is better compared to GP and is technically simpler in comparison to PI. The fueling efficiency of SMBI is around 30 ~ 60 %, about three to four times higher than the conventional gas puffing (GP) [1-3]. SMBI can also enhance the penetration depth and the fueling efficiency. A higher fueling efficiency and a reduced gas-wall interaction and therefore a lower wall saturation limit have been observed in L-mode plasma experiments in limiter tokamaks. In divertor tokamak experiments with the supersonic gas jet, a fueling efficiency of 10-30% has been reported [4,5].

Through a Laval nozzle, stagnation gas (P_0, T_0) is accelerated by imposing pressure difference so that it enters into vacuum chamber with a definite speed. When the background pressure (P_b) of vacuum chamber is low enough and the ratio (P_0/P_b) exceeds a critical value (G) which is less than 2.1 for all gases, the flow reaches a sonic speed or Mach number equals to 1 at the nozzle throat. The critical value is expressed as

$$G = ((\gamma + 1)/2)^{\gamma/(\gamma - 1)} \tag{1}$$

where γ is the ratio of heat capacity, C_p and C_v.

Beyond the nozzle throat, an isentropic expansion can produce a supersonic gas flow and the Mach number increases. The characteristic dimension of the supersonic area (X_M) i.e. the axial length from the nozzle to the position of Mach disk, measured in terms of nozzle diameter (d) is given by [6]

$$X_M/d = 0.67\,(P_0/P_b)^{1/2} \tag{2}$$

SST-1 Tokamak [7] is a medium size superconducting tokamak with major radius 1.1 m, minor radius 0.2 m having toroidal field of B_t = 3T at the center. Recently SST-1 machine has been upgraded [8] with graphite first wall for divertor experiment. This machine is designed for double-null D-shaped plasma operation of 1000 s duration. In order to maintain such long plasma duration and hence to enhance the plasma density, SMBI system is designed and installed in SST-1 machine. The design, experimental validation and the implementation of SMBI system is discussed in the subsequent sections.

2. Nozzle design and parameters

In SST-1 machine, two Laval nozzles made up of SS 304L with throat diameters of 0.5 mm and 1 mm were installed at HFS side at the port locations 7 and 10 respectively. The taper angle is 12°–15° for these two exit diameters of 5 mm & 10 mm with the lengths of 12.6 mm and 30 mm respectively as shown in figure 1(a) and (b). The plenum gas pressure P_0 can be varied from 0.5–1MPa. If the background pressure P_b is 10^{-3} Pa, then the corresponding X_M is 7.5–10.5 m and 15–21 m respectively. The ratio of reservoir stagnation pressure to the background pressure P_0/P_b must be very high to avoid the Mach disk from being too close to the nozzle exit. To avoid the emergence of Mach shock and guarantee the beam convergence, the distance from the nozzles to edge plasma should be as short as possible. The schematic of SMBI inside the vacuum vessel as well as its PID is shown in figure 2. A Series-99 (Parkar) solenoid valve is used as a fast response valve (FSRV) with a response time of 1 ms which can be operated up to 5 MPa. To avoid the influence of toroidal field, this solenoid valve is mounted outside the toroidal filed coils.

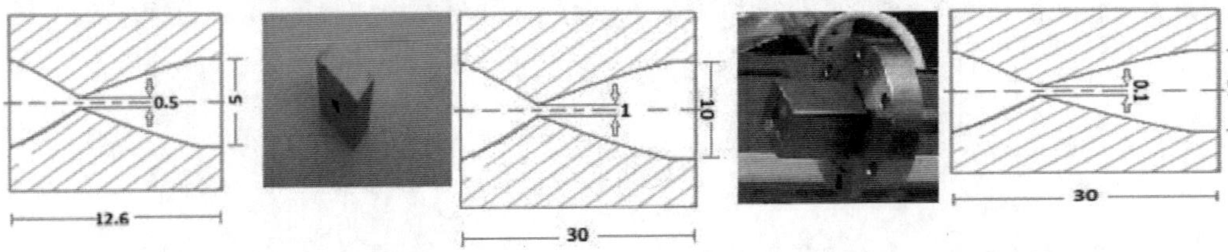

(a) (b) (c)

Figure 1. Nozzle design having throat diameters (a) Ø 0.5 mm, (b) Ø 1.0 mm, (c) Ø 0.1 mm.

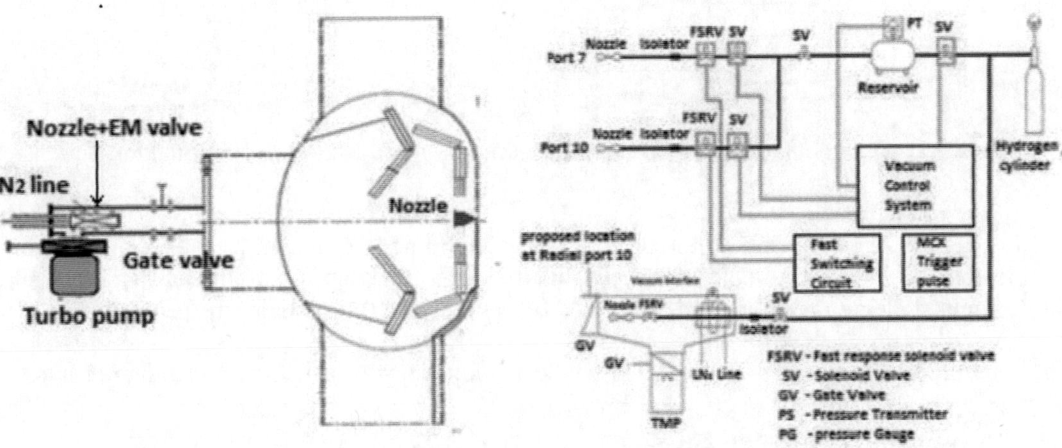

(a) (b)

Figure 2. (a) Schematic of SMBI nozzle inside SST-1 vacuum vessel, (b) PID of SMBI system.

Similarly, at LFS side, a nozzle of 0.1 mm diameter with the plenum gas temperature at 80 K, maintained using liquid nitrogen (LN_2) as shown in figure 1(c) shall be used during next experimental campaign. The onset of clustering can be described by an empirical scaling parameter referred as the Hagena parameter [9] which is given by

$$\Gamma^* = k\,\frac{(d/\tan\alpha)^{0.85}}{T_0^{2.29}}\,P_0 \qquad (3)$$

where d is the nozzle diameter in μm, α is the expansion half angle (α = 45° for sonic, α < 45° for supersonic), P_0 is the backing pressure behind the valve in mbar, T_0 is the pre-expansion temperature in K, and k is a constant related to bond formation (for hydrogen $k = 184$). Clustering generally begins for $\Gamma^* > 100$–300. For the both type of nozzles, $\Gamma^* \sim 200$ for high plenum pressure operation and hence onset of cluster may occur. The distance between the nozzle and the plasma column is maintained at ~ 1.2 m for LFS side while it is maintained at 70 mm from HFS side.

3. Experimental set-up

The initial characterization of the nozzle for flow rate and injection rate was carried out in the laboratory using a chamber of 50 mm diameter and 400 mm length. The cylinder is connected to a turbo-molecular pump having an effective pumping speed of 150 ls/. The pressure inside the chamber was maintained at ~ 1.0×10^{-2} Pa. A laser source (405 nm, 5 mW) was mounted horizontally with a fast imaging camera mounted perpendicular to the line of sight of laser as shown in figure 3.

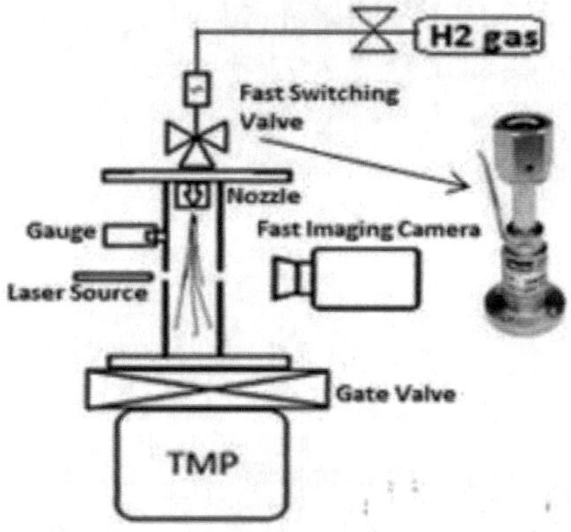

Figure 3. Schematic of experimental set-up.

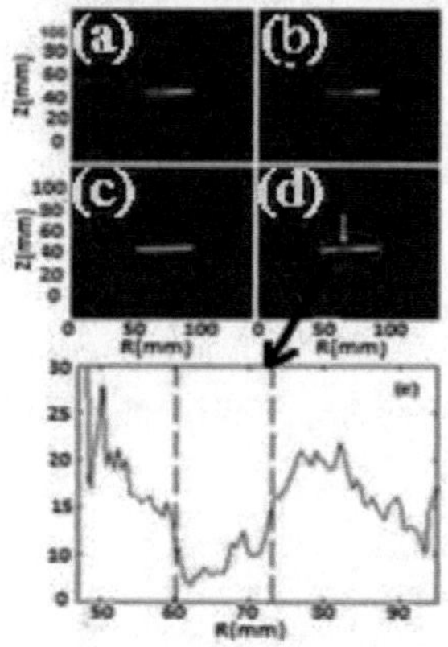

Figure 4. Fast imaging data.

3.1. Flow rate measurement

The flow rates (Pa m^3s^{-1}) of both the nozzles were determined at plenum pressure ranging from 0.25 MPa to 1 MPa as shown in table 1. Particle flux measured for nozzle with throat diameter Ø 0.5 mm and Ø 1 mm at plenum pressure 1 MPa is 2.6×10^{22} and 8.0×10^{22} particle/s respectively.

Table 1. Hydrogen gas flux (Pa m^3/s) measured for nozzle at different plenum pressure.

Sr. No.	Plenum pressure (MPa)	Flux (Pa m^2/s)	
		Throat diameter (0.5 mm)	Throat diameter (1.0 mm)
1	0.25	34	133
2	0.50	49	200
3	0.75	67	266
4	1.00	110	334

3.2. Beam divergence measurement

The divergence of beam for the nozzle with throat diameter 1.0 mm is measured from scattering of light using a fast imaging camera as shown in figure 4. Figure 4(a) and figure 4(b) are the raw images before and after the firing of supersonic molecular beam nozzle. Clear gradient in intensity can be seen along the laser path as light intensity gets attenuated along the path as laser is fired from left to right. Figure 4(c) and figure 4(d) are the intensity calibrated images corresponding to figure 4(a) and figure 4(b) showing the uniform intensity before firing the nozzle. A clear discontinuity in intensity after SMBI firing is shown by white solid arrow in figure 4(d) while figure 4(e) is the portion of laser path circled by white broken oval. The two vertical black broken lines denotes the actual beam width at the location. Actual beam width at laser path location is 12.88 mm at a distance of 130.45 mm from the nozzle.

3.3. Injection rate per pulse

The injection rates for both the nozzles have been determined at different plenum pressure ranging from 0.25 MPa to1.0 MPa) for the pulse lengths of 3 ms and 4 ms respectively. The measured injection rate per pulse of nozzle with throat diameter Ø 1.0 mm is shown in figure 5.

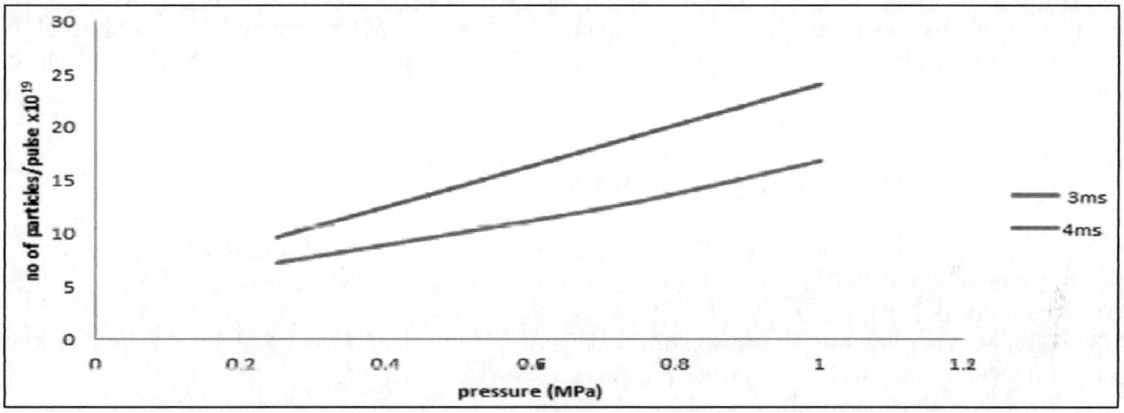

Figure 5. Injection rate per pulse measured for nozzle with throat diameter Ø 1.0 mm at different plenum pressure.

4. Implementation of SMBI system in SST-1 tokamak

The SMBI system will be used for gas puffing alongside conventional GP system. Also it will be used for density control as shown in figure 6. The pulse length can be varied from 1-20 ms. Pulse duration and pulse numbers will be controlled from main control system and can be adjusted according to the experimental requirement. Heterodyne radiometry will be used as a feedback for density control. The feedback system will start when the plasma current reaches 80% and the plasma density reaches to 6.0×10^{12} /cm^3 with 1 ms on-time and 20 ms off-time. Output of control will be amplified and applied to SMBI solenoid valve. Once the plasma density reaches a maximum value of 2.4×10^{13} /cm^3 or the plasma current start decaying, the feedback system will accordingly stop the pulse for gas puffing.

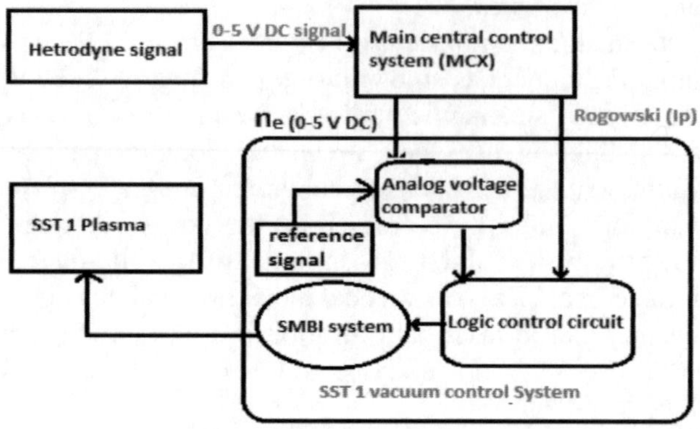

Figure 6. Schematic of density control system for SST-1 tokamak.

5. Conclusion

The SMBI system for SST-1 machine has been successfully characterised experimentally for its flow rate and injection rate. Two nozzles of different diameters has been successfully installed in SST-1 tokamak at two locations near HFS. In next experimental campaign, the same type of nozzles will be mounted in LFS with 80 K Plenum temperature and will be used for gas fuelling with density feedback control.

References

[1] Pégourié B et al. 2003 *J. Nucl. Mater.* **313,** 539
[2] Pégourié B et al. 2007 *Plasma Phys. Control .Fusion* **49**, R87
[3] Yao L et al. 1993 *Proc. 20th EPS Conf. on Fusion & Plasma Physics* **17C(I)** 303
[4] Bucalossi J et al. 2004 *Proc. 31st EPS Conf. on Plasma Physics* ECA **28G**
[5] Yao L et al. 2004 *Nuc. Fusion* **44** 420
[6] Hagena O F 1992 *Rev Sci Instrum* **63** 2374
[7] Pradhan S et al. 2012 *IEEE Trans. Plasma Sci.* **40** 614
[8] Khan Z et al. *Procedia Materials Science* **6** 216
[9] Smith R A, Ditmire T and Tisch W G 1998 *Rev. Sci. Instrum.* **69** 3798

RF assisted Glow Discharge Condition experiment for SST-1 Tokamak

Dilip Raval[1], Ziauddin Khan, Siju George, Kalpeshkumar R Dhanani, Yuvakiran Paravastu, Pratibha Semwal, Prashant Thankey, Mohammad Shoaib Khan, Bharat Kakati, Subrata Pradhan

[1]Institute for Plasma Research, Near Indira Bridge, Bhat, Gandhinagar 382 428, India

E-mail: raval@ipr.res.in

Abstract. Impurity control reduces the radiation loss from plasma and hence enhances the plasma operation. Oxygen and water vapors are the most common impurities in tokamak devices. Water vapour can be reduced with extensive baking while in order to have a significant reduction in oxygen it is necessary to use glow discharge condition (GDC). RF assisted glow discharge cleaning system will be implemented to remove low z impurities at PFC installed SST-1 vacuum vessel. A RF assisted Glow discharge conditioning is studied at laboratory to find the optimum operating parameters in a view to implement at SST-1 tokamak. Helium is used as a fuel gas in the present experiment. It is observed that the ultimate impurity level is reduced significantly below to the accepted level for plasma operation after RF assisted GDC. The experimental findings of RF assisted Glow discharge conditioning is discussed in details in this paper.

1. Introduction

Vessel wall conditioning plays a very important role for fusion grade plasma performance of tokamak device. It helps to prepare vessel wall for plasma operation and controls plasma generated low and high-Z impurities which desorbs from the wall during the tokamak discharge. The edge plasma properties strongly depend on these impurities concentration. It is well known that the edge plasma properties play a vital role to produce high confinement mode (H-mode) tokamak plasma [1–4]. The two main effects related to plasma impurities are the radiative power loss and the dilution of the hydrogenic species [1, 5]. For achievement of reproducible and clean plasma conditions as well as to improve the overall plasma performance [1–5], an efficient wall conditioning is very necessary. A wide variety of conditioning techniques are implemented in tokamak devices during the last few decades [1, 3, 4, 5–8] to reduce the impurities concentration. Some of the widely used conditioning techniques are DC glow discharge cleaning (GDC), RF assisted GDC, Taylor discharge cleaning (TDC), electron cyclotron resonance (ECR) and ion cyclotron resonance (ICR) discharge cleaning. Generally, H-GDC and He-GDC are used for desorption of impurities either by a chemical process or ion induced desorption process [3-5]. He-GDC is preferred over hydrogen in such devices where large parts of the wall are covered with carbon material [1, 3, 4 and 6]. It is also observed that the wall cleaning efficiency is much higher at RF assisted GDC than normal GDC. Also, RF assisted GDC can be operated at lower working pressure compare to the normal GDC. As a result, the recombination probability of the impurities with background gas and vessel wall decreases which helps to improve the cleaning efficiency during RF assisted GDC. Steady-state Superconducting Tokamak (SST-1) is a medium sized tokamak with a major radius of 1.1 m and a minor radius of 0.2 m. The top view of SST-1 device is shown in figure 1. SST-1 machine has two vacuum chambers such as (i) Vacuum vessel for plasma discharge and confinement and (ii) Cryostat for enclosing cold masses along with their associated components. Vacuum vessel is a toroidally continuous single wall metallic structure made of SS 304L material. In order to prevent the vacuum vessel directly getting exposed to high energetic neutrons, it is covered with graphite tiles of 20 mm thickness as a plasma facing surfaces (PFC) [9-12]. Since GDC is very proven technique to prepare vessel and PFC for plasma operation, an experiment is carried out in laboratory to evaluate the optimum conditioning parameters of RF assisted GDC so that it could be implemented for conditioning of SST-1. In this paper, the experimental set-up with methodology, the results and their physical interpretations are discussed in subsequent sections.

Figure 1. Top view of SST-1 machine.

2. Experimental set up and methodology

The experimental set-up is shown in figure 2. The vacuum chamber was fabricated from SS 304L having a surface area of 1.0 m^2. Since graphite to SS surface area for SST-1 is 0.46, hence in order to maintain SST-1 machine like surface condition, few graphite tiles having total surface area of 0.5 m^2 were kept inside the experimental chamber. Before installation, the graphite tiles were properly cleaned and baked at 1000 °C for 24 hours in a dedicated vacuum furnace. The chamber was evacuated using a turbo-molecular pump with an effective pumping speed of 110 lit/sec.

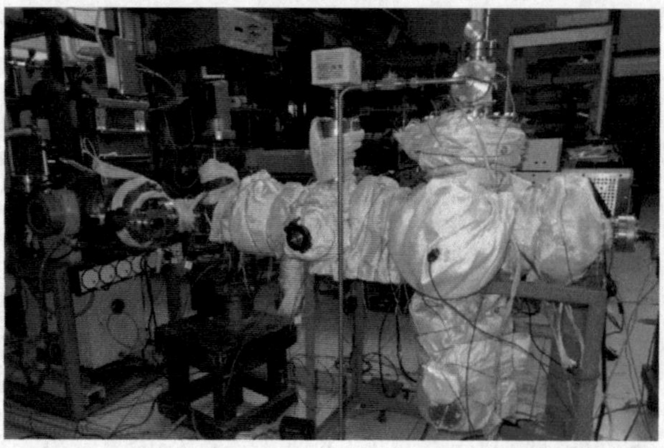

Figure 2. Snap shot of the experimental set up.

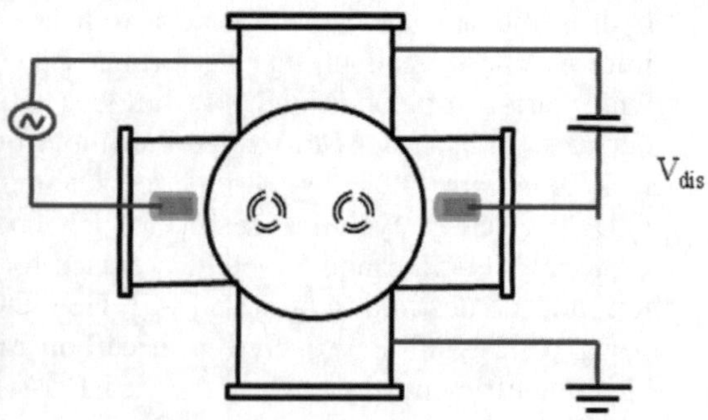

Figure 3. Schematic of discharge circuit.

Ultra-high pure helium gas was fed into the chamber using two manual leak valves for GDC. RF and DC power supply is used to produce RF assisted DC plasma. The DC discharge voltage is applied between chamber wall and a cylindrical anode. Another cylindrical electrode mounted inside the chamber was used as a power electrode for RF power supply. The RF discharge voltage was applied between chamber wall and power electrode. The schematic of the discharge circuit for RF assisted DC GDC is shown in figure 3.

GDC plasma was produced using a DC power supply of 1000 V and 1.0 A while RF generator with maximum power 300 watt and frequency 13.56 kHz was used to produce RF assisted GDC. A residual gas analyser (Make: SRS; Model: RGA200) was used to measure the partial pressure of impurities. The plasma density, electron temperature and the floating potential were measured with the help of a double Langmuir probe made of Tungsten wire of 1.0 mm dia and 3.0 mm length. Before GDC, the chamber was baked continuously at 250 °C for 16 hours to reduce the water vapour.

The experiment was carried out in two different phases. During 1st phase, the desorption yield of different impurities at different operating pressures was studied for constant RF power (63 watt). During 2nd phase, the desorption yield of different impurities at different RF powers was studied at constant operating pressure of 1.5×10^{-3} mbar. For each condition, the experiment was carried out for 4.0 hours keeping the constant discharge current density at 0.1 A/m^2. After each experimental cycle, the chamber was also exposed to air for 24.0 hours to maintain the same initial condition.

3. Results and discussions

3.1. Effect of operating pressure on desorption of impurities

In order to study the effect of operating pressure, the operating pressure was varied from 1.5×10^{-3} mbar to 6.5×10^{-3} mbar. During GDC, the plasma density and the electron temperature were found to be of the order of 10^{16}/m^3 and ~ 2.6 – 3.7 eV respectively. QMA scan for the graphite containing SS chamber during GDC for 1.5×10^{-3} mbar and 4.0×10^{-3} mbar is shown in figure 4 (a) and (b). With increase of working pressure, the reduction in partial pressure of some impurities was observed which indicates the degradation of cleaning efficiency.

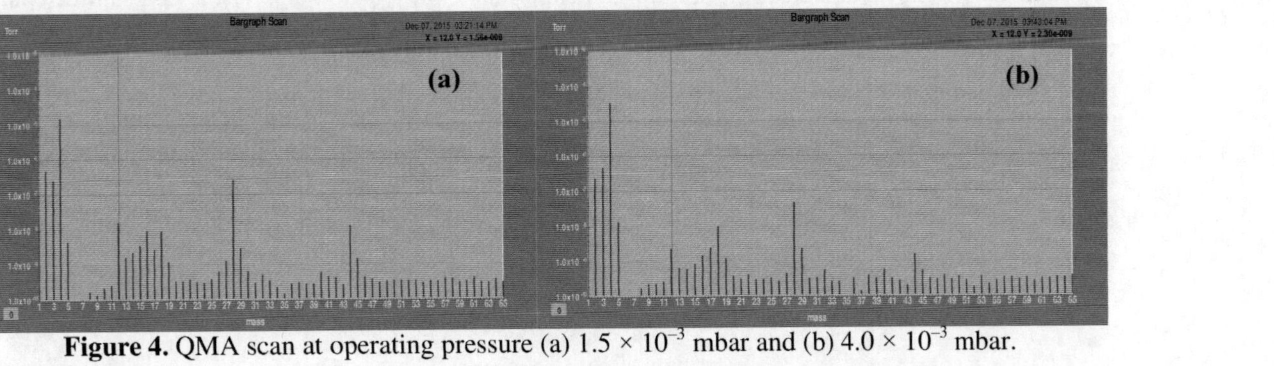

Figure 4. QMA scan at operating pressure (a) 1.5×10^{-3} mbar and (b) 4.0×10^{-3} mbar.

The gas load of different impurities at different working pressures is shown in figure 5(a). This figure shows that at higher operating pressure, the gas load of the impurities M2, M28, M44 and M16 slightly decreases whereas the gas load of M18, M32 and M17 remains constant. Such behaviour of impurities desorption is due to the reduction of recombination probability at lower working pressure. As the operating pressure increases, the discharge current changes which in turn affects the plasma parameters. Hence, it is necessary to maintain the same discharge current for the entire operating pressure regime so that the effect of working pressure can be studied. It is observed that as the operating pressure increases, the required DC discharge voltage to maintain the constant discharge current density 0.1 A/m^2 need to be decreased. The variation of discharge voltage at different operating pressure is shown in figure 5(b). The required discharge voltage at higher operating pressure is less as compared to the lower operating pressure. The decrease in discharge voltage decreases the sheath electric field. Since, the required energy of +ve ion for desorption of impurities during GDC depends on the sheath electric field formed around the vessel walls, the lower discharge voltage at higher operating pressure is also another reason for the reduction of impurities desorption.

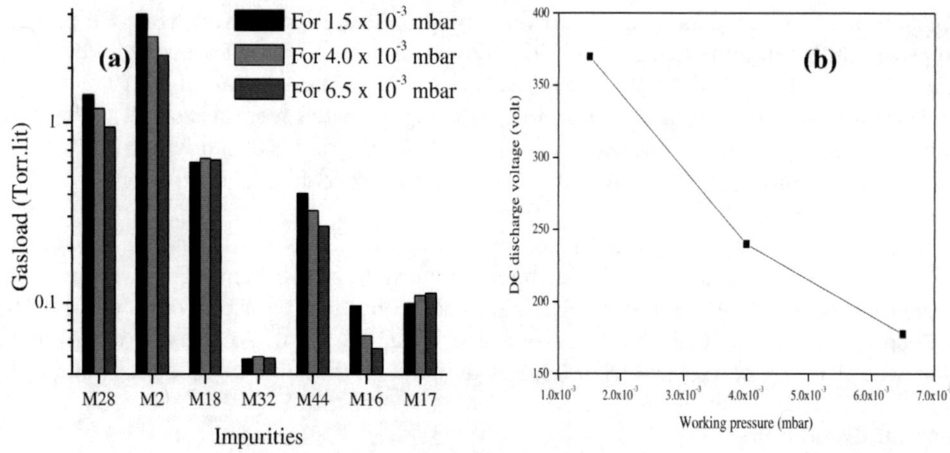

Figure 5. (a) Gas load at different operating pressures (b) DC discharge voltage at different operating pressure.

3.2. Effect of RF power on desorption of impurities

The effect of RF power on desorption of different impurities was studied at a constant operating pressure of 1.5 × 10^{-3} mbar. The partial pressure of M2 and M44 impurity at different RF power is shown in figure 6(a). The RF power is varied from 0–63 watt. At each RF power, the partial pressures of impurities were recorded for 2.0 hours. The experiment was carried out for the constant discharge current density 0.1 A/m^2 for which the discharge voltage was decreased at higher RF power. The variation of DC discharge voltage at different RF power is shown in figure 6 (b). The partial pressure of different impurities mainly M2, M28, M44, M18, M32, M16 and M17 were investigated using RGA. It was observed that there was no effect of RF power on desorption of M18, M32, M16 and M17 impurities while there was a slight increase of desorption rate for M2 and M44 impurities at higher RF power. Also it was observed that the electron temperature increases with increase of RF power.

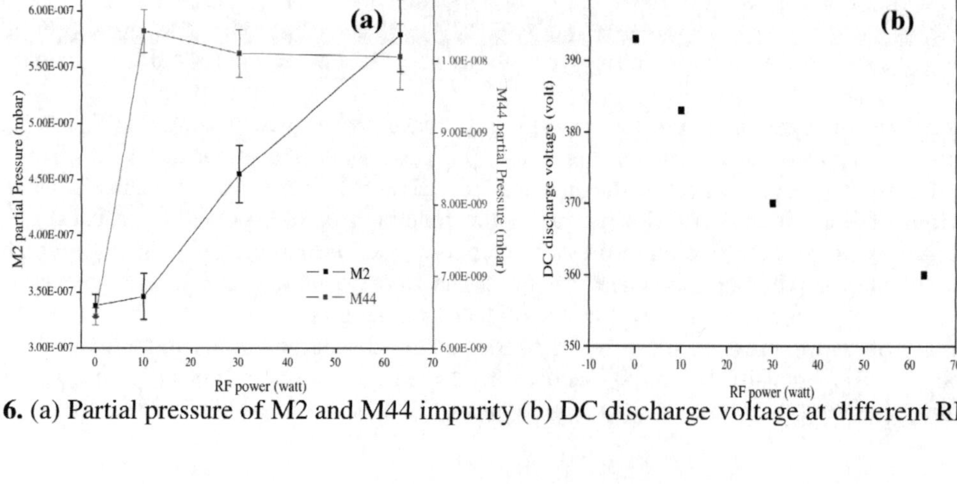

Figure 6. (a) Partial pressure of M2 and M44 impurity (b) DC discharge voltage at different RF power

4. Conclusion

The results show that the lower operating pressure is beneficial for wall conditioning. At lower working pressure, the recombination probability of the impurities with fuel gas decreases. Also, it is observed that the higher discharge voltage is required at lower operating pressure to sustain the discharge. The increase in discharge voltage increases the cathode sheath electric field which in turn increases the impact energy of +ve ions. The effect of RF power on desorption of different impurities indicates that the desorption of M2 and M44 impurity slightly increases at higher RF power whereas the effect of RF power on desorption of M18, M32, M16 and M17 impurities is negligible.

References

[1] Dylla H F 1980 *J. Nuclear Materials* **93-94** 61
[2] Federici G, Skinner C H, Brooks J N, Coad J P, Grisolia C, Haasz A A, Hassanein A, Philipps V, Pitcher C S, Roth J, Wampler W R and Whyte D G 2001 *Nucl. Fusion* **41** 1967
[3] Waelbroeck F, Winter J and Wienhold P 1984 *J. Vac. Sci. Technol.* **A2** 1521
[4] Winter J 1989 *J. Nucl. Materials* **161** 265
[5] Li J, Shimada M, Zhao Y, Hu J, Gong X, Yu Y W and Zhuo G Z 2011 *J. Nucl. Materials* **415** S35–S41
[6] Oren L, Taylor R J 1977 *Nucl. Fusion* **17** 1143
[7] Yu Y W, Gao X, Hu J S, Li J G, Shan J F and The HT-7 Team 2009 *J. Nucl. Materials* **390–391** 1051
[8] Ushigusa K, Seki M, Suganuma K, Toyoshima N and Ikeda Y 1999 *Fusion Engineering and Design* **45** 137
[9] Khan Z, Paravastu Y and Pradhan S 2014 *Procedia Materials Science* **6** 216
[10] Paravastu Y, Khan Z and Pradhan S 2015 *Fusion Engineering and Design* **98–99** 1375
[11] Chaudhuri P, Parashar S K S, Santra P and Reddy D C 2014 *Int. J. Thermal Sciences* **86** 299
[12] Saxena Y C and SST-1 Team 2000 *Nucl. Fusion* **40** 1069

IOP Conf. Series: Journal of Physics: Conf. Series **823** (2017) 012065 doi:10.1088/1742-6596/823/1/012065

Gas Fuelling System for SST-1Tokamak

Kalpesh Dhanani[1], D C Raval[1], Ziauddin Khan[1], Pratibha Semwal[1], Siju George[1], Yuvakiran Paravastu[1], Prashant Thankey[1], M S Khan[1] and Subrata Pradhan[1]

[1]Institute for Plasma Research, Near Indira Bridge, Bhat, Gandhinagar 382 428, India

E-mail: kdhanani@ipr.res.in

Abstract. SST-1 Tokamak, the first Indian Steady-state Superconducting experimental device is at present under operation in the Institute for Plasma Research. For plasma break down & initiation, piezoelectric valve based gas feed system is implemented as a primary requirement due to its precise control, easy handling, low construction and maintenance cost and its flexibility in the selection of the working gas. Hydrogen gas feeding with piezoelectric valve is used in the SST-1 plasma experiments. The piezoelectric valves used in SST-1 are remotely driven by a PXI based platform and are calibrated before each SST-1 plasma operation with precise control. This paper will present the technical development and the results of the gas fuelling system of SST-1.

1. Introduction

Magnetic fusion devices require fast-responding, versatile and accurate gas injection system in order to initiate the plasma breakdown and subsequently to maintain and control the plasma density. A number of techniques like external gas feed, neutral beam injection, pellet injection and their combinations are employed in fusion devices to maintain the plasma density at desired levels for long periods. Main functions of the gas fuelling system are (a) pre-filling of the vessel to the required pressure, (b) feeding of required amount of gas during plasma current and density build up phase, (c) feeding of required amount of gas to maintain plasma density at the desired value, (d) feeding of gases into the vessel during wall conditioning and (e) feeding of required amount of gases in the divertor region to remove the heat and exhausted particles. For all these operations, a very efficient gas fuelling system is essential and therefore developments of effective fuelling methods has been one of the key issues in the plasma study. Hydrogen gas feed, Supersonic molecular beam injection and pellet injection are the techniques implemented for the SST-1 tokamak.

The steady-state superconducting tokamak (SST-1) has a minor radius of 0.2 m and a major radius of 1.1 m and was designed for a steady state plasma operation of 1000 s with both single null and double null configuration [1,2]. Commissioning of SST-1 was completed in 2013 and the machine was under operation for initial plasma break down up to 100 kA limiter-assisted circular plasma for more than 100 ms using hydrogen gas. Graphite plasma facing components were installed inside the vessel in mid 2015 and a maximum plasma current of 110 kA forms has been achieved with a toroidal field of 1.5 T using ohmic transformer (OT) and ECRH system of 250 kW & 42.6 GHz. A pair of vertical field coils provides equilibrium during this phase. The vacuum system of SST-1 comprises of an ultra-high vacuum vessel (VV) and a high vacuum cryostat. SST-1 VV is fabricated from SS 304L material in torus D-shaped form. It has 16 rectangular radial ports (RPs), 16 triangular top vertical ports, 16 triangular bottom vertical ports, 16 interconnecting rings and 16 vessel sector rings. It has two radial pumping lines connected to turbo-molecular (TMP) pumping stations with a total effective pumping peed of 4675 l/s (N_2 gas) [4]. The total pressure is monitored in the vessel radial port as well as in the pumping lines using a hot cathode ionization gauge. The vacuum vessel and plasma facing components are baked at 150 °C and 250 °C respectively using hot nitrogen gas heating & supply system [4]. DC and RF assisted glow discharge cleaning using H_2/He gas and their mixture was carried out both before and in between plasma shots. For this purpose, two rectangular SS electrodes mounted at diagonally opposite ports and two power supplies each having a rating of 0-1000 V, 0-15 A and an RF supply of 300 W at 13.56 MHz were used. Few of the SST-1 machine as well as operating parameters are given in Table 1.

Table 1. Major SST-1 Machine & Operational Parameters.

Parameter	Value
Major radius	1.1 m
Minor radius	0.2 m
Toroidal field	3.0 T
Plasma current	220 kA
Plasma species	Hydrogen
Pulse length	1000 s
Configuration	Double null
Vessel base pressure	$< 5.0 \times 10^{-8}$ mbar
Effective pumping speed (Hydrogen)	6500 l/s

2. SST-1 Gas Fuelling System

Three types of fuelling systems have been installed on SST-1 vacuum vessel such as (a) Supersonic molecular beam injection system (SMBI) (b) Pellet injection system (PI) and (c) Fast gas injection system (FGI). A schematic of the SST-1 tokamak with the pumping lines and the injection points of these three fuelling systems are shown in figure 1.

2.1. Supersonic Molecular Beam Injection system (SMBI)

The SMBI was designed for feedback based density control. Main advantage is that the neutrals are fed at supersonic velocity and therefore have higher probability of penetrating into the core as compared to ordinary gas puffing. Two (02) numbers of M/s Parker make solenoid valves have been installed on the high field side at radial port nos. R-7 & R-10. An inlet pressure of 10 bar(g) was maintained at the valve inlet and a rectangular voltage pulse of 20 VDC with varying time duration was applied to feed the desired number of neutrals into the plasma chamber. Each of these two valves has a molecular injection rate of 2.8×10^{22} H-atoms/s.

2.2. Pellet Injection System (PI)

A single barrel hydrogen pellet injection system has been recently integrated with SST-1 vessel which is capable of producing pellets of 1.6 mm length having 1.8 mm diameter with a velocity in the range of 700 to 900 m/s. This injector is an in-situ gun type injector, in which, a solid hydrogen pellet is formed and accelerated to high speed using high pressure propellant gas.

2.3. Fast Gas Injection Fuelling system (FGI)

Out of the three fuelling systems installed on SST-1 vessel, only the fast gas injection method has been used extensively in SST-1 plasma campaigns. The objective of fast gas injection system is to feed the desired amount of hydrogen gas during pre-filling and density build-up phase during the plasma operation as well as to feed fuel gas during the wall conditioning, Lower Hybrid Current Drive (LHCD) and Ion Cyclotron Resonance Heating (ICRH) assisted plasma shots. The key element of this system is piezoelectric valve which was chosen due to its fast response time of ~ 2.0 ms, capability to work in the presence of magnetic field, low cost and reliable performance.

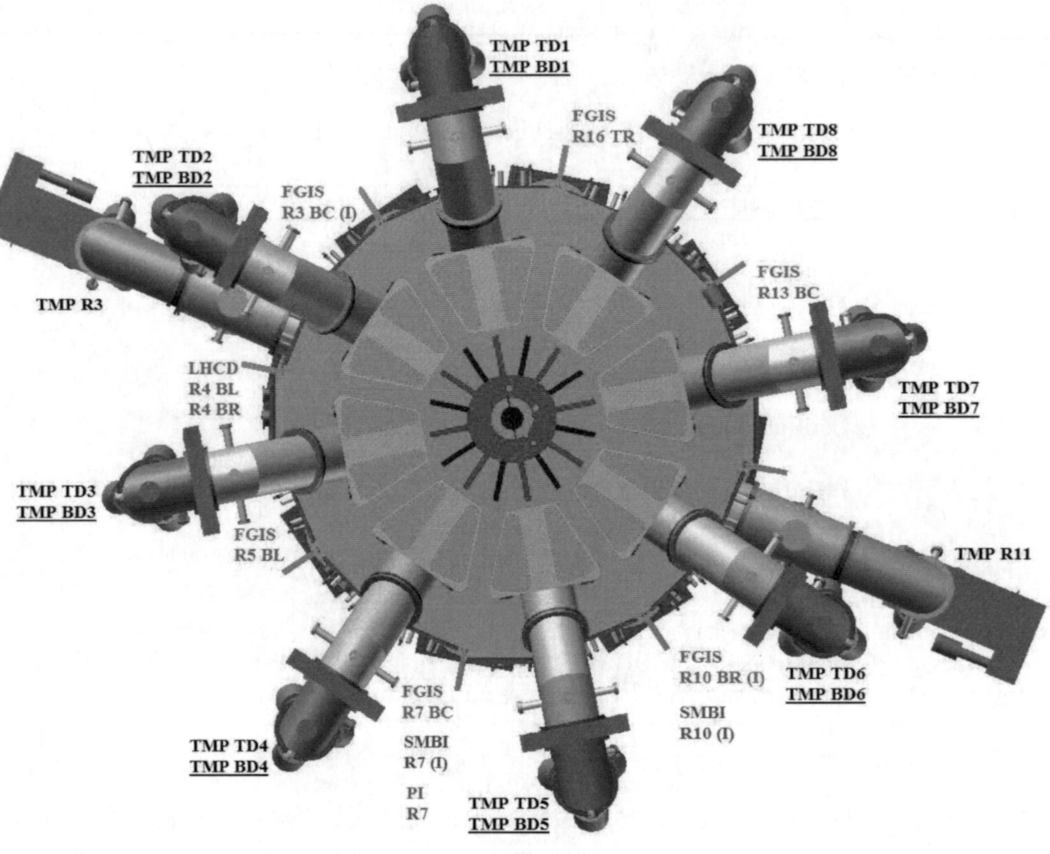

Figure 1. Schematic of the SST-1 tokamak showing the pumping and fuel injection locations.

This valve has a maximum throughput of 625 sccm at 100 V, 25 °C and 1.0 bar(g) inlet pressure. The throughput of this piezo valve is linear with respect to its inlet pressure and has a capacity to withstand a maximum inlet pressure of 3 bar(g). FGI system consists of reliable hardware and software necessary to control and monitor the flow of gases required for the plasma experiments remotely from the control room. Four (04) numbers of M/s. Maxtek make MV-112 piezo valves (02 nos. on the high field and 02 nos. on the low field side) are used to feed gas into SST-1 vessel during pre-filling. Another two (02) numbers of valves were used during the density build-up phase. These valves were mounted in radial ports diagonally opposite to each other to maintain the overall balance of gas distribution inside the vessel. In addition, two (02) more numbers of valves were mounted on either side of LHCD antenna and one (01) number of valves near ICRH antenna to feed the desired amount of neutrals during LHCD and ICRH assisted plasma shots. For Glow discharge cleaning (GDC) and boronization experiments, two (02) numbers of valves were mounted at diagonally opposite radial ports R-3 and R-11.

Research grade ultra-high pure hydrogen, helium and their mixtures were used in SST-1 plasma experiments. A palladium based hydrogen purifier was connected in hydrogen inlet line which provides 99.99995% pure hydrogen gas at its outlet. Since the variation in inlet pressure to the valve

IOP Conf. Series: Journal of Physics: Conf. Series **823** (2017) 012065 doi:10.1088/1742-6596/823/1/012065

changes its flow rate, this ultra-high pure gas was stored in buffer chamber and then fed into the vessel. The buffer chamber was maintained at 2.0 bar(g) with closed loop feedback system. A schematic of closed loop feedback control for auto refilling of the reservoir is shown in figure 2.

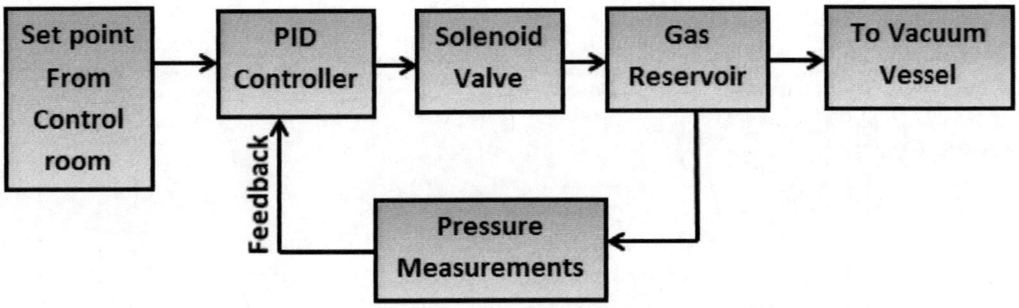

Figure 2. Schematic of closed loop feedback control for auto refilling the gas reservoir.

The entire gas feed line was fabricated using Swagelok connectors and electro-polished SS pipes. The entire piping and the connectors were leak tested in sniffer mode at 3.5 bar(g) helium pressure for the leak tightness at a background of 1.0×10^{-6} mbar l/s. A square voltage pulse of 100 VDC with different time duration was applied to the piezo-valves for gas puffing. All these piezoelectric valves were electrically isolated both from high-pressure side and low-pressure side using isolators to avoid the grounding interference which may changes the applied signal. The Piezoelectric valves were operated according to the requirement of the different gas puffing modes such as: (1) Pulsed flow through software, (2) Continuous flow at constant pressure and (3) Pulsed flow using density feedback.

3. Results and analysis

3.1. Calibration of the fuelling system

Four numbers of piezo valves were used for prefilling the vessel from 7.0×10^{-6} mbar to 3.0×10^{-5} mbar for plasma discharge operation. During operation, the total amount of gas puffed in to the vessel was varied from phase to phase by varying the amplitude and time duration of applied DC voltage to the piezo valve. The injection rate of 1.92×10^{21} H-atoms/sec was estimated when four numbers of the piezo valves were used. With an inlet pressure of 2.0 bar(g) when a pulse of 100 VDC and a width of 7 ms was applied to the piezovalve, an estimated 1.3×10^{19} H-atoms were injected inside the chamber. This was experimentally verified by the pressure-rise method. The pressure vs. time plot acquired is shown in figure 3. From the plot the number of hydrogen atoms injected into the vessel is found to be 1.0×10^{20}. This difference may be due to the inaccuracy in the total pressure measurement and also due to the difference in the inlet pressures.

3.2. Effect of reservoir pressure

The effect of reservoir pressure on piezo valve characteristics was studied for the reservoir pressure range varying from 0.5 bar(g) to 2.5 bar(g). The maximum vessel pressure achieved for a 7 ms pulse width is shown in table 2. It is observed that a change of 25 % in the reservoir pressure results in 10-20 % change in the injected gas molecules.

IOP Conf. Series: Journal of Physics: Conf. Series **823** (2017) 012065 doi:10.1088/1742-6596/823/1/012065

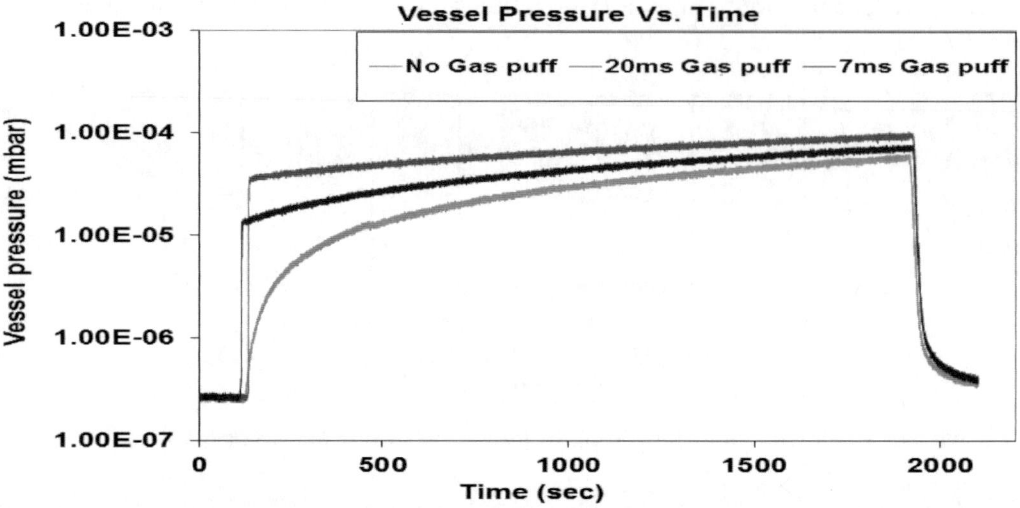

Figure 3. Pressure-time plot for 7 ms and 20 ms pulse width.

Table 2. Effect of reservoir pressure on piezo valve characteristics.

Tank Pressure (bar-g)	Peak Vessel Pressure (mbar)
0.5	6.5×10^{-6}
1.0	7.5×10^{-6}
1.5	9.2×10^{-6}
2.0	1.0×10^{-5}
2.5	1.1×10^{-5}

3.3. Gas puff with and without plasma shot

A screenshot of the vessel pressure is shown for a pulse width of 7 ms in the presence and absence of plasma shot in figure 4.

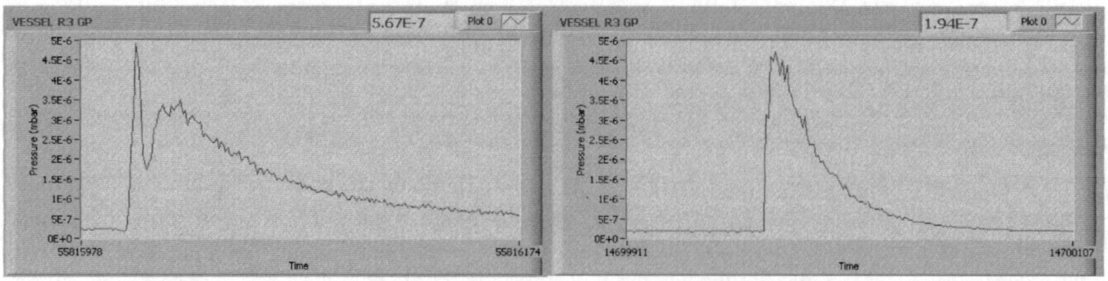

Figure 4. Pressure-time plot for 7 ms (a) During plasma shot (b) Without plasma shot.

IOP Conf. Series: Journal of Physics: Conf. Series **823** (2017) 012065 doi:10.1088/1742-6596/823/1/012065

4. Conclusion

From SST-1 campaigns, it is observed that the amount of gas injected through piezo valve plays an important role in achieving a repetitive plasma shots. Also experimental measurements show that a change of 25 % in the reservoir pressure results in 10-20 % change in injected gas molecules. In order to reduce such effect, auto refilling of reservoir with closed loop feedback controlled is very important. Experimental measurements validate that the theoretical estimate of the injection rate is more than ~ 10^{21} H-atoms/s.

References

[1] Pradhan S and SST-1 Mission Team 2010 *J. Plas. Fus. Res.* **9** 650
[2] Pradhan S, Khan Z, Tanna V L et al. 2015 *Nucl. Fusion* **55** 104009
[3] Pradhan S, Tanna V L, Sharma A N, Khan Z et al. 2015 *Physics Procedia* **67** 756
[4] Khan Z, Paravastu Y and Pradhan S 2014 *Proc. Mat. Sci.* **6** 216

IOP Conf. Series: Journal of Physics: Conf. Series **823** (2017) 012066 doi:10.1088/1742-6596/823/1/012066

Nuclear Analyses of Indian LLCB Test Blanket System in ITER

H L Swami[1], A K Shaw[1], C Danani[1] and Paritosh Chaudhuri[1]

[1]Institute for Plasma Research, Near Indira Bridge, Bhat, Gandhinagar 382 428, India

E-mail: hswami@ipr.res.in

Abstract. Heading towards the Nuclear Fusion Reactor Program, India is developing Lead Lithium Ceramic Breeder (LLCB) tritium breeding blanket for its future fusion Reactor. A mock-up of the LLCB blanket is proposed to be tested in ITER equatorial port no.2, to ensure the overall performance of blanket in reactor relevant nuclear fusion environment. Nuclear analyses play an important role in LLCB Test Blanket System design & development. It is required for tritium breeding estimation, thermal-hydraulic design, coolants process design, radioactive waste management, equipment maintenance & replacement strategies and nuclear safety. The nuclear behaviour of LLCB test blanket module in ITER is predicated in terms of nuclear responses such as tritium production, nuclear heating, neutron fluxes and radiation damages. Radiation shielding capability of LLCB TBS inside and outside bio-shield was also assessed to fulfill ITER shielding requirements. In order to supports the rad-waste and safety assessment, nuclear activation analyses were carried out and radioactivity data were generated for LLCB TBS components. Nuclear analyses of LLCB TBS are performed using ITER recommended nuclear analyses codes (i.e. MCNP, EASY), nuclear cross section data libraries (i.e. FENDL 2.1, EAF) and neutronic model (ITER C-lite v.1). The paper describes a comprehensive nuclear performance of LLCB TBS in ITER.

1. Introduction

One of the major technical objectives in ITER will be to test tritium breeding concepts [1] that would lead in a future reactor to tritium self-sufficiency, extraction of high-grade heat, and electricity production. LLCB breeding blanket for fusion DEMO reactor is being developed by India [2], [3] and a mockup of it is planned to be tested in ITER equatorial port no. 2 through TBM program [3], [5]. It is necessary to evaluate the overall nuclear performance of LLCB TBS in ITER nuclear environment to conceptualize the design & to proceed with further development of LLCB Test Blanket System (TBS). There are three major nuclear aspects of LLCB TBS in ITER. First is the neutronic performance of LLCB TBM in ITER in terms of tritium production, nuclear heat deposition, neutron flux and radiation damages. Second, safety concern in terms of shutdown dose rates in human accessible zones belongs to LLCB TBS and third, the handling & managing the irradiated TBM & associated systems in terms of activation responses. The paper describes a comprehensive nuclear performance of LLCB TBS in ITER, carried out using the ITER recommended codes and guidelines [6].

2. Description of LLCB TBS & its Neutronic Model

LLCB TBM contains lithium titanate Li_2TiO_3 as Ceramic Breeder (CB) material in the form of packed pebble beds with 60 % ^6Li enrichment and PbLi eutectic as tritium breeder with 90 % enriched ^6Li, neutron multiplier and coolant for the CB zones. India specific Reduced Activation Ferritic Martensitic Steel (IN-RAFMS) is used as the structural material and TBM first wall is actively cooled by high-pressure helium (He) gas. A shield block, made of steel and water, is also associated with TBM for radiation shielding to port inter-space area. LLCB TBM along with shield block, also called TBM set will be placed in one half of equatorial port no. 2. The coolant pipes will route through the shield block and port inter space, also known as pipe forest (PF) area and will be connected to the other associated systems which are placed outside bio-shield. LLCB TBS [7] consists of TBM set and its different ancillary systems, such as First Wall Helium Cooling System (FWHCS), Lead-Lithium Helium Cooling System (LLHCS), Lead-Lithium Cooling Systems (LLCS), Coolant Purification System (CPS) and Tritium Extraction System (TES) [8], [9]. A schematic view of LLCB TBS is shown in figure 1.

In order to evaluate the nuclear performance, it is required to prepare a neutronic model of necessary systems. Neutronic model of LLCB TBS are prepared using MCAM [9] and integrated into ITER 40^0 sector model C-lite. The integrated neutronic model up to bio-shield is shown in figure 2 and LLCB TBM set is shown in figure 3. Model of port cell no. 2 ancillary equipments unit (AEU) are shown in figure 4. In order to include the effect of neighboring test blanket system, China HCCB TBS [10] neutronics models which are received from China TBS team, are also integrated into ITER neutronic model.

3. Neutronic Performance of LLCB TBM set

Neutronic performance of LLCB TBM set [12] in ITER has been evaluated using radiation transport code MCNP6 [13] and nuclear cross section data library FENDL 2.1 [14]. All nuclear responses have been estimated considering 500 MW D-T ITER fusion neutron source. The radial profile of total neutron flux with contributions of high energy and low energy is shown in figure 5. Neutron flux attenuation in LLCB TBM is around 1 order; however attenuation in shield block is around 5 orders. The flux level just after TBM set flange is around ~10^8 n/s.cm^2. Tritium production rate in LLCB TBM is 1.66×10^{17} tritons/s. For a full power day operation of ITER (continues D-T operation), total tritium production will be 71.8 mg; contribution from PbLi in the total tritium production is around 51.5 % whereas Li_2TiO_3 is around 48.5 %. Total heat deposited in LLCB TBM is around 0.46 MW and shield block is 7.13 kW. The highest power density in LLCB TBM set is 5.2 Watt/cc at TBM first wall. Two main sources of radiation damage through transmutation in a fusion reactor are the atomic displacement (DPA) and gas production and knowledge of these is essential to determine the component's useful lifetime. Estimations of DPA, Helium and Hydrogen production estimation have been carried out, assuming the neutron fluence in ITER 0.3MWa/m^2 accumulated during full ITER life time. Calculations predict the maximum DPA value 2.72 in TBM First wall IN-RAFM steel. Helium and hydrogen gas productions are also higher at same location with value 33.4 appm and 145 appm respectively.

4. Estimation of shutdown dose rate at human accessible zones

There are two human accessible areas allocated for assembly, maintenance and dismantling of the LLCB TBS in port no. 2. One is in port inter-space (PI) area (pipe forest area) as shown in figure 6 and second is in the port cell (PC) corridor as given in figure 7. As per the ITER safety guidelines, limits of shutdown dose rates (SDDR) are 100 μSv/hr after 12 days cooling in PI area and 10 μSv/hr after 1 day cooling in PC area. We have adopted MCNP D1S [15], [15] (direct 1 step) method for estimating the dose rate in PI area. Radial Profile of dose rates in PI area is shown in figure 8. Maximum estimated shutdown dose rate at PI area is 370 μSv/hr 12 days after shutdown. Dose rate in PC area has been estimated considering two radiation sources, one is activated structural materials and the second is activated PbLi in the dump tank. Calculation of dose rate in PC area has been performed in two steps. In the first step we calculate the radiation source 1 day after shutdown and in second step transport of radiations at interested locations. Since irradiated PbLi has high activation level, a 12 cm thick lead shield on the tank has been placed. Radial profile of dose rates in PC area is shown figure 9.

The maximum dose rate at PC area is 1.38 μSv/hr 1 day after shutdown. Estimated dose rate at PI area is higher than limit which indicates that improvement in shielding is required at PI area. Dose estimation in PC area are under the limit but there are other dose sources like Activated Corrosion Products (ACPs) in coolant (PbLi) and activated PbLi residual film present in pipes which are not yet included. Contributions in dose rate in PC area from ACPs & residual film have to be added and equatorial port no. 2 bio-shield plug (BSP) geometry model needs to be improved in nuclear analyses of TBS during next design phase for revised estimation of dose rates.

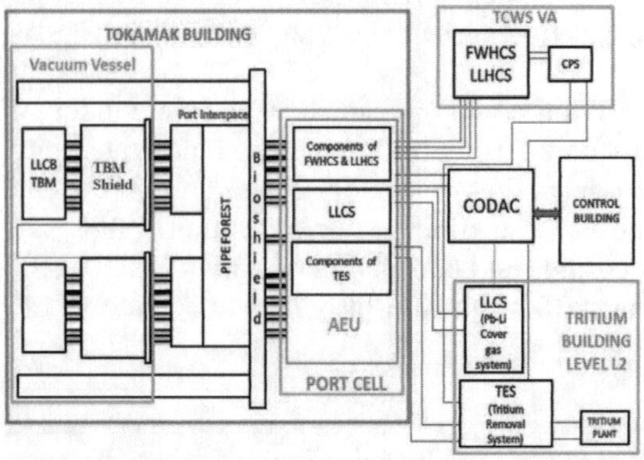

Figure 1. Schematic of LLCB Test Blanket System

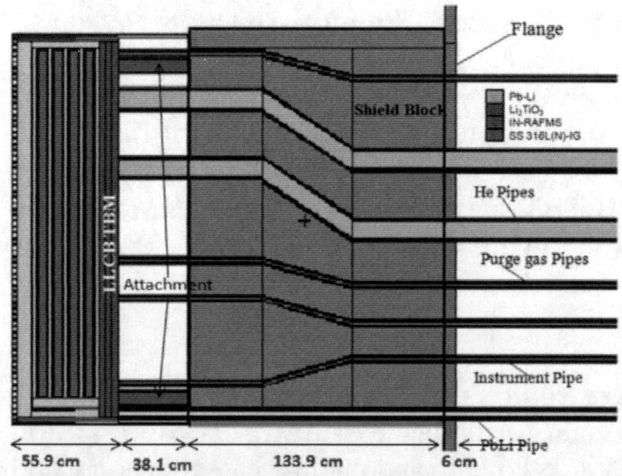

Figure 3. LLCB TBM set Neutronic model

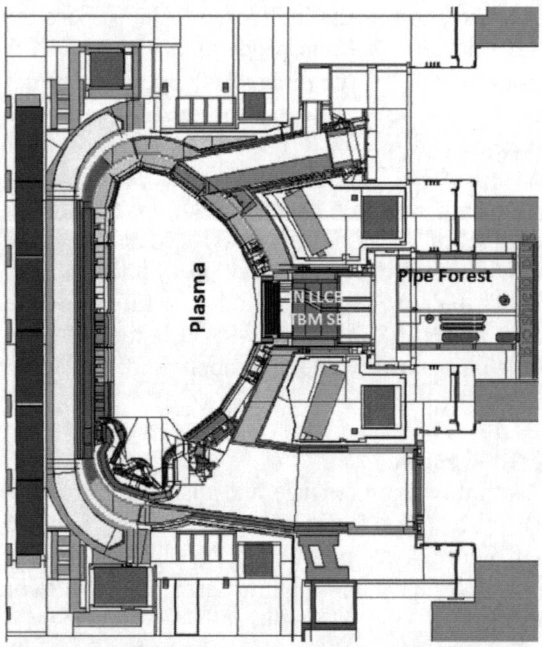

Figure 2. ITER Neutronic model with TBM port no. 2

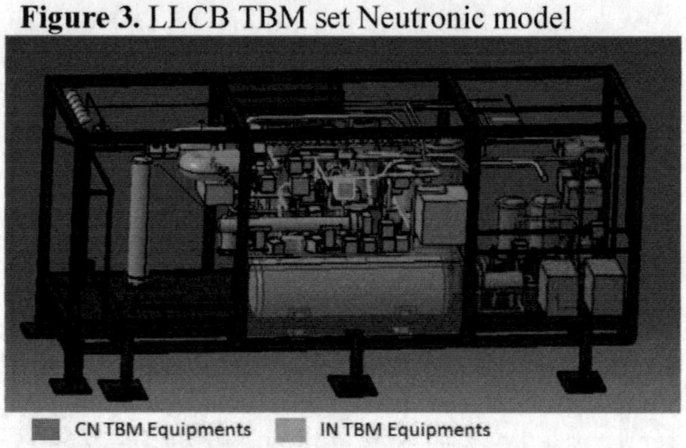

Figure 4. AEU in port cell #2

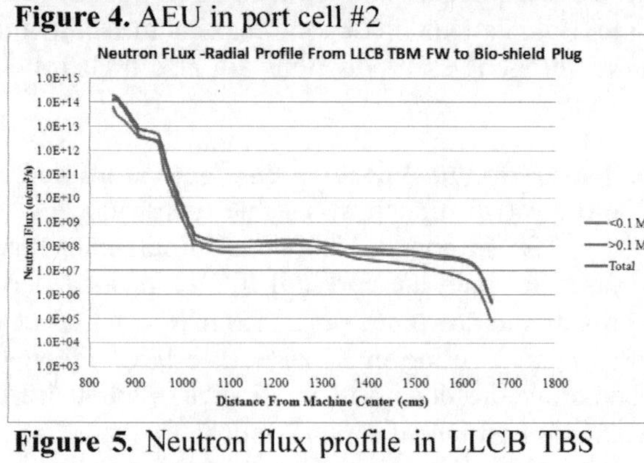

Figure 5. Neutron flux profile in LLCB TBS

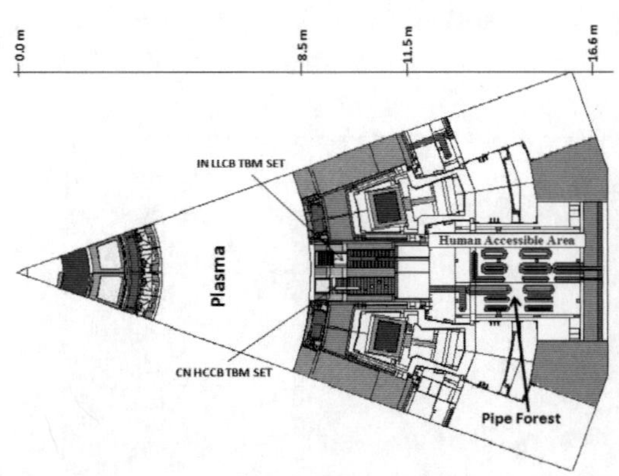

Figure 6. Top view of ITER Neutronic model

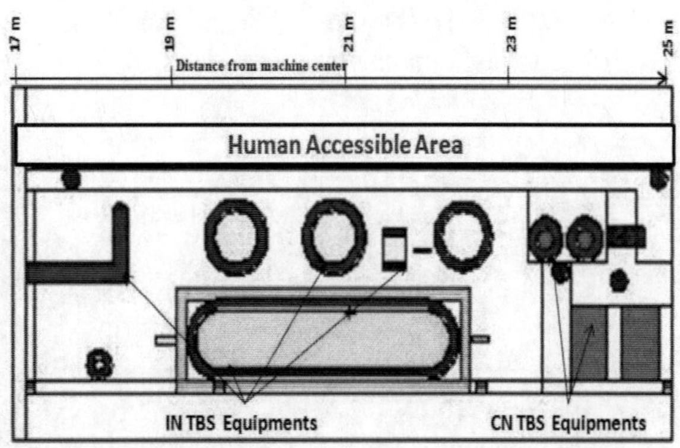

Figure 7. Top view of ITER port no. 2 PC area

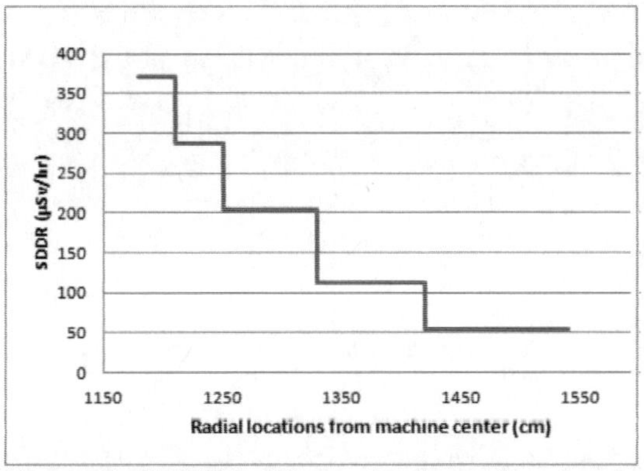

Figure 8. SDDR in PI area 12 days after shutdown

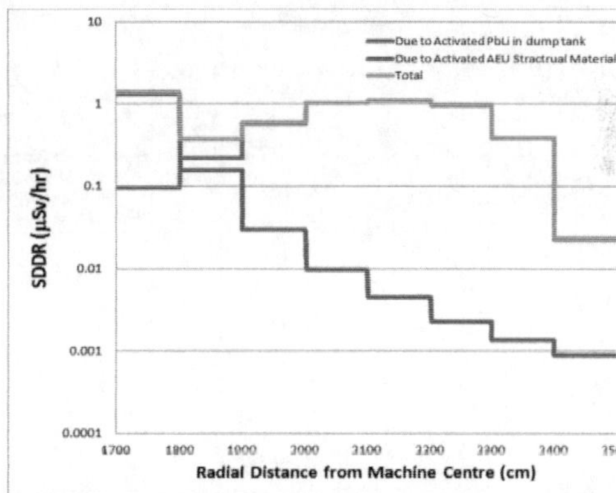

Figure 9. SDDR in PC area 1 day after shutdown

5. Activation Analyses of LLCB TBS

Neutron induced activation calculations are needed to assess the radioactive hazards from irradiated LLCB TBS. Activation analyses provide the contact dose rate, decay heat and radioactivity estimations required for the maintenance plan, assembling-dismantling plan and rad-waste management of LLCB TBS. Activation analysis has been performed for LLCB TBS by making use of inventory code FISPACT-2007 [17]. ITER TBM safety irradiation scenario having 4 years of D-T neutron irradiation (2 ITER D-T operation cycles) and 0.17 MWa/m^2 fluence has been assumed for activation analyses. The radioactive inventories of LLCB TBM materials as a function of cooling times are shown in figure 10. The total radioactive inventory after shutdown in LLCB TBM is 1.41 MCi. The contribution in total activity from the structural material, ceramic breeder, and lead lithium are 0.094 MCi, 0.055 MCi and 1.26 MCi respectively. Figure 11 gives the total decay heat and the contribution from each material. The total decay heat just after the shutdown is 13.4 kW with a contribution of 11.7 kW from the lead lithium, 1.02 kW from Ceramic Breeder and 0.678 kW from IN-RAFMS, respectively. Estimation of contact dose rate is required to perform maintenance and replacement activities for TBM set. Contact dose rate from TBM materials as a function of the cooling time is shown in figure 12. TBM First wall is the main contributor to contact dose rate. The ^{59}Co present in IN-RAFMS produces ^{60}Co ($T_{1/2}$ = 5.271y) which is a major contributor to dose rate for up to 10^5 years of cooling time. The maximum contact dose rates at TBM first wall just after shutdown is 1.61×10^4 Sv/hr and after 1 month cooling is 1.01×10^3 Sv/hr.

Specific activities were also estimated for components of LLCB TBS in other areas like Pipe Forest (PF), Bio-shield Plug (BSP), and AEU. Specific activities of PF, BSP and AEU just after shutdown are 2.79×10^6 Bq/kg, 1.12×10^6 Bq/kg and 1.93×10^5 Bq/kg, respectively. After 100 years of cooling, the value drops to 2.1×10^4 Bq/kg, 1.2×10^3 Bq/kg and 1.44×10^3 Bq/kg. These estimations support the LLCB TBS rad-waste management in ITER.

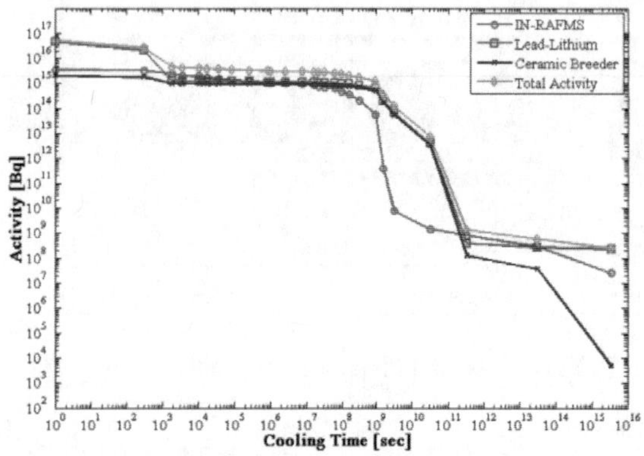

Figure 10. Radioactivity of irradiated LLCB TBM

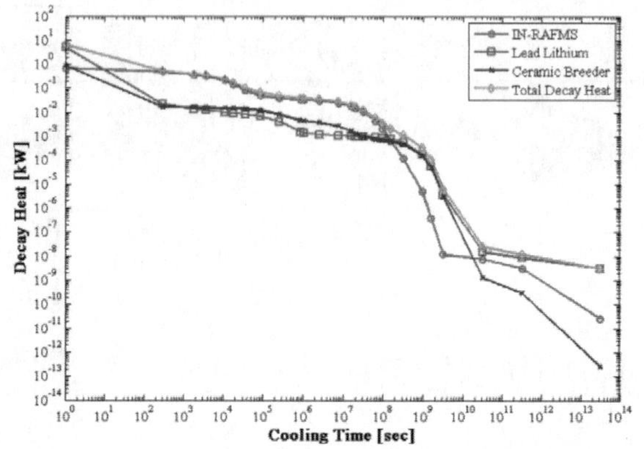

Figure 11. Decay heat of irradiated LLCB TBM

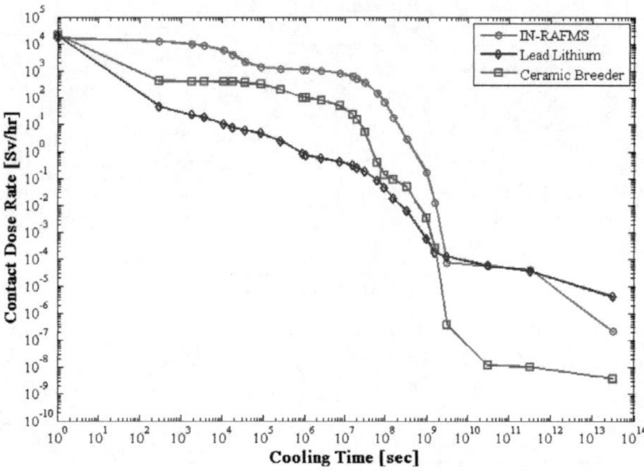

Figure 12. Contact dose rate of irradiated LLCB TBM

6. Conclusion

The nuclear analyses play a key role in LLCB test blanket systems design for testing in ITER neutron radiation environment. The nuclear performance of LLCB TBS in ITER is evaluated as per ITER recommended codes and guidelines. Integrated neutronic models of LLCB TBS were constructed and the comprehensive nuclear performance was carried out. LLCB TBM set neutronic performances outcome i.e. tritium production, nuclear heating and radiation damages have been estimated and provided for LLCB TBM set engineering design. The total tritium produced in LLCB TBM in a full power day is 71.8 mg and total heat deposited inside LLCB TBM is 0.46 MW. The estimated shutdown dose rate at PI area is 370 μSv/hr 12 days after shutdown. Evaluation of dose rates indicates that shielding in PI area needs to be improved. The maximum dose rate in PC area is 1.38 μSv/hr 1 day after shutdown. The dose rates from ACPs and PbLi residual film in pipes have not been included in estimation total dose rate inside PC area which will have an impact on dose rate. It will be assessed during next design phase of TBS. A large database on activation characteristic of LLCB TBS has been prepared to support the rad-waste management and handling of irradiated TBS. The presented work has supported the conceptual design of LLCB TBS in ITER. Results of the LLCB TBS nuclear analyses are utilized in engineering analyses, safety analyses & process design analyses of TBM set and associated systems.

Acknowledgement

Authors are very much thankful to Oak Ridge National Laboratory, U.S., for providing the MCNP access through the cloud server. Authors are also thankful to Dr. Michael Loughlin ITER Organization France and his colleagues for supporting to get the MCNP access and also provide the facility to run D1S code for dose rate estimation. Authors acknowledge the China HCCB TBM team for providing the model of Chinese TBM set for completeness of the analyses.

This work was carried out using an adaptation of the C-LITE MCNP model which was developed as a collaborative effort between the FDS team of ASIPP CHINA, University of WISCONSIN-MADISON, ENEA Frascati, CCFE UK, JAEA NAKA, and the ITER Organization.

Authors wish to thank Mr. Rajendra Kumar, Indian TBM Project Leader for his continuous support and encouragement. We also acknowledge the contribution of other LLCB TBM colleagues.

References

[1] Giancarli L M et al. 2006 *Fusion Eng. Des.* **81** Issues 1–7 393
[2] Srinivasan R, Deshpande S P et al. 2008 *Fusion Eng. Des.* **83** 889
[3] Paritosh Chaudhuri et al. 2009 *Fusion Eng. and Des.* **84** Issues 2–6 573
[4] Rajendra Kumar E et al. 2012 *Fusion Eng. Des.* **87** 461
[5] Giancarli L M, et al. 2012 *Fusion Eng. Des.* **87** Issues 5–6 395
[6] Quality Assurance in Neutronic Analyses 2006 *ITER IDM Document, ITER_D_23H9A4*
[7] Rajendra Kumar E et al. 2016 *Fusion Eng. Des.* **109–111**, Part B, 1 1522
[8] Chaudhuri P et al. 2012 *Fusion Eng. Des.* **87** 1009
[9] Bhattacharyay R et al 2014 *Fusion Eng. Des.* **89** Issues 7–8 1107
[10] Y Wu, FDS Team 2009 *Fusion Eng. Des.*, **84** 1987
[11] Feng K M et al. 2012 *Fusion Eng. Des.* **87** Issues 7–8 1138
[12] Swami H L et al. 2016 *Fusion Eng. Des.* **113** 71
[13] Goorley J T et al 2013, "Initial MCNP6 Release Overview - MCNP6 version 1.0", LA-UR-13-22934
[14] Aldama D L, Trkov A 2004 FENDL-2.1: Update of an Evaluated Nuclear Data Library for Fusion Applications, IAEA Report INDC (NDS)-467, FENDL-2.1 Library Home: http://www-nds.iaea.org/fendl21/
[15] Villari R, *et al.* 2012 *Fusion Eng. Des.* **87** 1095
[16] Valenza D *et al.* 2001 *Fusion Eng. Des.* **55** (4) 411
[17] Forrest R A, FISPACT-2007: User manual, EASY Documentation Series UKAEA FUS 534

IOP Conf. Series: Journal of Physics: Conf. Series **823** (2017) 012067 doi:10.1088/1742-6596/823/1/012067

Quality control of FWC during assembly and commissioning in SST-1 Tokamak

Hitesh Patel[1], Prosenjit Santra[1], Tejas Parekh[1], Prabal Biswas[1], Snehal Jayswal[1], Pradeep Chauhan[1], Yuvakiran Paravastu[1], Siju George[1], Pratibha Semwal[1], Prashant Thankey[1], Gattu Ramesh[1], Arun Prakash[1], Kalpesh Dhanani[1], D C Raval[1], Ziauddin Khan[1] and Subrata Pradhan[1]

[1]Institute for Plasma Research, Near Indira Bridge, Bhat, Gandhinagar 382428, India

E-mail: hpatel@ipr.res.in

Abstract. First Wall Components (FWC) of SST-1 tokamak, which are in the immediate vicinity of plasma, comprises of limiters, divertors, baffles, passive stabilizers designed to operate long duration (~1000 s) discharges of elongated plasma. All FWC consist of copper alloy heat sink modules with SS cooling tubes brazed onto it, graphite tiles acting as armour material facing the plasma, and are mounted to the vacuum vessels with suitable Inconel support structures at inter-connected ring & port locations. The FWC are very recently assembled and commissioned successfully inside the vacuum vessel of SST-1 undergoing a rigorous quality control and checks at every stage of the assembly process. This paper will present the quality control aspects and checks of FWC from commencement of assembly procedure, namely material test reports, leak testing of high temperature baked components, assembled dimensional tolerances, leak testing of all welded joints, graphite tile tightening torques, electrical continuity and electrical isolation of passive stabilizers from vacuum vessel, baking and cooling hydraulic connections inside vacuum vessel.

1. Introduction

After eleven successful experimental campaigns since the commissioning of Steady-state Super conducting tokamak (SST-1) [1, 2], First wall components (FWC) were installed in SST-1 vacuum vessel as one of the up-gradation objective during October, 2014 to May 2015. FWC of the SST-1 Tokamak consists of Inboard divertor plates (IDP), Outboard divertor plates (ODP), Inboard passive stabilizers (IPS), Outboard passive stabilizers (OPS), main baffle (MBAF) [3]. Assembly requirements of FWCs are proper alignment with the machine centre, leak tightness of hydraulic connections, Electrical isolation of IPS and OPS from other in-vessel components of vacuum vessel and baking of all graphite tiles up to 1000 °C [4,5] before assembly inside SST-1 machine.

2. Baking of graphite tiles

FWCs of the SST-1 machine consist of copper alloy back plates with brazed stainless steel tube. Plasma facing side of these back plates were covered with graphite tiles which are mounted mechanically. These graphite tiles have been manufactured by CNC machining. During the machining graphite tiles absorb many impurities i.e. atmospheric dust, gases and metallic tool particle. In order to remove all these impurities from porous graphite tiles, baking of these tiles was carried out at 1000 °C for 10 hours in vacuum condition. A dedicated vacuum furnace with graphite as electrode having a capacity to reach temperature up to 1200 °C was used for the baking of ~ 3800 PFCs graphite tiles. As per loading capacity of furnace, each batch consists of ~ 200 tiles were baked to 1000 °C in a vacuum condition of 1.0×10^{-5} mbar as shown in figure 2. Residual gas analyser (RGA) has been used for analysing the release of the different gas species. A snap shot of RGA scan shown in figure 1 depicts the low level of the hydrogen and oxygen residuals in range of 1.0×10^{-6} mbar at room temperature (RT) and at the flat top condition. After baking, all the graphite tiles were packed in vacuum tight sealed bags.

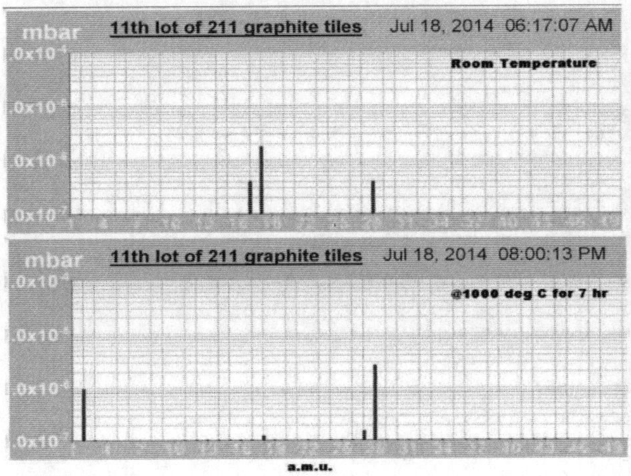

Figure 1. RGA Spectra of graphite tiles at RT and 1000 °C.

Figure 2. Vacuum Furnace loaded with graphite tiles.

3. Leak testing and baking of FW modules

All FW modules copper alloy back plates were required to be baked to achieve ultra high vacuum inside vacuum vessel and also they have to be cooled during the plasma experiments for the extraction of incident heat flux. A common hydraulic circuit was used for heating and cooling of these modules (figure 3). During the heating, hot nitrogen gas was passed through these piping and while for cooling demineralised water at RT was used. In order to ensure the leak tightness of ~ 10^{-9} mbar l/s, all FW modules brazed piping along with their hydraulic circuit including flexible bellows & isolators were tested for leaks at RT as well as at baking temperature using a dedicated vacuum chamber. The vacuum baking chamber along with loaded copper back plates, flexible bellows & isolators is shown in figure 4 and the schematic P&I diagram is shown in figure 5. At RT, the entire circuit was leak tested and then the chamber was pumped down to high vacuum of ~ 1.0×10^{-5} mbar. At this vacuum, it was heated up to 250 °C and at this condition leak tightness was carried out by pressurizing it with helium gas. The holding time and temperature of baking were 8 hours and 250 °C respectively. A snapshot of the all copper plate baking temperature with respect to time is shown in figure 6 which depicts the uniform temperature rise of all FW modules up to 250 °C.

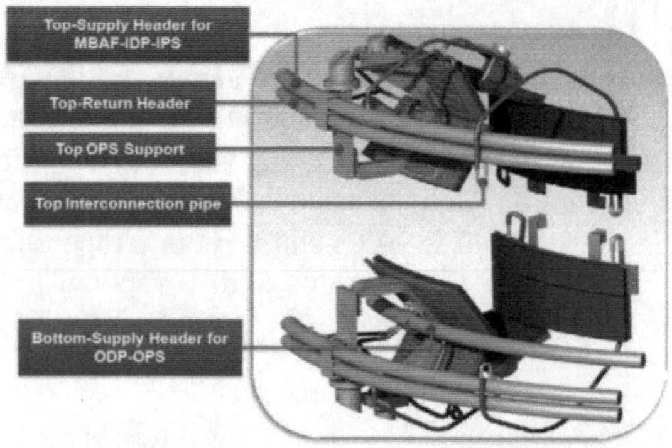

Figure 3. 3D model of PFC with Hydraulics

Figure 4. Baking facility for copper back plate

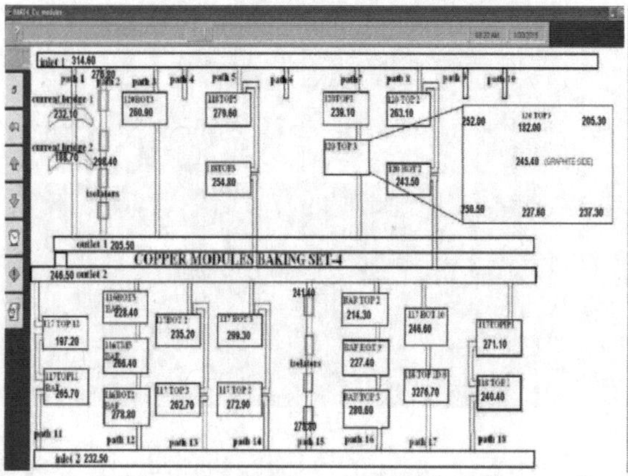

Figure 5. Schematic diagram of backing flow path

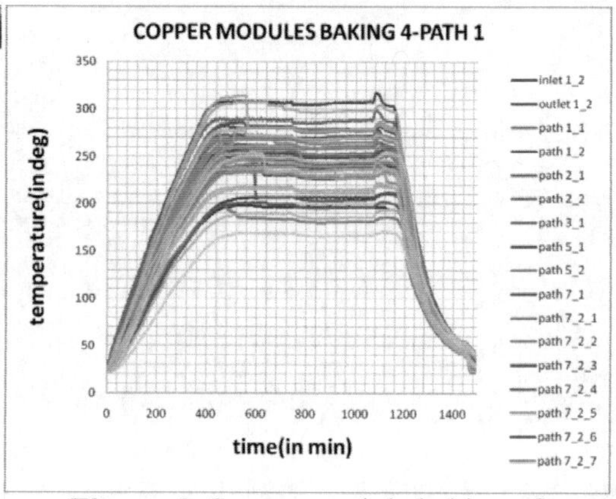

Figure 6. Copper module baking of set 4

The hydraulic circuit of FWCs consists of 2 supplies headers and 1 common return header each at top and bottom sides. MBAF, IDP and IPS (top/bottom) are connected in series and fed through two supply headers each having 11 parallel sub-connections while ODP and OPS (top/bottom) are connected in series and fed through another two supply headers each having 16 parallel sub-connections. Due to space constraints in vacuum vessel, all the possible joints were welded outside. Minimum joints were kept for the in-situ welding. The joints made outside of the vessel were leak tested in vacuum condition before their welding inside the vacuum vessel. Final in-situ welding connections of each path were made in parallel to the final assembly of respective back plates. All the connections of each hydraulic path except the connection with supply and return header were leak tested in vacuum mode. During the leak testing of these 54 parallel paths, leaks of ~ 10^{-2} mbar l/s to 10^{-6} mbar l/s were found in 8 welding joints. These joints were repaired and re-tested for leaks. At the end of the assembly of all the back plates, two joints of each hydraulic path were remained to be leak tested. Leak testing of these connections was done in the final testing of entire hydraulic circuit in vacuum and sniffer mode at 8 bar pressure.

4. QC aspect of metrology of PFC assembly

FWC are required to be assembled with high positional accuracy as plasma is to interact with FWC at certain angle. To insure the same, strict metrology protocols were generated and followed during the assembly of FWC. Typical module of FWC comprises 4 components which includes, welded part of support, bolted part of support, back plates and graphite tiles. The profiles of all the back plates were checked with accurate templates and through the laser scanning by coordinate measurement machine on ground (figure 7) before their final assembly inside vacuum vessel. IPS and OPS back plates were assembled on the ground for the identification of flaws in lap joints. For the accurate measurement of coordinates inside vacuum vessel, the combination of theodolite based laica ECDS (electronic co-ordinate determination system) measurement along with photogrammetry measurement (figure 8) was implemented. An inboard side limiter was mastered with respect to machine centre using both the

IOP Conf. Series: Journal of Physics: Conf. Series **823** (2017) 012067 doi:10.1088/1742-6596/823/1/012067

techniques and aligned within an accuracy of 1 mm. The surface of the limiter was then considered as a reference and other back plates were assembled with respect to it. Precise templates of respective profiles were also used for reconfirmation. Suitable fixtures were developed for handling and alignment of back plates. The acceptance criterion was kept as ± 2 mm in deviation of elevation, poloidal and toroidal gaps between various FWC and surface profiles. Electrical isolation between IPS and OPS circuits were also ensured. Since the headers and in-vessel coils were supported on the OPS supports, thermal isolation between them was also insured. For insuring the isolation amongst the components, isolation checks were done continuously at various voltages (0-1 kV) at each stage of component assembly. The minimum value of resistance achieved amongst the components was 1 MΩ.

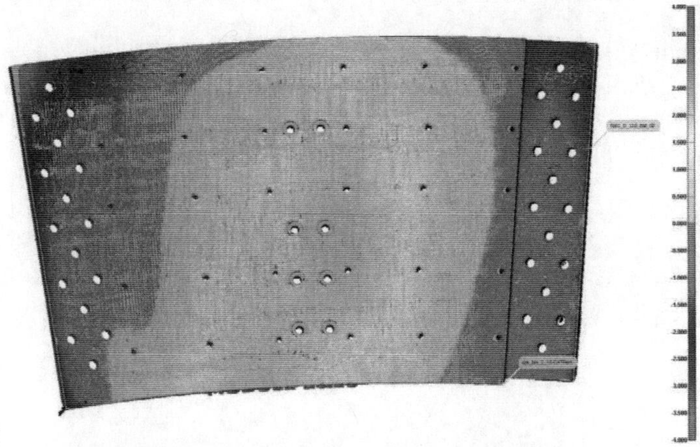

Figure 7. Scan analysis of Main baffle.

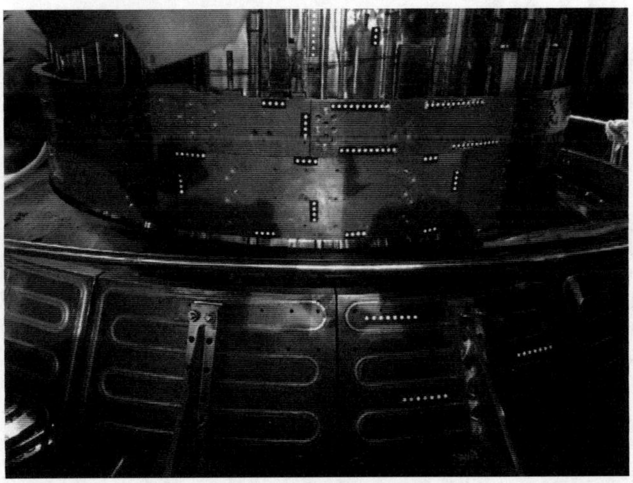

Figure 8. Photogrammetry of PFC components during assembly.

5. Conclusion

After successful assembly of PFC components in May 2015 and its rigorous testing in terms of pressure testing, vacuum mode and sniffer mode testing and thermal cycling testing, the successful result of plasma breakdown in excess of 110 kA plasma current with duration of 450 ms was achieved.

IOP Conf. Series: Journal of Physics: Conf. Series **823** (2017) 012067 doi:10.1088/1742-6596/823/1/012067

Even after four successive campaigns of plasma operation, none of the components of the PFC assembly along with its hydraulics has been failed.

Acknowledgement

I acknowledge the sincere effort of Mr. Kirit Vasava for his valuable contribution towards the assembly and integration of PFCs which leads to successful operation of tokamak for achieving desired plasma parameters.

References

[1] Pradhan S, Sharma A N, Tanna V L, Khan Z, Prasad U et al. 2012 *IEEE Trans. Plasma Sci.* **40** 614
[2] Pradhan S, Khan Z, Tanna V L et al. 2015 *Nucl. Fusion* **55** 104009
[3] Jacob S et al. 1996 *Journal of Nuclear Materials* **233-237** 655
[4] Paravastu Y, Khan Z and Pradhan S 2015 *Fusion Engineering and Design* **98-99** 1375
[5] Babu G R, Khan Z and Pradhan S 2014 *IPR internal report* IPR/TR-297/2014

IOP Conf. Series: Journal of Physics: Conf. Series **823** (2017) 012068 doi:10.1088/1742-6596/823/1/012068

Measurement of electron energy probability function in weakly magnetized plasma

D Kalita[1], B Kakati[2], B K Saikia[1], M Bandyopadhyay[3] and S S Kausik[1]

[1]Centre of Plasma Physics-IPR, Nazirakhat, Sonapur, Kamrup -782 402, Assam, India
[2]Institute for Plasma Research, Bhat, Gandhinagar- 382 428, India
[3]ITER-India, Institute for Plasma Research, Bhat, Gandhinagar- 382 428, India

E-mail: bharatkakati15@gmail.com, dkalita@cppipr.res.in

Abstract. Knowledge of the real EEDF is of great importance in understanding the underlying physics of processes occurring at the magnetized plasma, such as the formation of transport barriers, cross-field diffusion coefficients and plasma–substrate interactions. In the present experiment, the application of Langmuir probe to evaluate EEPF in presence of magnetic fields within the range 594 –32 G is investigated. The data recorded for EEPFs in magnetic fields and in dust is acquired using current–voltage characteristics measured in low pressure hydrogen plasma. The values of plasma density, electron temperature and EEPF are evaluated with a single cylindrical Langmuir probe at different axial positions (1 cm to 6 cm) from the magnet. From the recent EEPF observations in presence of magnetic field, it shows a bi-Maxwellian EEPF structure at different magnetic fields. But at different magnetic field, it is observed that the low energy electron population changes whereas the high-energy electron population remains almost constant. EEPF measurement shows almost identical behaviour with the unmagnetized plasma when the larmour radius of electron is greater than or equal to 10 times of the probe radius.

1. Introduction

The importance of the knowledge of the electron energy distribution function (EEDF) is of great interest in different branches of plasma physics ranging from laboratory to fusion plasmas [1,2]. The EEDF is a measure of the number of electrons within a unit energy interval and provides information on other internal parameters of the plasma. The EEDF is obtained using the Druyvesteyn procedure [3] by the second derivative of *I–V* characteristic measured by the probe which is proportional to the electron energy probability function (EEPF) [4,5].

where, $F(\varepsilon) = \varepsilon^{1/2} \times f_p(\varepsilon)$ (1)

Here ε is the electron energy in eV.

In most of the laboratory plasma devices, a magnetic field is practically observed in all situations either in ambient or is applied in the industrial/ fusion plasma research. It is known that the magnetic field can influence the anisotropy of the plasma properties. Hence, the magnetic field affects the plasma properties by altering the transport of ions and electrons to the chamber wall. However, due to many difficulties in measuring of the EEDF magnetized plasmas [6], the evolution of the EEDF and associated plasma parameters across magnetic fields still remains a topic of important research.

Also, the understanding of the dust behavior in plasma is of great importance for experimental and industrial applications [6,7]. The dust charging behaviour and its controlling in filamentary hydrogen plasma are reported by our group in our earlier publications [8-11]. The presence of dust (grains) changes the plasma parameters by redistributing the low and high energy electron populations [12]. The density of the plasma electrons in presence of dust particles decreases by capturing the low energy plasma electrons on the surface of the dust grain, which in turn increases the energy of the electrons. However, the study on EEPF for dust containing hydrogen plasma in presence of magnetic field is very rare [13].

The present experimental work is largely dealt with the study of the effect of external magnetic field on the plasma parameters such as plasma density, electron temperature, EEPF in absence/presence of dust in low pressure hydrogen plasma. Special emphasis is given on the measurement of EEPF by Langmuir probe for "high and low magnetic field case" for pristine and dust containing hydrogen plasma. The electron energy distributions functions measured in this work are presented in terms of electron energy probability function EEPF, in units of $cm^{-3}V^{-3/2}$ rather than the EEDF itself.

2. Experimental set up

The experiments are carried out in the existing dusty plasma device consisted of two stainless steel chambers placed one above the other as shown in figure 1. The lower chamber is the plasma chamber that is placed horizontally and vertical upper chamber facilitates the holding of dust dropper. The plasma discharge chamber was made out of a cylindrical stainless steel chamber of 100 cm in length and 30 cm in diameter. The dust unit consists of cylindrical stainless steel chamber of height 72 cm and 15 cm in diameter with a dust dropper, fitted inside the dust chamber. For the confinement of the plasma elements, a full line cusped magnetic cage made of strontium ferrite of 1.3 kG is used. Hydrogen plasma is produced by hot cathode filament discharge technique by striking a discharge between incandescent tungsten filaments and grounded magnetic cage. Here filaments are acting as cathode and the grounded chamber as an anode. A permanent magnet having magnetic field strength 1.2 kG is placed inside the plasma chamber to observe the effect of magnetic field on plasma parameters.

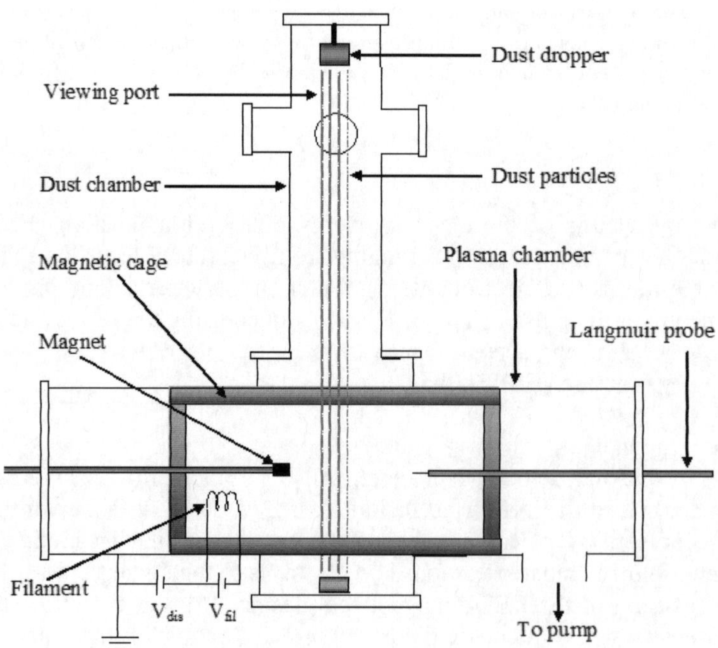

Figure 1. Schematic diagram of the experimental device.

Plasma parameters are measured by a single cylindrical Hiden Analytical Limited's ESPION advanced Langmuir probe system, consisting of a tungsten wire of 0.15 mm in diameter and a length of 10 mm. For probe measurement validity in magnetic field, the probe tip is oriented normally to the magnetic field.

In general, the probe radius should be much smaller than the Larmor radius of the electron or ion. If the Larmor radius of the electron or ion is smaller than the probe radius, the electron or ion current to the probe is reduced, and as a result the data acquired in presence of magnetic field are not accurate. In presence of strong magnetic field, the charge collections by the Langmuir probe is limited due to orbital effect and the reduction of collected charged particle can be represented by a parameter β. The parameter β is estimated by the ratio of the probe radius (r_p) to the mean Larmor gyroradius (r_L) of charged particle. In the present experiment for different magnetic field strength, the parameter β is found out to be 1.16 (max) and 0.05 (min) for electron and for ion, it is 0.20 (max) and 0.009 (min). Thus the electron collection by the probe in presence of magnetic field is affected more compared to the ion collection.

In the present work, tungsten dust particles (having dia ~ 4-6 μm) are used to study the effect on EEDF in presence of dust grains which are dropped from the top of the dust chamber with the help of an electrically controlled dust dropper.

3. Results and Discussions

The presence of the magnetic field lines is generally thought to reduce the electron or ion current drawn from the plasma by a Langmuir probe because the charged particles tend to follow the magnetic field lines. As the magnetic field strongly restrict the motion of the electrons than the ions, so the electron density measured from the electron saturation current with the help of a Langmuir probe deviates from the ion density measurements. In presence of magnetic field, the electron saturation current takes the form of [14]

$$I_e = I_0 \delta \tag{2}$$

Here $I_0 = -\frac{1}{4} n_{e0} \vec{v}_e e A_p$ is the undisturbed electron current, where A_p is the probe surface area, e is the elementary charge, $\bar{v}_e$ is the electron's mean velocity, n_{e0} is the electron density of the undisturbed plasma. The δ is the reduction factor due to the magnetic field which is given by;

$$\delta = \left(1 + \frac{\pi}{8} \frac{r_p}{\lambda} \left[1 + \left(\lambda / r_L^e \right)^2 \right]^{1/2} \right)^{-1} \tag{3}$$

where r_p is the probe radius, λ is the Debye length and r_L^e is the Larmor radius of the electron.

The reduction factor δ at different magnetic field strengths for our present experiment is shown in figure 2.

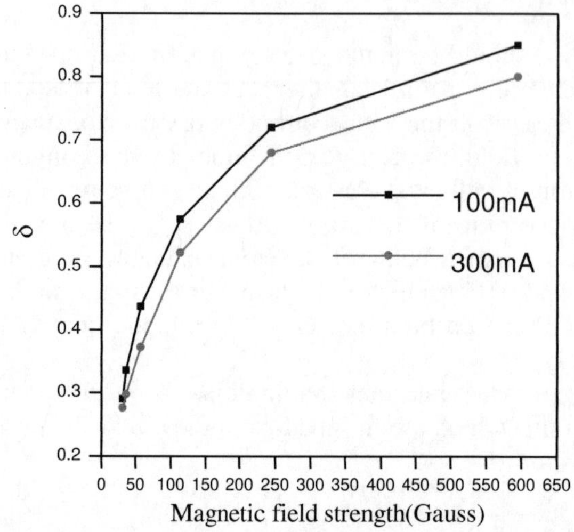

Figure 2. Reduction factor at different magnetic field strengths.

The maximum reduction factor δ is around 0.85 at magnetic field, B = 594 Gauss for I_{dis} = 100 mA and it reduces to 0.2806 at magnetic field, B = 32 Gauss for I_{dis} = 300 mA.

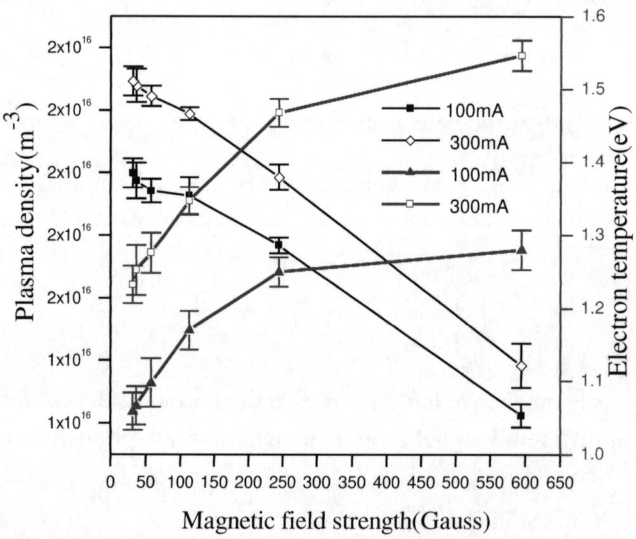

Figure 3. Plasma density (Black colour lines) and electron temperature (Blue colour lines) at different magnetic field strengths.

The influence of magnetic field on the plasma parameters are strongly deviates from the unmagnetized plasma [10]. The plasma parameters are calculated from the I-V characteristics of the Langmuir probe. Figure 3 shows the plasma density and electron temperature measurement at different magnetic field strengths for constant working pressure ~ 2×10^{-4} mbar. From figure 3, it is seen that as the magnetic field strength increases, the electron temperature decreases from ~ 1.55 eV to ~ 1.10 eV for discharge currents 300mA. Because, at higher field strength, only high energetic electrons are collected by the probe and the lower energetic electrons are trapped by the magnetic field lines. As the field strength decreases, the electrons with lower energy can also reach the probe. Thus, the electron temperature decreases with the decrease in field strengths.

Also, the plasma density decreases with the increase of magnetic field strength because higher magnetic field constraints the motion of electrons and ions. Hence, the charge collection by the probe is reduced from the actual value. It is observed that the plasma density and the electron temperature measured by Langmuir probe shows almost consistent result at lower magnetic field strength (at B ≤ 37) as the influence of magnetic field on the collection of charge particle is almost negligible. At this field strength, r_L^i and $r_L^e \geq 10 r_P$.

The shape of EEPF for molecular gases (like N_2, O_2) is completely different and complicated than atomic gases [8]. It is reported that the EEPF for molecular gases (N_2 and O_2) plasma shows nearly Maxwellian distribution whereas the atomic gases (like noble gases) shows Druyvesteyn distribution at low pressure. Godyak *et al.* [6] suggested that if the electron-electron collision frequency is of the order of 10^{10} $(\text{cmV}^{1/2})^{-3}$, then the EEPF shows Maxwellian structure. The electron-electron collision frequency is proportional to the value of $N(\langle\varepsilon\rangle/e)^{-3/2}$. It is found that the the value of $N(\langle\varepsilon\rangle/e)^{-3/2}$ is very less compare to 10^{10} $(\text{cmV}^{1/2})^{-3}$ at the operating pressure 2 x 10^{-4} mbar. Thus, a bi- Maxwellian EEPF is observed for the entire magnetic field range (figure 5). In the present work $N(\langle\varepsilon\rangle/e)^{-3/2}$ is presented in the form of $n^*(\langle\varepsilon\rangle/e)^{-3/2}$ $(V^{-3/2})$ where $n^* = n_e/n_i$. The value of $n^*(\langle\varepsilon\rangle/e)^{-3/2}$ $(V^{-3/2})$ at different magnetic field is shown in figure 4.

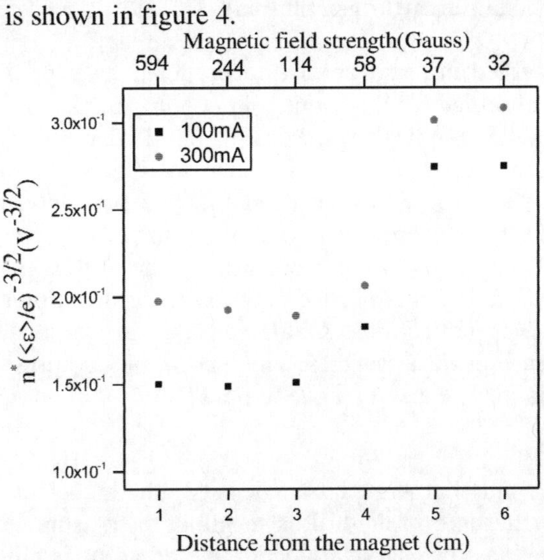

Figure 4. The values of $n^*(\langle e\rangle/e)^{-3/2}$ $(V^{-3/2})$ at different magnetic field strengths.

Figure 5(a) and (b) shows the EEPFs of hydrogen plasma, in absence and presence of dust, for higher field strength (B = 594G) and lower field strength (B = 37 G). A bi-Maxwellian EEPF is observed in both the cases for pristine and dust containing hydrogen plasma. From the observations, it is found that the low energy electron population changes whereas the high-energy electron population remains almost constant in presence of magnetic field. At higher magnetic field, the probe does not collect the lower energetic electrons due to the higher value of δ. So, the probe collects mostly the higher energetic electrons. So, the peak of bulk electrons i.e. the low energy electrons are very small at higher magnetic field compared to the lower magnetic field. As the magnetic field reduces the bulk electron peak of the EEPF becomes steeper and reaches a saturated value. It reflects the decrease in electron temperature at lower magnetic field strength, which is also observed in the electron temperature measurements (figure 2).

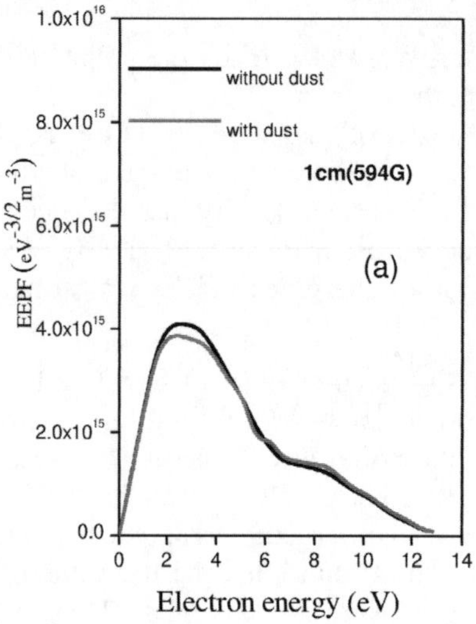

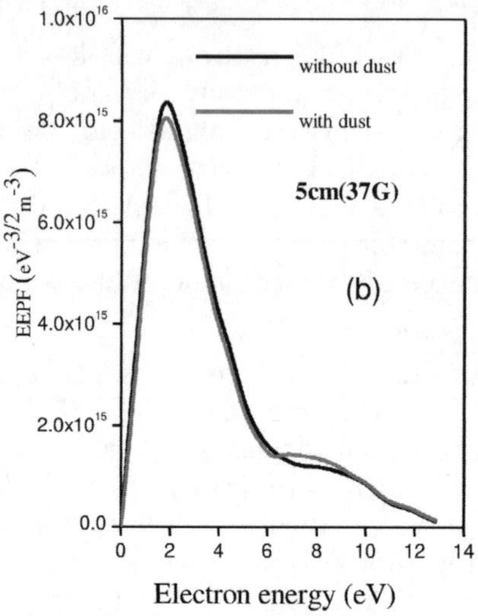

Figure 5. EEPF plots at B = 594 G and B = 37 G (black line represent for pristine hydrogen plasma and red line represent for dust containing hydrogen plasma).

Additionally, the EEPF is studied in presence of dust particles. Like pristine hydrogen plasma, a similar bi-Maxwellian EEPF is observed for dust containing hydrogen plasma in presence of magnetic field. It is observed that the bulk electron peak of the EEPF slightly decreases and the high-energy tail becomes slightly larger in presence of dust grains than the pristine plasma due to the collection of electrons and ions by the dust particles. It is seen from figure 5 that only the low energy electron populations are reduced and the high energy electrons are enhanced in presence of dust particles. Because of collection of the bulk electrons by the dust particles, the overall electron loss increases and the system then self-organizes to maintain a balance between the production and loss of electrons by increasing the number of high-energy electrons. As a result, the fraction of high-energy electrons in the EEPF tail enhances in presence of dust particles. The present observation shows similar EEPF for dust containing hydrogen plasma in absence of magnetic field [8].

4. Conclusion

From the observations, it is found that the plasma density and the electron temperature, shows consistent results at lower magnetic field strength. It is seen that the plasma density increases with the increase of magnetic field strength as the charge collection by the probe is highly restricted for higher field strengths and correspondingly the electron temperature gradually decreases with the reduction of magnetic field strength as both the higher and lower energetic electrons are collected by the probe. We have found from the measured EEPFs, a bi-Maxwellian behaviour at a very low pressure (~ 2×10^{-4} mbar) in presence of magnetic field. In the EEPF graphs, the peak of bulk electrons is very small at higher magnetic field compared to that of lower magnetic field. As the magnetic field decreases, the bulk electron peak in the EEPF increases as both the higher energetic and lower energetic electrons are collected by the probe. In presence of dust particles, the bulk electron peak of the EEPF slightly decreases due to the collection of electrons by the dust particles and the high-energy tail becomes slightly larger compared to pristine hydrogen plasma in presence of magnetic field.

Acknowledgement

The authors would like to thank Mr. G. D. Sharma for his technical support all throughout the work.

References

[1] Fischery R and Dose V 1999 *Plasma Phys. Control.Fusion* **41** pp 1109-1123
[2] Popov Tsv K , Ivanov P, Dimitrova M , Kovacic J, Gyergyek T and Cercek M 2012 *Plasma Sources Sci. Technol.* **21** 025004

IOP Conf. Series: Journal of Physics: Conf. Series **823** (2017) 012068 doi:10.1088/1742-6596/823/1/012068

[3] Druyvesteyn M J 1930 *Z. Phys.* **64** 781
[4] Toader E I 2004 *Plasma Sources Sci. Technol.* **13** pp 646–653
[5] Abdel-Fattah E, Bazavan M and Sugai H 2012 *Phys. Plasmas* **19** pp 113503
[6] Godyak V A and Demidov V I 2011 *J. Phys. D: Appl. Phys.* **44** 233001
[7] Denysenkoa I , Yu M Y , Ostrikov K , Azarenkov N A and Stenflo L 2004 *Physics of Plasmas* **11** 4959
[8] Kakati B , Kalita D , Kausik S S, Bandyopadhyay M and Saikia B K 2014 *Physics of Plasmas* **21** 083704
[9] Kakati B , Kausik S S , Bandyopadhyay M , Saikia B K and SaxenaY C 2014 *J. Appl. Phys.* **116** 163302
[10] Kakati B, Kausik S S, Saikia B K and Bandyopadhyay M 2011 *Phys.Plasmas* **18** 033705
[11] Kakati B, Kausik S S , Saikia B K and Bandyopadhyay M 2011 *Phys. Plasmas* **18** 063704
[12] Denysenko I, Yu M Y, Ostrikov K, and Smolyakov A 2004 *Phys. Rev. E* **70**, 046403
[13] Liu J -Y , Gao Y and Wang G 2012 *Pramana–J. Phys* .**79** 113
[14] Stangeby P C 1982 *J. Phys. D* **15** 1007-29

IOP Conf. Series: Journal of Physics: Conf. Series **823** (2017) 012069 doi:10.1088/1742-6596/823/1/012069

Engineering design and integration of in-vessel single turn segmental coil in vacuum vessel of SST-1

Snehal Jayswal[1], P Chauhan[1], P Santra[1], K Vasava[1], T Perekh[1], H Patel[1], P Biswas[1] and S Pradhan[1]

[1]Institute For Plasma Research,Bhat, Gandhinagar, Gujarat State, INDIA

Email: snehal@ipr.res.in

Abstract: SST-1 tokamak is having the error field due to unsymmetrical positioning of Toroidal field coils which push the plasma to inner side from its major radius of 1100 mm. hence it is required to install the In-vessel Coil (PF6) at a location of 1350 mm radius and elevation of 350 mm above and below the mid plane of the toroidal field coils. The In-Vessel coil was decided to make in eight segments for futuristic use, to control the individual localized error field correction by supplying the different current. A single turn, eight segments, copper conductor with 18 mm diameter with GFRP insulation and in housed in SS304 L casing to carry 8000 A current for 10 s was designed and installed in vacuum vessel of SST-1. This paper will present the design drivers, material selection, advantages and constraints of the in-vessel coils, its conceptual and engineering design, CAD models, finite element analysis using ANSYS, its fabrication, quality assurance/control and assembly/integration aspects inside vacuum vessel of SST-1.

1. Introduction

During the refurbishment of the SST-1, the error field was mapped with the help of the hall probe and mirnov's probe magnetic field mapping methods. Which in turn revels the shifting of plasma null to one side due error in magnetic field generated due to unsymmetrical positioning of the Toroidal field coils. The unsymmetrical positioning was due to error in TF supporting ring design and partially due to uneven tightening of Toroidal field coils. These error field causes plasma to form inboard side of the vacuum vessel from its major radius of 1100 mm. the simulation of the exact positioning of the TF coil and correcting of this field lead to evolution of the design of the in-vessel coil. To counter the effect of the magnetic field error, it was required to install pair of In-vessel Coil (PF6) at a location of 1300 mm radius and elevation of 350 mm above and below the mid plane of the toroidal field coil (figure 1). The In-Vessel coil was decided to make in eight segments (each segment with 3000 mm and one coil of 6900 mm length) for futuristic use, to control the individual localized error field correction by supplying the different current. The design of the in-vessel coil has many critical aspects for considerations like terminals of two successive copper conductor along with stainless steel casing must be come out from 46 mm diameter nozzle, secondly casing should be welded to radial port such to vacuum integrity of vessel remains intact, thirdly copper conductor can able to withstand maximum current of 8000 A for 10 s with maximum temperature rise of 50° C, there must be close tolerance between copper conductor and SS casing to reduce any movement of copper conductor inside the casing for keeping geometrical stability within vessel. Hence The in-vessel coil was made from single solid ETP copper conductor encased inside the prefabricated SS 304 L piped casing in eight segment to form full circle.. The In-vessel coil is open to atmosphere and experiencing the vacuum inside the vacuum vessel as shown in figure 2. To maintain the circular shape of the coil copper rod inside the SS casing, very close tolerances are maintained e.g. copper conductor has outer diameter of 18mm and after FRP insulation and Kepton tap wrapping the outer diameter reaches to 19mm (figure 3). This coil is made from solid copper rod which under forces while charging can vibrate and change the shape. Hence it is required to keep in circular shape by encasing in steel pipe casing with close tolerances.

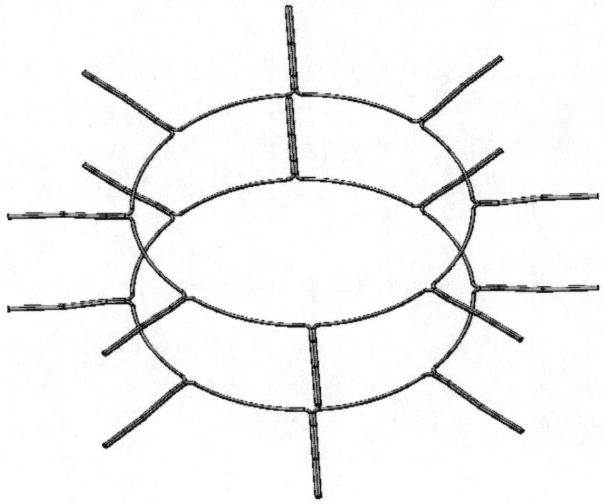

Figure 1. An open view of segmental In-vessel coil

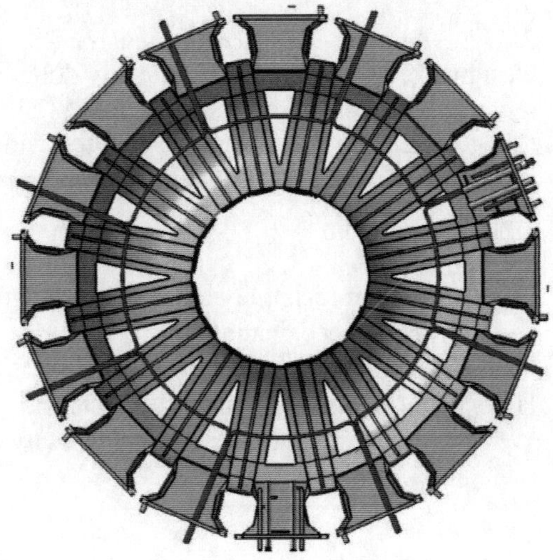

Figure 2. In-vessel coil inside the vessel

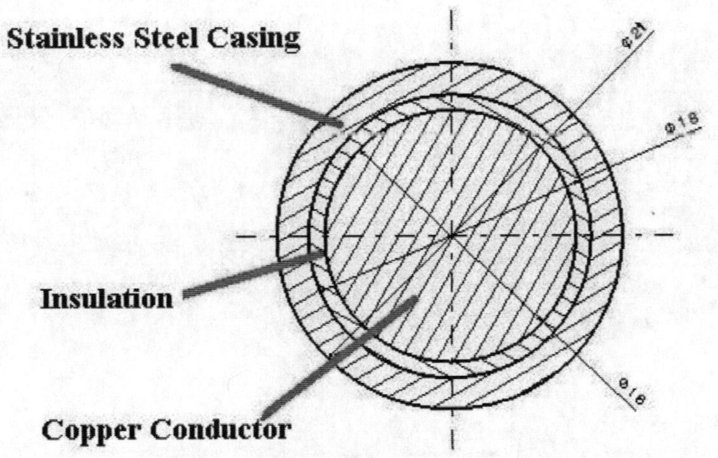

Figure 3. Cross section detail of In-vessel coil

2. Design of components

2.1. Design of SS Casing

The Stainless Steel pipe is experiencing internal vacuum pressure of vessel and is open to atmosphere. Though entire casing house the copper conductor along with insulation in tight tolerances, the entire assembly is act as solid rod. Also all pipes has to go through hydraulic testing at 70Psi pressure, there isn't need to design check for bursting pressure. Even though it has been verified by the following equation (2.1.1) and value derived for hoop stress is 0.6 MPa much below the yield stress value of SS 304L casing material (170 MPa)

$$\text{Hoop stress} = \frac{Pressure(p) \times InnerDiameter(d)}{2 \times Pipethickness \times weldefficiency} \quad (2.1.1)$$

2.2. *Design of In-vessel coil conductor*

Design of copper conductor is mainly done with input rating of 8000 A current carrying capacity in 10 seconds and with maximum differential temperature rise to 50° C with following equations (2.2.1,2.2.2,2.2.3) and result is validated with ansys analysis software and is plotted below.

(a) Resistance of the single copper segment

$$R = \frac{\rho \times l}{A} \quad (2.2.1)$$

Here ρ = copper resistivity in Ω/m, L = length of the copper conductor, A= Area of copper conductor

(b) Mass of the conductor

$$m = V \times \delta \quad (2.2.2)$$

Here V = Volume of conductor, δ = Density of copper

(c) Maximum current carrying capacity of conductor

$$I^2 = \frac{m \times C_p \times \Delta T}{R \times t} \quad (2.2.3)$$

Here C_p = Specific heat of copper, ΔT = Temperature rise difference, t = Charging time

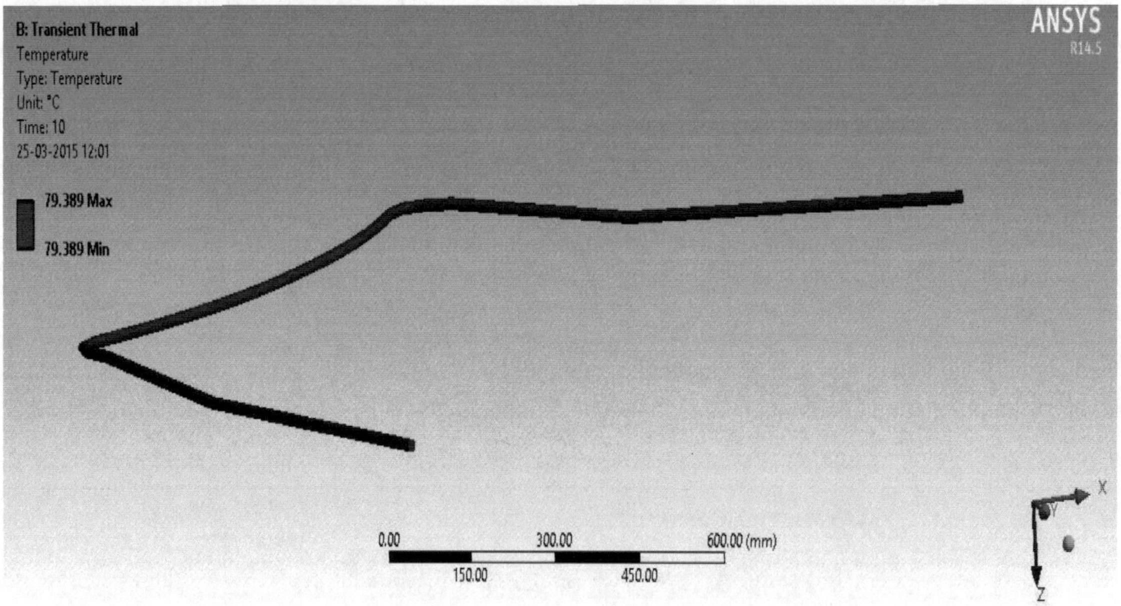

Figure 4. Thermal Ansys analysis of Rise of temperature during charging of In-vessel coil.

The above analysis plot shows the rise of the temperature of about 79° C from the ambient temperature while charging to maximum current of 8000 A for 10 s. The cool down time for the in-vessel coil after maximum current charging condition is also calculated to find out time lapse required between two successive plasma shots as per equation (2.2.4). The below mentioned formula used to calculate the cooling time and also analysis plot at the end validated the calculation as the cool down time is 10° C for 10 min. of cooling by radiation. The copper conductor duly in contact with ss casing leaves the heat with casing through conduction which in case dissipated through radiation to vessel in vacuum.

$$\text{Time of cooling} = \frac{Nk}{2\varepsilon\sigma A}\left[\frac{1}{Tfinal^{3}} - \frac{1}{Thot^{3}}\right] \quad (2.2.4)$$

Here N = number of particles of copper = $\frac{mN_A}{M} = \frac{6.72\times 6.022\times 10^{23}}{63.54} = 0.636\times 10^{26}$

M = molecular mass of copper, m= mass of copper, N_A = avagrado number, K = thermal conductivity of copper (401 W/m K), ϵ = emissivity of copper (0.65) σ= Stephen Boltzman constant = 5.67x 10^{-8} w/m^2 k^4, A= area of radiating surface. $Tfinal = 343.18\,\text{K}$ = **70.03 °C**

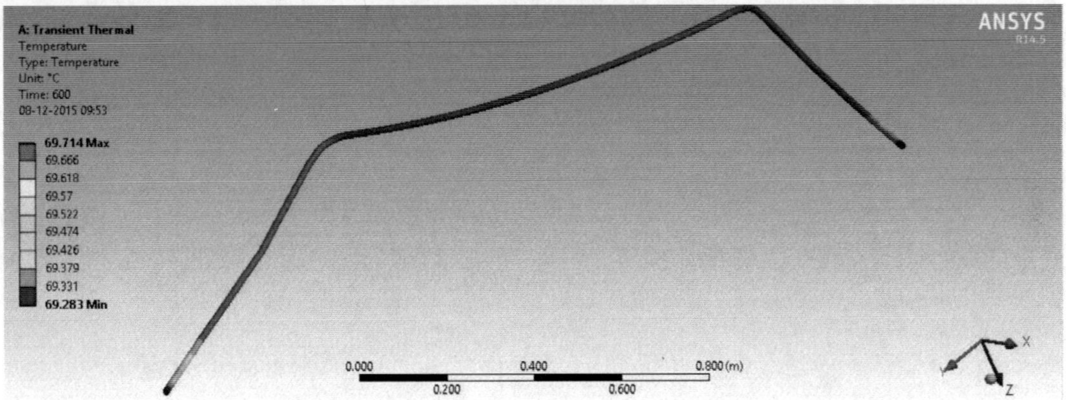

Figure 5. Thermal Ansys analysis of cool down of in-Vessel coil after maximum charge

2.3. *Design and Analysis of support structure*

The pair of In-vessel coil experiencing the vertical force from each other which is very much negligible, hence design is done from force generated by radial control coils which experience both vertical and radial forces. The support for holding the In-vessel coil and Radial control coil is taken from Outer passive stabilizer support at eight Inter Connecting Ring locations at top and bottom as shown in modelled figure 6 and figure 7.

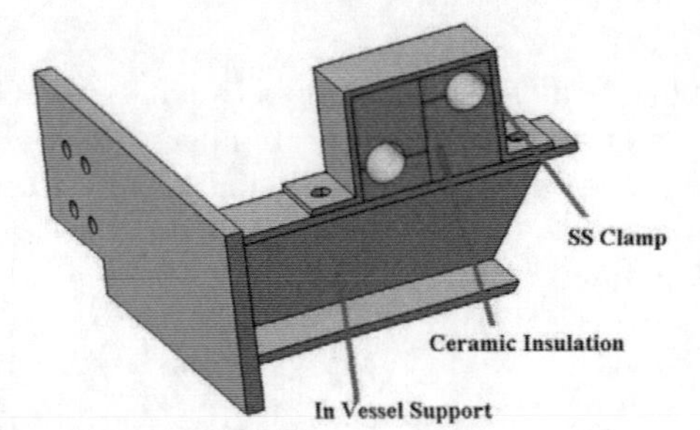

Figure 6. Conceptual support with ceramic insulation

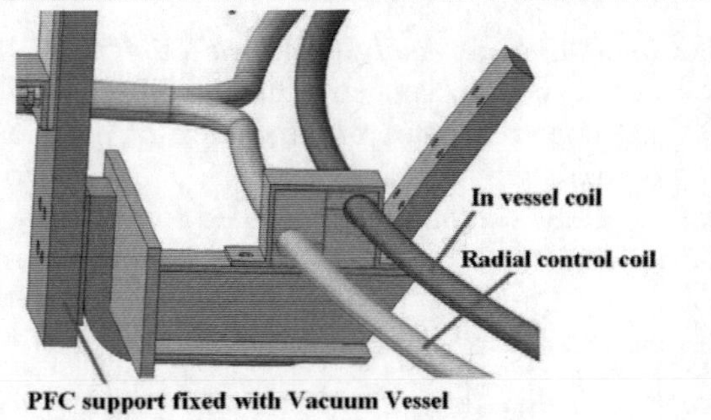

Figure 7. Fixing of support inside the Vacuum vessel

These supports experiences the Radial force of 30000 N and Vertical force = 1300 N with total 8 nos. of support for holding 8 segments of in-vessel coil (Load data provided by simulation group). A simplified free body force diagram is shown in figure 8 below showing the vertical and radial force acting on single support and the stresses and deflection found which are around 49.14 MPa and 0.58

IOP Conf. Series: Journal of Physics: Conf. Series **823** (2017) 012069 doi:10.1088/1742-6596/823/1/012069

mm deflection respectively. The same is validated by the ansys analysis software and its image plotted here in figure 9.

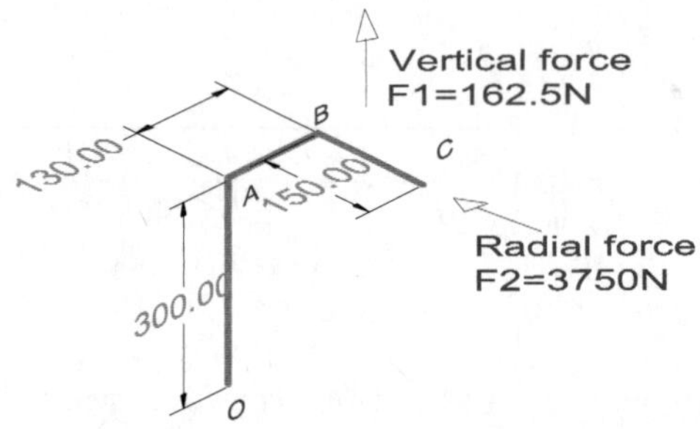

Figure 8. Free Body diagram of the forces and unknown reactions in problem

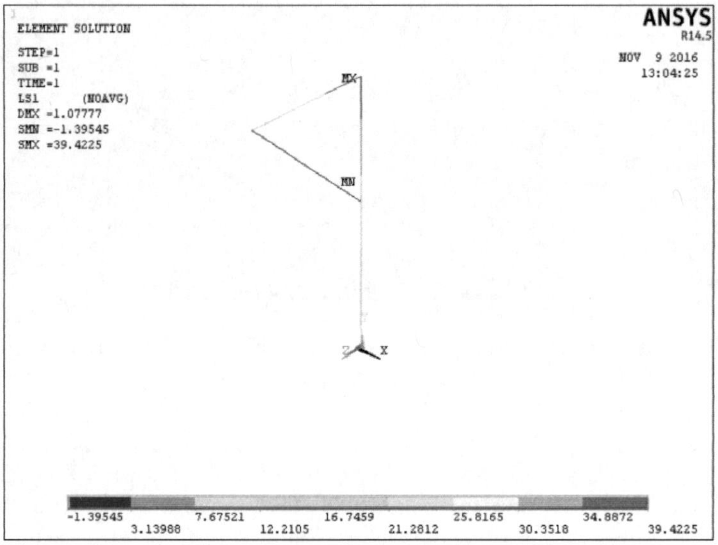

Figure 9. Bending stress

2.4. Calculation and Analysis of Insulation

The in-vessel coil is clamped with supports along with alumina ceramic as isolation between In-vessel coil and support for stopping the loop of eddy current formation. 99 % purity alumina with high compressive strength was used in four pieces to hold the In-vessel coil and Radial control coil together. These ceramics are designed for crushing strength as per equation (2.4.1) and the values are validated with ansys analysis software as shown in figure 10 below.

$$\text{Crushing stress} = \frac{P}{d \times l} = 4.07\ \text{N/mm}^2 \tag{2.4.1}$$

Here P = Radial force experience by ceramic, d = Projected diameter of ceramic, l = Length of ceramic

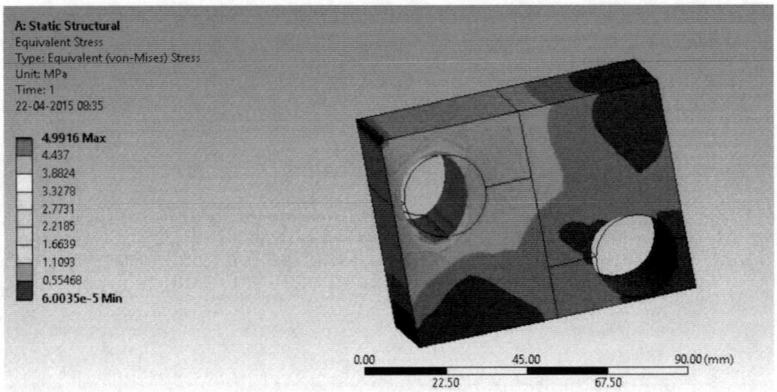

Figure 10. Stress on ceramic

3. Fabrication and Installation

The In-vessel coil copper conductor is duly insulated with fiberglass tap and kepton inserted within close tolerances to SS casing. These pair of coil is installed on outboard passive stabilizer support mounted in Inner connecting ring at the radius of 1300mm and elevation of 350mm above and below the mid-plane of Toroidal field coil with ECDS system with vary close tolerances as shown in image. The Stainless steel casing are welded from outside with vessel radial port in order to make the coil vacuum proof. The copper conductor leads comes out form the vessel is brazed with successive terminal to form full circle. In future, these terminals can be separated for individual segment operation for localized field error correction (N mode correction). All the weld joint of the radial control coil is tested at the leak rate of 1x 10^{-8} mbarl/s. The final megger test is done between In-vessel coil casing and vacuum vessel to find resistance in order of giga-ohms.
The installed in-vessel coil inside SST-1 VV is shown in Figure 11 and 12.

Figure 11. Zoomed view of In-vessel coil inside Vacuum Vessel

Figure 12. Terminal of two segments going out of Vacuum vessel

4. Characterization of the coils

In vessel coil has been tested with giving 200 A, 500 mHz square signal current and checked the magnetic field effect at the plasma radius of 1100 mm with putting hall probe at plasma radius at different location which measured value of average 3.5 Gauss. This measured value comes to vary close to the theoretical value as well with standard 1st & 2nd order elliptical equations for determining magnetic fields of circular loop. A measured signal wave form is also shown in below image.

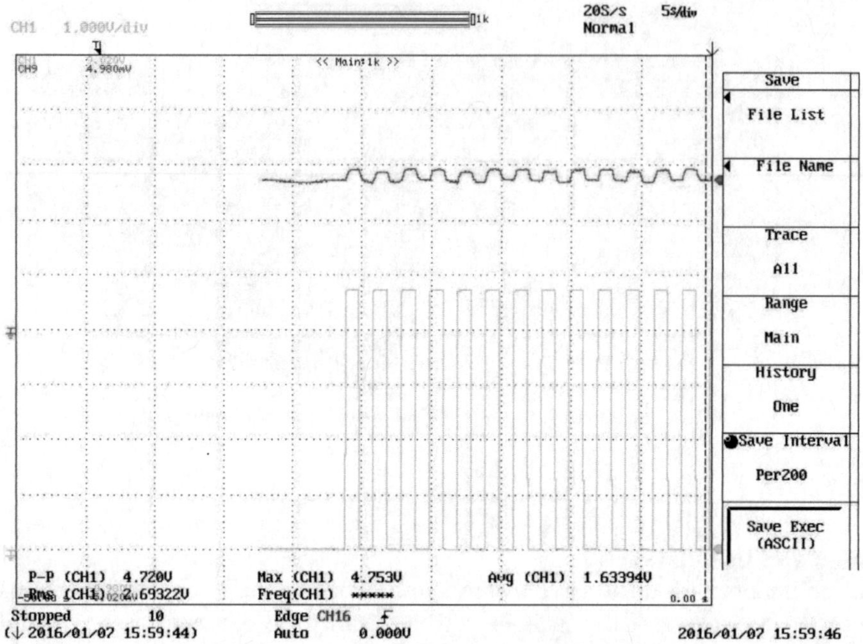

Figure 13. An oscilloscope wave form image

5. Conclusions

The In-vessel coil support design, analysis, fabrication aspects and assembly inside SST-1 vacuum vessel are presented in detail. The analytical stress calculations and ANSYS FE results match closely. In-vessel coil and its supports are successfully assembled inside SST-1 vacuum vessel. The initial characterization of the In-vessel coil performance has found satisfactorily and the operation of the coil will commence in short duration.

6. References

[1] Pradhan S, Sharma A N, Tanna V L, Khan Z, Prasad U, Doshi K, Raval D C, Pathan F, Gupta N C, Tank J, Gupta M K, Santra P, Biswas P, Parekh T, Masand H, Sharma D, Srivastava A and Patel H 2012 *IEEE Transaction on Plasma Science* **40** Number 3 614
[2] ANSYS Software Version 14.5, ANSYS Inc. PA, USA
[3] Khurmi R S 2005 *A text book of machine design* (Eurasia publication house) pp 96-265
[4] Bansal R K 2009 *A text book of strength of material* (Laxmi publication pvt ltd) p 746

Institute of Physics
Dirac House, Temple Back
Bristol BS1 6BE UK

ISSN: 1742-6588
ISBN 978-1-5108-3962-5

CURRAN ASSOCIATES INC.
proceedings.com